Lecture Notes in Computer Science **16851**

The series Lecture Notes in Computer Science (LNCS), including its subseries Lecture Notes in Artificial Intelligence (LNAI) and Lecture Notes in Bioinformatics (LNBI), has established itself as a medium for the publication of new developments in computer science and information technology research, teaching, and education.

LNCS enjoys close cooperation with the computer science R & D community, the series counts many renowned academics among its volume editors and paper authors, and collaborates with prestigious societies. Its mission is to serve this international community by providing an invaluable service, mainly focused on the publication of conference and workshop proceedings and postproceedings. LNCS commenced publication in 1973.

Faisal Shafait · Adrian Ulges ·
Momina Moetesum · Xu-Cheng Yin
Editors

Document Analysis Systems

17th IAPR International Workshop, DAS 2026
Vienna, Austria, September 3–4, 2026
Proceedings

Springer

Editors
Faisal Shafait
National University of Sciences and Technology (NUST)
Islamabad, Pakistan

Adrian Ulges
RheinMain University of Applied Sciences (HSRM)
Wiesbaden, Germany

Momina Moetesum
National University of Sciences and Technology (NUST)
Islamabad, Pakistan

Xu-Cheng Yin
National University of Sciences and Technology (NUST)
Beijing, China

ISSN 0302-9743 ISSN 1611-3349 (electronic)
Lecture Notes in Computer Science
ISBN 978-3-032-36206-3 ISBN 978-3-032-36207-0 (eBook)
https://doi.org/10.1007/978-3-032-36207-0

This Springer imprint is published by the registered company Springer Nature Switzerland AG
The registered company address is: Gewerbestrasse 11, 6330 Cham, Switzerland

Preface

It is our pleasure to welcome you to the proceedings of the Seventeenth IAPR International Workshop on Document Analysis Systems (DAS 2026). Vienna provided an inspiring setting to continue the long-standing tradition of the DAS workshop series, bringing together researchers and practitioners from around the world to share advances, exchange ideas, and foster collaboration in the field of document analysis and recognition.

DAS 2026 built on a distinguished series of previous workshops held in Kaiserslautern, Germany (1994); Malvern, PA, USA (1996); Nagano, Japan (1998); Rio de Janeiro, Brazil (2000); Princeton, NJ, USA (2002); Florence, Italy (2004); Nelson, New Zealand (2006); Nara, Japan (2008); Boston, MA, USA (2010); Gold Coast, Australia (2012); Tours, France (2014); Santorini, Greece (2016); Vienna, Austria (2018); Wuhan, China (2020); La Rochelle, France (2022); and Athens, Greece (2024). Like these previous editions, this year's DAS focused on system-level challenges and holistic approaches to document analysis and recognition. With the rapid emergence of generative AI and agentic systems as key enablers of real-world document understanding applications, this perspective is gaining renewed importance.

Since 2024, DAS has adopted a revised format and is now organized in conjunction with the International Conference on Document Analysis and Recognition (ICDAR) as its largest associated workshop, while retaining its own proceedings and two-day structure. We consider this change successful, and have put considerable effort into making DAS a high-quality workshop in the spirit of its successful predecessors. We ensured a single-track format, a rigorous peer-review process, and complete participation, complemented by invited talks, contributed papers, and discussion sessions. With this format, we hope to bring together practitioners and researchers from industry with academics in a unique environment.

Following DAS 2024, this year's edition featured two separate calls for papers, an earlier one corresponding to regular papers to be included in the proceedings, and a second call for short papers presenting preliminary results or research in progress. Of the 50 submissions received for the regular paper track, 29 were accepted for presentation at the workshop, resulting in an acceptance rate of 58%. All submissions received at least two double-blind reviews from our team of 49 Program Committee members. The accepted regular papers are published in this volume of the Springer Lecture Notes in Computer Science (LNCS) series. DAS also received nine short paper submissions and which are made available in PDF format on the DAS conference website. The short papers were also reviewed in a single-blind form by the Program Committee. The final program included four oral sessions, two poster sessions, one demo session, and one discussion group. In addition, DAS hosted industry keynote talks delivered by leading experts: Jordy van Landeghem, Nibal Nayef, Brandon Smock and Sheraz Ahmed. Outstanding contributions were recognized through the Best Paper Award and the Best Student Paper Award.

We extend our sincere gratitude to everyone who contributed their time and effort to making DAS 2026 a leading event for the community. The success of the workshop program was made possible through the dedication and hard work of many individuals. We sincerely thank the Program Committee members for their hard work reviewing submissions, and especially those last-minute, emergency reviewers who were instrumental in avoiding notification delays. We also thank the organizing committee for their important contributions. We acknowledge the efforts of Industrial and Sponsorship Chairs Yves-Noel Weweler and Lukasz Borchmann in coordinating sponsorship activities and keynote arrangements. Special thanks go to Carlos Moreno-García for creating and maintaining the DAS website and social media presence; to Publication Chair Liangrui Peng for overseeing the proceedings preparation; and to Ajoy Mondal for serving as Demo and Discussion Group Chair.

Finally, we would like to express our sincere appreciation to the authors for their outstanding contributions and active participation in DAS 2026. We hope that the workshop program encouraged future research, inspired new collaborations, and provided practitioners with useful methods, algorithms, and tools. It is both an honor and a privilege to present the latest advances in document analysis systems through these proceedings, and we trust that readers will find them informative and inspiring. hope you enjoyed DAS 2026 in Vienna!

September 2026

Faisal Shafait
Adrian Ulges
Momina Moetesum
Xu-Cheng Yin

Organization

General Chairs

Faisal Shafait	National University of Sciences and Technology, Pakistan
Adrian Ulges	RheinMain University of Applied Sciences, Germany

Program Chairs

Momina Moetesum	National University of Sciences and Technology, Pakistan
Xu-Cheng Yin	University of Science and Technology Beijing, China

Publication Chair

Liangrui Peng	Tsinghua University, China

Industrial and Sponsorship Chairs

Yves-Noel Weweler	Buildsimple, Germany
Lukasz Borchmann	Snowflake Inc., Poland
Tobias Alt-Veit	Proalpha, Germany

Publicity and Website Chair

Carlos Moreno-García	Robert Gordon University, UK

Demo and Discussions Group Chair

Ajoy Mondal	International Institute of Information Technology Hyderabad, India

Local Organizers

Marco Peer	Université de Fribourg, Switzerland
Rafael Sterzinger	TU Wien, Austria
Marvin Burges	TU Wien, Austria

Program Committee

Adrian Ulges	RheinMain University of Applied Sciences, Germany
Ajoy Mondal	IIIT Hyderabad, India
Anna Scius-Bertrand	University of Fribourg, Switzerland
Ashok Popat	Google, Inc., USA
Bin Liu	National Laboratory of Pattern Recognition, Institute of Automation, Chinese Academy of Sciences, China
Biying Fu	RheinMain University of Applied Sciences, Germany
Camille Kurtz	Université Paris Cité, France
Carlos Mello	Universidade Federal de Pernambuco, Brazil
Chiranjoy Chattopadhyay	FLAME University, India
Daniel Benniah John	University of California, Berkeley, USA
Dirk Krechel	RheinMain University of Applied Sciences, Germany
Eliott Thomas	Laboratoire Informatique, Image et Interaction, La Rochelle University, France
Elisa Barney	Boise State University, USA
Ernest Valveny	Computer Vision Center, Spain
Florence Cloppet	Université Paris Cité, France
Florian Kleber	TU Wien, Austria
Frédéric Rayar	LIFAT - University of Tours, France
Hung Tuan Nguyen	Tokyo University of Agriculture and Technology, Japan
Irina Rabaev	Shamoon College of Engineering, Israel
Jianjuan Liang	Guizhou University, China
Jin Wei	Lenovo Research, China
Josep Llados	Computer Vision Center, Universitat Autònoma de Barcelona, Spain
Kaspar Riesen	University of Bern, Switzerland
Kengo Terasawa	Future University Hakodate, Japan
Kenny Davila	DePaul University, USA

Maham Jahangir	Guangdong CAS Cogniser Information Technology Co. Ltd., China
Marco Peer	University of Fribourg, Switzerland
Mark Clement	Brigham Young University, USA
Maroua Mehri	University of Sousse, Tunisia
Mickaël Coustaty	L3i Laboratory, La Rochelle University, France
Nam Ly	Tokyo Metropolitan University, Japan
Olumayowa Onabanjo	University of Oviedo, Spain
Oriol Ramos Terrades	Universitat Autònoma de Barcelona - Computer Vision Centre, Spain
Rina Buoy	Techo Startup Center, Cambodia
Robert Sablatnig	TU Wien, Austria
Ronaldo Messina	Mitek Systems, USA
Salvatore-Antoine Tabbone	IDMC-Université Lorraine, France
Shivakumara Palaiahnakote	University of Salford, UK
Tayyibah Khanam	Kensho Technologies (S&P Global), USA
Thierry Paquet	University of Rouen Normandie, France
Verónica Romero	Universitat de València, Spain
Vincent Christlein	Friedrich-Alexander-Universität Erlangen-Nürnberg, Germany
Vincent Poulain d'Andecy	Yooz, France
Vlad Atanasiu	University of Bern, Switzerland
Xiangdong Su	Inner Mongolia University, China
Xugong Qin	Nanjing University of Science and Technology, China
Yaping Zhang	National Laboratory of Pattern Recognition, Institute of Automation, University of Chinese Academy of Sciences, China
Yuchen Zheng	Shihezi University, China
Yugo Kubota	Kyushu University, Japan

Sponsors/Organizers

IAPR, SEECS NUST, HSRM, USTB, IIIT, Robert Gordon University, Tsinghua University, University of Fribourg, TU Wien, Insiders Technologies GmbH, Snowflake Inc.

International Association for Pattern Recognition (IAPR)

Insiders Technologies GmbH

National University of Sciences and Technology (NUST)

Hochschule RheinMain University of Applied Sciences and Arts (HSRM)

University of Science and Technology Beijing (USTB)

International Institute of Information Technology Hyderabad (IIIT Hyderabad)

Robert Gordon University

Tsinghua University

University of Fribourg

Vienna University of Technology (TU Wien)

Snowflake Inc.

Fig. 1. Sponsors and Organizers of DAS 2026

Contents

Handwriting, Writer Identification and Music Documents

OCR, Layout and Recognition Systems

Tables, Forms and Structured Extraction

POTATR: A Lightweight Image-to-Graph Model for Page-Level Table Extraction

Brandon Smock(✉), Libin Liang, Maxim Sokolov, Amrit Ramesh, Valerie Faucon-Morin, Tayyibah Khanam, and Maury Courtland

Kensho Technologies, Cambridge, USA
{brandon.smock,maury.courtland}@kensho.com

Abstract. Large-scale document processing requires contextually aware table extraction (TE) that is both accurate and efficient. Yet current approaches require billions of parameters, hundreds of autoregressive steps, or costly API inference. Motivated by this, we introduce the Page-Object Table Transformer (POTATR), a 29M parameter image-to-graph model that extends the Table Transformer for contextualized page-level TE. POTATR outperforms all models tested on PubTables-v2 Single Pages—including frontier MLLMs—achieving $\text{GriTS}_{\text{Con}}$ of 0.964 while running over 130× faster at 300× lower cost. Further, POTATR's output is spatially grounded: every recognized element has a bounding box, enabling visual verification and geometric text assignment. As a result, POTATR performs unified page-level TE while composing with other models, enabling extension to scanned documents via external OCR and to full-document TE via techniques like cross-page merging. Code and models will be released at https://github.com/kensho-technologies/POTATR.

Keywords: Page-level table extraction · Image-to-graph

1 Introduction

Vision-language models (VLMs) [3,10,14,26] have emerged as the dominant paradigm for contextually aware table extraction (TE). Such models frame TE as autoregressive text generation—producing HTML or markdown token-by-token. This has the advantage of leveraging vast amounts of structured text data for training, which has been shown to lead to strong recognition performance.

However, this paradigm has several downsides. These include inherently sequential generation, lack of structural validity guarantees, and often a lack of reliable spatial grounding for page elements. Moreover, despite their increasing effectiveness, frontier multi-modal large language models (MLLMs) can be prohibitively costly to deploy at scale [16]. Task-specific systems, in contrast, are orders of magnitude less expensive than multi-purpose generative models [11].

F. Shafait et al. (Eds.): DAS 2026, LNCS 16851, pp. 3–17, 2027.
https://doi.org/10.1007/978-3-032-36207-0_1

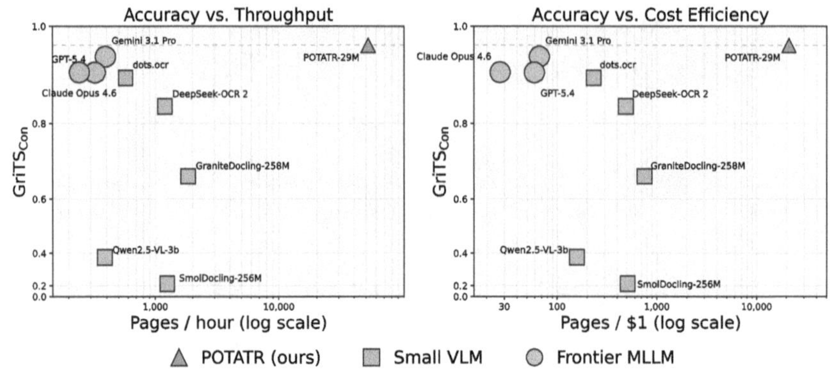

Fig. 1. Accuracy vs. efficiency on PubTables-v2 Single Pages. POTATR (29M parameters) achieves the highest page-level table extraction accuracy (GriTS$_{Con}$) while being over 130× faster and 300× cheaper than frontier MLLMs. The y-axis uses a nonlinear scale to better separate high-accuracy models. (Color figure online)

Motivated by these limitations, we explore an alternative paradigm. We introduce the Page-Object Table Transformer (POTATR), a lightweight 29M-parameter image-to-graph model for contextually aware page-level TE. Rather than autoregressive text generation, POTATR frames page-level TE as parallel-decoded spatial graph prediction. It extends TATR [22,23]—a proven architecture for table structure recognition (TSR) on cropped tables—to the page level, adding page-specific output classes and a relation head for graph prediction. Despite being orders of magnitude smaller than VLMs, POTATR outperforms all tested models on page-level TE—including frontier MLLMs—at a fraction of the cost (Fig. 1).

Our contributions:

- We introduce POTATR, a 29M-parameter model that frames page-level TE as image-to-graph prediction, jointly detecting tables, their structures, captions, and footers, with hierarchical relationships between them, and every element spatially grounded.
- We show that initializing from pre-trained TSR weights provides a significant advantage over training image-to-graph models from scratch, validating the choice to extend an existing TSR model for page-level TE.
- We show that POTATR outperforms all tested models on PubTables-v2 Single Pages—including frontier MLLMs—achieving GriTS$_{Con}$ of 0.964 and caption text F1 of 0.979, while running over 130× faster at 300× lower cost.
- Finally, we show that POTATR's modular, spatially-grounded design composes naturally with other specialized models, which enables system-level improvements on downstream tasks such as multi-page table extraction. Altogether, we believe this shows that small, non-autoregressive models remain a compelling option for deploying real-world document understanding at scale.

2 Related Work

2.1 Table Extraction

Table extraction has traditionally decomposed into table detection on individual pages followed by table structure recognition (TSR) on cropped table images [13,17,19,22]. Several works have attempted to unify these stages in a single model, including segmentation-based approaches [15,17], detection-based hierarchies [29], and corner-edge segmentation [2]. However, many of these methods rely on spatial heuristics or rule-based post-processing to associate detected elements with their parent tables, rather than learning these relationships directly.

The Table Transformer (TATR) [22,23] is a DETR-based [4] model that achieves strong TSR on cropped tables, but is limited to processing individual tables in isolation. POTATR extends TATR to full-page TE, leveraging its pre-trained weights while adding the capabilities needed for contextualized extraction.

Most closely related to our work, the End-to-End Table Transformer (ETT) [5] extends Deformable DETR with a symmetric adjacency matrix to group detected structures by table membership. POTATR differs in several key respects: it predicts *directed* parent-child relations rather than symmetric same-table adjacency, leverages pre-trained TATR weights rather than training from scratch, includes caption and footer classes as first-class page elements, handles rotated tables directly in page context, and uses a lighter architecture (29M vs. ~41M+ parameters).

2.2 Image-to-Graph Models

Image-to-graph models jointly predict objects and their relationships from images. DocParser [18] applied this paradigm to document understanding, predicting hierarchical relationships between page elements. Relationformer [20] proposes a general-purpose framework based on Deformable DETR [31], using a dedicated relation token and relation head to predict edges between detected objects. EGTR [8] repurposes Deformable DETR's internal query-key interactions for scene graph generation.

Page-level TE is inherently hierarchical: tables contain structures (rows, columns, headers) and are associated with captions and footers. Unlike prior work, POTATR produces a full directed hierarchical graph for page-level TE, predicting not only parent-child edges between tables and their structures, but also between tables and their captions and footers.

2.3 Vision-Language Models for Document Parsing

VLMs [1,3,25] combine visual input with text prompts for multi-modal understanding. A recent trend is toward smaller, specialized VLMs for document parsing [7,10,14,26–28], which are less expensive to train and deploy. However, even these smaller models typically contain hundreds of millions to billions of parameters and are evaluated only on isolated cropped tables.

In practice, document parsing often involves modular pipelines combining multiple specialized models (e.g., Docling [14]). Our work demonstrates that a much smaller specialized model (29M parameters) can outperform VLMs on page-level TE when paired with accurate text extraction. Unlike VLMs that entangle structure detection with text generation, POTATR's modular design separates these concerns—yielding a component that can be integrated into broader document parsing pipelines while remaining independently upgradeable (Fig. 2).

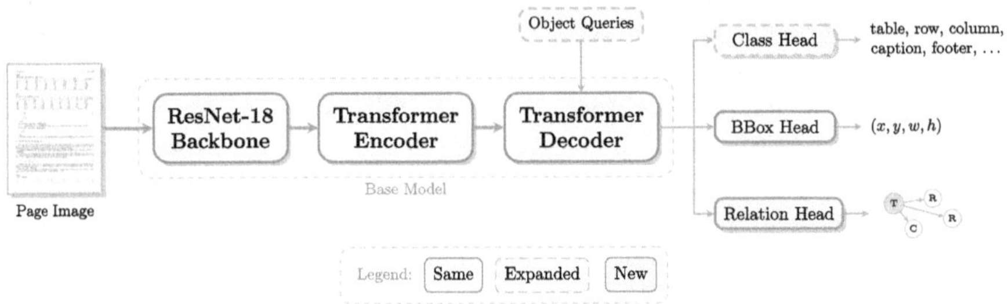

Fig. 2. POTATR architecture. POTATR extends TATR with 125 additional object queries, 10 new page-level object classes, and a relation head that predicts parent-child relationships between detected objects. *Same* (gray): modules unchanged from pre-trained TATR. *Expanded* (orange, dashed): modules with pre-trained components plus additional randomly initialized parts. *New* (teal): modules added in POTATR and randomly initialized. The relation head takes pairs of decoder embeddings and predicts directed edges, producing a hierarchical graph over detected page objects. (Color figure online)

3 Method

3.1 POTATR Architecture

We extend the Table Transformer (TATR) [23] from a model for table structure recognition on individual cropped tables to a model for full-page table extraction, which we call the Page-Object Table Transformer (POTATR).

The original TATR model predicts 6 object classes: *table*, *table column*, *table row*, *table column header*, *table projected row header*, and *table spanning cell*. POTATR adds 2 additional page-level classes: *table caption* and *table footer*. Because a table can be rotated within a page, POTATR also adds 8 additional classes that are the rotated counterparts of the first 8 (*i.e. table rotated* and *table column rotated*), for a total of 16 classes. Since page images can have more objects to detect, we also double the number of object queries from 125 to 250.

POTATR adds a relation head similar to Relationformer [20], making it an image-to-graph model. We frame relation prediction as a binary classification problem that predicts whether object 1 is the parent of object 2. We define the table object to always be the parent, while its structures (rows, columns, etc.), caption, and footer objects are children. This serves the purpose of associating captions and footers with their table, as well as grouping table structures that belong to the same table. The grouping of table structures using explicit relation prediction departs from the approach used by TATR, which instead uses bounding box overlap with a parent table object to determine the grouping, similar to DocParser [18].

POTATR as an image-to-graph architecture differs from Relationformer in two key ways. First, POTATR is based on the original DETR architecture [4], same as TATR. Relationformer and other image-to-graph models adopt Deformable DETR [31] as their base architecture. Second, POTATR simplifies the relation head, dropping the use of a relation token. We hypothesize that the relation token is unnecessary when using full self-attention as in the original DETR architecture. The relation head is implemented as a three-layer MLP. The total parameter count for POTATR is 29M.

The main purpose for extending TATR into POTATR rather than adopting an existing image-to-graph model such as Relationformer or EGTR [8] is to leverage TATR's pre-trained weights. We hypothesized that this would significantly boost model performance and confirm this with a small-scale experiment in Sect. 4.2.

3.2 Training

We train and evaluate on PubTables-v2 [21], a large-scale dataset for page- and document-level TE sourced from over one million PubMed articles. PubTables-v2 (PTv2) contains multiple collections, including two collections with full pages as input: PTv2 Single Pages (468k pages with 548k tables annotated with hierarchical relationships between tables, captions, and footers), and PTv2 Full Documents (9k documents containing 137k pages and ∼25k tables, including 9.5k multi-page tables, annotated with HTML for TSR and bounding boxes for table detection).

We train POTATR in two stages. In the first stage, we train for 100 epochs on the Single Pages collection without any change to the data distribution. In the second stage, we train for an additional 35 epochs where each epoch contains an additional 2,000 table-free document pages from the Full Documents collection, as well as repeats the 10k largest tables in the Single Pages collection with different augmentation. These additions are intended to address the long tail of the data distribution.

Training is conducted on 8 Nvidia T4 GPUs with a batch size of 2 per GPU, for an effective batch size of 16. The initial learning rate is 0.00005 (5e-5) with a learning rate gamma of 0.9 applied every 2 epochs. We use the default hyperparameters from TATR [22], except the weight on the "no object" class, `eos_coef`, which we set to 0.1. There is also an additional relation loss with a weight of 0.05.

Most weights are initialized from `TATR-v1.1-Pub` [23], a pre-trained model for TSR on cropped tables. The additional relation head, expanded class head, and the 125 additional object queries are randomly initialized.

4 Experiments

4.1 Metrics

For page-level table extraction, we evaluate using the page-level generalizations of GriTS [24] and TEDS [30] for TE, proposed by Smock *et al.* [21]. At the page-level for TE, it is no longer assumed that there is exactly one predicted table evaluated against one ground truth table, as in the traditional TSR task. Instead, each metric compares two *sets* of tables (predicted and ground truth) and uses the Hungarian algorithm to determine the one-to-one correspondence that maximizes their aggregate F1 score, with GriTS or TEDS used to determine the score for all candidate matches between tables. This flexibility enables evaluation for methods that do not produce a bounding box prediction for a table's location. In our case, we apply the Hungarian matching within a single page, although the same approach can be applied across an entire document.

For object detection we use the standard COCO metrics and for evaluating graph prediction, we use the edge F1 metric from DocParser [18]. We use an IoU threshold of 0.8 to determine if the predicted nodes for each edge are a true positive. For caption text accuracy, we use character-level F1 based on the longest common subsequence (LCS) between predicted and ground truth caption strings.

Table 1. Image-to-graph model comparison. To validate the choice to base POTATR on the pre-trained TATR model, we train each candidate image-to-graph model for 10 epochs on the PubTables-v2 Single Pages collection. Interestingly, the benefits of leveraging pre-trained weights from `TATR-v1.1-Pub` [23] extend not only to classifying the new additional object classes but also to the novel relation prediction task (edge F1), which is not part of the original TATR model.

Model	AP50	AP75	AP	$\text{GriTS}_{\text{Top}}$	$\text{GriTS}_{\text{Con}}$	Edge F1
Relationformer [20]	0.808	0.523	0.484	0.852	0.840	0.339
EGTR [8]	0.791	0.599	0.549	0.850	0.837	0.707
POTATR (ours)	**0.904**	**0.755**	**0.698**	**0.902**	**0.892**	**0.746**

4.2 Image-to-Graph Model Comparison

To validate which modeling approach to adopt for our full-scale training experiments, we compare POTATR against Relationformer [20] and EGTR [8] in a small-scale experiment. We match the parameters of Relationformer and EGTR to those of POTATR and use each model's default losses and loss weights. All models are trained on 8 Nvidia T4 GPUs with a batch size of 2 per GPU (effective batch size 16) for 10 epochs on the PubTables-v2 Single Pages collection, with an initial learning rate of 0.00005 and a learning rate gamma of 0.66.

POTATR outperforms both alternatives across all metrics (Table 1). Interestingly, POTATR outperforms Relationformer and EGTR not only on object detection but also on relation prediction, despite the relation head being an entirely new component. This validates the choice to extend pre-trained TATR with a relation head rather than adopting an existing image-to-graph architecture to train from scratch.

Table 2. Page-level table extraction. Evaluation results on the PubTables-v2 Single Pages collection. POTATR is a 29M-parameter model; all other models are evaluated zero-shot. The best scores among VLMs/MLLMs are underlined; the best overall scores are bolded.

Model	$\text{GriTS}_{\text{Top}}$	$\text{GriTS}_{\text{Con}}$	Acc_{Top}	Acc_{Con}	TEDS_{S}	TEDS
SmolDocling-256M [14]	0.2502	0.2177	0.2800	0.0298	0.5980	0.5250
GraniteDocling-258M [7]	0.7252	0.6670	0.3794	0.0829	0.6685	0.6155
Qwen2.5-VL-3b [3]	0.4672	0.3803	0.3651	0.0619	0.2859	0.2632
Granite-Vision-3.2-2b [26]	0.8015	0.7480	0.4245	0.0331	0.8665	0.8092
DeepSeek-OCR [27]	0.8629	0.8207	0.6184	0.1671	0.7996	0.7639
DeepSeek-OCR 2 [28]	0.8839	0.8386	0.5258	0.1120	0.7688	0.7263
dots.ocr [10]	0.9241	0.8991	0.6425	0.1872	0.9168	0.8743
Claude Opus 4.6	0.9146	0.9106	0.4934	0.2713	0.9073	0.8945
GPT-5.4	0.9208	0.9102	0.5461	0.2074	0.9084	0.8874
Gemini 3.1 Pro	0.9500	0.9418	0.6348	0.3524	0.9392	0.9250
POTATR + EasyOCR [9]	0.9555	0.7819	0.6595	0.0022	0.9468	0.8162
POTATR + PaddleOCR [6]	0.9607	0.8758	0.6915	0.0074	0.9528	0.8759
POTATR + docTR [12]	0.9601	0.8799	0.6707	0.0683	0.9500	0.8952
POTATR + DT	**0.9665**	**0.9636**	**0.6992**	**0.6454**	**0.9565**	**0.9502**

4.3 Page-Level Table Extraction

In the next experiment, we evaluate the fully-trained version of POTATR on page-level table extraction. We compare against seven specialized document VLMs and three frontier MLLMs, all evaluated zero-shot.

Unlike VLMs, POTATR separates structure detection from text recognition: its structure predictions are combined with text either extracted directly from the document (DT), as would be present in digitally-native documents, or from a separate OCR model, as needed for scanned documents. In addition to text extracted directly from the document, we evaluate POTATR with text extracted by three lightweight OCR models (EasyOCR [9], PaddleOCR [6], and docTR [12]).

Results are in Table 2. POTATR with direct text achieves $GriTS_{Con}$ of 0.964—outperforming even the strongest frontier MLLM despite being a 29M-parameter model. The best frontier MLLM (Gemini 3.1 Pro, $GriTS_{Con}$ 0.942) outperforms POTATR paired with traditional OCR. The OCR progression ($0.782 \rightarrow 0.876 \rightarrow 0.880 \rightarrow 0.964$) demonstrates that POTATR's structure recognition is already highly accurate; remaining errors are dominated by text extraction quality and are independently fixable.

Table 3. Caption text content accuracy. Character-level LCS-based F1 on PubTables-v2 Single Pages. POTATR detects caption bounding boxes and assigns text via direct extraction (DT) or OCR; MLLMs are prompted to output caption text alongside table structure.

Model	F1	Precision	Recall
Claude Opus 4.6	0.9611	0.9866	0.9370
GPT-5.4	0.9658	0.9743	0.9575
Gemini 3.1 Pro	0.9774	**0.9882**	0.9669
POTATR + EasyOCR	**0.9789**	0.9830	**0.9749**
POTATR + PaddleOCR	0.9764	0.9799	0.9728
POTATR + docTR	0.9786	0.9828	0.9744
POTATR + DT	0.9781	0.9817	0.9746

4.4 Caption Text Accuracy

Caption extraction is a natural byproduct of POTATR's spatial detection: detected caption bounding boxes are populated with text from the same source used for table cells. We evaluate caption text accuracy using character-level LCS-based F1 (Table 3). POTATR with direct text achieves F1 of 0.978 with balanced precision and recall, matching the best frontier MLLM (Gemini 3.1 Pro, F1 0.977). MLLMs tend toward higher precision but lower recall—they are accurate when they output captions but occasionally omit them. POTATR's detection-based approach ensures systematic caption coverage without requiring explicit prompting.

Interestingly, caption recognition for POTATR with EasyOCR very slightly outperforms caption recognition with text extracted directly from the document itself. While the difference is very slight, it may be due to occasional whitespace captured within document-extracted text bounding boxes or minor reading order ambiguity.

4.5 Inference Cost and Throughput

We compare inference cost and throughput for page-level TE (Table 4). POTATR processes over 52k pages per hour on a single A10G GPU—over 130× the

Table 4. Inference cost and throughput. Measured on the PubTables-v2 Single Pages test set. All local models use a single NVIDIA A10G GPU (24 GB); VLMs use batched inference. Frontier MLLMs are priced via API. Local cost is based on an on-demand cloud instance priced at \$2.45/hr.

Model	Params	Pages/hr	\$/1M pages
Claude Opus 4.6	—	244	\$36,680
GPT-5.4	—	326	\$16,800
Gemini 3.1 Pro	—	393	\$15,000
Qwen2.5-VL-3b	3B	386	\$6,336
dots.ocr [10]	3B	572	\$4,281
DeepSeek-OCR 2 [28]	3B	1,201	\$2,038
SmolDocling-256M	256M	1,247	\$1,963
GraniteDocling-258M	258M	1,847	\$1,325
POTATR	29M	**52,031**	**\$47**

throughput of the fastest frontier MLLM (Gemini 3.1 Pro at 393 pages/hour) and 28× faster than the fastest local VLM (GraniteDocling at 1,847 pages/hour). At under \$50 per million pages, POTATR is roughly 300× cheaper than the most economical frontier MLLM and 28× cheaper than the most economical local VLM. These differences make POTATR practical for large-scale batch processing where even locally-deployed VLMs would be costly.

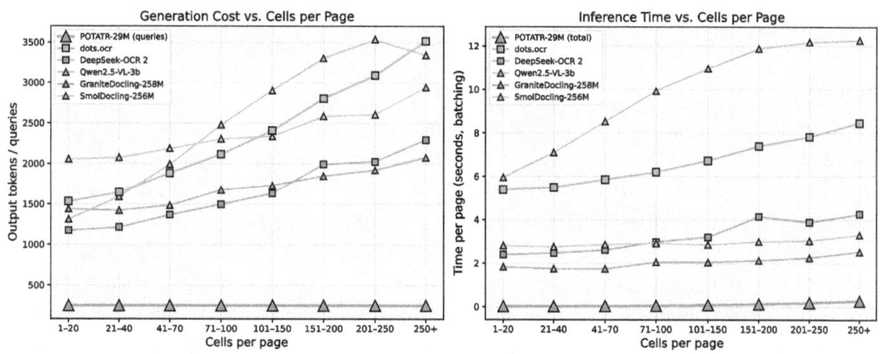

Fig. 3. Generation cost vs. page table cell volume. Left: output tokens (or fixed queries for POTATR) as a function of cells per page. Right: inference time per page with batching. POTATR uses a constant 250 queries regardless of page content; autoregressive VLMs generate tokens proportional to table size, leading to linearly growing cost and latency.

This efficiency gap widens as a page's table cell count increases (Fig. 3). POTATR's inference time is constant at ∼21 ms regardless of cells per page.

Only post-processing (text assignment)—a relatively trivial operation—scales with page table cell count. In contrast, autoregressive VLMs generate output tokens proportional to table size, with inference time growing from 1.8 s to 12 s across the same range. This makes POTATR's throughput advantage even more pronounced on table-heavy document types, even compared with the most economical VLMs running in batch mode.

5 Discussion

5.1 Modularity as a Feature

POTATR's modular design separates structure detection from text recognition, making errors decomposable and each component independently improvable.

POTATR with direct text achieves $\text{GriTS}_{\text{Con}}$ of 0.964, leaving a total error of 0.036. The gap from POTATR + docTR (0.880) to POTATR + DT (0.964) is 0.084—more than twice the remaining error with perfect text. This indicates that text recognition error dominates structure recognition error in the POTATR + docTR result, and that structure recognition is near its ceiling. The OCR progression (EasyOCR 0.782 → PaddleOCR 0.876 → docTR 0.880 → DT 0.964) confirms that a system for page-level table extraction with POTATR automatically benefits from better text extraction without retraining.

VLMs entangle structure and text in a single autoregressive generation process. When a VLM produces incorrect output, it can be difficult to fully attribute the underlying cause of the error. It is also challenging to remedy each issue in isolation, as improving structure recognition risks harming text recognition and vice versa. POTATR allows each component to be improved, replaced, or upgraded independently.

5.2 From Page-Level to Document-Level TE

In practice, documents—not individual pages—are the unit of processing. Assessing page-level models in a document-level context is therefore important for understanding real-world applicability.

PubTables-v2 includes a Full Documents benchmark. But this benchmark is not intended to assess TE performance across a typical distribution of documents. Instead, it is designed specifically to augment the Single Pages benchmark by focusing on documents that contain multi-page tables, which is the case where page-level models are most likely to fail.

Since POTATR is trained on single pages, it has never seen tables that are split across page boundaries. Optimizing POTATR for full documents with multi-page tables is beyond the scope of this work. However, we do think it is of interest to assess the gap between fully single-page TE and complicated document-level TE involving multi-page tables.

We evaluate POTATR in this context in the Appendix (Appendix C). Within a system designed to incorporate a page-level model, POTATR performs well even on multi-page tables, closing 60–73% of the gap to frontier MLLMs at roughly two orders of magnitude lower inference cost (Table 4). Further optimization of POTATR for out-of-distribution multi-page content, and extending the pipeline to handle within-page table part merging, could narrow the remaining gap.

6 Conclusion

We introduced POTATR, a 29M-parameter image-to-graph model extending the Table Transformer for page-level table extraction, which aims to address some of the shortcomings and prohibitive costs associated with autoregressive VLMs. POTATR outperforms all tested models on the PubTables-v2 Single Pages benchmark, including frontier MLLMs, achieving a $\text{GriTS}_{\text{Con}}$ of 0.964 while running over 130× faster at 300× lower cost. POTATR also achieves similar performance to frontier MLLMs on caption text recognition, with a text F1 score of 0.979. Notably, its modular design, paired with its spatially-grounded recognition output, composes well with other small specialized models. Overall, these results suggest small, non-autoregressive models remain a compelling option for real-world page-level table extraction, where their constant inference time regardless of table size offers a fundamental scaling advantage.

7 Limitations

POTATR is trained and evaluated on a dataset of English scientific articles from PubMed; generalization to other document types and languages remains untested. However, TATR—the model that POTATR extends—has been demonstrated to generalize across a variety of document types [23], suggesting that domain adaptation is feasible.

Disclosure of Interests. The authors have no competing interests to declare that are relevant to the content of this article.

Appendix

A Licenses

We plan to open source models and code under the MIT license.

B Accuracy vs. Page Table Cell Count

We break down $\text{GriTS}_{\text{Con}}$ by the number of table cells per page (Fig. 4). Interestingly, no major trends in recognition performance emerge across models. In fact, the frontier MLLMs show somewhat increased table extraction performance as the number of table cells grows. This is despite the fact that the number of tokens

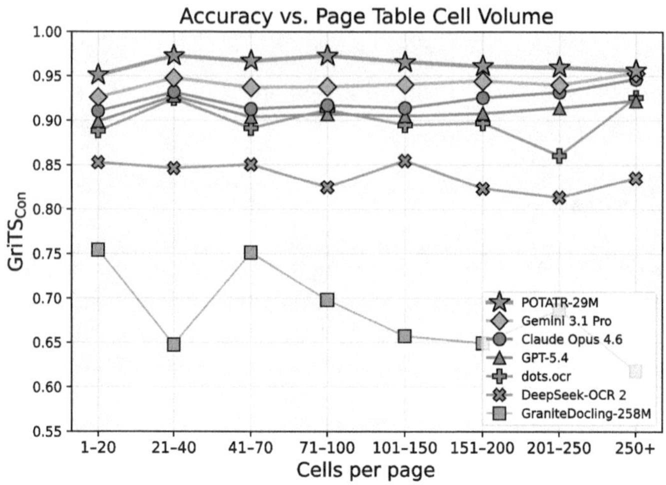

Fig. 4. Accuracy vs. Page table cell count. GriTS$_{Con}$ for each model broken down by cells-per-page on PubTables-v2 Single Pages.

generated, and thus the cost of output, significantly increases with page table cell count (Fig. 3). Furthermore, as the number of previous tokens in a model's context window grows with page table cell count, the amount of previous—potentially distributed—information on which the current token's prediction might need to be conditioned grows. We speculate that this apparent paradox in performance could be because the page layout becomes more uniform (simpler) overall as table cell count on the page increases. Thus any increase in the probability of error due to the increased token output is offset by the increased regularity in structure. On the other hand, POTATR, which is not autoregressive, exhibits potentially more expected behavior, with performance peaking on the most typical tables in the data distribution (Table 5).

C Multi-page Table Extraction

In this experiment, we assess the gap in performance when applying POTATR, optimized for single-page TE, to the most challenging scenario for document-level TE—documents containing multi-page tables.

Initially we apply POTATR to these documents using page-by-page processing only. This leads to an overall GriTS$_{Con}$ score of 0.671. While POTATR can recognize single-page tables in these documents, any predicted table corresponds to at most only a fraction of any multi-page ground truth table, significantly lowering the achievable GriTS$_{Con}$ score. For reference, we also compare to the best-performing small VLM, dots.ocr, which is also single-page only and achieves a GriTS$_{Con}$ score of 0.577. Frontier MLLMs with full-document context, on the other hand, perform strongly: the best performing frontier model, Gemini 3.1 Pro, achieves GriTS$_{Con}$ of 0.931.

Table 5. Multi-page table extraction. Evaluation on PubTables-v2 Full Documents containing multi-page tables. Full-document models receive all pages at once, whereas page-by-page models process each page independently. The last two rows show the effect of applying cross-page merging [21] to the page-by-page models' output.

Model/Context	$\text{GriTS}_{\text{Top}}$	$\text{GriTS}_{\text{Con}}$	Acc_{Top}	Acc_{Con}	TEDS_{S}	TEDS
Full document						
Claude Opus 4.6	0.9116	0.9079	**0.5799**	**0.2452**	0.8914	0.8798
GPT-5.4	0.8865	0.8847	0.5038	0.1636	0.8563	0.8341
Gemini 3.1 Pro	**0.9365**	**0.9309**	0.4854	0.2025	**0.9048**	**0.8905**
Pages processed individually						
dots.ocr	0.6054	0.5768	0.3958	0.1180	0.6179	0.5876
POTATR	0.6973	0.6710	0.3285	0.2038	0.6353	0.6128
Pages processed individually + cross-page merging						
dots.ocr + merging	0.7772	0.7495	0.5059	0.1180	0.8022	0.7731
POTATR + merging	0.8507	0.8269	0.4234	0.2176	0.8058	0.7824

Next, we pair dots.ocr and POTATR with a ViT-B-16 table continuation classifier [21]. This model takes as input a pair of page images and predicts whether a table spans a page boundary. When a single-page model predicts a table on both pages and the continuation model predicts a positive continuation, we merge the two table segments vertically.

POTATR with cross-page merging achieves $\text{GriTS}_{\text{Con}}$ of 0.827, a 47.4% reduction in error over processing pages without merging (0.671). Applying the same merging strategy to `dots.ocr` yields a 40.8% reduction in error. Notably, this improvement requires no retraining of the page-level model; it is achieved entirely through composition with a lightweight classifier (ViT-B-16, 86M parameters) and a simple vertical concatenation rule. This addresses a substantial proportion of the gap between single-page models and frontier MLLMs on documents with multi-page tables—60% for POTATR versus Gemini 3.1 Pro, and 73% for POTATR versus GPT-5.4.

Cross-page table continuation prediction is near-optimal (0.995 recall), so the remaining gap is attributable to page-level extraction quality on multi-page table content (pages that POTATR never encounters during its current single-page training) and to within-page table part merging, which the current pipeline does not address. Optimizing POTATR for these out-of-distribution pages, and extending the merging strategy to handle table parts split within a single page, are clear directions for future work.

References

1. Alayrac, J.B., et al.: Flamingo: a visual language model for few-shot learning. Adv. Neural. Inf. Process. Syst. **35**, 23716–23736 (2022)
2. Baek, Y., Nam, D., Surh, J., Shin, S., Kim, S.: TRACE: table reconstruction aligned to corner and edges. In: International Conference on Document Analysis and Recognition, pp. 472–489. Springer (2023)
3. Bai, S., et al.: Qwen2.5-VL Technical report. arXiv preprint arXiv:2502.13923 (2025)
4. Carion, N., Massa, F., Synnaeve, G., Usunier, N., Kirillov, A., Zagoruyko, S.: End-to-end object detection with transformers. In: ECCV, pp. 213–229 (2020)
5. Choi, Y.Y., Kim, T., Kim, N., Lee, T., Joe, S.: End to end table transformer. In: International Conference on Document Analysis and Recognition, pp. 331–345. Springer (2024)
6. Cui, C., et al.: PaddleOCR 3.0 Technical report. arXiv preprint arXiv:2507.05595 (2025)
7. IBM-Granite: Granite Docling (2025). https://huggingface.co/ibm-granite/granite-docling-258M
8. Im, J., Nam, J., Park, N., Lee, H., Park, S.: EGTR: extracting graph from transformer for scene graph generation. In: Proceedings of the IEEE/CVF Conference on Computer Vision and Pattern Recognition, pp. 24229–24238 (2024)
9. JaidedAI: EasyOCR (2024). https://github.com/JaidedAI/EasyOCR
10. Li, Y., Yang, G., Liu, H., Wang, B., Zhang, C.: dots.ocr: Multilingual document layout parsing in a single vision-language model. arXiv preprint arXiv:2512.02498 (2025)
11. Luccioni, A.S., Jernite, Y., Strubell, E.: Power hungry processing: Watts driving the cost of ai deployment? In: Proceedings of the 2024 ACM Conference on Fairness, Accountability, and Transparency (FAccT), pp. 85–99. ACM (2024)
12. Mindee: docTR: Document text recognition (2021). https://github.com/mindee/doctr
13. Nassar, A., Livathinos, N., Lysak, M., Staar, P.: Tableformer: table structure understanding with transformers. In: Proceedings of the IEEE/CVF Conference on Computer Vision and Pattern Recognition, pp. 4614–4623 (2022)
14. Nassar, A., et al.: SmolDocling: an ultra-compact vision-language model for end-to-end multi-modal document conversion. arXiv preprint arXiv:2503.11576 (2025)
15. Paliwal, S.S., Vishwanath, D., Rahul, R., Sharma, M., Vig, L.: Tablenet: deep learning model for end-to-end table detection and tabular data extraction from scanned document images. In: 2019 international conference on document analysis and recognition (ICDAR), pp. 128–133. IEEE (2019)
16. Poznanski, J., et al.: olmocr: unlocking trillions of tokens in pdfs with vision language models. arXiv preprint arXiv:2502.18443 (2025)
17. Prasad, D., Gadpal, A., Kapadni, K., Visave, M., Sultanpure, K.: CascadeTabNet: an approach for end to end table detection and structure recognition from image based documents. In: Proceedings of the IEEE/CVF Conference on Computer Vision and Pattern Recognition Workshops, pp. 572–573 (2020)
18. Rausch, J., Martinez, O., Bissig, F., Zhang, C., Feuerriegel, S.: Docparser: hierarchical document structure parsing from renderings. In: Proceedings of the AAAI Conference on Artificial Intelligence. vol. 35, pp. 4328–4338 (2021)
19. Schreiber, S., Agne, S., Wolf, I., Dengel, A., Ahmed, S.: DeepDeSTR: deep learning for detection and structure recognition of tables in document images. In: 2017 14th

IAPR International Conference on Document Analysis and Recognition (ICDAR), vol. 1, pp. 1162–1167. IEEE (2017)
20. Shit, S., et al.: Relationformer: a unified framework for image-to-graph generation. In: European Conference on Computer Vision, pp. 422–439. Springer (2022)
21. Smock, B., et al.: PubTables-v2: a new large-scale dataset for full-page and multi-page table extraction. arXiv preprint arXiv:2512.10888 (2025)
22. Smock, B., Pesala, R., Abraham, R.: PubTables-1M: towards comprehensive table extraction from unstructured documents. In: Proceedings of the IEEE/CVF Conference on Computer Vision and Pattern Recognition (CVPR), pp. 4634–4642 (2022)
23. Smock, B., Pesala, R., Abraham, R.: Aligning benchmark datasets for table structure recognition. In: International Conference on Document Analysis and Recognition, pp. 371–386. Springer (2023)
24. Smock, B., Pesala, R., Abraham, R.: GriTS: grid table similarity metric for table structure recognition. In: International Conference on Document Analysis and Recognition, pp. 535–549. Springer (2023)
25. Team, G., et al.: Gemini: a family of highly capable multimodal models. arXiv preprint arXiv:2312.11805 (2023)
26. Team, G.V., et al.: Granite vision: a lightweight, open-source multimodal model for enterprise intelligence. arXiv preprint arXiv:2502.09927 (2025)
27. Wei, H., Sun, Y., Li, Y.: Deepseek-OCR: contexts optical compression. arXiv preprint arXiv:2510.18234 (2025)
28. Wei, H., Sun, Y., Li, Y.: Deepseek-OCR 2: visual causal flow. arXiv preprint arXiv:2601.20552 (2026)
29. Zheng, X., Burdick, D., Popa, L., Zhong, X., Wang, N.X.R.: Global table extractor (GTE): a framework for joint table identification and cell structure recognition using visual context. In: Proceedings of the IEEE/CVF winter conference on applications of computer vision, pp. 697–706 (2021)
30. Zhong, X., ShafieiBavani, E., Jimeno Yepes, A.: Image-based table recognition: data, model, and evaluation. In: European Conference on Computer Vision, pp. 564–580. Springer (2020)
31. Zhu, X., Su, W., Lu, L., Li, B., Wang, X., Dai, J.: Deformable DETR: deformable transformers for end-to-end object detection. In: ICLR (2021)

Structured Handwritten Form Synthesis via Unsupervised Template Recovery and Field-Level Segmentation

Natalia Bottaioli[1(✉)], Yanhao Li[1], Thomas Gorges[2], Solène Tarride[3], Vincent Christlein[2], Gabriele Facciolo[1,5], and Jean-Michel Morel[4]

[1] Centre Borelli, ENS Paris-Saclay, Université Paris-Saclay, CNRS, Gif-sur-Yvette, France
natalia.bottaioli@ens-paris-saclay.fr
[2] Pattern Recognition Lab, Department of Computer Science, Friedrich-Alexander-Universität Erlangen-Nürnberg, Erlangen, Germany
[3] Teklia, Paris, France
[4] Lingnan University, Tuen Mun, Hong Kong
[5] Institut Universitaire de France, Paris, France

Abstract. Structured handwritten forms are common in large digitization projects and are often the starting point for downstream tasks such as Handwritten Text Recognition (HTR) and Named Entity Recognition (NER). Out-of-the-box solutions for these tasks typically perform poorly on such data. Although fine-tuning is a common strategy, privacy regulations often limit the direct use of real-world documents due to sensitive entries. Toward the goal of privacy-preserving dataset generation, we propose a pipeline to synthesize sets of handwritten forms that share the same template. In this work, we validate the foundational synthesis process. Our method first recovers the underlying empty template using hierarchical homography-based registration and quantile-based aggregation, in a fully unsupervised manner. Next, it surgically segments handwritten fields. Finally, it generates reconstructed versions of the original documents by pasting these fields onto the recovered empty template. Our evaluation is twofold. First, using image processing metrics on the public NAF dataset and a private civil registry dataset, we demonstrate robust template recovery and accurate field segmentation across varying levels of degradation and handwriting styles. Second, we evaluate the downstream utility of the reconstructed documents by training a key-value information extraction model both on real and synthesized RIMES forms. When evaluated on 200 real unseen RIMES documents, the model trained with our synthesized data achieves slightly lower CER and WER (4.79% and 16.15%) than a model trained on real RIMES data (5.00% and 17.05%).

Keywords: Document Analysis · Unsupervised Template Recovery · Image Registration · Handwritten Form Synthesis · Data Anonymization

F. Shafait et al. (Eds.): DAS 2026, LNCS 16851, pp. 18–35, 2027.
https://doi.org/10.1007/978-3-032-36207-0_2

1 Introduction

Structured handwritten forms are common in administrative and historical archives, particularly in civil registry collections and institutional records. Large-scale digitization initiatives have produced vast collections of such documents [2,4]. Many of them share pre-printed templates that were filled out by hand. Automatic processing of these forms typically involves layout analysis, field segmentation, and Handwritten Text Recognition (HTR) [3,7]. In practice, the use of pre-trained systems without adaptation often leads to poor results, both because form structure is template-specific and handwriting varies strongly across writers.

Despite recent improvements in HTR transcription performance in unconstrained settings [5,28], structured forms still raise significant challenges: handwriting is spatially constrained within predefined fields, templates may vary between document families, and degradation or scanning artifacts can affect layout consistency.

In order to reach adequate levels of accuracy, a frequent approach is to fine-tune existing models. However, in many real-world scenarios, digitized documents contain personal or sensitive data and cannot be directly used for training. Regulations such as the General Data Protection Regulation (GDPR) can also hinder the direct publication of original documents for research purposes [6]. In such contexts, artificially generating documents that resemble the targeted ones becomes essential to fine-tune existing models or to train new ones.

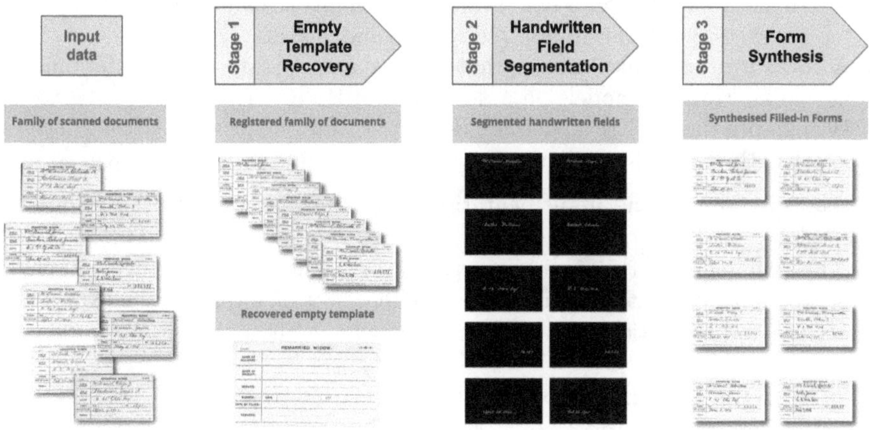

Fig. 1. An overview of the proposed pipeline showing the main outputs of each of its three stages.

To address these limitations and move toward fully privacy-preserving data generation, we propose a modular framework for synthesizing handwritten forms starting from a real family of handwritten forms sharing the same printed template. Our framework consists of three core stages: 1) recovering the underlying empty template from a collection of filled forms, 2) "surgically" segmenting the handwritten fields, and 3) recombining these fields onto the recovered clean template.

Ultimately, this framework is designed to enable robust data anonymization by randomly shuffling segmented fields across different documents in a family. Recombining fields from multiple documents onto a single template can break the linkability of the original personal data and generate privacy-compliant, multi-writer documents for training.

Our purpose here is to design and validate the crucial foundational synthesis process itself. If the layout geometry is distorted during that process, or if writer-specific characteristics are degraded during template recovery and segmentation, the downstream utility of the generated documents will fail, regardless of the quality of the ensuing anonymization. Therefore, in this work, we focus on establishing and evaluating the core pipeline up to the single-document synthesis step—reconstructing documents. Our validation is performed on reconstructed documents obtained by pasting the original fields of a document back onto the recovered clean template. We then show that the resulting synthesized forms retain sufficient structural and textual integrity to successfully train downstream Document Understanding models.

Specifically, our main contributions are:

- A robust fully unsupervised template recovery method that mitigates the effects of handwriting variation, degradation, and scanning artifacts;
- A "surgical" field-level segmentation mechanism enabling document-level reconstruction without annotated layout information.
- A form synthesis mechanism that proves to generate documents that are usable for training models on a downstream HTR/NER task.
- We explore the quality of shuffled documents for our private dataset and also for the RIMES [15] subset used in our experiments.

The source code and implementation details for our framework are publicly available.[1]

The remainder of this paper is organized as follows: Sect. 2 reviews related work. Section 3 presents the proposed pipeline. Section 4 reports the experimental results and limitations. Lastly, Sect. 5 concludes the paper.

2 Related Work

Form understanding [1] and template-based document processing have been studied extensively. Early approaches relied on rule-based systems and structural matching techniques to align filled forms with predefined templates [18]. Subsequent work introduced image registration and homography-based alignment methods to normalize documents within the same family [31]. As for template extraction from aligned documents, most methods assume access to a clean reference [10] or require supervised annotations of layout components.

[1] https://github.com/natibottaioli/structured-handwritten-form-synthesis.

Classic approaches for empty template recovery usually register multiple filled-in documents and then extract the template from the aligned set. Hutchison *et al.* [16,17] utilized Fourier-based and local-feature methods for this purpose, but these methods can be sensitive to local variations caused by handwriting. Other works align a known empty template with filled instances: Mahajan *et al.* [23] used character keypoints, Kessi *et al.* [21] adapted non-local means for registration and separation, and Narayan and Gowda [25] subtracted common foreground after alignment to isolate handwriting.

Regarding handwritten text segmentation, most traditional methods typically work at the line or word level but not at the field level, and they provide bounding boxes which are either rectangles or more generic polygons. Modern instance segmentation models such as YOLOv8 [20] offer pixel-level segmentation. However, in order to produce this kind of mask, they typically need annotations on the pixel level.

Beyond standard geometric transformations [14], the generation of new document instances is often explored through deep generative models [27]. However, these approaches typically require explicit layout annotations to enforce complex spatial constraints. Alternatively, recombination-based methods offer a highly realistic and practical way to synthesize documents [29]. Although existing recombination techniques rely on pristine reference templates and annotated layout components, our framework achieves this in a fully unsupervised manner using only collections of filled-in documents.

To the best of our knowledge, the availability of public handwritten preprinted form datasets is very limited. The NAF dataset [11] contains a variety of templates. However, the number of documents sharing the same template is usually small, which limits certain experiments. The RIMES dataset [15] contains more than 5,000 images of scanned forms, but there are only two templates in the database. Moreover, there are not many forms filled in by a single writer. Several additional datasets for form understanding, such as FUNSD [19] and FormNLU [12], are discussed in [1]. Nevertheless, the presence of handwritten-filled forms is scarce. For these reasons, we also use a private dataset containing more than 20 different templates, some of which feature multiple forms completed by the same writer.

3 Methodology

Our pipeline, shown in Fig. 1, includes three stages: *empty template recovery*, *handwritten field segmentation*, and *form synthesis*. First, the blank printed template is recovered from a set of scanned documents sharing a fixed layout. Next, we perform field-level segmentation to isolate handwritten content. Finally, new forms are synthesized by pasting the segmented fields onto the clean template. In this work, when we segment all fields from a single original document and paste them onto an empty template, we refer to the output as a reconstructed document. Table 1 summarizes the mathematical symbols and structural conventions used in this section.

Table 1. Summary of mathematical notations.

Notation	Description
$\mathcal{I}_i$, $\mathcal{M}_i$	The i-th original document image and its corresponding mask
$\mathcal{I}_{reg,i}$, $\mathcal{M}_{reg,i}$	The i-th original document image after registration and its corresponding mask
$\mathcal{I}^T$, $\mathcal{M}^T$	The (recovered) empty template and its corresponding mask
$\mathcal{I}^H_{i,k}$, $\mathcal{M}^H_{i,k}$	The k-th handwritten connected curve within the i-th document after removing the empty template and its corresponding mask
$\mathcal{I}^F_j$, $\mathcal{M}^F_j$	The j-th handwritten field of the document family and its corresponding mask

3.1 Empty Template Recovery

The key challenge to recovering empty templates from a set of handwritten-filled documents is to align scans robustly despite handwriting and scanning artifacts. This stage consists of the following three steps: (1) create an averaged reference image; (2) register all images to the averaged reference; (3) extract the empty template by quantile aggregation.

We adopt the parameter-free homographic registration method of Moisan *et al.* [24], which estimates a homography between two images using local feature matches based on SIFT [22], and robustly removes outliers with RANSAC (Random Sample Consensus) algorithm [13] coupled with an *a contrario* criterion. This approach is well suited to forms, as most stable keypoints originate from the printed template, while handwriting is mostly treated as outliers.

Step 1.1: Create an Averaged Reference Image. Let $\mathcal{I}_0, \ldots, \mathcal{I}_{p-1}$ be p handwritten document images created using the same form. To compute an averaged registered image, we adopt the following hierarchical registration scheme. The function `register` estimates the homography between two images and warps the first image into the coordinate frame of the second.

(1) **Initialization**: set $n = \lfloor \log_2 p \rfloor$; set $A_i^{(0)} = I_i$, for $i = 0, \ldots, 2^n - 1$.
(2) **Iterative registration**: for each iteration $k = 1, \ldots, n$, register the averaged pairs of images from the previous iteration:

$$A_i^{(k)} = \frac{A_{2i}^{(k-1)} + \texttt{register}(A_{2i+1}^{(k-1)}, A_{2i}^{(k-1)})}{2}, \tag{1}$$

for $i = 0, \ldots, \lfloor \frac{p}{2^k} \rfloor - 1$.
(3) **Final result**: the final averaged reference image is $A_0^{(n)}$.

Figure 2 illustrates the generation of an averaged registered image from eight documents that share the same underlying template. When two documents are stitched together, the result resembles a single form filled out twice (with potentially different handwritten content). Since both handwritings are overlaid, the strokes appear lighter than in the original images. As more documents are combined, the handwriting regions become more even, while the underlying template remains visible.

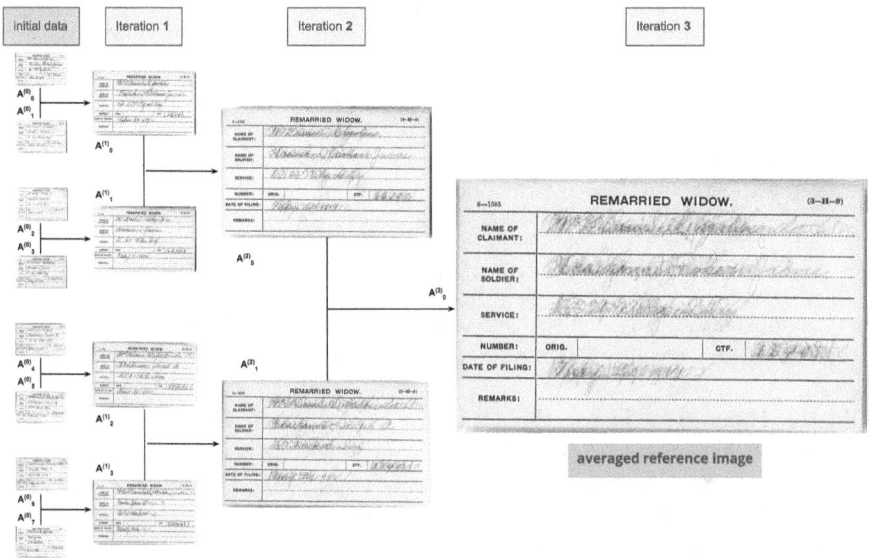

Fig. 2. Hierarchical generation of an averaged reference image using eight documents from the NAF dataset [11]. At each iteration, pairs of images are registered and aggregated, reinforcing the stable template structure while progressively attenuating variable handwritten content. The alignment remains sharp across iterations, leading to a clean reference for global registration.

Step 1.2: Register All Images to the Averaged Reference. Once the averaged reference image is established, we again use the registration method from [24] to align all documents within the same family to this reference. Each registered image $\mathcal{I}_{reg,i}$ is resampled to the same size as the reference image to facilitate subsequent operations.

Step 1.3: Extract the Empty Template by Quantile Aggregation. Following the previous step, the printed template is aligned across all registered document images $\{\mathcal{I}_{reg,i}\}_{i=0}^{p-1}$. *A priori*, images belonging to the same document family after registration have approximately the same values at pixels belonging to the underlying empty template. For pixels not belonging to the printed form, the values of different images can vary, depending on whether handwritten text is present or not.

We assume that handwritten and printed pixels are darker than the blank paper background. In ideal settings, where the printing and scanning conditions are identical for all documents, we could recover the empty template $\mathcal{I}^T$ from the pixel-wise maximum over the registered images: $\mathcal{I}^T(x, y) = \max_i \mathcal{I}_{reg,i}(x, y)$ for all pixel positions (x, y).

However, in practice, some template strokes can be missing or degraded due to printing defects or document wear. In these cases, taking the maximum may result in unwanted gaps in the empty template result, as shown in Fig. 8e. To improve reconstruction robustness, instead of using the maximum, we obtain the empty template image $\mathcal{I}^T$ by computing the pixel-wise q-quantile over the registered document images: $\mathcal{I}^T(x, y) = \text{quantile}_q(\{\mathcal{I}_{reg,i}(x, y)\})$.

In our experiments, $q = 0.9$ produces a clean empty template.

3.2 Handwritten Field Segmentation

The previous stage provides (i) the recovered empty template $\mathcal{I}^T$ and (ii) a set of registered document images $\{\mathcal{I}_{reg,i}\}_{i=0}^{p-1}$, where all documents are spatially aligned.

In this stage, we will use these outputs to automatically separate the field areas and extract handwritten content from each field. It consists of the following steps: (1) estimate field masks from the family mean, (2) compute a handwriting-only image for each registered document, and (3) extract per-document field instances using the estimated field masks.

Step 2.1: Identify Field Regions (Fig. 3). The initial step identifies distinct fields within the form. We generate a handwritten zone mask for each field by averaging the family of documents and then removing the template. Specifically, we compute the pixel-wise mean of the registered documents, $\bar{\mathcal{I}} = \frac{1}{p} \sum_{i=0}^{p-1} \mathcal{I}_{reg,i}$, and isolate the handwritten content by subtracting the recovered template $\mathcal{I}^T$. The resulting residual image is binarized using Otsu's method [26]. Stable handwritten regions are identified by extracting connected components and filtering out noise using a minimum area threshold. Adjacent components within the same text line are merged to form a unified field mask. The result is a set of field masks $\{\mathcal{M}_j^F\}_{j=1}^m$ corresponding to different handwritten field regions, as can be seen in the right-most part of Fig. 3.

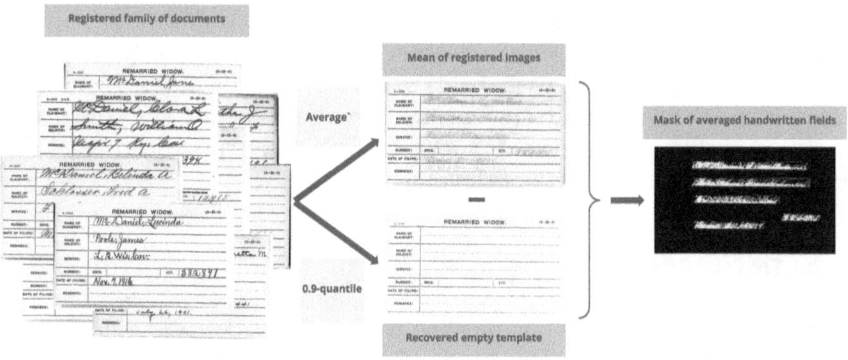

Fig. 3. Process of identifying field regions based on the average of all registered documents.

Step 2.2: Compute Handwriting-only Images (Fig. 4). For each registered document $\mathcal{I}_{reg,i}$, we remove the printed template by masking template pixels using the binarized recovered template $\mathcal{M}^T$. Masked pixels are set to background prior to inpainting. Because handwritten strokes may overlap the printed template, this removal can erase parts of characters; we therefore apply biharmonic inpainting [9] in the masked regions to restore missing content. This yields a handwriting-dominant image $\mathcal{I}_i^H$ and a corresponding binarized mask $\mathcal{M}_i^H$ (as can be seen in Fig. 4, under the label "Mask of handwritten text"). We then use the connected components of the mean document as a structural reference to localize the corresponding field regions in each document, enabling the construction of a handwritten mask for every field.

Step 2.3: Extract Field Instances (Fig. 5). Given the mean field masks $\{\mathcal{M}_j^F\}_{j=1}^m$ and a handwriting-only mask $\mathcal{M}_i^H$ for document i, we divide the mask $\mathcal{M}_i^H$ into K separate handwritten curves $\{\mathcal{M}_{i,k}^H\}_{k=1}^K$ by connected component segmentation, then extract the handwritten contents of field j by intersecting the handwritten components with the field mask:

$$\mathcal{M}_{i,j}^F = \bigcup_{k'} \{\mathcal{M}_{i,k'}^H : \mathcal{M}_{i,k'}^H \cap \mathcal{M}_j^F \neq \emptyset\}. \tag{2}$$

Intuitively, the mean field mask functions as a spatial prior to isolate the relevant handwriting components for each field. The extracted handwritten field $\mathcal{I}_{i,j}^F$ is obtained through element-wise multiplication: $\mathcal{I}_{i,j}^F = \mathcal{I}_{reg,i} \odot \mathcal{M}_{i,j}^F$.

3.3 Form Synthesis

Form synthesis can use segmented contents at any order. For reconstructed documents, we take a document of the given family after registration, $\mathcal{I}_{reg,i}$, we extract all its handwritten fields $\mathcal{I}_{i,j}^F$ and integrate them into the recovered (empty) template $\mathcal{I}^T$. An example of the result of such process applied to a document of the NAF dataset can be seen in Fig. 9. Alternatively, to generate shuffled documents, we can paste fields extracted from different original documents within the same family onto a single empty template $\mathcal{I}^T$. As discussed in Sect. 1, our evaluation in this paper focuses exclusively on reconstructed documents to validate the integrity of the pipeline.

4 Experiments

In this section, we evaluate our pipeline from two perspectives: the structural fidelity of the image processing stages, and the practical utility of the reconstructed documents for downstream Document Understanding tasks (HTR and NER).

To evaluate the image processing pipeline, we utilize two datasets: a subset of the public NAF dataset (noisy historical scans from the late 1800s and early 1900s) and a private dataset of 1,000 documents from Peru's National Civil Registration and Identification Office (RENIEC). Due to data protection constraints, we only show original images for the public NAF dataset. For the private dataset, we present derived visualizations that do not reveal individuals' records. Finally, to evaluate downstream HTR/NER utility, we use a subset of the RIMES dataset.

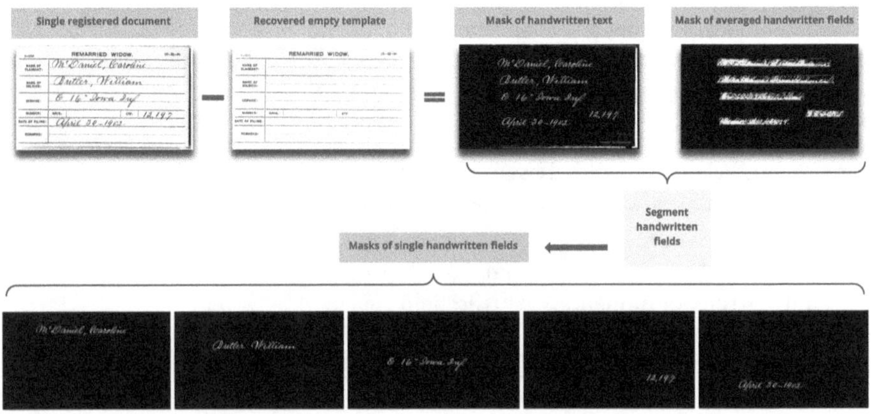

Fig. 4. Handwritten field segmentation applied to one single document of the NAF dataset. Note that these are the masks that are then used to reconstruct the given document that appears in the top-left part of the figure.

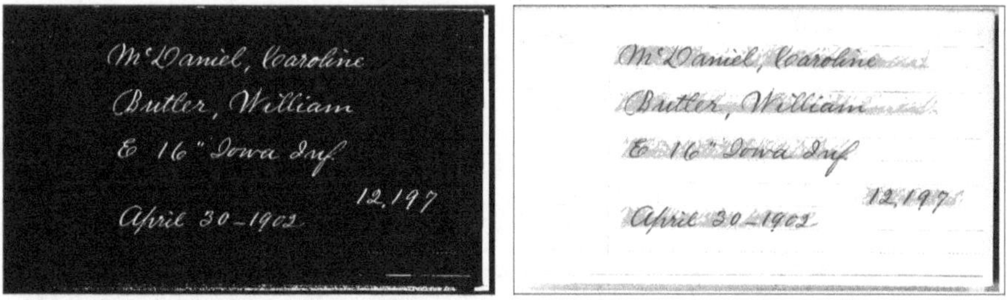

(a) Mask of one single document after erasing the empty template.

(b) Mean fields plus handwritten text intersected by mean fields.

Fig. 5. Intermediary steps of the process of obtaining a mask of the handwritten fields of one single document (shown in dark red) by using the masks of the handwritten fields of the mean of the registered documents (shown in pink). Note that, in this example, the comma between 'Butler' and 'William' and a tiny part of the '9' in the last field are not shown in dark red. This means that they will be ignored when cutting handwritten fields in future steps.

4.1 Registration Quality

First, we evaluate the alignment quality by calculating an overlap score between the registered scan $\mathcal{I}_{reg,i}$ and the recovered template $\mathcal{I}^T$. We binarize both the recovered template and each registered scan using OpenCV's adaptive Gaussian thresholding. To reduce border artifacts, we ignore a 5% margin in the image by putting to zero the corresponding border region in the template mask. The alignment score is defined as the fraction of template-foreground pixels that are also foreground in the registered scan:

$$q_i = \frac{\sum_x \mathbb{1}[\mathcal{M}^T(x) = 1 \wedge \mathcal{M}_{reg,i}(x) = 1]}{\sum_x \mathbb{1}[\mathcal{M}^T(x) = 1] + \varepsilon}, \tag{3}$$

where x are the pixel locations and ε is a small constant for numerical stability. A value close to 1 indicates that the printed template foreground is well aligned with the registered scan. Importantly, because this metric only evaluates pixels belonging to the template mask, extra foreground in scan-such as handwriting and noise-is not penalized. The lower values, in contrast, suggest geometric misalignment or missing template strokes.

On the NAF subset, overlap scores are consistently high, with 0.9244 ± 0.0113 (USA Declaration of Intention), 0.9258 ± 0.0052 (Alien Crew Vessel), and 0.9456 ± 0.0214 (Remarried Widow). These results show that the registration process produces high-quality alignment despite handwritten variability and scan artifacts.

Figure 6 shows the alignment score distribution and visualizes three randomly selected examples (one of each template family from the NAF dataset). In these visualizations, pixels in green represent where the template foreground is supported by the scan, while cyan pixels represent missing foreground pixels in the scan. For readability, we dilate the visualization overlay by a 3×3 morphological dilation. The scores are computed from the undilated masks. Some cyan pixels are visible, which can be attributed to small local misalignment. These results suggest that field extraction operates on reliably registered data.

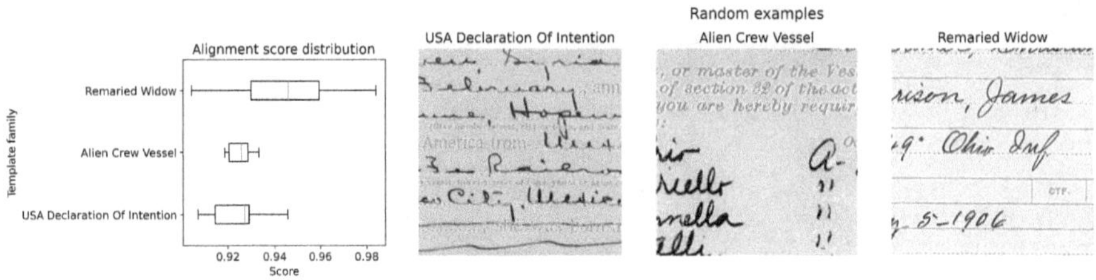

Fig. 6. Registration quality on a subset of the NAF dataset [11]. The distribution of the alignment quality is shown in the box plot, as well as zoomed-in overlap visualizations for three randomly selected examples. Supported template pixels are shown in green, while missing template pixels are shown in cyan. Overlay is morphological dilated 3×3 for readability.

4.2 Empty Template Recovery

We evaluate two factors that affect the quality of the recovered template: the form layout and the quantile parameter.

Robustness to Template Diversity: The second row of Fig. 7 shows the final empty templates recovered from the three existing document families. These families vary in noise and degradation, the number of available documents, and the handwriting styles. Despite these variations, all recovered templates are sufficiently clear, producing synthetic documents. It is also visible that the recovered templates preserve visible global appearance properties from the family of documents, such as overall intensity.

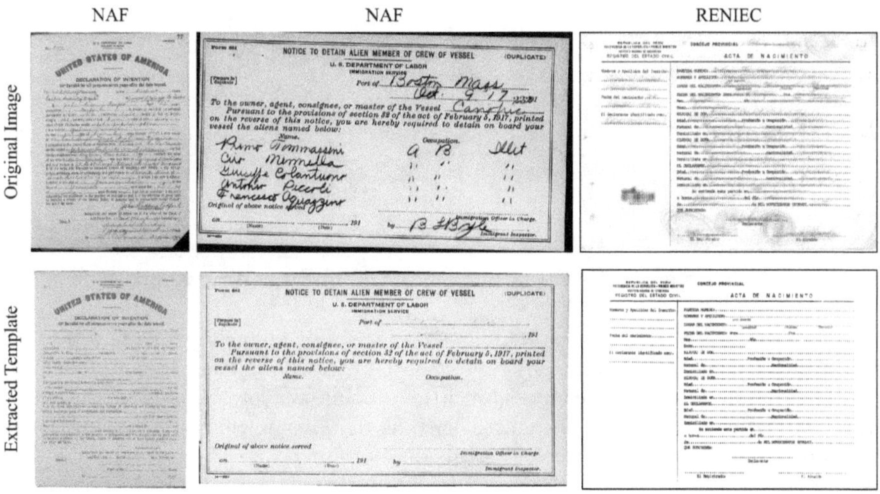

Fig. 7. The first row shows two original images belonging to the NAF dataset [11], and an image taken from the process of finding an averaged reference image for a family of birth certificates belonging to the private (RENIEC) dataset. The empty templates shown in the second row were obtained using 10, 9, and 78 images, respectively. The original documents of the first family were filled in by two different writers, while all original documents of the second family were written by a single writer. As a result, visible handwritten text residuals remain in the second extracted template because the consistent writing habits of a single writer are not fully attenuated during averaging.

Impact of the Quantile. We employ a quantile-based approach in the template recovery stage described in Sect. 3.1 because the printed templates are not always evenly printed or scanned. If a template pixel is missing or faint in some documents, a pixel-wise maximum can select an unusually bright value or introduce artifacts. Using the quantile provides a robust alternative. Figure 8 compares different quantile values. In our experiments, the 0.9-quantile gave the best trade-off between template clarity and suppression of handwriting traces.

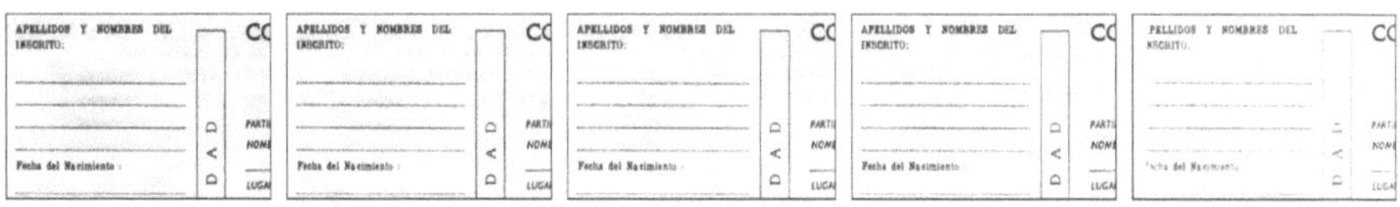

(a) 0.5 quantile (b) 0.7 quantile (c) 0.8 quantile (d) 0.9 quantile (e) maximum

Fig. 8. Zoomed-in examples of empty-template recovery from a family of 78 documents with different quantile values. Lower quantiles ((a), (b)) leave visible handwritten traces, whereas near-maximum quantiles ((c), (d)) make the template less visible as light pixels dominate. With more than 16 documents, the 0.9 quantile provides a good trade-off between template clarity and suppression of handwritten content, as illustrated both in (d) and in the second row of Fig. 7.

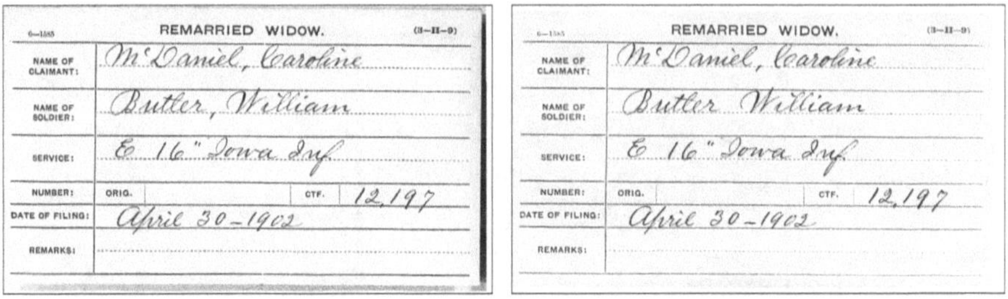

(a) Registered image. (b) Reconstructed image.

Fig. 9. Registered image from the NAF dataset (a) obtained by Step 1.2 (Sect. 3.1) and reconstructed image (b) obtained by Steps 2.1–2.3 (Sect. 3.2). Two types of differences are visible: template-related changes, such as missing shadow borders in the reconstruction, and handwritten-field changes, such as the missing comma between 'Butler' and 'William' and the slightly cut '9' in the date of filing.

4.3 Handwritten Field Segmentation and Form Synthesis

We evaluate the second and third stages of our pipeline, which extract handwritten fields and synthesize a reconstructed version of the given documents, respectively (see Fig. 9). The fidelity of this reconstruction is quantified using the Structural Similarity Index (SSIM) [32]. This metric measures the structural similarity between the original and reconstructed images (a value of 1 indicates a perfect match, while 0 indicates no similarity).

In order to have an SSIM evaluation that is more capable of expressing the structural (geometric) differences between the reconstructed and original documents, we also evaluate this metric after applying denoising and standardization by making the data zero-mean and the standard deviation equal to one. We use the Peak Signal-to-Noise Ratio (PSNR) metric in both scenarios.

Table 2 shows experiments on 16 NAF and 64 RENIEC documents, demonstrating that denoising consistently improves PSNR and SSIM. A robust registration stage prevents geometric distortions, meaning reconstruction errors primarily arise from segmentation and intensity normalization. Intragroup SSIM variations are explained by structural differences: the first 8 higher-performing NAF documents use identical templates and fields, unlike the second half. For RENIEC, all documents share one template, but the first group benefits from significantly lower noise.

Table 2. Comparative analysis of PSNR and SSIM metrics in two templates, one from the NAF dataset (public) and one from the RENIEC dataset (private).

	PSNR (dB)			SSIM		
Dataset / Group	Before	After	Δ	Before	After	Δ
NAF Experiment						
Docs 0–7	22.26	23.32	+1.06	0.8853	0.9477	+0.0624
Docs 8–15	18.58	19.12	+0.54	0.8594	0.9049	+0.0455
All (16 documents)	**20.42**	**21.22**	**+0.80**	**0.8723**	**0.9263**	**+0.0540**
RENIEC Experiment						
Docs 0–29	22.60	22.81	+0.21	0.9357	0.9472	+0.0115
Docs 30–63	18.33	21.35	+3.02	0.8676	0.9223	+0.0547
All (64 documents)	**20.33**	**22.03**	**+1.70**	**0.8995**	**0.9340**	**+0.0345**

(a) Crop of the registered version of birth certificate #19.

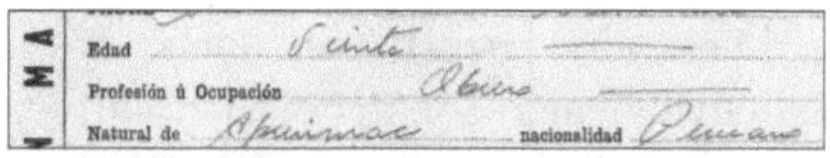

(b) Crop of the registered version of birth certificate #59.

(c) Crop of the reconstructed version of birth certificate #19.

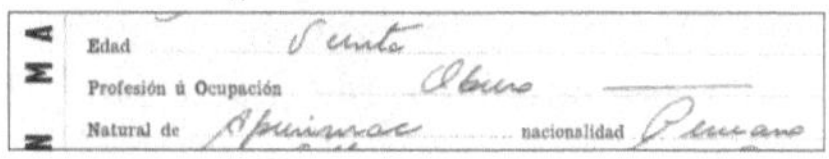

(d) Crop of the reconstructed version of birth certificate #59.

Fig. 10. Crops of two birth certificates (#19 and #59) belonging to a family of 64 birth certificates, together with the corresponding crop of their reconstructed versions (using our pipeline). Although both registered documents share the same underlying template, they have very different equalization profiles. Since handwritten fields are segmented without carrying any background pixels, no evidence of the gray background behind the handwritten text in (b) is pasted on the empty templates, as (d) shows.

Limitations. Our field extraction relies on mean-derived field masks as spatial priors and on the overlap-based selection of handwriting components. This design is effective when handwritten content is consistently located within fields, but has the following limitations.

(1) *Missing disconnected tails in long fields.*
If a long field contains disconnected leftmost (or rightmost) components that do not overlap the mean mask, the overlap criterion in Step 2.2 may fail to select them, causing missing characters or punctuation (as shown in Fig. 5). This typically occurs with faint ink, strong binarization artifacts, or large intra-word gaps.

(2) *Optional or rare fields are under-represented in the mean.* Fields that contain handwriting in only a small fraction of documents may be suppressed when computing the family mean in Step 2.1. As a result, these regions do not produce a stable mean mask $\mathcal{M}_j^F$, and the pipeline cannot reliably extract or synthesize them.

(3) *Sensitivity to differences in intensity across documents.* Large differences in scan equalization across the family can affect binarization and connected component structure in Steps 2.1 and 2.2. This effect is visible in the synthesis errors observed for documents with markedly different contrasts, as shown in Fig. 10.

4.4 Evaluation on Downstream HTR

To evaluate the extent to which reconstructed documents can be used to train a model on a downstream HTR task, we employ a version of the DAN model [8] already adapted for NER [30] and fine-tune two models for key-value information extraction of four fields (first name, last name, address and city): Model A is fine-tuned on 180 original documents from the RIMES database, while Model B is fine-tuned on reconstructed versions of the same 180 documents reconstructed with our pipeline.

The main objective of the experiment is to determine whether models trained exclusively on reconstructed documents can generalize effectively to real-world data without degrading HTR/NER performance. Therefore, we test both models on 200 original documents from the RIMES dataset. Since reconstructed RIMES documents are registered by design, we also evaluate both models on the same RIMES documents after registering them using the first stages of our pipeline.

Table 3 shows that Model B (trained on reconstructed RIMES forms) achieves slightly lower CER and WER than Model A (trained on real RIMES forms) when no registration is applied. Furthermore, registering documents before HTR/NER improves Model A but not Model B. This indicates that, although Model B was fine-tuned on registered documents, it does not require test documents to be registered at inference time in order to achieve competitive results.

Due to the limitations of the field segmentation method discussed in Sect. 4.3, manual correction of the annotated texts was needed on reconstructed documents.[2]

Table 3. CER and WER (%) evaluation of Models A and B on 200 RIMES documents, and on the same 200 RIMES documents after applying registration using our pipeline.

Model (fine-tuning data)	RIMES		RIMES registered	
	CER	WER	CER	WER
Model A (original RIMES)	5.00	17.05	4.84	16.60
Model B (reconstructed RIMES)	**4.79**	**16.15**	4.82	15.96

[2] In total, 36.5% of the reconstructed documents were perfect and required no correction. At the field level, 74.7% of the extracted fields required no correction. Most errors were suffix truncations, while prefix truncations and internal character errors were rare. The total manual correction effort on the 180 reconstructed documents amounted to 1.5 work hours.

5 Conclusion

We presented a fully unsupervised pipeline for recovering empty templates from filled structured handwritten forms, segmenting handwritten text at the field level, and synthesizing realistic documents. By combining hierarchical homographic registration with quantile-based aggregation, our approach extracts clean templates from filled pages alone. These templates enable precise field-level segmentation and the synthesis of structurally and stylistically consistent documents.

Experiments on public historical forms and private civil registries demonstrate robust registration and template recovery despite degradation and writer variability, with SSIM analysis confirming that minor reconstruction differences stem from field segmentation and intensity normalization, rather than geometric misalignment. Furthermore, downstream experiments on the RIMES dataset demonstrate that our synthesized forms can effectively train page-level HTR and NER models, achieving slightly better error rates (CER and WER) than models trained on real data.

Looking ahead, we plan to diversify field extraction in three directions: 1) to replace mean-derived masks with robust probabilistic spatial models to better handle rare or optional fields; 2) to extend mean-derived masks so that they capture longer fields whose tails our current method cuts; and 3) to shrink mean masks in order to capture shorter fields for data augmentation purposes. Ultimately, we will conduct experiments on shuffled documents and further explore the generation of anonymized forms.

Acknowledgments. This work was partially supported by the IRL-CNRS IFUMI-2030.

Disclosure of Interests. No competing interests.

References

1. Abdallah, A., Eberharter, D., Pfister, Z., et al.: A survey of recent approaches to form understanding in scanned documents. Artif. Intell. Rev. **57**(4), 342 (2024)
2. Antonacopoulos, A., Karatzas, D.: Document image analysis for world war ii personal records. In: Proceedings of the First International Workshop on Document Image Analysis for Libraries, p. 336. IEEE Comput. Soc. (2004)
3. Ares Oliveira, S., Seguin, B., Kaplan, F.: dhsegment: a generic deep-learning approach for document segmentation. In: 2018 16th International Conference on Frontiers in Handwriting Recognition (ICFHR), pp. 7–12. IEEE (2018)
4. Barrett, W., Hutchison, L., Quass, D., Nielson, H., Kennard, D.: Digital mountain: from granite archive to global access. In: Proceedings of the First International Workshop on Document Image Analysis for Libraries (DIAL'2004). p. 104. IEEE Computer Society (2004)
5. Bluche, T., Messina, R.O.: Gated convolutional recurrent neural networks for multilingual handwriting recognition. In: 2017 14th IAPR International Conference on Document Analysis and Recognition (ICDAR), vol. 1, pp. 646–651 (2017)

6. Bottaioli, N., et al.: Normalized vs diplomatic annotation: a case study of automatic information extraction from handwritten Uruguayan birth certificates. In: Mouchère, H., Zhu, A. (eds.) Document Analysis and Recognition – ICDAR 2024 Workshops, pp. 40–54. Springer Nature Switzerland (2024)
7. Clausner, C., Pletschacher, S., Antonacopoulos, A.: Aletheia - an advanced document layout and text ground-truthing system for production environments. In: 2011 International Conference on Document Analysis and Recognition, pp. 48–52 (2011)
8. Coquenet, D., Chatelain, C., Paquet, T.: DAN: a segmentation-free document attention network for handwritten document recognition. PAMI (2023)
9. Damelin, S.B., Hoang, N.S.: On surface completion and image inpainting by biharmonic functions: numerical aspects. Int. J. Math. Math. Sci. **2018**(1), 3950312 (2018)
10. Dar, M., Nagabhushan, P., Mir, A.H.: Pre-printed form recognition and extraction of data. In: AIP Conference Proceedings. vol. 1060, pp. 233–236. American Institute of Physics (2008)
11. Davis, B., Morse, B., Cohen, S., Price, B., Tensmeyer, C.: Deep visual template-free form parsing (2019)
12. Ding, Y., et al.: Form-NLU: dataset for the form natural language understanding. In: Proceedings of the 46th International ACM SIGIR Conference on Research and Development in Information Retrieval, pp. 2807–2816. Association for Computing Machinery (2023)
13. Fischler, M.A., Bolles, R.C.: Random sample consensus: a paradigm for model fitting with applications to image analysis and automated cartography. Commun. ACM **24**(6), 381–395 (1981)
14. Groleau, A., Chee, K.W., Larson, S., Maini, S., Boarman, J.: Augraphy: a data augmentation library for document images. In: International Conference on Document Analysis and Recognition, pp. 384–401. Springer (2023)
15. Grosicki, E., Carre, M., Brodin, J.M., Geoffrois, E.: RIMES evaluation campaign for handwritten mail processing. In: Proceedings ICFHR 2008: 11th International Conference on Frontiers in Handwriting Recognition, pp. 1–6. Concordia University (2008)
16. Hutchison, L., Barrett, W.: Fast registration of tabular document images using the fourier-mellin transform. In: First International Workshop on Document Image Analysis for Libraries, 2004. Proceedings, pp. 253–267. IEEE (2004)
17. Hutchison, L., Barrett, W.: Fourier-Mellin registration of line-delineated tabular document images. IJDAR **8**(2), 87–110 (2006)
18. Ishitani, Y.: Model matching based on association graph for form image understanding. In: Proceedings of 3rd International Conference on Document Analysis and Recognition, vol. 1, p. 287. IEEE Computer Society (1995)
19. Jaume, G., Kemal Ekenel, H., Thiran, J.P.: FUNSD: a dataset for form understanding in noisy scanned documents. In: 2019 International Conference on Document Analysis and Recognition Workshops (ICDARW). vol. 2, pp. 1–6 (2019)
20. Jocher, G., Chaurasia, A., Qiu, J.: YOLO by ultralytics (2023)
21. Kessi, L., Le Bourgeois, F., Garcia, C.: An efficient image registration method based on modified nonlocal-means: application to color business document images. In: 10th International Joint Conference on Computer Vision, Imaging and Computer Graphics Theory and Applications, pp. 166–173. SciTePress-Science and Technology Publications (2015)
22. Lowe, D.G.: Distinctive image features from scale-invariant keypoints. International Journal of Computer Vision **60**(2), 91–110 (2004)

23. Mahajan, K., Sharma, M., Vig, L.: Character keypoint-based homography estimation in scanned documents for efficient information extraction. In: 2019 International Conference on Document Analysis and Recognition Workshops (ICDARW), vol. 4, pp. 25–30. IEEE (2019)
24. Moisan, L., Moulon, P., Monasse, P.: Automatic homographic registration of a pair of images, with a contrario elimination of outliers. Image Processing On Line **2**, 56–73 (2012)
25. Narayan, S., Gowda, S.D.: Document image registration for imposed layer extraction. ICTACT J. Image Video Process. **7**(3) (2017)
26. Otsu, N.: A threshold selection method from gray-level histograms. IEEE Transactions on Systems, Man, and Cybernetics **9**(1), 62–66 (1979)
27. Pippi, V., et al.: Quo vadis handwritten text generation for handwritten text recognition? In: Proceedings of the IEEE/CVF International Conference on Computer Vision (ICCV) Workshops, pp. 7473–7483 (2025)
28. Puigcerver, J.: Are multidimensional recurrent layers really necessary for handwritten text recognition? In: 2017 14th IAPR International Conference on Document Analysis and Recognition (ICDAR). vol. 1, pp. 67–72. IEEE (2017)
29. de Sousa Neto, A.F., Bezerra, B.L.D., de Moura, G.C.D., Toselli, A.H.: Data augmentation for offline handwritten text recognition: a systematic literature review. SN Comput. Sci. **5**(2), 258 (2024)
30. Tarride, S., Boillet, M., Kermorvant, C.: Key-value information extraction from full handwritten pages. In: Fink, G.A., Jain, R., Kise, K., Zanibbi, R. (eds.) Document Analysis and Recognition - ICDAR 2023, pp. 185–204. Springer Nature Switzerland (2023)
31. Wang, J., Brown, M.S., Tan, C.L.: Automatic corresponding control points selection for historical document image registration. In: 2009 10th International Conference on Document Analysis and Recognition, pp. 1176–1180. IEEE (2009)
32. Wang, Z., Bovik, A., Sheikh, H., Simoncelli, E.: Image quality assessment: from error visibility to structural similarity. TIP **13**(4), 600–612 (2004)

Nutri-TabNet: Detection of Non-planar Nutrition Information Panels in Real-World Food Packaging

Shiza Maham[1(✉)], Abdullah Tariq[1], Claire Pulker[2,3], Lauren Blekkenhorst[2], Martin Masek[1,2], Faisal Shafait[4], and Syed Zulqarnain Gilani[1,2,5]

[1] Centre of AI & ML (CAIML), School of Science, Edith Cowan University, Joondalup, Australia
smaham@our.ecu.edu.au, {a.abdullah,m.masek,s.gilani}@ecu.edu.au
[2] Nutrition and Health Innovation Research Institute, Edith Cowan University, Joondalup, Australia
{c.pulker,l.blekkenhorst}@ecu.edu.au
[3] School of Population Health, Curtin University, Perth, Australia
[4] School of Electrical Engineering and Computer Science (SEECS), National University of Sciences and Technology (NUST), Islamabad, Pakistan
faisal.shafait@seecs.edu.pk
[5] Department of Computer Science, University of Western Australia, Perth, Australia

Abstract. Accurate detection and localization of Nutrition Information Panels (NIPs) on food packaging is critical for enabling automated dietary assessment and assistive retail applications. Unlike conventional table detection in scanned documents or PDFs, NIPs in real-world settings present significant challenges: images are captured in-the-wild, packaging surfaces are often non-planar (e.g., pouches, cylindrical containers, and boxes), and panels frequently exhibit deformation, perspective distortion, partial borders, and incomplete ruling lines. These factors limit the effectiveness of existing document-centric table detection methods. We propose Nutri-TabNet, a multimodal framework for robust NIP detection and localization in such challenging environments. The model integrates textual descriptions to complement visual representations using a Geometry-Guided Cross-Attention block, enabling better reasoning about geometric distortions caused by non-planar packaging. This multimodal domain-informed design allows the model to distinguish NIPs from other structured text regions on packaging. We evaluate Nutri-TabNet on NIP-Pack, a custom-collected dataset of real-world food packaging images. Our method achieves an mAP of 63% and outperforms strong baselines by 13% at mAP@0.75-0.95. The proposed work introduces NIP localization in food packaging as a new application domain for table detection.

F. Shafait et al. (Eds.): DAS 2026, LNCS 16851, pp. 36–53, 2027.
https://doi.org/10.1007/978-3-032-36207-0_3

Keywords: Table Detection · Deep Learning · Vision Transformer · Cross-Attention · Nutrition Information Panels · Non-planar Surfaces · Scene Text Detection · Multimodal Learning · Food Packing Dataset

1 Introduction

Table detection is a fundamental task in document analysis, enabling downstream applications such as structured information extraction, document understanding, and semantic parsing [5]. Accurate detection and localization of tables allow visually structured information to be transformed into machine-readable representations, which is essential in domains such as financial document processing, scientific article analysis, and automated data digitization. Recent advances in deep learning have substantially improved table detection performance through convolutional neural networks [9,22], object detection [28] and specialized frameworks [24] designed for joint table detection and structure recognition. More recently, transformer-based architectures [20,32,36] and graph neural network [26] approaches have further enhanced detection capabilities by modeling long-range spatial dependencies and hierarchical structures within tabular data. These approaches have demonstrated strong performance on widely used benchmark datasets such as ICDAR table competitions [6,8] TableBank [15], and PubLayNet [39], where tables are embedded in planar document images.

Despite some progress in table detection under unconstrained settings [18,37] (such as long parsing tables and table detection in the natural scenes), Nutrition Information Panels (NIPs) on food packaging remain particularly challenging due to the highly variable packaging visuals and geometric properties of consumer packaging. NIPs exhibit diverse layouts and are often embedded within real-world product imagery affected by specular highlights, surface curvature, crumpling, and complex branding elements. Furthermore, these panels are commonly printed on non-planar or deformable packaging surfaces such as bottles, flexible pouches, and cans. Consequently, table regions frequently undergo curvature-induced deformation, perspective variation, and non-uniform scaling, making reliable localization considerably more difficult.

To address these challenges, we propose *Nutri-TabNet*, a multimodal framework for NIPs detection and localization in unconstrained food packaging conditions. Our approach incorporates geometry-guided priors through textual prompts, allowing the model to explicitly reason about packaging geometries such as cylindrical and pouch-like surfaces. In addition, domain-specific textual cues are used to rank candidate regions, acting as weak semantic anchors that guide the selection of the most plausible table instance. This design enables robust detection even when structural cues are incomplete or visually ambiguous.

A key challenge in developing and evaluating such methods (i.e. Nutri-TabNet) is the scarcity of publicly available datasets capturing tables under real-world consumer packaging conditions. Existing benchmarks largely consist of scanned or digitally generated documents and thus fail to reflect the geometric variability and visual complexity of packaging imagery. To bridge this gap and effectively train and evaluate our proposed framework, we curate a custom food packaging dataset, named *NIP-Pack*, comprising diverse packaging types, layouts, and design variations under practical real-world conditions.

Experimental results demonstrate that our proposed framework surpasses several table detection methods including documented images and scene text methods, such as CascadeTabNet [25], TATR [32], PyramidTabNet [35], SwinText-v2 [11] and HybridTabnet [21], by up to 10% to 13% mAP points on our NIP-Pack dataset, underscoring the severity of the domain gap on non-planar surfaces. These results indicate that our approach not only addresses the challenges of NIP detection on curved and deformable surfaces, but also generalizes effectively to conventional table detection benchmarks, highlighting its broader applicability.

To the best of our knowledge, this is among the first works to specifically study NIP localization on non-planar food packaging using geometry-aware vision-language fusion. The major contributions of our work are as follows:

- We address NIP detection on food packaging, a challenging domain where table-like panels appear under substantial geometric and visual variability, including curved and deformable non-planar surfaces, specular reflections, perspective distortions, and cluttered packaging designs.
- We introduce NIP-Pack, a custom dataset of 1,100 food packaging images spanning diverse shapes, including cylindrical, pouch, rectangular, partially curved, and irregular forms, annotated at both polygon and bounding box levels. This dataset enables evaluation of NIP detection under realistic, non-planar conditions. The dataset will be made public after obtaining necessary approvals.
- We propose Geometry-Guided Cross-Attention (GGCA), a fusion block that conditions ViT visual queries on BERT-encoded shape prompts via cross-attention, injecting geometry-aware priors into the detection features to improve robustness under perspective distortions and surface curvature.

2 Related Work

Table detection aims to identify tabular regions in documents, with methods evolving from heuristic pipelines to deep learning-based detection and segmentation. Early approaches relied on handcrafted cues such as ruling lines, whitespace distribution, and layout regularity. Although effective on structured pages, they often degraded on complex layouts and borderless tables. Shafait and Smith [29] proposed a practical algorithm for heterogeneous multi-column documents and released an open-source implementation within the Tesseract OCR engine. Their method builds on Tesseract's layout analysis via tab-stop detection and leverages the resulting column layout and column partitions for table spotting.

With deep learning, robustness improved as the models learned hierarchical visual features directly from document images. Gilani et al. [7] preprocess document images prior to a region proposal and classification pipeline. In a related direction, Arif and Shafait [1] encode explicit foreground and background cues prior to detection by color-coding numeric/textual content and applying distance transformation, then using Faster R-CNN for table localization.

Beyond box detection, unified pipelines began coupling detection with structural understanding. DeepDeSRT [28] combines table detection with structure recognition, while segmentation-driven methods such as TableNet [23] cast table localization as semantic segmentation. Instance-segmentation and cascade-based detectors (e.g., CascadeTabNet [25], HybridTabNet [21], CasTabDetectoRS [10]) improve localization through iterative refinement and multi-scale feature reuse. More recently, one-stage detectors and transformer-based models (e.g., YOLO variants [12,38] and deformable transformer detectors [31]) have gained popularity due to efficiency and their ability to model long-range layout dependencies. In parallel, a few studies such as [18,37] have extended table detection toward natural-scene and unconstrained environments beyond scanned document settings. Moreover, recent vision-language foundation models (e.g., CLIP [27], BLIP [13], Grounding DINO [17]) have popularized prompt-conditioned localization, motivating multimodal detectors that inject textual priors into visual representations.

Despite these advances, reliable localization of Nutrition Information Panels (NIPs) on food packaging remains challenging, as existing methods are not explicitly designed to handle the combined effects of packaging-specific geometric distortions, deformable surfaces, specular reflections, and highly variable branding layouts. Motivated by these challenges, we propose Nutri-TabNet, a geometry-aware multimodal framework for robust NIP localization under non-planar food-packaging conditions.

3 Proposed Method

We present **Nutri-TabNet** as shown in Fig. 1, a geometry-guided detection framework for localizing Nutrition Information Panels (NIPs) on real-world food packaging and tables in document images. Nutri-TabNet builds upon established components, including ViT-B/16 [4], BERT-base encoder [3], and YOLOv8m [33] detection neck. These components provide the backbone feature extraction and detection infrastructure. Building on these components, our main contributions are: (i) a geometry-descriptive prompt design that encodes packaging surface types such as cylindrical, pouch, rectangular, and irregular forms, (ii) a Geometry-Guided Cross-Attention (GGCA) block for injecting non-planar surface priors, (iii) dilated C2f blocks within the YOLOv8m neck, and (iv) a domain knowledge based semantic re-ranking module.

Unlike GLIP [14] and Grounding DINO [17], which use text to specify *what* to detect, our prompts encode *how* the target appears by describing packaging surface geometry.

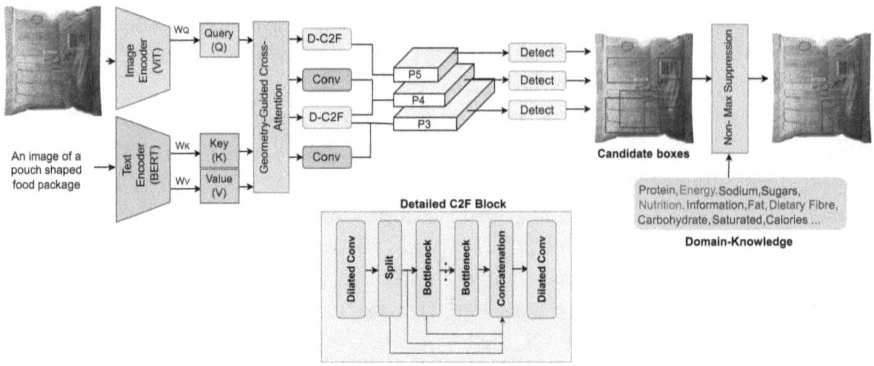

Fig. 1. Overview of Training Pipeline of Nutri-TabNet. The framework combines visual features from a ViT encoder with shape-descriptive text prompts encoded by BERT. A cross-attention module fuses visual and textual representations, followed by a multi-scale detection head for table/NIP localization. A nutrition-aware semantic re-ranking module refines predictions before final non-maximum suppression.

3.1 Problem Formulation

Given an input image $I \in \mathbb{R}^{3\times H\times W}$ and an associated shape description prompt $\mathcal{B}$, we predict bounding boxes $\mathcal{D} = \{(x_i, y_i, w_i, h_i, s_i)\}$, where (x_i, y_i, w_i, h_i) denotes box coordinates and $s_i \in [0, 1]$ the confidence. We formulate the task as single-class detection (table vs. background) targeting two regimes: (i) non-planar packaging with cylindrical, pouch, and box surfaces exhibiting barrel distortion, perspective warping, reflections, and clutter; and (ii) planar documents with flat, axis-aligned content. The prompt $\mathcal{B}$ serves as a geometric prior conditioning the detector on expected surface characteristics, enabling a single architecture to operate across both regimes.

3.2 Geometry-Descriptive Prompt Design

Each packaging image is manually categorized into one of the following geometric categories: cylindrical, rectangular, pouch, and partially curved/irregular. Based on this taxonomy, we define a generic prompt template with a placeholder {}:

- P_{generic}: "An image of a {} shaped food package."

where the placeholder is replaced with the corresponding geometry label from a controlled vocabulary (cylindrical, rectangular, pouch, partially curved/irregular). For instance, when the geometry label is cylindrical, the instantiated prompt becomes: "*An image of a cylindrical shaped food package.*" This yields category-level prompts with a consistent linguistic structure while encoding coarse geometric priors. In contrast, for planar document benchmarks, we use:

- P_{flat}: "An image of a flat document page."

Importantly, prompts are assigned at the category level and remain identical for all samples within a given category, i.e., no instance-specific prompt customization is performed. Furthermore, these category-level prompts are introduced only during training to encode coarse geometric priors and are not used during inference (see Fig. 2 for the inference pipeline).

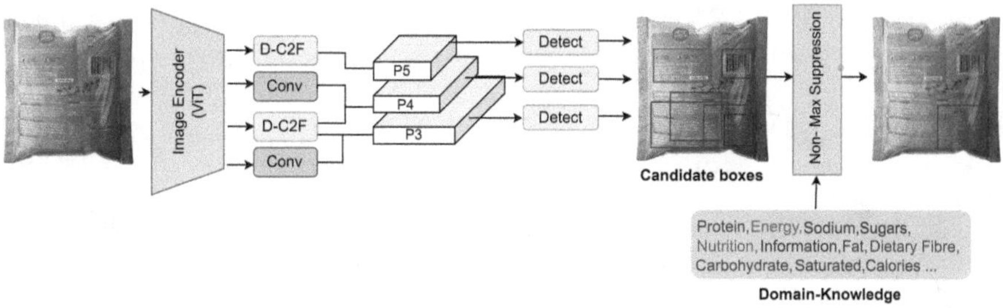

Fig. 2. Overview of Testing Pipeline of Nutri-TabNet. A packaging image is encoded by the ViT backbone and decoded by the multi-scale detector to produce NIP candidates; a nutrition-aware semantic re-ranking module refines these candidates, and non-maximum suppression yields the final localization. Note that Geometry-guided text prompts are not used at test time.

3.3 Vision Language Feature Extraction

Visual Encoder: We employ ViT-B/16 [4] initialized from ImageNet-21k pre-trained weights. The input image is resized to 640×640 and partitioned into $N = 40 \times 40 = 1600$ non-overlapping patches of size 16×16. Each patch is linearly projected and augmented with learnable positional embeddings, producing:

$$X = [x_1, \ldots, x_N] \in \mathbb{R}^{N \times d}, \quad d = 768. \tag{1}$$

The ViT encoder processes X through 12 transformer layers with multi-head self-attention. We extract the output patch tokens (excluding [CLS]) as the visual feature sequence, preserving full spatial resolution.
Text Encoder: We employ BERT-base [3] (12 layers, 768 hidden, 12 heads) to encode shape prompts. Given a tokenized prompt $\mathcal{B}$ of M tokens ($M \in [12, 14]$ after WordPiece tokenization), BERT produces:

$$Y = [y_1, \ldots, y_M] \in \mathbb{R}^{M \times d_t}, \quad d_t = 768. \tag{2}$$

Since $d = d_t = 768$, no additional projection is required.

3.4 Geometry-Guided Cross-attention

We introduce a Geometry-Guided Cross-Attention (GGCA) module that enables image patches to query geometry-descriptive prompts, producing geometry-aware visual embeddings. We adopt a *visual-queries-text* (image→text) attention direction: image patch tokens query the prompt tokens, producing N prompt-conditioned patch descriptors that preserve spatial granularity for downstream dense detection. Formally, given ViT patch tokens $X \in \mathbb{R}^{N\times d}$ and BERT token embeddings $Y \in \mathbb{R}^{M\times d}$, we compute linear projections

$$Q = XW_Q, \qquad K = YW_K, \qquad V = YW_V, \tag{3}$$

where $W_Q, W_K, W_V \in \mathbb{R}^{d\times d_h}$ and $d_h = d/h$ for h attention heads. For each head j, we obtain attention weights over prompt tokens for every image patch,

$$A^{(j)} = \text{softmax}\left(\frac{Q^{(j)}(K^{(j)})^\top}{\sqrt{d_h}}\right) \in \mathbb{R}^{N\times M}, \qquad Z^{(j)} = A^{(j)}V^{(j)} \in \mathbb{R}^{N\times d_h}, \tag{4}$$

and concatenate the head outputs to form $Z = \text{Concat}(Z^{(1)}, \ldots, Z^{(h)})W_O \in \mathbb{R}^{N\times d}$. We fuse the text-conditioned features with the original visual tokens through a residual connection followed by normalization:

$$X' = \text{LayerNorm}(X + Z) \in \mathbb{R}^{N\times d}. \tag{5}$$

Finally, we reshape X' back to the $H_p \times W_p$ patch grid to obtain a dense feature map $F_{\text{fused}} \in \mathbb{R}^{H_p\times W_p\times d}$, and apply a lightweight 1×1 convolutional adapter (two layers with BN and SiLU) to reduce the channel dimension from d=768 to C=256, yielding $F_0 \in \mathbb{R}^{H_p\times W_p\times C}$ which interfaces with our multiscale detection neck.

3.5 Multi-scale Detection Neck

We adopt the YOLOv8-medium (YOLOv8m) [33] detection neck operating on F_0. The neck constructs a three-level feature pyramid $\{P3, P4, P5\}$ via a top-down feature payramid network (FPN) and bottom-up Path Aggregation Network (PAN) pathway, producing features at strides 8, 16, and 32, which enables robust localization under large scale variation across packaging and document images. To better capture the tables on distorted and curved surfaces, we replace standard C2f bottlenecks with *dilated* convolutions to enlarge the receptive field without additional downsampling.

Dilated C2f Blocks: Standard C2f bottlenecks use 3×3 convolutions with unit dilation, providing local receptive fields insufficient for elongated structures distorted by surface curvature. We replace standard convolutions with dilated convolutions at rates $r \in \{1, 2, 4\}$ cycling across successive bottlenecks.A dilated C2f block first applies a 3×3 dilated convolution (dilation $r = 2$) to expand the receptive field, then splits the feature channels into two equal branches. One branch is processed by a cascade of n bottleneck modules, where each bottleneck uses a 1×1 convolution to reduce channels, followed by a 3×3 dilated convolution with dilation cycling through $r \in \{1, 2, 4\}$, and a final 1×1 convolution to restore channels, with residual skip connections. The outputs of all bottlenecks are then concatenated with the split branch, and a final 3×3 dilated convolution (again with $r = 2$) is applied for feature refinement. The mixed dilation provides effective receptive fields of 3, 5, and 9 pixels within a single block without increasing parameters, benefiting cylindrical packaging where table borders span long arc lengths but appear thin due to foreshortening.

3.6 Nutrition-Aware Semantic Re-ranking

Dense textual regions on food packaging (e.g., ingredient lists, marketing claims, allergen warnings and barcode) frequently produce high-confidence false positives. To improve discrimination, we introduce a nutrition-aware semantic re-ranking module that incorporates domain-specific textual cues.
Phrase Template Bank: We construct a bank $\mathcal{K}$ of 15 nutrition-related phrase templates capturing characteristic nutrient-name and unit patterns (e.g., "protein (g)", "sodium (mg)", "per 100 g", "nutrition facts"). Each phrase $k \in \mathcal{K}$ is encoded using a frozen BERT encoder to obtain a semantic embedding $\mathbf{e}_k \in \mathbb{R}^d$. When adapting the framework to other public table-detection datasets, we replace the nutrition-specific templates with generic table-centric cues, such as "table", "results", "summary", and related structural terms.
Region-Text Consistency Scoring: For each candidate box $\mathbf{b}_i$, we extract a region feature from the detection head using RoIAlign and pass it through a lightweight projection head comprising two linear layers with ReLU activation that maps the visual feature into the same d-dimensional space as the BERT phrase embeddings, yielding an aligned region embedding $\mathbf{r}_i \in \mathbb{R}^d$. We compute a semantic consistency score as the maximum cosine similarity between $\mathbf{r}_i$ and the phrase embeddings:

$$c_i = \max_{k \in \mathcal{K}} \frac{\langle \mathbf{r}_i, \mathbf{e}_k \rangle}{||\mathbf{r}_i|| \, ||\mathbf{e}_k||}. \tag{6}$$

The final detection confidence is computed as:

$$s_i' = \alpha s_i + (1 - \alpha)\sigma(c_i) \tag{7}$$

where $\alpha = 0.7$. Refined scores are then used during Non-Maximum Suppression. Unlike OCR keyword matching, this approach operates directly in embedding space, making it robust to partial occlusion, curved surfaces, and text distortion.

4 Dataset

To evaluate the performance of our proposed model, we utilized both publicly available table detection and structure recognition datasets, as well as our custom collected food packaging dataset. The different datasets used for experimentation in this study include:
TableBank: TableBank [15] is a large-scale table detection and recognition dataset constructed from Word and LaTeX documents. It contains over 417K annotated table instances generated through weak supervision by leveraging the document source files.

PubTables-1M: PubTables-1M [32] is a large-scale dataset curated from PubMed Central open-access articles. It is widely adopted for both table detection and structure recognition tasks. The dataset contains approximately 1 million tables, with detailed structural annotations including table bounding boxes, rows, columns, and cell segmentation. We used just table detection samples.
ICDAR 2019 cTDaR: The ICDAR 2019 cTDaR dataset [6] was introduced as part of the competition on Table Detection and Recognition, providing separate evaluation tracks for table detection and table structure recognition. In this work, we utilize Track A, which focuses on the table detection task. This track is further divided into archival and modern document subsets. We conducted our experiments on the modern document subset, where table region annotations are available for each image. It consists of 600 training images and 240 testing images, covering a diverse collection of document types derived from PDF sources.
UNLV: The UNLV dataset [30] is an early benchmark in document layout analysis research. It consists of 2,889 pages of scanned document images. We have utilized only a subset of 427 images that contain tables.
DocBank: DocBank [16] is a large-scale document layout dataset with over 500k pages, annotated across multiple semantic categories, including tables, and constructed via weak supervision from LaTeX sources.
NIP-Pack Dataset: We collected a food-packaging dataset of 1,100 images. For the table detection task, NIPs are annotated as the target tables. A total of 900 images were captured using multiple mobile devices across different grocery stores under unconstrained conditions (varying illumination, viewpoint, background clutter, and packaging material), while 200 images were scraped from online retail sources. To characterize geometric variability, the dataset includes cylindrical, pouch, rectangular/box-shaped, partially curved, and irregular packages. Table 1 shows the sample count per category.

Table 1. Distribution of Packaging Geometry Types in the NIP-Pack dataset.

Geometry Type	No. of Samples	Examples
Pouch	150	Sachets, Flexible Packs
Cylindrical	425	Bottles, Cans
Rectangular/Box-shaped	387	Boxes, Cartons
Partially Curved/Irregular	138	Tubs, cups
Total	**1,100**	–

All NIPs are annotated using polygon vertices to capture non-planar boundaries caused by perspective distortion and surface curvature. For compatibility with standard object-detection evaluation, an axis-aligned bounding box are also available. Representative examples across geometry types are shown in Fig. 3. Overall, the dataset is designed to support evaluation of NIPs detection under non-planar geometry and layout complexity, which are prevalent in food packaging but underrepresented in conventional document benchmarks.

Fig. 3. Sample images from the NIP-Pack dataset show cylindrical, rectangular, pouch, partially curved, and irregularly shaped packages, highlighting geometric variations and real-world challenges like surface curvature, perspective distortion, and complex layouts for table detection.

To ensure a consistent distribution between training and evaluation data, the proposed dataset is randomly divided into training and testing subsets using a stratified image-level split. Approximately 75% of the images are allocated for training, while the remaining 25% are reserved for testing and evaluation. This random split is performed at the image level to maintain diversity across packaging geometries, layout variations, and challenging cases in both subsets. Following this protocol, the final dataset consists of 825 training images and 275 testing images.

5 Results and Discussion

This section presents implementation details and a comprehensive evaluation of the proposed Nutri-TabNet framework across five widely used public benchmarks and custom real-world dataset.

5.1 Implementation Details

We use AdamW [19] with weight decay 5×10^{-4}. Learning rates are 1×10^{-4} for cross-attention and detection modules and 1×10^{-5} for pre-trained ViT and BERT. We use cosine annealing with linear warmup over 5 epochs. Training uses batch size 16 on 2× NVIDIA A6000 GPUs (effective 32 with gradient accumulation). We train for 100 epochs on the NIP dataset and 50 epochs on public benchmarks. All the images were resized to 640×640. Augmentations include mosaic ($p = 0.5$), random affine ($\pm 10°$, scale 0.5–1.5), HSV jittering, and horizontal flipping. Pre-trained weights initialize ViT/BERT; Kaiming (conv) and Xavier (linear) initialization are used for newly added modules.

Following standard object detection protocols, we report Precision (P), Recall (R), mAP@50, mAP@50–95, and mAP@75–95. The latter two metrics are included to assess localization performance under stricter IoU thresholds, which is critical for accurate table boundary detection.

5.2 Results on NIP-Pack Dataset and Cross-Dataset Evaluation

Table 2 compares the unimodal baseline (Only ViT) with the full Nutri-TabNet configuration (+ Text Cues + Shape Guidance) on our NIP-Pack dataset, and further reports performance across five widely used benchmarks to assess cross-dataset generalization.

On our NIP-Pack dataset, mAP@50 improves from 61.72% to 79.65%, while mAP@50–95 increases from 48.30% to 70.22%. On the other hand, mAP@75–95 rises from 40.17% to 63.21%, reflecting substantial gains under stricter IoU thresholds. Precision improves from 61.0% to 76.31% and recall from 69.31% to 83.03%. The larger gains at stricter IoU thresholds indicate that the proposed conditioning improves not only detection coverage but also box tightness under non-planar distortions such as curvature, perspective warping, specular highlights, and clutter.

On TableBank, the baseline already achieves strong performance (91.19% mAP@50). The proposed method further improves results to 93.91% mAP@50 and 93.73% mAP@50–95. On PubTables-1M, which contains scientific tables, mAP@50 increases from 93.77% to 98.87%, while mAP@75–95 improves from 82.69% to 84.03%. Consistent gains at stricter IoU thresholds indicate improved structural alignment in tightly packed layouts. For ICDAR 2019 cTDaR, the proposed method achieves 93.01% mAP@50 and 90.83% mAP@50–95, outperforming the baseline across all metrics. Similarly, on the UNLV dataset, mAP@50 improves from 92.01% to 97.30%, and mAP@75–95 increases from 78.48% to

Table 2. Comparison of detection performance between standalone ViT [4] and the Nutri-TabNet framework across five public datasets

Dataset	Method	P	R	mAP@50	mAP@50–95	mAP@75–95
NIP-Pack	Only ViT	61.00	69.31	61.72	48.30	40.17
	+ Text Cues + Shape Guidance	**76.31**	**83.03**	**79.65**	**70.22**	**63.21**
TableBank [15]	Only ViT	90.19	91.17	91.19	90.91	85.31
	+ Text Cues + Shape Guidance	**93.83**	**94.10**	**93.91**	**93.73**	**92.01**
PubTables-1M [32]	Only ViT	92.03	92.18	93.77	87.10	82.69
	+ Text Cues + Shape Guidance	**98.65**	**94.21**	**98.87**	**90.65**	**84.03**
cTDaR [6]	Only ViT	87.35	84.64	88.41	84.87	80.01
	+ Text Cues + Shape Guidance	**92.96**	**90.79**	**93.01**	**90.83**	**87.97**
UNLV [30]	Only ViT	92.12	91.73	92.01	83.11	78.48
	+ Text Cues + Shape Guidance	**97.83**	**96.50**	**97.30**	**86.99**	**81.83**
DocBank [16]	Only ViT	87.35	87.19	85.62	78.89	76.21
	+ Text Cues + Shape Guidance	**91.10**	**90.31**	**88.83**	**83.70**	**80.01**

81.83%, reflecting improved robustness to scanned and legacy document artifacts. On DocBank, which contains diverse document elements and layout noise, the proposed approach improves mAP@50 from 85.62% to 88.83% and mAP@75–95 from 76.21% to 80.01%. Significant improvement under stricter localization criteria demonstrates enhanced discrimination between tables and the surrounding layout components. Overall, the model improves both detection accuracy and localization quality across a wide range of domains, spanning from real-world packaging to structured document benchmarks.

5.3 Comparison with Different Detection Frameworks

Table 3 reports the results against several detectors fine-tuned and evaluated on our NIP-Pack dataset: DETR [2], YOLOv8 [33], YOLOv11 [34], CascadeTabNet [25], HybridTabNet [21], TATR [32], PyramidTabNet [35], and SwinText-v2 [11]. Although SwinText-v2 was originally proposed for scene text spotting, NIPs are visually text-dense and have few common challenges with scene text layouts; we therefore fine-tuned it using only its detection branch on NIP-Pack. Among competing methods, the best mAP@50 is 67.99% achieved by CascadeTabNet, while PyramidTabNet achieves 57.97% mAP@50–95. In contrast, our method attains 79.65% mAP@50 and 70.22% mAP@50–95, outperforming the strongest competitor by more than 11% and 12%, respectively. The performance margin is substantial, demonstrating that the integration of textual and geometric cues provides complementary information beyond purely visual detection frameworks. Experimental comparison with the WTW Cycle-CenterNet [18] model was not conducted because the official implementation is not publicly available.

Table 3. Comparison of generic object detectors, scene-text detectors, document table detection methods, and multimodal grounding models on the proposed NIP-Pack. All methods are evaluated under the same experimental protocol such as identical train/test splits, input resolution and evaluation metrics used by our method.

Method	P	R	mAP@50	mAP@50–95	mAP@75–95
DETR [2]	60.94	68.00	59.21	45.07	38.85
YOLOv8 [33]	58.01	66.43	57.72	47.84	39.65
YOLOv11 [34]	64.98	70.04	64.83	56.68	48.05
CascadeTabNet [25]	64.27	<u>71.46</u>	<u>67.99</u>	52.11	46.56
HybridTabNet [21]	<u>65.63</u>	69.98	67.36	53.85	47.01
PyramidTabNet [35]	62.01	65.39	62.50	<u>57.97</u>	46.98
TATR [32]	64.98	61.87	62.80	56.76	48.87
Grounding DINO [17]	65.19	68.35	67.09	57.43	<u>50.13</u>
SwinText-v2 [11]	61.20	68.97	65.13	55.79	48.61
Ours	**76.31**	**83.03**	**79.65**	**70.22**	**63.21**

5.4 Qualitative Analysis of Partial Success and Failure Cases

In this section, we analyze the NIP detection outputs of Nutri-TabNet and a few table detection methods. As shown in Fig. 4(a), ground truth annotations mark the nutritional information panel across diverse packaging types. In Fig. 4(b) and Fig. 4(c), CascadeTabNet and PyramidTabNet produce multiple bounding boxes that partially cover the NIP region on some samples. Notably, on the black deformed rectangular packaging, PyramidTabNet incorrectly localizes a non-NIP region as a table, while CascadeTabNet partially detects the NIP across multiple boxes with an additional spurious detection outside the NIP boundary. Although both methods are able to detect the NIP on the flat, non-deformed rectangular green packaging, PyramidTabNet still produces a slightly oversized bounding box enclosing non-NIP regions, suggesting these methods perform better under planar and regular surface conditions. However, both methods fail to detect NIP on green tin packaging, suggesting that their architectures are not consistently robust to surface curvature and deformations inherent in food packaging. In contrast, Fig. 4(d) shows that our proposed Nutri-TabNet generally produces tighter, more precise bounding boxes closely aligned with the ground truth across varying packaging geometries, though it produces only a partial detection on the green tin packaging due to surface reflectance and high curvature.

5.5 Ablation Study

Table 4 analyses the contribution of each proposed component on the NIP-Pack dataset. The unimodal baseline (Only ViT) attains 61.00 precision, 69.31 recall, and 61.72 AP@50, with a marked drop under stricter localization criteria (AP@75 = 48.30, AP@95 = 40.17), highlighting the difficulty of NIP detection

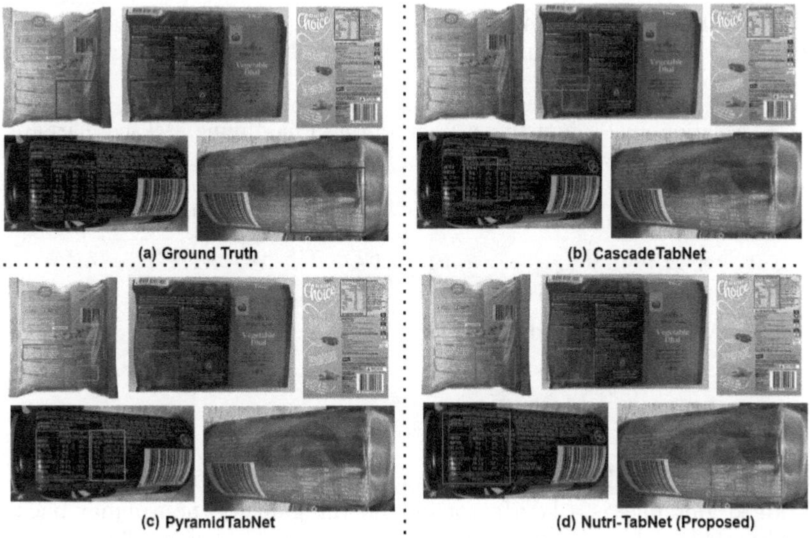

Fig. 4. Qualitative visualization of NIP detection on the NIP-Pack dataset. (a) shows ground-truth NIP annotations in blue. (b) and (c) show CascadeTabNet and PyramidTabNet predictions, where partial localization, false positives, and missed detections occur under cluttered and non-planar packaging conditions. (d) shows Nutri-TabNet predictions across diverse packaging surfaces. (Color figure online)

Table 4. Component Ablation on the NIP-Pack dataset. 'Shape Guidance only' indicates the model without the re-ranking module, while 'Text Cues only' refers to the model without shape-conditioned prompts.

Method	P	R	AP@50	AP@75	AP@95
Only ViT	61.00	69.31	61.72	48.30	40.17
+ Shape Guidance only	68.21	74.87	66.93	56.03	46.98
+ Text Cues only	70.53	69.31	70.41	63.29	54.86
+ Text Cues + Shape Guidance	76.31	83.03	79.65	70.22	63.21

on in-the-wild packaging with perspective and surface deformation. Incorporating shape guidance improves both precision and recall (68.21/74.87), suggesting better proposal coverage and fewer missed panels under geometric distortion. In contrast, semantic re-ranking ("text cues only") yields a larger boost in precision and high-IoU AP (AP@50 = 70.41, AP@75 = 63.29, AP@95 = 54.86) while leaving recall essentially unchanged, consistent with re-ranking operating on detector candidates by promoting nutrition-consistent boxes and suppressing false positives rather than generating new proposals. Combining both modules provides complementary gains, achieving the best overall performance (76.31 precision, 83.03 recall, 79.65 AP@50) with the strongest improvements at higher

IoU thresholds (AP@75: 48.30→70.22; AP@95: 40.17→63.21), indicating more reliable localization and robustness under severe geometric variability.

Table 5. Ablation on Prompt Supervision during Fine-tuning. All modules are kept identical; only the prompt text used during fine-tuning is changed.

Fine-tuning Prompt Setting	P	R	AP@50	AP@75	AP@95
No Shape Prompts	62.8	70.5	64.1	50.9	42.2
Shape Prompts	**76.3**	**83.0**	**79.6**	**70.2**	**63.2**
Synonyms (Can, Round, Bottle, Box)	75.9	82.6	79.1	69.7	62.5

Effect of Prompt Selection. Overall, these results validate that prompt-conditioned guidance is a key factor behind the gains on non-planar packaging while remaining robust to reasonable variations in prompt wording. To isolate the role of prompt supervision, Table 5 keeps all modules and hyperparameters fixed and varies only the prompt text used during fine-tuning. Removing shape prompts degrades performance to 64.1 AP@50 and 42.2 AP@95, which confirms that the detector benefits from explicit geometric priors rather than relying purely on visual appearance. Using the shape prompts produces the best overall results having 79.65 AP@50, 70.22 AP@75 and 63.21 AP@95. Replacing the prompts with close synonyms (e.g., can/round/bottle/box) maintains performance with only a minor drop which suggests that the model is not overfitting but instead leverages semantically equivalent shape descriptions.

6 Conclusion and Future Work

In this paper, we present Nutri-TabNet, a multimodal framework for localizing Nutrition Information Panels (NIPs) on real-world food packaging while remaining effective for conventional table detection in planar document images. By coupling a ViT–BERT dual-encoder with an image-to-text cross-attention fusion block, Nutri-TabNet injects geometric surface priors through shape-descriptive prompts, while a nutrition-aware phrase-template re-ranking module suppresses text-driven false positives. Through extensive experiments, we have shown that Nutri-TabNet consistently improves over a strong unimodal baseline across multiple benchmarks. It also delivers particularly large gains at stricter IoU thresholds, indicating more accurate and robust localization under curvature, perspective distortion, and clutter packaging environments. For future work, manually defined shape prompts can be replaced with learnable prompt embeddings, enabling a systematic evaluation of their impact on detection accuracy and generalization. Beyond localization, Nutri-TabNet can be extended further to table structure recognition, including row–column decomposition, cell adjacency modeling, and field-level extraction (e.g., nutrient names, values, and units).

Acknowledgments. SZG acknowledges the support of Future Health Research and Innovation Fund, Department of Health WA through the WANMA grant.

References

1. Arif, S., Shafait, F.: Table detection in document images using foreground and background features. In: Digital Image Computing: Techniques and Applications (DICTA). IEEE (2018)
2. Carion, N., Massa, F., Synnaeve, G., Usunier, N., Kirillov, A., Zagoruyko, S.: End-to-end object detection with transformers. In: European Conference on Computer Vision. Springer (2020)
3. Devlin, J., Chang, M.W., Lee, K., Toutanova, K.: BERT: pre-training of deep bidirectional transformers for language understanding. In: Proceedings of the 2019 Conference of the North American Chapter of the Association for Computational Linguistics: Human Language Technologies (2019)
4. Dosovitskiy, A., et al.: An Image is Worth 16x16 Words: Transformers for Image Recognition at Scale (2020). arXiv:2010.11929 arXiv preprint
5. Fernandes, J., Xiao, B., Simsek, M., Kantarci, B., Khan, S., Alkheir, A.A.: TableStrRec: framework for Table Structure Recognition in Data Sheet Images. International Journal on Document Analysis and Recognition (IJDAR) (2024)
6. Gao, L., et al.: ICDAR 2019 Competition on Table Detection and Recognition (cTDaR). In: International Conference on Document Analysis and Recognition (ICDAR). IEEE (2019)
7. Gilani, A., Qasim, S.R., Malik, I., Shafait, F.: Table detection using deep learning. In: International Conference on Document Analysis and Recognition (ICDAR). IEEE (2017)
8. Göbel, M., Hassan, T., Oro, E., Orsi, G.: International Conference on Document Analysis and Recognition (ICDAR), IEEE (2013). ICDAR 2013 Table Competition
9. Hao, L., Gao, L., Yi, X., Tang, Z.: A table detection method for pdf documents based on convolutional neural networks. In: 12th IAPR Workshop on Document Analysis Systems (DAS), IEEE (2016)
10. Hashmi, K.A., Pagani, A., Liwicki, M., Stricker, D., Afzal, M.Z.: CasTabDetectoRS: cascade network for table detection in document images with recursive feature pyramid and switchable atrous convolution. J. Imaging (2021)
11. Huang, M., et al.: Swintextspotter v2: towards better synergy for scene text spotting. Int. J. Comput. Vision **133**(8), 5281–5301 (2025)
12. Huang, Y., Yan, Q., Li, Y., Chen, Y., Wang, X., Gao, L., Tang, Z.: A YOLO-based table detection method. In: International Conference on Document Analysis and Recognition (ICDAR). IEEE (2019)
13. Li, J., Li, D., Xiong, C., Hoi, S.: BLIP: bootstrapping language-image pre-training for unified vision-language understanding and generation. In: International Conference on Machine Learning. PMLR (2022)
14. Li, L.H., et al.: Grounded language-image pre-training. In: Proceedings of the IEEE/CVF Conference on Computer Vision and Pattern Recognition (CVPR) (2022)
15. Li, M., Cui, L., Huang, S., Wei, F., Zhou, M., Li, Z.: TableBank: table benchmark for image-based table detection and recognition. In: Proceedings of the Twelfth Language Resources and Evaluation Conference (2020)

16. Li, M., Xu, Y., Cui, L., Huang, S., Wei, F., Li, Z., Zhou, M.: DocBank: a benchmark dataset for document layout analysis. In: Proceedings of the 28th International Conference on Computational Linguistics (2020)
17. Liu, S., et al.: Grounding-DINO: marrying DINO with grounded pre-training for open-set object detection. In: European Conference on Computer Vision. Springer (2024)
18. Long, R., et al.: Parsing table structures in the wild. In: Proceedings of the IEEE/CVF International Conference on Computer Vision, pp. 944–952 (2021)
19. Loshchilov, I., Hutter, F.: (2017). arXiv:1711.05101 Decoupled Weight Decay Regularization. arXiv preprint
20. Nassar, A., Livathinos, N., Lysak, M., Staar, P.: TableFormer: table structure understanding with transformers. In: Proceedings of the IEEE/CVF Conference on Computer Vision and Pattern Recognition (CVPR) (2022)
21. Nazir, D., Hashmi, K.A., Pagani, A., Liwicki, M., Stricker, D., Afzal, M.Z.: Hybridtabnet: towards better table detection in scanned document images. Appl. Sci. (2021)
22. Nguyen, D.D.: TableSegNet: a fully convolutional network for table detection and segmentation in document images. Int. J. Doc. Anal. Recogn. (2022) IJDAR
23. Paliwal, S.S., Vishwanath, D., Rahul, R., Sharma, M., Vig, L.: TableNet: deep learning model for end-to-end table detection and tabular data extraction from scanned document images. In: international conference on document analysis and recognition (ICDAR), IEEE (2019)
24. Polewczyk, M., Spinaci, M.: ClusterTabNet: supervised clustering method for table detection and table structure recognition. In: International Conference on Document Analysis and Recognition (ICDAR). Springer (2024)
25. Prasad, D., Gadpal, A., Kapadni, K., Visave, M., Sultanpure, K.: CascadeTabNet: an approach for end-to-end table detection and structure recognition from image-based documents. In: Proceedings of the IEEE/CVF Conference on Computer Vision and Pattern Recognition Workshops (2020)
26. Qasim, S.R., Mahmood, H., Shafait, F.: Rethinking table recognition using graph neural networks. In: International Conference on Document Analysis and Recognition (ICDAR). IEEE (2019)
27. Radford, A., et al.: Learning transferable visual models from natural language supervision. In: International Conference on Machine Learning, PmLR (2021)
28. Schreiber, S., Agne, S., Wolf, I., Dengel, A., Ahmed, S.: DeepDeSRT: deep learning for detection and structure recognition of tables in document images. In: International Conference on Document Analysis and Recognition (ICDAR). IEEE (2017)
29. Shafait, F., Smith, R.: Table detection in heterogeneous documents. In: Proceedings of the 9th IAPR International Workshop on Document Analysis Systems (2010)
30. Shahab, A., Shafait, F., Kieninger, T., Dengel, A.: An open approach towards the benchmarking of table structure recognition systems. In: Proceedings of the 9th IAPR International Workshop on Document Analysis Systems (2010)
31. Shehzadi, T., Azeem Hashmi, K., Stricker, D., Liwicki, M., Zeshan Afzal, M.: Towards end-to-end semi-supervised table detection with deformable transformer. In: International Conference on Document Analysis and Recognition (ICDAR). Springer (2023)
32. Smock, B., Pesala, R., Abraham, R.: PubTables-1M: towards comprehensive table extraction from unstructured documents. In: Proceedings of the IEEE/CVF Conference on Computer Vision and Pattern Recognition (CVPR) (2022)
33. Ultralytics: YOLOv8 (2023). https://github.com/ultralytics/ultralytics

34. Ultralytics: YOLOv11 (2024). https://github.com/ultralytics/ultralytics
35. Umer, M., Mohsin, M.A., Ul-Hasan, A., Shafait, F.: PyramidTabNet: transformer-based table recognition in image-based documents. In: International Conference on Document Analysis and Recognition (ICDAR). Springer (2023)
36. Wang, Y., et al.: TransTab: a transformer-based approach for table detection and tabular data extraction from scanned document images. Machine Learning with Applications (2025)
37. Yang, F., Hu, L., Liu, X., Huang, S., Gu, Z.: A large-scale dataset for end-to-end table recognition in the wild. Sci. Data **10**(1), 110 (2023)
38. Zhang, D., Mao, R., Guo, R., Jiang, Y., Zhu, J.: Yolo-table: Disclosure Document Table Detection with Involution: D. Zhang et al, International Journal on Document Analysis and Recognition (IJDAR) (2023)
39. Zhong, X., Tang, J., Yepes, A.J.: PubLayNetÃćâĆňÂŕ: largest dataset ever for document layout analysis. In: International Conference on Document Analysis and Recognition (ICDAR). IEEE (2019)

Document-Centric AI, Retrieval and Reasoning

Criticality-Aware Evaluation of Table Understanding Systems on Synthetic ESG Documents

Giorgia Bertacchini[1], Marcello Pietri[2,3], Nicola Bicocchi[1,3], Enrico Rossini[1], and Marco Mamei[2,3](✉)

[1] Department of Science and Methods for Engineering, University of Modena and Reggio Emilia, Reggio Emilia, Italy
{giorgia.bertacchini,nicola.bicocchi,enrico.rossini}@unimore.it
[2] Department of Engineering "Enzo Ferrari", University of Modena and Reggio Emilia, Modena, Italy
{marcello.pietri,marco.mamei}@unimore.it
[3] Artificial Research Innovation Center, University of Modena and Reggio Emilia, Modena, Italy

Abstract. Corporate Environmental, Social, and Governance (ESG) reports are visually rich documents where key evidence is often embedded in complex tables with multi-level headers, merged cells, footnotes, graphical indicators, and heterogeneous values. These features make table understanding a challenge for document analysis and LLM-based applications such as question answering and automated reporting. This paper presents a synthetic benchmark for evaluating table understanding systems on ESG-style documents. The pipeline generates PDF reports paired with complete JSON ground truth, including text, table structure, cell values, units, formulas, semantic types, graphical indicators, and criticality annotations. The benchmark is paired with a criticality-aware hierarchical evaluation framework measuring structural fidelity, content accuracy, and semantic completeness. We compare rule-based parsers, layout-aware pipelines, table-specific models, and multimodal LLMs. Results show that recent multimodal LLMs perform best overall, but table extraction remains the main bottleneck, especially for complex headers, merged cells, footnotes, and large tables. This highlights the need for domain-specific benchmarks for ESG document understanding.

Keywords: ESG reporting · synthetic benchmark · table understanding · criticality-aware evaluation · document parsing · multimodal LLMs

1 Introduction

Environmental, Social, and Governance (ESG) reporting has become a major operational and regulatory requirement for many organizations. Regulatory instruments such as the European Union's Corporate Sustainability Reporting

F. Shafait et al. (Eds.): DAS 2026, LNCS 16851, pp. 57–73, 2027.
https://doi.org/10.1007/978-3-032-36207-0_4

Directive (CSRD) require companies to collect, validate, and disclose increasingly detailed sustainability information, across heterogeneous sources and templates, a process that still relies heavily on costly and error-prone manual inspection. This is particularly challenging for visually rich ESG documents, where key sustainability indicators are often embedded in tables and depend on both textual content and structural cues such as layout, row–column organization, units, notes, and aggregate values [16]. Unlike the regular layouts often found in academic benchmarks [12,20], tables in sustainability and financial reports frequently contain multi-level headers, merged cells, hierarchical row groups, footnotes, formula-based aggregation rows, graphical indicators, and heterogeneous value types, whose interpretation depends on the preservation of row–column relations, header semantics, and local visual structure—further complicated in PDFs by page layout and reading order. Table extraction is not merely a low-level recognition task, but a prerequisite for preserving the evidential value of ESG indicators in downstream analytical and reporting workflows. This issue has become more relevant with the growing adoption of Large Language Models (LLMs) and Retrieval-Augmented Generation (RAG). Although RAG can improve grounding by retrieving external evidence, its reliability depends critically on the fidelity of upstream parsing, table reconstruction, and semantic normalization [7,17,21]. Without preserving the structural and semantic organization of tables, retrieval and reasoning over ESG indicators remain unreliable. Recent benchmarking studies on heterogeneous PDFs confirm that robust end-to-end table extraction remains challenging, particularly because performance varies across datasets and current methods still face limitations in generalizability and robustness [14] (Fig. 1).

Remediation Type	FY 2024 Risk Assessment		Coverage Status
	Allocation	Estimated Exposure	
Total Environmental Remediation	$220,000,000	$278,000,000	✓
Site Contamination Cleanup	$90,000,000	$116,000,000	⚠ ‡
Emissions Control Retrofits	$75,000,000	$94,000,000	✓
Water and Effluent Treatment	$55,000,000	$68,000,000	✓
Total Social Remediation	$64,000,000	$80,000,000	✗
Supply Chain Compliance Audits	$38,000,000	$47,000,000	✗
Workforce Safety Upgrades	$26,000,000	$33,000,000	✗

‡ Remediation milestone rescheduled to Q3 2025 pending regulatory approval.

Total ESG Financial Commitments 2024			
$468,000,000	Environmental Commitments		
	$227,000,000	Climate Action Investments	
		$168,000,000	Renewable Energy *
			$120,000,000
			Energy Efficiency Retrofits
			$48,000,000
		Sustainable Operations	
		$32,000,000	Waste and Water
			$32,000,000
	Social and Governance Commitments		
	$89,000,000	Workforce Development and Inclusion	

Fig. 1. Examples of ESG-style tables containing hierarchical headers, merged cells, numerical indicators, and visual or textual annotations.

A further limitation concerns the available evaluation resources. Existing benchmarks have advanced table detection, structure recognition, and financial document reasoning, but they only partially capture the visual complexity, semantic richness, and end-to-end extraction requirements of ESG-style documents [4,9,11,19]. Consequently, they provide limited support for evaluating whether extraction systems preserve the structural, semantic, and criticality-related information required in ESG reporting scenarios.

In this paper, we address this gap through a synthetic benchmark for ESG-style table understanding. Given a set of manually designed templates, the proposed pipeline automatically generates PDF reports paired with complete JSON ground truth under controlled and reproducible conditions. The generated tables include multi-level headers, merged cells, formula-based rows, footnotes, graphical indicators, and heterogeneous value types, making them more representative of sustainability and financial reporting practice. The resulting benchmark is used to evaluate rule-based parsers, layout-aware pipelines, table-specific models, and multimodal LLM-based systems through a criticality-aware hierarchical framework that decomposes extraction quality into structural, content, and semantic dimensions.

Our study asks: *to what extent can current document understanding systems, including multimodal LLMs, reliably recognize and reconstruct complex tabular information from ESG-style PDF documents?* We address this question using a controlled synthetic corpus with exact structural and semantic ground truth. The experimental results show that recent multimodal LLMs form the strongest-performing group among the evaluated systems, but table extraction remains the main bottleneck, particularly for complex layouts. These findings motivate realistic and semantically grounded benchmarks for ESG reporting and related high-stakes document analysis tasks. The main contributions of this paper are[1]:

- a synthetic ESG-style document generation pipeline producing PDF reports with complex tabular structures and paired JSON ground truth;
- a criticality-aware hierarchical evaluation framework for assessing structural fidelity, content accuracy, and semantic completeness;
- an empirical benchmark comparing rule-based parsers, layout-aware pipelines, table-specific models, and multimodal LLM-based systems.

2 Related Work

Table extraction is a central task in document understanding and is particularly relevant for ESG reporting, where evidence is often embedded in visually complex business documents. In this setting, values depend not only on text, but also on layout, row–column relations, headers, units, footnotes, formulas, graphical indicators, and surrounding context. This makes ESG-style tables challenging for end-to-end understanding, especially when they include merged cells, multi-level headers, irregular alignments, and corporate-style layouts.

Early approaches mainly adapted object detection architectures to localize tables in document pages, while more recent methods jointly address table detection and structure recognition. In particular, Table Transformer (TaTR) showed that DETR-style transformer models can achieve strong performance when trained on large-scale and carefully canonicalized table annotations [3,12]. Practical parsing frameworks have also become important. Tools such as pdfplumber, Camelot, Tabula, PyMuPDF, and Unstructured provide access to text

[1] https://github.com/giorgiaBertacchini/document_ai_for_ESG_reporting.

spans, coordinates, and table extraction, although comparative studies show that performance strongly depends on layout variability and extraction goals [1]. More recent end-to-end systems such as Docling combine layout analysis and table structure recognition in a unified pipeline [2], while industrial open-source tools such as OpenDataLoader targets structured PDF extraction for RAG-oriented workflows, including reading order, tables, and layout-aware representations.

Multimodal Large Language Models offer a direction for visually rich document understanding, since they can process rendered pages and reason over textual, spatial, and visual cues. OCR-oriented multimodal systems such as GLM-OCR illustrate this convergence between document extraction and multimodal reasoning [5]. However, it remains necessary to evaluate whether these capabilities translate into reliable understanding of complex ESG-style tables.

Benchmark datasets have played an important role in the development of table extraction and document-understanding methods. TableBank provided a large-scale weakly supervised dataset for table detection and recognition, automatically generated from Word and LaTeX documents [8]. Later datasets such as PubTabNet and PubTables-1M improved the study of table-structure recognition by introducing image-to-HTML annotations and richer structural labels for scientific documents [12, 20]. Although these resources are widely used, they mainly represent scientific documents, whose tables are often cleaner and more regular than those found in business reports.

Financial datasets such as FinTabNet reduce the gap between scientific-table benchmarks and real business documents by focusing on complex tables extracted from annual reports [19]. However, their annotations are mainly HTML structural representation, capturing table layout rather than richer semantic information about the reported indicators, entities, or financial meaning. SynthTabNet further shows the potential of synthetic data by automatically generating annotated table images for table-recognition tasks [9], but it remains limited to table-level generation. These resources therefore support progress in table-structure recognition, but leave complete domain-specific document generation less explored. ESG reports represent a broader setting, where realistic synthetic data must combine narrative sections, sustainability indicators, heterogeneous layouts, and semantically meaningful tables. In parallel, FinQA and DocFinQA focus on downstream question answering and numerical reasoning over financial documents, with DocFinQA extending the task to full-document contexts [4, 11]. However, their focus remains on question answering rather than on the broader challenges of document-level information extraction.

Beyond document extraction itself, recent research has also investigated how AI-based infrastructures can support broader sustainability and ESG-related objectives. For instance, federated and privacy-preserving approaches have been proposed for collaborative forecasting in local energy communities, showing how AI can support sustainability-oriented data ecosystems [15]. Similarly, Digital Twins and Federated Learning have been explored in Industrial IoT scenarios for worker-centric safety, which is closely related to the Social dimension of ESG [10]. These works highlight the growing role of intelligent data infrastructures in ESG-relevant contexts, while the present paper focuses on a complementary prerequisite: the reliable extraction of structured ESG information from complex corporate documents.

Existing evaluation metrics for document extraction capture different aspects of system performance, but each provides only a partial view. OCR-oriented metrics such as Character Error Rate (CER) and Word Error Rate (WER) measure transcription errors at character or word level, while precision, recall, F1-score and Exact Match are commonly used to evaluate whether expected fields or cells are correctly extracted. More flexible string-based metrics, such as Normalized Levenshtein Similarity, can also assign partial credit when predictions differ from the reference only because of minor textual or formatting variations. However, complex tables require metrics that account not only for textual content but also for structure. Tree-Edit-Distance-based Similarity (TEDS) addresses this by representing a table as a tree, typically derived from its HTML structure, and comparing predicted and reference tables through structural edit operations [20]. This allows TEDS to capture errors in cell content, row and column organization, and merged-cell structures. Grid Table Similarity (GriTS) extends this perspective by evaluating tables in their natural grid representation, comparing topology, cell locations, and cell content, which makes it particularly relevant for complex tabular layouts [13].

Taken together, prior work leaves three related gaps. Existing table benchmarks have advanced table detection and structure recognition, but only partially reflect ESG-style documents, where tables are embedded in realistic report layouts and include hierarchical headers, merged cells, footnotes, formulas, graphical indicators, and heterogeneous value types. At the same time, practical parsers and multimodal LLMs are increasingly used for such documents, but their behavior is rarely compared under controlled conditions with complete structural and semantic ground truth. Finally, existing metrics often treat extraction errors as equally important and only partially account for the semantic role or criticality of extracted information. These gaps motivate our synthetic ESG-style benchmark and hierarchical criticality-aware evaluation of both conventional parsers and multimodal LLM-based systems along structural, content, and semantic dimensions.

3 Methods

This section describes the synthetic benchmark, the document generation pipeline, the evaluated systems, and the criticality-aware evaluation framework.

3.1 Synthetic ESG Document Generation

Reliable evaluation of table extraction systems requires benchmarks that combine realistic document complexity with exact ground truth. The automated benchmark generation pipeline (Fig. 2) addresses this by producing visually varied PDFs with fully controlled annotations, difficult to combine in manually annotated datasets. Starting from a set of manually designed ESG-style table

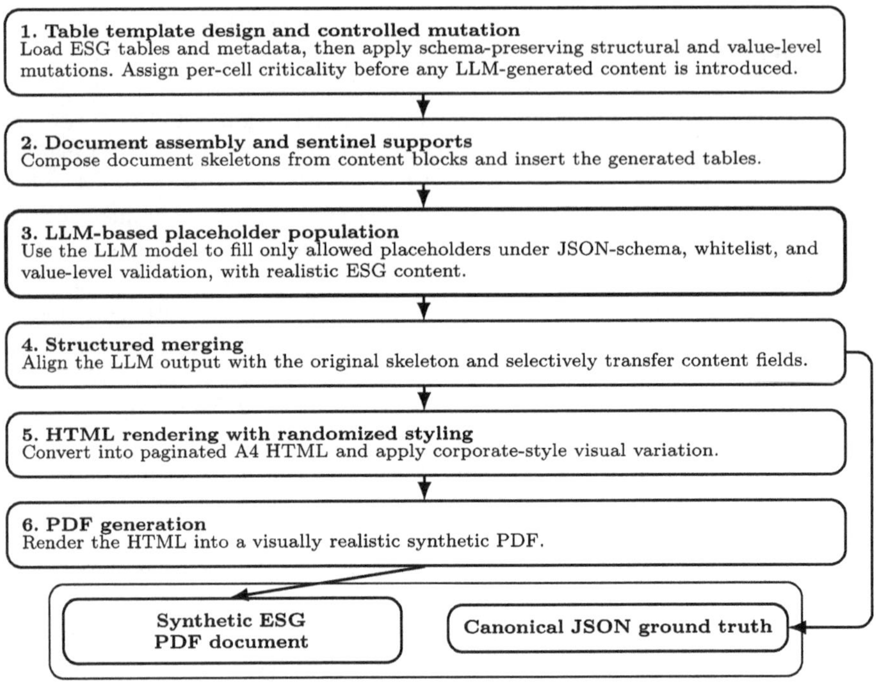

Fig. 2. Synthetic dataset generation pipeline. Automated generation of PDF documents paired with complete JSON ground truth for evaluation.

templates, the pipeline generates multiple document instances through deterministic stages that define structure—controlled structural mutations, randomized visual styling, and deterministic PDF rendering—while an LLM only fills constrained content, preserving stable structural and semantic ground truth.

Each sample contains a rendered PDF, used as system input, and a canonical JSON document, used as ground truth. The JSON schema is designed not only for evaluation, but also for downstream use: it represents documents as ordered blocks and tables as structured objects with cells, groups, spans, footnotes, graphical indicators, and formulas. This structure can be flattened into relational tables or converted into a graph representation, making the benchmark usable for scoring extraction systems, but also for downstream document intelligence pipelines. Table 1 summarizes the main characteristics of the generated corpus, including the number of documents, pages, tables, cells, merged cells, formula cells, indicator cells, and footnotes.

Seed Tables and Controlled Structural Variation. The generation process starts from curated ESG-style seed tables stored as Excel workbooks and paired with JSON sidecars. The benchmark is built from nine curated seed tables modeled on recent ESG and financial disclosures of major Italian banking groups, giving the dataset a realistic reporting profile. Each seed defines a distinct logical

Table 1. Summary statistics of the generated benchmark.

Property	Value	Property	Value
Documents	92 PDFs	Total pages	1,680
Seed tables	9	Total tables	1,597
Total cells	32,052	Narrative text	204,574 words
Pages/doc.	med. 12, mean 18.26	Tables/doc.	med. 11, mean 17.36
Cells/table	med. 18, mean 20.07	Rows/cols/table	mean 7.41 / 3.32
Page range	5–41	Table range/doc.	3–39
Cell range/table	4–82	Mutations/table	med. 6, mean 5.97
Merged cells	4,660 / 741 tables	Multi-level headers	417 tables (26.1%)
Formula cells	3,339 / 642 tables	Indicator cells	10,513 (32.8%)
Footnotes	1,749 / 1,421 tables	Critical cells	25.8% critical

structure, differing in header depth, row group nesting, formula-based aggregation rows, and footnote and indicator usage.

Controlled mutation operators are then applied to increase variability, for example by adding or removing rows and columns, changing value types, varying units and decimal formats, attaching footnotes, or modifying header regions. These operators are sampled using user-defined weights, which bias transformations such as structural edits, value changes, footnotes, KPI markings, and graphical indicators, controlling the balance between structural, content-level, and semantic variations. All mutations are checked through schema validation and consistency rules, ensuring that the resulting tables remain structurally valid and semantically coherent.

Every cell uses one canonical schema. *Content* fields carry the payload: `raw` (display text), `formula`, and `label`. *Metadata* fields describe structure and formatting like: `valueType`, `unit`, `format`, `note`, `colspan`, `rowspan`, `symbol` (graphical indicators), `role`, and a `criticality` level. Cell criticality is assigned before any LLM-generated content is introduced. Criticality labels distinguish low-importance decorative or contextual cells from high-importance and critical cells. Criticality annotations, structural spans, layout metadata, and formula definitions are inherited from the source templates and preserved deterministically, keeping the ground truth stable and controlled. This enables the evaluator to weight extraction errors according to their downstream relevance.

Document Assembly and Placeholder-Based LLM Population. After mutation, tables are embedded into synthetic ESG report skeletons represented as ordered blocks, including titles, narrative paragraphs, tables, and side-by-side layouts. Although this paper focuses on table analysis, this surrounding context is preserved because headings, captions, and nearby text often support table interpretation. The structure builder replaces writable fields with sentinel placeholders, while non-writable fields such as cell identifiers, row–column structure, merged spans, units, formats, formulas, and criticality labels remain fixed.

The generator treats the LLM as a replaceable content-population module to generate realistic ESG-related text and cell values, without modifying the underlying structural schema. In this benchmark, the population stage uses the local Ollama model `gpt-oss120b`, but the surrounding pipeline does not depend on this specific model. The prompt provides the document theme, JSON skeleton, indicator-field constraints, and strict output contract. The model must return valid JSON, preserve identifiers and list lengths, and fill only placeholders with ESG-plausible titles, paragraphs, labels, numerical leaves, footnotes, and graphical indicators selected from the predefined allowed set. Graphical indicators are represented directly in the JSON through three fields: `meaning`, `symbol`, and `category`. The `meaning` captures the semantic interpretation, such as positive, negative, stable, warning, or trend-related; `symbol` stores the visible mark; and `category` groups it into classes such as trend, status, or alert.

Invalid generations are handled through schema and whitelist validation. Only content-bearing fields may overwrite the original template, while protected fields such as list lengths, criticality annotations, layout metadata, and formula expressions remain unchanged. Numerical values, indicator fields, and text are then checked for validity; failed outputs can be retried with correction hints, and unresolved cases are preserved as partial documents for audit and recovery.

Formula Representation. Formula cells are not generated freely by the LLM. They are stored in the ground-truth JSON as structured formula objects that reference cells within the same table. Conceptually, each formula is represented as a normalized excel-like expression or abstract syntax tree, with operators and operands defined over stable cell identifiers rather than visual text. This preserves both the displayed formula value and the dependency structure needed to assess whether an extractor recovers the correct computational relation.

After LLM population, the formula evaluator computes derived cells per table and in dependency order, checking references, units, and invalid divisions. Values are then formatted with the declared unit and decimal format, keeping formula outputs consistent with the ground truth.

Rendering and Output Artifacts. The final JSON document is rendered to paginated A4 HTML and then to PDF with controlled corporate-style variations, including typography (font families), colour/accent palette, spacing, header treatment, and table styles (horizontal-only, minimal, boxed, striped); the renderer also supports cross-page tables linked to a single ground-truth table object. The pipeline outputs ESG-like PDFs paired with exact JSON ground truth, plus provenance artifacts for mutation history and seed-template usage, supporting reproducibility and auditability. Because generation is fully procedural, the same pipeline can also emit arbitrarily large *paired* training data, making dedicated train/validation splits for fine-tuning models a natural extension.

3.2 Benchmark Setting

Our goal is to evaluate not only content recovery, but also preservation of the structure that gives values meaning, including rows, columns, headers, merged cells, footnotes, and formula relations. We evaluated representative systems from four methodological families.

Rule-based PDF Extractors. We included lightweight parsers that operate directly on the PDF text layer and convert the extracted content into an LLM-friendly representation. In particular, we considered `PyMuPDF4LLM`, which produces Markdown while preserving headings, paragraphs, and tables. These systems are computationally efficient and simple to integrate, but they often flatten table structure, making them less suitable for hierarchical headers, spanning cells, and visually complex layouts.

Layout-aware Document Pipeline. We evaluated end-to-end document processing frameworks that integrate OCR, layout detection, and table extraction. These include `Unstructured`, `Docling`, and `OpenDataLoader`. Among them, Docling is particularly relevant because it combines RT-DETR for layout analysis and TableFormer for table structure recognition, making it well suited to structured business documents [9,18]. OpenDataLoader was used in a hybrid mode that delegates complex pages to a Docling backend while preserving a unified hierarchical output.

Table-specific Transformer Models. To provide a strong structure-oriented baseline, we implemented a two-stage pipeline based on Table Transformer (TaTR), using separate models for table detection and structure recognition. Detected tables were cropped from page images, processed by the structure recognizer, and finally combined with OCR-based cell content extraction.

Vision-language and Multimodal LLM Systems. Finally, we evaluated page-level generative systems that parse each document page from its image representation. This group includes the frontier multimodal LLMs analyzed in the results, namely `Sonnet 4.6`, `Gemini 3`, `GPT-5.2`, `Sonnet 4.5`, `Gemini 2.5`, and `GPT-5.1`, together with the open-weight `Gemma 3` and `GLM-OCR`. For the LLM-based systems, each page was processed with a structured prompt specifying the expected canonical JSON schema, aligned with the ground truth representation. The prompt described the required document blocks, table structure, cell fields, units, formulas, footnotes, and semantic types, and encouraged step-by-step reasoning before emitting the final JSON. `GLM-OCR` was treated separately, since it exposes a dedicated OCR/document parsing interface rather than a schema-constrained prompting setup.

Canonical Output Representation. Except about the structured LLM results that are inalterated, the evaluated systems return heterogeneous outputs, including Markdown, HTML, OCR text and parser-specific JSON. To enable a fair comparison, every prediction is converted into a common canonical JSON representation aligned with the benchmark ground truth. The document is represented as an ordered sequence of blocks, where text blocks are labeled as title or narrative, and table blocks are encoded through columns, optional header rows, body rows, cells, footnotes, and typed values. Markdown and HTML outputs are parsed from explicit row and cell delimiters, while parser-specific outputs are mapped by transferring the available row, column, span, and cell metadata into the same schema. The canonicalization layer is deterministic and intentionally conservative. Its role is limited to format normalization and schema alignment: it does not use ground truth, infer missing rows or columns, correct wrong cell values, or add structural relations that are not present in the original system output. This ensures that score differences reflect extraction quality rather than the corrective power of post-processing. The same rule-based procedures are applied across systems to remove tool-specific metadata, normalize identifiers, collapse trivial whitespace artifacts, and map equivalent output conventions to a shared representation. For example, at the cell level, numerical strings are converted into `numerical` values with raw numbers, explicit units such as currency symbols, and detected locale formats by detecting decimal and thousands separators. Graphical indicators are mapped to a controlled icon vocabulary only when directly visible, such as status marks or trend arrows, while footnote markers are separated from cell text and linked to table-level definitions when extracted.

3.3 Domain-Specific Hierarchical Evaluation

After canonicalization, each prediction P is compared with its ground truth G by a hierarchical evaluator returning an overall score $S \in [0, 1]$, separate text and table scores (the table score split into precision, recall, F_1), the three diagnostic dimension scores. Beyond the scalar scores, the evaluator emits a *detailed per-document mismatch report* that records every mismatched element: each cell-level difference is tagged with its field, its criticality, and expected/observed values, alongside table-level column, header, and footnote differences. Because systems may omit, add, split, or reorder blocks, alignment is content-aware rather than positional: text blocks are matched by normalized similarity and table blocks by structural signatures over columns and rows (Fig. 3).

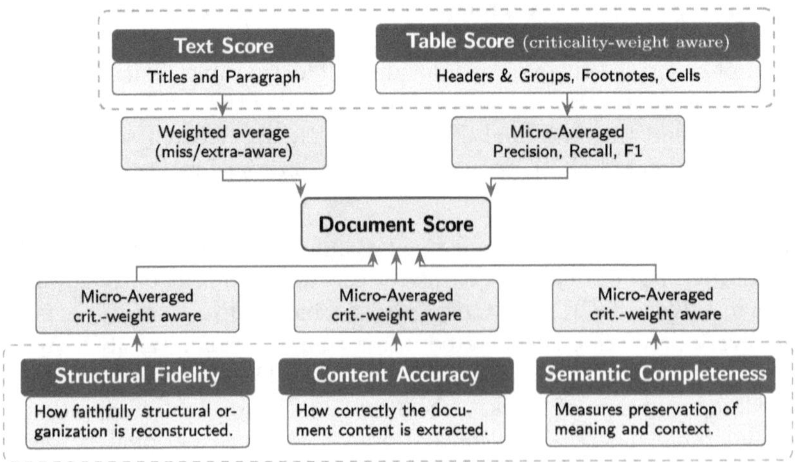

Fig. 3. Domain-specific hierarchical scoring schema of the evaluation framework, showing the metrics aggregation for the overall document score.

Text Block. The main element in scoring an aligned textual pair is the normalized Levenshtein similarity [6] of its contents—the same measure used for block alignment—which is bounded in $[0, 1]$, tolerating minor OCR and formatting variation rather than requiring exact equality.

Criticality-weighted Table Scoring. Tables are evaluated as structured objects composed of columns, headers, footnotes, rows, and cells. Each component is compared separately and then aggregated into table-level precision, recall, and F1. This distinction captures both omissions, such as missing indicators, and hallucinations, such as extra rows.

The score is also criticality-aware. Errors on more important elements, such as KPIs, reporting years, aggregation totals, or business dimensions, receive stronger penalties than errors in secondary fields. Rows are aligned through normalized row signatures, while cells are compared by considering both *content* and *metadata*, listed above. *Metadata* fields are scored by exact matching, *content* fields by normalized Levenshtein similarity with partial credit, except numeric values (exact match) and formulas (token-level exact match). Each ground-truth cell/header level maps to a weight through one spread parameter $\kappa \geq 1$:

$$\left(\omega_{\text{low}}, \omega_{\text{med}}, \omega_{\text{high}}, \omega_{\text{crit}}\right) = \left(1,\ 1 + \tfrac{\kappa-1}{3},\ 1 + \tfrac{2(\kappa-1)}{3},\ \kappa\right) \approx (1.0, 2.3, 3.7, 5.0) \quad (1)$$

Formula-cell Evaluation. Formula cells require a dedicated evaluation because their meaning is not limited to the displayed value. In ESG and financial tables, totals, subtotals, percentages, and aggregation rows often depend on other cells. A system may correctly extract the final number without losing the computational relation that produced it. For this reason, formula cells are treated as semantic units that encode dependencies between cells. Since predicted and ground-truth JSON files may use different internal cell identifiers, formulas are first compared after cell-ID resolution. The evaluator builds a mapping between predicted and ground-truth identifiers from the already established table, row, and positional cell alignment.

After this step, the evaluator applies *AST-Based Canonical Normalization.* The formula is parsed into an operator-aware representation, so that the comparison focuses on the computational structure rather than superficial formatting. This normalization also avoids penalizing harmless variations, such as reordered operands in commutative operations, while preserving the order of non-commutative operations. Finally, the normalized formulas are compared by using a *Canonical Token-Level Exact Match.* This verifies whether the predicted formula preserves the same referenced cells and arithmetic structure as the ground truth, distinguishing systems that only recover the displayed result from those that also preserve the underlying dependency.

Document Aggregation and Diagnostic Dimensions. After block and table evaluation, text and table scores are reported separately and combined into the overall document-level score, a weighted mean that gives greater importance to table extraction, since structured tabular evidence is the main focus of the benchmark.

The framework also reports three diagnostic dimensions, each computed as a criticality-weighted ratio over a disjoint set of evaluation units. Structural fidelity measures whether document and table organization are reconstructed correctly, covering columns, header cells, merged-cell spans, and cell and value types. Content accuracy measures the correctness of extracted text and raw cell values. Semantic completeness measures whether higher-level information is preserved, including units, numerical formats, notes, graphical indicator metadata, and formula dependencies. This decomposition makes the metric more informative than a single aggregate score, because it identifies whether errors mainly affect, supporting a more precise comparison between the evaluated systems.

4 Results

This section reports the results of the benchmark on the ESG-style synthetic documents. The analysis first considers the overall ranking of the evaluated systems, then separates text and table extraction, and finally examines table precision, recall, and the diagnostic dimensions of structure, content, and semantics.

All multimodal LLM-based systems were evaluated with the same input strategy. Each document was divided into chunks of five consecutive pages, with a one-page overlap between adjacent chunks. The outputs were then merged through overlap-aware deduplication. This setting reduces the effect of model-specific context limits and increases the probability that tables spanning page boundaries are processed completely.

4.1 Overall Ranking and Score Distribution

Table 2 reports the overall score statistics for all evaluated systems. The results show a clear separation between frontier multimodal LLMs and the other approaches. Sonnet 4.6 and Gemini 3 achieve the highest mean overall scores, 0.833 and 0.831 respectively, and also the lowest variability, indicating both high accuracy and strong consistency. A second tier is formed by Sonnet 4.5 and GPT-5.2, followed by Gemini 2.5 and GPT-5.1.

Table 2. Overall score summary statistics by evaluated system.

Category	System	Mean	Median	Std.	IQR
Frontier LLM	Sonnet 4.6	0.833	0.841	0.134	0.144
Frontier LLM	Gemini 3	0.831	0.852	0.138	0.174
Frontier LLM	Sonnet 4.5	0.766	0.790	0.184	0.166
Frontier LLM	GPT-5.2	0.729	0.760	0.179	0.202
Frontier LLM	Gemini 2.5	0.687	0.687	0.146	0.171
Frontier LLM	GPT-5.1	0.659	0.662	0.175	0.200
Parser/OCR	OpenDataLoader	0.636	0.651	0.230	0.377
Parser/OCR	Docling	0.631	0.621	0.195	0.237
Parser/OCR	Unstructured	0.514	0.528	0.158	0.265
Parser/OCR	PyMuPDF4LLM	0.482	0.441	0.197	0.256
Parser/OCR	GLM-OCR	0.462	0.455	0.187	0.211
Open LLM	Gemma 3	0.459	0.448	0.135	0.156
Table model	TATR	0.340	0.274	0.178	0.198

Document parsing pipelines occupy the intermediate range. OpenDataLoader and Docling are the strongest non-frontier systems, both slightly above 0.63 mean overall score, whereas Unstructured, PyMuPDF4LLM, and GLM-OCR remain substantially lower. The open-weight Gemma 3 and the table-specific TATR baseline close the ranking.

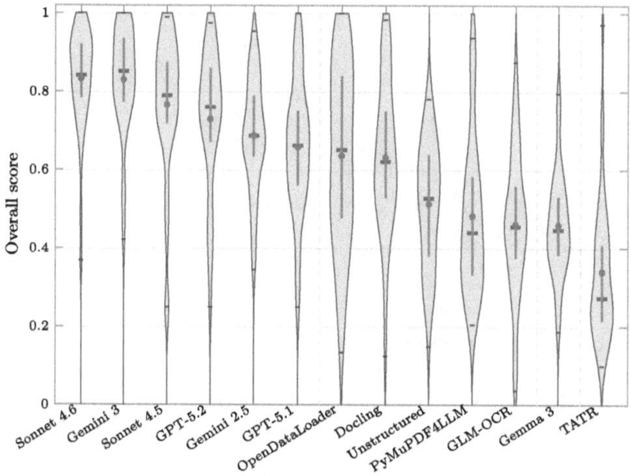

Fig. 4. Distribution of per-document overall scores across the evaluated systems. The width of each violin reflects the kernel density of the score distribution; horizontal markers indicate the median and interquartile range.

Figure 4 confirms this pattern. The best frontier LLMs have compact distributions concentrated in the upper part of the scale, while parsing pipelines are more dispersed. This is relevant in practice because, in ESG reporting, a system must not only perform well on average, but also behave predictably across different document layouts.

4.2 Text vs Table Extraction

To better understand the source of the performance differences, we decompose the overall score into text and table components. Most systems achieve high text scores. All frontier LLMs are above 0.89, with Sonnet 4.6 reaching 0.992. Some weaker systems also perform well on text: Gemma 3 reaches 0.920, and PyMuPDF4LLM reaches 0.901, mainly because it can exploit the native PDF text layer. The main differences emerge on table extraction. Gemini 3 obtains the highest mean table score (0.782), followed closely by Sonnet 4.6 (0.779). Sonnet 4.5 remains competitive at 0.700, whereas GPT-5.2, Gemini 2.5, and GPT-5.1 fall in the 0.58–0.66 range. Among the non-frontier systems, OpenDataLoader and Docling achieve the strongest table performance (0.556 and 0.544), but remain roughly 23 points below the leaders. PyMuPDF4LLM, Gemma 3, and TATR all score near or below 0.34 on tables (Fig. 5).

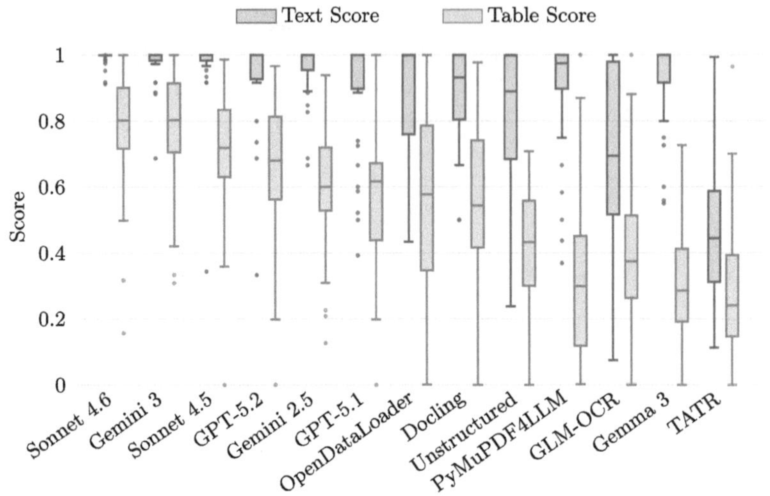

Fig. 5. Distribution of per-document text score and table score across the evaluated systems. Boxes show the interquartile range, the central line marks the median, whiskers follow the 1.5 IQR rule, and dots indicate outliers.

A similar trend emerges when table performance is separated into precision and recall. Gemini 3 reaches 0.781 precision and 0.783 recall, while Sonnet 4.6 obtains 0.775 precision and 0.786 recall, showing that their higher table scores come from balanced and reliable reconstruction. In contrast, GPT-5.1 shows a more conservative profile, with higher precision than recall, suggesting that it more often omits part of the tabular content than introduces incorrect additional structures. Overall, the text–table gap shows that the benchmark is not primarily limited by plain text recovery. The central challenge is the reconstruction of tables with multi-level headers, merged cells, formula rows, footnotes, graphical indicators, and heterogeneous value types.

4.3 Diagnostic Decomposition

The evaluation framework decomposes performance into *structure*, *content*, and *semantic* diagnostic dimensions. The frontier LLMs obtain the best structural and content scores. Gemini 3 leads on structure with 0.844, followed by Sonnet 4.6 with 0.828, while Sonnet 4.6 achieves the highest content score, 0.808. This shows that the strongest systems recover both table organization and cell values more effectively. Semantic completeness remains more difficult. Even the best LLMs score lower on this dimension, showing that units, value types, formulas, footnotes, and graphical indicators are still challenging. OpenDataLoader and Docling show a different behaviour: their semantic scores, 0.720 and 0.710, but their structural scores are lower (Table 3).

Table 3. Diagnostic scores, reporting mean, median, and interquartile range (IQR), grouped by methodological category.

Model	Structure			Content			Semantic		
	Mean	Median	IQR	Mean	Median	IQR	Mean	Median	IQR
Sonnet 4.6	0.828	0.880	0.236	0.808	0.878	0.275	0.723	0.735	0.273
Gemini 3	0.844	0.880	0.264	0.797	0.817	0.250	0.714	0.671	0.397
Sonnet 4.5	0.757	0.847	0.327	0.722	0.801	0.334	0.581	0.578	0.350
GPT-5.2	0.721	0.787	0.400	0.704	0.772	0.362	0.503	0.527	0.294
Gemini 2.5	0.674	0.652	0.324	0.686	0.708	0.244	0.471	0.479	0.349
GPT-5.1	0.706	0.742	0.285	0.668	0.700	0.387	0.432	0.424	0.404
OpenDataLoader	0.577	0.592	0.471	0.682	0.722	0.352	0.720	0.792	0.407
Docling	0.607	0.638	0.409	0.640	0.627	0.261	0.710	0.779	0.387
Unstructured	0.474	0.475	0.286	0.535	0.556	0.279	0.568	0.661	0.389
PyMuPDF4LLM	0.396	0.324	0.369	0.455	0.423	0.351	0.416	0.397	0.592
GLM-OCR	0.517	0.553	0.309	0.414	0.386	0.268	0.515	0.620	0.526
Gemma 3	0.329	0.336	0.205	0.454	0.526	0.345	0.115	0.000	0.268
TATR	0.385	0.359	0.373	0.357	0.327	0.248	0.414	0.481	0.734

4.4 Discussion

Overall, frontier multimodal LLMs provide the best end-to-end results, with Sonnet 4.6 and Gemini 3 achieving the highest and most stable scores. However, table extraction remains the main bottleneck: text is recovered reliably, while hierarchical headers, merged cells, formulas, footnotes, and graphical indicators still cause errors.

5 Conclusion

This paper addressed table understanding in ESG-style PDF documents, where relevant information is often encoded not only in plain text, but embedded in complex and ambiguous tables. Such characteristics make reliable table extraction a necessary prerequisite for downstream ESG reporting, auditing, and retrieval-augmented analysis.

We presented a synthetic benchmark for ESG-like reports, generated as PDF documents paired with complete JSON ground truth. The generation pipeline combines controlled structural mutations with realistic corporate-style visual variation, while preserving exact information about document blocks, table structure, cell values, units, formulas, footnotes, and semantic annotations. This makes the benchmark suitable for evaluating extraction systems under realistic conditions. The benchmark was used to compare conventional parsing pipelines, table-specific models, open-weight vision-language models, and frontier multimodal LLMs within the same evaluation setting. The results show that frontier multimodal LLMs achieve the strongest overall performance, with Sonnet 4.6 and Gemini 3 leading both in accuracy and stability. However, the experiments also confirm that table extraction remains the main bottleneck.

Another contribution of the work is the domain-specific hierarchical evaluation framework. By separating structural fidelity, content accuracy, and semantic completeness, and by weighting errors according to criticality, the metric captures failures that would be weakly reflected by conventional text-oriented evaluation. This is especially important in ESG scenarios, where an omitted unit, an incorrect formula, or a misinterpreted indicator can have a stronger downstream impact than a minor textual mismatch.

Future work will extend the comparison to additional document-understanding systems, including table-specialized VLMs such as Table-LLaVA, suitable Qwen vision-language models, and recent parsers such as PaddleOCR-VL or MonkeyOCR. The dataset generation process will also be improved to produce increasingly realistic ESG-style documents, including more diverse layouts, stronger dependencies between tables and surrounding narrative text, and controlled noise elements, and a broader catalog of graphical indicators beyond the limited set currently supported. In parallel, while the current evaluation provides a consistent basis for comparing systems, the framework will be further refined to better handle edge cases and ambiguous scoring situations, leading to more accurate and precise assessment. This may also include more semantic forms of comparison, for example by assessing whether extracted indicators preserve the intended meaning rather than only matching their surface form. Finally, an important next step is to connect table extraction with downstream table-aware RAG and structured reasoning pipelines, where preserving relational semantics is essential for trustworthy ESG analysis.

Acknowledgments. The project has received funding within the Chips Joint Undertaking in collaboration with the Horizon Europe Framework Programme under project Ecomobility. Grant agreement number 101112306, CUP: E83D23000150006.

References

1. Adhikari, N.S., Agarwal, S.: A comparative study of pdf parsing tools across diverse document categories. arXiv:2410.09871 arXiv preprint (2024)
2. Auer, C., et al.: Docling technical report. arXiv:2408.09869 arXiv preprint (2024)
3. Carion, N., Massa, F., Synnaeve, G., Usunier, N., Kirillov, A., Zagoruyko, S.: End-to-end object detection with transformers. In: European Conference on Computer Vision, pp. 213–229. Springer (2020)
4. Chen, Z., et al.: Finqa: a dataset of numerical reasoning over financial data. In: Proceedings of the 2021 Conference on Empirical Methods in Natural Language Processing, pp. 3697–3711 (2021)
5. Duan, S., et al.: GLM-OCR technical report. arXiv preprint arXiv:2603.10910 (2026)
6. Levenshtein, V.I.: Binary codes capable of correcting deletions, insertions, and reversals. In: Soviet Physics Doklady, vol. 10, pp. 707–710. Soviet Union (1966)

7. Lewis, P., et al.: Retrieval-augmented generation for knowledge-intensive NLP tasks. Adv. Neural. Inf. Process. Syst. **33**, 9459–9474 (2020)
8. Li, M., Cui, L., Huang, S., Wei, F., Zhou, M., Li, Z.: Tablebank: table benchmark for image-based table detection and recognition. In: Proceedings of the Twelfth Language Resources and Evaluation Conference, pp. 1918–1925 (2020)
9. Nassar, A., Livathinos, N., Lysak, M., Staar, P.: Tableformer: table structure understanding with transformers. In: Proceedings of the IEEE/CVF Conference on Computer Vision and Pattern Recognition, pp. 4614–4623 (2022)
10. Pietri, M., Martinelli, M., Turazza, F., Bertacchini, G., Picone, M., Mamei, M.: Digital twins and federated learning in industrial IoT: worker-centric safety perspectives. In: 2026 IEEE 23rd Consumer Communications & Networking Conference (CCNC), pp. 1–6. IEEE (2026)
11. Reddy, V., Koncel-Kedziorski, R., Lai, V.D., Krumdick, M., Lovering, C., Tanner, C.: Docfinqa: a long-context financial reasoning dataset. In: Proceedings of the 62nd Annual Meeting of the Association for Computational Linguistics (Volume 2: Short Papers), pp. 445–458 (2024)
12. Smock, B., Pesala, R., Abraham, R.: Pubtables-1m: towards comprehensive table extraction from unstructured documents. In: Proceedings of the IEEE/CVF Conference on Computer Vision and Pattern Recognition, pp. 4634–4642 (2022)
13. Smock, B., Pesala, R., Abraham, R.: Grits: grid table similarity metric for table structure recognition. In: International Conference on Document Analysis and Recognition, pp. 535–549. Springer (2023)
14. Soric, M., Gracianne, C., Manolescu, I., Senellart, P.: Benchmarking table extraction from heterogeneous scientific extraction documents. arXiv:2511.16134 arXiv preprint (2025)
15. Turazza, F., Pietri, M., Hadjidimitriou, N.S., Picone, M., Burgio, P., Mamei, M.: Empowering local energy communities with blockchain-based federated forecasting and zero-knowledge proof verification. SN Comput. Sci. **6**(8), 970 (2025)
16. Wang, Z., Zhou, Y., Wei, W., Lee, C.Y., Tata, S.: VRDU: a benchmark for visually-rich document understanding. In: Proceedings of the 29th ACM SIGKDD Conference on Knowledge Discovery and Data Mining, pp. 5184–5193 (2023)
17. Yu, X., Jian, P., Chen, C.: Tablerag: a retrieval augmented generation framework for heterogeneous document reasoning. In: Proceedings of the 2025 Conference on Empirical Methods in Natural Language Processing, pp. 14074–14093 (2025)
18. Zhao, Y., et al.: DETRS beat yolos on real-time object detection. In: Proceedings of the IEEE/CVF Conference on Computer Vision and Pattern Recognition, pp. 16965–16974 (2024)
19. Zheng, X., Burdick, D., Popa, L., Zhong, X., Wang, N.X.R.: Global table extractor (GTE): a framework for joint table identification and cell structure recognition using visual context. In: Proceedings of the IEEE/CVF Winter Conference On Applications Of Computer Vision, pp. 697–706 (2021)
20. Zhong, X., ShafieiBavani, E., Jimeno Yepes, A.: Image-based table recognition: data, model, and evaluation. In: European Conference on Computer Vision, pp. 564–580. Springer (2020)
21. Zou, Y., et al.: Esgreveal: an LLM-based approach for extracting structured data from ESG reports. J. Clean. Prod. **489**, 144572 (2025)

A Feasibility Study of Concept-Based XAI Classification Methods on Chart Images

Josh Knize(✉), Kenny Davila, and Min Song

School of Computing, DePaul University, Chicago, IL 60604, USA
{jknize1,kdavilac,msong13}@depaul.edu

Abstract. Explainable AI is a rapidly growing field that is closely linked to the critical effort to build trustworthy AI systems. Concept-based XAI methods like Concept Relevance Propagation (CRP) and Label-free Concept Bottleneck Models (LF-CBMs) aim to extract and provide human-understandable explanations for the behavior of computer vision classifiers. We evaluated the feasibility of these methods by adapting them for chart image classification, a particularly difficult vision task due to limited non-synthetic datasets, class similarity, class imbalance, and overall chart heterogeneity. We generated, analyzed, and decomposed CRP explanations, and we evaluated LF-CBM concept generation and prediction using a combination of real-world and synthetic chart image data. We conclude that these XAI techniques are not practically useful out of the box for several reasons, including a wide interpretability gap, concept entanglement, poor concept generation, and poor concept prediction. We suggest future work to enable more complete explanations and better evaluation of XAI methods.

1 Introduction

Recent growth and progress in AI has led to more powerful models and increased adoption across countless domains [34]. As AI becomes more integral to our society and infrastructure, we risk giving up control to systems that we are unable to trust or understand [5,8,34]. Longo et al. [20] define explainability as one of the four core pieces of trustworthy AI. Progress in Explainable AI (XAI) enables improved model accuracy, better alignment between progress and human values, and responsible adoption of AI across different domains [1,8,20,26].

Deep learning architectures are especially powerful but remain particularly difficult to thoroughly understand. Concept-based XAI techniques are one of several popular frameworks for improving explainability and interpretability of these models [20]. Concepts can be defined as human-understandable features that a model might use for prediction or discrimination. For example, a model might learn a representation from a set of pixels and use it to predict images of the class "dog". A human, however, can recognize different parts of the image as *concepts* relating to a dog, such as "snout", "fur", or "tail". Concept-based techniques seek to explain model behavior by breaking it down into these intuitive

F. Shafait et al. (Eds.): DAS 2026, LNCS 16851, pp. 74–90, 2027.
https://doi.org/10.1007/978-3-032-36207-0_5

concepts, which make these methods particularly intriguing for bridging the gap between black-box model behavior and trustworthy insights. This paper studies the effectiveness and generalizability of two concept-based XAI frameworks for computer vision: Concept Relevance Propagation (CRP) [1] (Sect. 2.2) and Concept Bottleneck Models (CBMs) [19,26] (Sect. 2.3).

We employ the model from our previous work, *Optimizing Chart Image Classification* (OCC) [18], to build CNN classifiers on the CHART-Info 2024 dataset [9]. These models are used to investigate CRP. We use OCC's fine-tuned DINOv2 [28] model as a backbone to investigate Label-free CBMs (LF-CBMs) [26].

Ultimately, we did not find any satisfactory out-of-the-box configurations of CRP or LF-CBMs to apply to chart image classification. CRP suffers from a layer selection problem, concept entanglement, concept dispersion, framework portability issues, and some more fundamental challenges. LF-CBMs were not able to generate reasonable concept sets or accurately predict concepts. The findings of this work highlight the need for improved explainability metrics and datasets to critically evaluate new XAI techniques. Additionally, we stress the need for XAI frameworks that allow for contrastive [23] explanations.

2 Background

2.1 Explainable AI

The rapid growth of XAI has led to many different approaches and frameworks with different aims and strategies [20,38].

Global vs. Local. XAI frameworks can typically be described as global or local [20,32,38]. Global XAI aims to explain a model's overall behavior, whereas local explanations focus on a specific input or individual prediction. For instance, a local explanation might use a heatmap on an image to highlight the regions that most influenced the model's prediction. In contrast, a global explanation might present a collection of dataset samples associated with a particular feature representation learned by the model. Figure 1 gives a clear example of each type of explanation.

Ante-hoc vs. Post-hoc. Ante-hoc XAI methods are designed to be intrinsically explainable [38], which can often lead to strong explanations [31]. Post-hoc methods differ by explaining models that were already trained end-to-end. In contrast to ante-hoc methods, post-hoc methods are usually more portable, often at the expense of clearer explanations [31].

Method Taxonomy. Assigning XAI methods to clear categories is a complex and challenging problem [32,38]. Here, it suffices to name a few relevant types of XAI methods. Note that these approaches are not mutually exclusive.

A popular, early, and relatively straightforward explainability method is an *attribution-based* approach [30,36]. These approaches, such as heatmaps (Fig. 1b), indicate important parts of an input for model prediction. They are often criticized for leaving too much room for human *interpretation* rather than model *explainability* [20,31]. We refer to this problem as the "interpretability gap".

Another relevant method is *example-based* explainability, which uses representative or prototypical examples to explain a model. For example, Fig. 1c uses representative samples from a dataset to give a global explanation of our model.

Concept-based learning [20] is the most relevant XAI technique to this paper. Concept-based algorithms can provide post-hoc explainability or a method for creating ante-hoc explainable models. Using these techniques, model behavior can be broken down into human-understandable ideas and abstractions.

Interdisciplinary Findings. Miller [23] shows that social science research can be applied to XAI research. This interdisciplinary approach is necessary because a purely computational approach will not be enough to build trustworthy AI systems [23]. Miller [23] argues that explanations in AI models should be contrastive, selective, social, and causal rather than probabilistic.

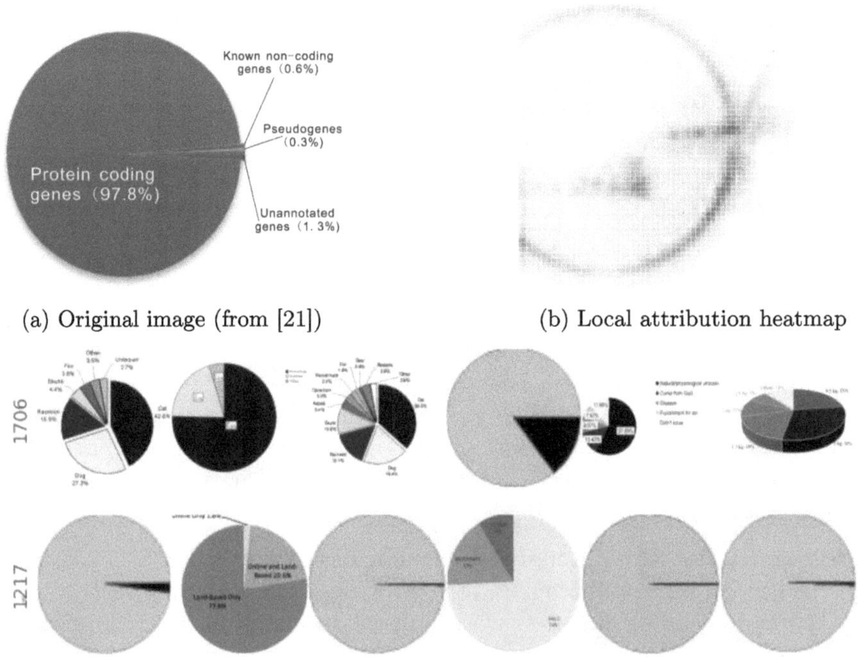

(a) Original image (from [21])

(b) Local attribution heatmap

(c) Global feature visualization for two concepts

Fig. 1. Overview of CRP for chart image classification.

2.2 Concept Relevance Propagation

CRP is a concept-based, post-hoc XAI technique that combines local and global perspectives [1]. CRP leverages Layer-wise Relevance Propagation (LRP) [4], a method to attribute relevance to the input as an algorithm propagates backward through a network. CRP extends this framework by conditionally masking neurons during propagation to create *channel-wise* relevance. This channel-wise relevance is then used to create heatmaps that correspond to different concepts learned in a model's layers and bridge the interpretability gap typical of traditional heatmaps.

Further, CRP introduces a measure, RelMax, used to select maximally relevant samples to latent model features. RelMax is a proposed alternative to a popular attribution measure, ActMax [24]. ActMax values, based solely on maximizing neuron activations, are not always consistent with the most useful parts of an input for model *prediction* [1,31]. RelMax offers an alternative method for attribution that searches samples for concepts relevant to prediction, thus linking the relevance score directly to model output. RelMax scores calculated from images can be used to generate a global, example-based explanation for a model's behavior.

2.3 Concept Bottleneck Models

Traditional CBMs. CBMs are an ante-hoc technique for building models with knowledge of human-understandable concepts. Traditional CBMs employ a learnable hidden layer of neurons, or concept bottleneck layer (CBL), that corresponds to user-specified concepts. A final classification layer is learned using the concept activations in this penultimate layer, which can also be extracted as an explanation for the model's decisions (Fig. 2).

There are two major drawbacks to traditional CBMs [26]. First, training a CBM requires ground truth labeling of concepts for each sample, which tends to be a tedious, manual process. Second, a drop in model performance can be expected due to the added bottleneck layer [26].

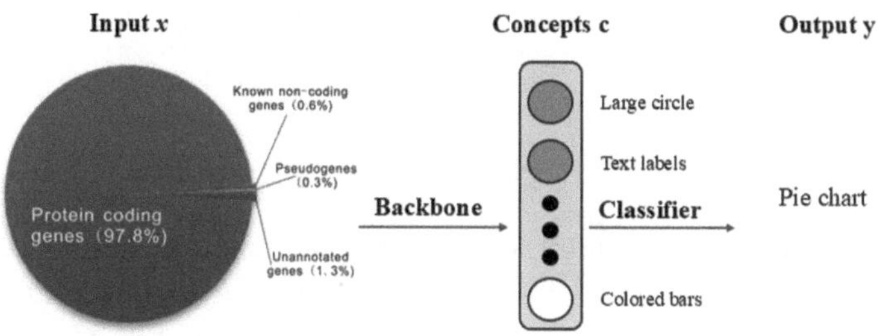

Fig. 2. Overview of CBM for chart image classification.

Label-Free CBMs. Label-free CBMs (LF-CBMs) build on the original CBM framework and ostensibly eliminate the need for concept ground truths while preventing the performance trade-off [26]. Rather than using labeled ground truths for each individual sample, LF-CBMs require only a list of concepts to be loaded before training a model. The framework can automatically generate these concept sets with various large language model (LLM) prompt options, or the concepts can be manually written and provided by the user. Concepts are generated for each class, combined into a single, class-agnostic set, and then filtered to a smaller set using criteria such as a cosine similarity calculation.

The LF-CBM CBL is learned automatically using CLIP-Dissect [27,29], which uses CLIP to provide scores of how closely individual neurons in a network represent a given concept. Another notable feature of LF-CBMs is that they seek to learn a *sparse* final layer, enabling more interpretable explanations.

2.4 Chart Image Classification

The ability to classify chart images can be seen as the first step of automated chart content extraction [11,12], a task with potential to build rich datasets and generate significant value [10,16]. A few recent surveys have explored this emerging subfield [10,13,39]. Chart images are a challenging domain for classification due to limited non-synthetic datasets, class similarity, class imbalance, and overall chart heterogeneity [11,18]. The CHART-Info 2024 dataset captures the diversity and challenges of this domain with over 50,000 real-world chart images categorized into 15 different classes [9]. There are several frameworks [14,17,18,35] that achieve strong results (>93% macro-averaged F1 score) on different versions of the CHART-Info datasets [9,11,12].

We used the model from our previous work, OCC [18], for our study because it is the most recent and highest scoring model on CHART-Info 2024 to our knowledge. The model reached an F1 score of 94.9% using a DINOv2 architecture [6,28] and domain-specific training strategies. In particular, OCC masks portions of each training image and uses gridline overlay augmentations to learn better feature representations and reduce overfitting. Further, the OCC model was trained with weighted random samples of the original dataset class distribution to give higher priority to underrepresented classes and ameliorate the class imbalance problem.

3 Methodology

In this section, we describe our choices and steps for adapting CRP [1] and CBMs [19,26] to the OCC [18] model and the CHART-Info 2024 dataset [9]. Also, we describe the strategies used to evaluate the effectiveness of the XAI frameworks.

3.1 Concept Relevance Propagation

We used CRP to visualize features from different layers of models trained on CHART-Info 2024. We employed ResNet-50 [15] for this experiment due to CRP compatibility challenges with other models (Sect. 4.1). Images were resized to the 224×224 input size of the ResNet-50's pre-training. The OCC framework uses square masks and gridline overlay augmentations of 30% and 15% of the image size, respectively. We found that our ResNet-50's [15] accuracy improved when reducing the training masks from 30% to 20% of the image area. This model was used for our CRP feature visualizations.

For each combination of chart type in our data and selected layer in our model, we used CRP to compute a normalized vector of cumulative concept relevance scores. We used this data to extract the top five most relevant concepts for each class and visualize the top 10 most representative dataset samples for that concept. We qualitatively evaluated CRP's effectiveness in extracting useful, human-understandable concepts.

As an example, we will break down a feature visualization (Fig. 3) from our model. The concept represented in the figure is the second most relevant concept in the final hidden layer, "layer4.2.conv3", for predicting the "venn" class. The label of 4.54 indicates that Channel 1355 is responsible for 4.54% of relevance flowing from layer4.2.conv3 to the final classification of a venn chart on average across the CHART-Info 2024 training set. We will call this "global" relevance. The five charts in the figure are the five most representative training set samples of Concept 1355 according to the averaged RelMax metric. The CRP output highlights pixels from the images that are most relevant for the concept.

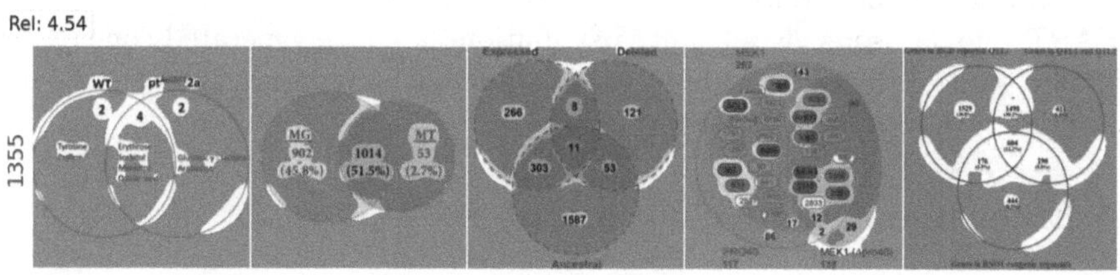

Fig. 3. An example CRP feature visualization.

3.2 Label-Free CBMs

An LF-CBM model was built using the OCC DINOv2 backbone and 2024 CHART-Info dataset. The hidden units of the CBL required a set of corresponding concepts. To match the LF-CBM paper, we created different concept sets using two provided GPT-3.5 Turbo Instruct prompts: "List the most important features for recognizing something as a {chart class} chart" and "List the things most commonly seen around a {chart class} chart". We tested the prompts and a combination of both. The generated concepts were then filtered by the LF-CBM framework (Sect. 2.3). Further, we experimented with parameters for stricter filters to remove concepts that are similar to each other (see Sect. 4.2).

The LF-CBM code repository natively supports feature extraction from four-dimensional data. We adapted the LF-CBM code repository to support feature extraction from three-dimensional data so we could obtain representations from

ViTs and OCC's DINOv2 model. We added custom code to track experiments, preprocess images, and evaluate custom metrics.

Quantitative analysis in the original LF-CBM paper relies on both final classification scores and a crowdsourced evaluation of the generated explanations. This type of evaluation provides some insight into the framework's usability, but might also suffer from noise expected in crowdsourced scenarios. Other than this, the authors of LF-CBM rely on qualitative analysis to evaluate interpretability, but this also has some limitations (see Sect. 4.2).

We propose an alternative strategy for quantitatively evaluating label accuracy using synthetic data. We created a simple synthetic dataset of five chart classes (line, pie, horizontal bar, vertical bar, and scatter) using Python and Matplotlib. These chart classes were given ground truth labels for sixteen of our CBM concepts that were feasible to accurately generate programmatically. Some concepts were labeled based on chart type, while others were conditionally labeled during chart generation. For example, "a large circle" is a concept present in all pie charts and none of the other synthetic classes. Alternatively, "colorful points" was labeled on some scatter charts if the random style generation selected to use a colorful palette when rendering the chart image. We generated 100 charts per class type for a total of 500 charts with 16 boolean concept labels each. Our objective with this dataset is only to evaluate the LF-CBM's ability to accurately predict concepts. For this reason, we did not need to match the CHART-Info dataset's class count, distribution, or entire generated concept set. We adapted the LF-CBM code to export concept predictions during inference, given different concept activation thresholds.

4 Experiments and Discussion

Our backbones were trained on NVIDIA A100 GPUs with 80 GB of VRAM and tuned using the OCC open-source code [18]. We used ResNet-50 [15] for CRP [1] and the OCC DINOv2 [28] backbone for the LF-CBM [26] experiments. Our base ResNet-50 model, using OCC's data augmentation parameters, achieved a 90.9% macro-averaged F1 score, which increased to 91.5% with our new parameters (Sect. 3.1). For comparison, the OCC model with a different DINOv2 backbone reached a 95.4% validation F1 score.

4.1 Concept Relevance Propagation

The open-source CRP repository provides extensive tooling for tailoring explanations to user preferences. We used the provided functions to visualize and quantify relevance flows for samples across different layers of our networks (Sect. 3.1). Also, we used CRP's functionality for visualization output options, model architecture tracing, and conditional channel masking.

A key appeal of post-hoc XAI methods such as CRP is their promise to enable quick adoption and support explanations for end-to-end trained models. Despite this, we found several critical drawbacks to CRP that would prevent us from using the method out of the box.

The Layer Selection Problem. The CRP framework allows the user to flexibly extract concepts from any layer of the model, conditioned on different channel masks. However, choosing one or more layers for feature extraction on a complex, modern architecture is not a trivial decision. In general, early layers of the model will represent basic features, while later layers will represent more complex ones [40]. In practice, building complete explanations might require extracting features from different depths across the model.

For example, we have a "surface" chart class in our dataset (Fig. 4). These charts typically contain distinctive, complex geometrical features likely to be represented in later layers of a model. Their three-dimensional axes, however, are a much simpler feature that will presumably be captured in an earlier layer. These features are both very likely to be useful concepts for classification, making layer selection complex.

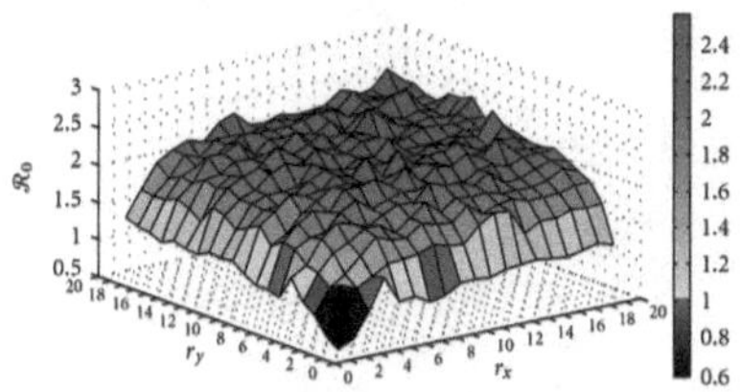

Fig. 4. Example surface chart [25]. These charts are often characterized by a complex geometry in the chart area and a three-dimensional axis.

Table 1. Surface Chart Concepts

Concept
a grid or lines to assist with reading...
a legend explaining the data
a legend or key explaining the symbols
a title describing the purpose of the chart
a title for the chart
data points plotted as symbols or markers
data points plotted on the chart
gridlines to aid in reading the data
horizontal and vertical axes with label...
x and y axes

Concept Entanglement. Even if a solution is found to the layer selection problem, the problem of concept entanglement persists. A single channel can contain multiple concepts, especially in later network layers. This introduces ambiguity when attempting to parse an explanation. Figure 3 shows the second most relevant channel for a venn chart classification in the penultimate layer of our ResNet-50. There clearly seem to be two concepts present in these samples: a circular edge and small text labels.

One option to decompose this higher-level feature is to propagate backward. Here, we take a channel (Fig. 3) and attempt to trace relevance back to the previous two consecutive input layers. Ideally, we hope to see the origin of the circular edge and small text label features. However, we found that Channel 1355 decomposes into more channels of entangled concepts in the previous layer (Fig. 5a); three of the five most relevant channels contain circular edges and small text labels (Channels #9, #37, #491). Because our desired features are most clear in Channel 37, we show a further trace to the next layer (Fig. 5b). Unfortunately, the desired concepts are not recoverable in this layer and condition's most relevant channels either. The visualizations ultimately show that similar concepts

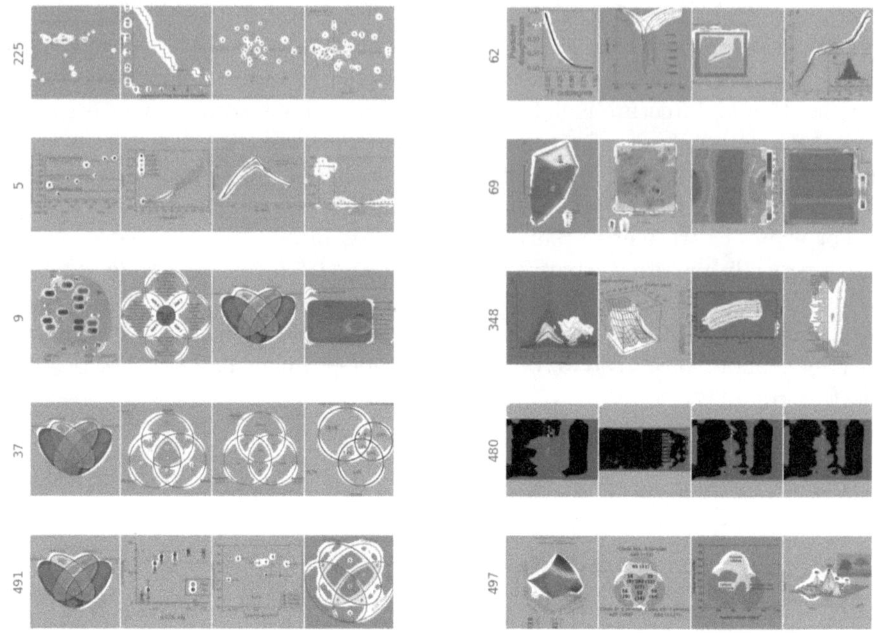

(a) Channel 1355 / layer4.2.conv2 (b) Channel 37 / layer4.2.conv1

Fig. 5. CRP venn chart concept decomposition.

can appear across both layers and channels, and this entanglement makes it challenging to extract clean, human-understandable concepts.

Concept Dispersion. Our experiments show that Channel 1355 is the second most relevant channel globally in the final hidden layer for predicting a venn chart. It represents 4.54% of the relevance in this layer (the most relevant was 5.60%). We found, however, that the most representative sample has a normalized relevance score of only 1.9% *locally* for Channel 1355, making it the third most prevalent concept in that sample. This suggests that the most relevant global concepts may not necessarily be particularly relevant locally due to highly dispersed relevance across the 2048 neurons in the last hidden layer.

Fundamental Challenges. The above findings lead us to reflect on the promising results presented in the CRP paper and on some of its claims. We believe that there are fundamental challenges with the current CRP output that lead to ambiguous explanations.

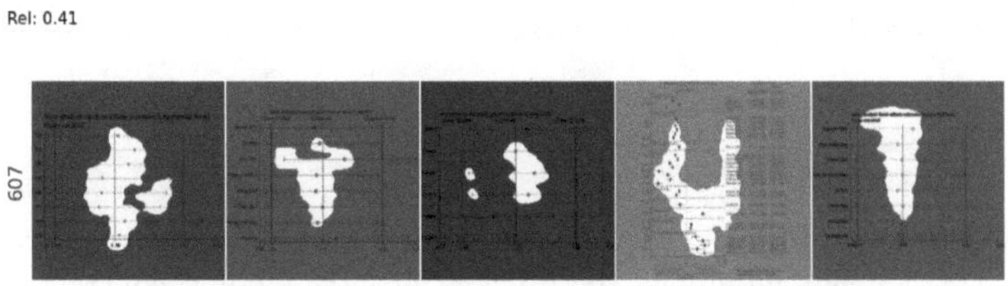

Fig. 6. Representative samples for a relevant horizontal interval chart feature.

We argue that the concepts visualized and represented in both local and global CRP outputs are not robust enough to prevent users from misinterpreting model behavior. We noted ambiguity in figures from the original CRP paper [1], where user interpretation would be required to define concepts in CRP's output and understand their locations. We reproduced some of this ambiguity in Fig. 1. For instance, does Concept 1706 represent circles, slices, text labels, or some combination of those features? Does Concept 1217 refer to only a thin slice of the pie chart, and is it specific to the middle, right-hand side of the circle?

The chart domain shows how the interpretability gap can widen with a different type of dataset. Chart images contain a wide range of colors, shapes, orientations, and data. The second most relevant concept across the horizontal interval chart provides a good example (Fig. 6). We can note from this feature visualization that all images contain a horizontal orientation, aligned horizontal lines, points with whiskers, and black and white colors. We found several other similar ambiguities and could not confidently extract concepts without leaning heavily on our own discretion.

The last fundamental issue to highlight is a lack of contrastive explanations. Miller [23] shows that an explanation must be contrastive to be useful, but CRP's explanations lack a contrastive interface. CRP's attribution heatmaps and representative samples show what parts of the image are used for prediction, but they do not show why another class was *not* predicted. A complete and satisfying explanation should contain an exploration of alternative classes and concepts. Figure 7 and 8 are useful examples of explainable visualization outputs that could enable more contrastive explanations.

Framework Portability. The last significant challenge with CRP is its out-of-the-box compatibility with different model architectures. CRP is designed to compute relevance scores while propagating backward through a network. The framework is built on *Zennit* (Zennit Explains Neural Networks In Torch) [3], a highly customizable attribution-based XAI framework that employs LRP [4]. While Zennit has extensive functionality for different types of networks, it's nontrivial to apply the method to transformers [2]. There is existing work that applies Zennit to transformers [2], and other work that builds attribution heatmaps using both transformer attention *and* MLP layers [7]. The problem, however, is that existing methods, including CRP, do not provide channel-wise (i.e. concept-based) explanations across both attention and MLP layers out of the box.

4.2 Label-Free CBMs

The LF-CBM framework offers a potential solution to easily train CBMs without extensive concept labeling. The authors provide strong, contrastive output options for explaining model behavior globally or locally (e.g., Fig. 7 and 8). The convenient setup and informative visualizations are not sufficient, though, and we found significant challenges in implementing the framework in practice.

The Concept Set Generation Problem. The first step of building an LF-CBM is to generate or provide a set of concepts to the model. We experimented with both approaches and found that the author-provided code for generating concept sets via an LLM (Sect. 3.2) was insufficient to produce a usable concept set in the chart image classification domain.

We found that generated concept sets often contain outright bad concepts. Upon closer review, we found these ambiguous concepts in the original LF-CBM paper. The authors provide a Sankey diagram (see Fig. 7) in their paper that shows concepts such as "a citrus fruit" and "a piece of fruit" as large contributors to predicting an orange or a lemon, but we argue that these concepts do not *entail* an orange or a lemon prediction.

In addition to predictive concepts, we need to generate contrastive concepts [23] to build a classifier that provides satisfactory explanations. We generated concepts by prompting GPT-3.5 Turbo Instruct using the default LF-CBM prompts with "chart" appended (Sect. 3.2). The framework, however, does not guarantee that useful concepts for discrimination and prediction are generated. We further used the prompt "List the most important features for recognizing something as a *surface* chart rather than other chart types". The results of this prompt are illustrated in Table 1. These were not particularly better than the results using the original prompt, therefore all results shown here use the default prompt.

We can see that most concepts (Table 1) are unlikely to be predictive; we would argue that all concepts generated are common across other chart types such as area, bar, line, scatter, scatter-line, box, and interval. A stronger concept set might have included "a three-dimensional axis" or "a rugged three-dimensional shape", for example. We generated our own Sankey chart to visualize the top seven weights for two of our classes, vertical box and horizontal bar, from our top-performing LF-CBM model (Fig. 7). We suggest that the simplest way to discriminate between these two classes is to recognize 1) vertical vs. horizontal orientation and 2) boxes vs. bars. These concepts are not clearly prioritized or present in a contrastive manner. It is possible that different prompting techniques could enable more contrastive concepts in future work. The problem, however, remains with the current framework.

The LF-CBM code enables concept filtering based on length, similarity to class names, similarity to one another, lack of presence in training data, and

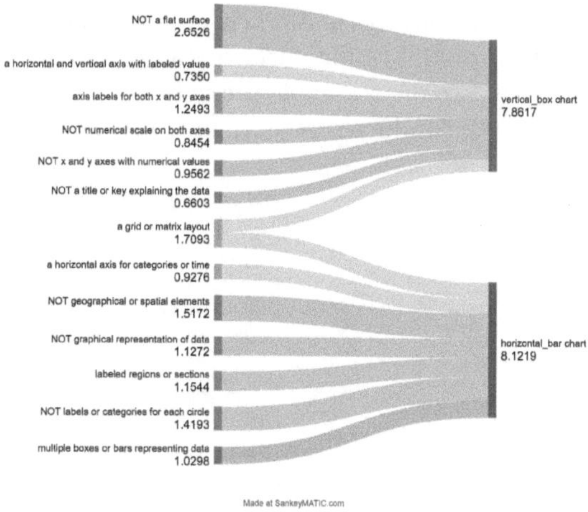

Fig. 7. Example Sankey diagram: concepts on the left correspond to lines, proportional to their labeled final layer weights, flowing into a class prediction.

interpretability. Despite the efforts detailed in the rest of this subsection, we were unable to leverage these filters and parameters to refine our concept sets and provide sufficient explanations.

In particular, the authors' cosine similarity method for filtering concepts that are too similar did not succeed in removing overlapping concepts. Our initial concepts, generated with default LF-CBM filtering parameters, except for adjusting the concept length threshold, returned 77 concepts. We found many overlapping concepts, such as the group of "axis labels and tick marks", "axis labels and titles", "axis labels for both x and y axes", and "axis labels for data points". We adjusted the cosine similarity parameter from 0.9 to 0.8, which then returned 24 concepts. Here, we still saw significant overlap, such as seven concepts focused on a "legend" or "key". Thus, adjusting the parameter reduced our concept set by 69% with 29% of the concepts now referring to a legend or key. This is not necessarily a problem for prediction, but it is a problem for explainability. Overlapping semantics in concept sets lead to unclear and excessive explanations, and it is important to filter explanations to key, contrastive information [23]. The example in Fig. 8 illustrates how the redundancy of "axis" concepts complicates the explanation.

Poor Concept Prediction. We trained custom LF-CBMs using the DINOv2 backbone from OCC and LLM-generated concept sets (Sect. 3.2). Some of our models' validation macro-averaged F1 classification scores retained state-of-the-art performance (up to 95.4%) on the CHART-Info 2024 dataset despite the challenges described in the previous section. Additionally, we found qualitative, local examples of problematic concept contributions that are not consistent with the actual features present in the sample. The list of concepts in Fig. 8 is predicted by our model as present in the line chart. Concepts such as "a linear or non-linear trend line (optional)", "a flat surface", "a circular shape", and "labels or categories for each circle" are questionable predictions at best. These concerning qualitative results led us to conduct further quantitative analysis presented later in this section.

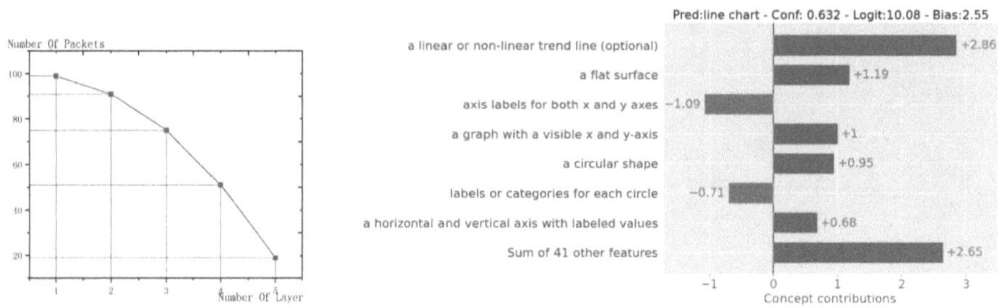

Fig. 8. An Explanation from our CBM for Line Chart.

Oikarinen et al. [26] evaluate their framework's concept predictions using a user survey. Task 1 of the survey asks whether the 10 images most strongly correlated with a concept are representative of that concept. In effect, the study asks users if the neuron in the CBL corresponds accurately to its labeled concept. The study yields evidence that the LF-CBM performs significantly better than a ResNet-50, however, the ResNet-50 performs only 5.2% worse on a user-rated 5-point scale despite having no concept-specific training at all. Also, the second half of the survey only includes a comparative evaluation and does not directly ask users whether they found the explanations themselves reasonable.

These concerns led us to devise our own quantitative strategy for evaluating LF-CBMs *concept* prediction. We evaluated our classifier (95.4% F1 score) on a set of synthetic chart data images with concept ground truths (Sect. 3.2). There is ambiguity regarding what concept activation threshold should be used to determine if a neuron in the CBL, and thus the concept that it corresponds to, is activated. Therefore, we evaluated the concept prediction accuracy of our LF-CBM at different concept activation thresholds (Table 2). For example, the third row of Table 2 indicates that our model correctly predicted 66.2% of concepts in our dataset when we classify a concept as present if its corresponding neuron's pre-activation value exceeds 1.0. The high accuracy and low F1 score at higher concept activation thresholds can be explained by a no-information rate of 82.2%.

Clearly, the model is not predicting concept ground truths well, as indicated by the F1 scores. The results support that it is dangerous to use concept classification performance as a metric for the overall effectiveness of an XAI method. The results also support the growing research consensus that CBMs are at risk of information leakage [22,33]. The hidden CBL is likely entangling concepts and using information outside of the specified concepts to make its final classification [22,33].

Table 2. Model Performance - Concept Prediction. Each concept is an independent classification problem. All scores are the micro-averages over all binary concept predictions.

Concept Act	Recall %	Precision %	F1 %	Accuracy %
0.0	89.6	24.5	43.3	48.5
0.5	84.7	28.1	46.0	57.3
1.0	73.0	31.4	45.9	66.2
1.5	52.8	34.1	40.4	74.7
2.0	28.9	34.5	29.1	79.6
2.5	15.4	36.0	18.5	82.6
3.0	5.3	26.9	8.6	82.6

5 Conclusion

Explainable AI is a critical, growing field [20,34]. In this paper, we explore two popular concept-based XAI techniques and evaluate the feasibility of applying them to a new domain: chart image classification.

Our findings highlight the difficulties of implementing CRP [1] and LF-CBMs [26] in our target domain using their original configurations. The adoption of CRP proved challenging due to the layer selection problem, concept entanglement, concept dispersion, other fundamental challenges, and framework portability. The LF-CBMs proved inadequate due to poor concept set generation and concept prediction within the concept bottleneck layer.

Our aim is not to disparage exciting new efforts in XAI, but rather to raise awareness of challenges and motivate progress towards better techniques. We believe this paper underscores a key issue in explainability research. We do not have standard, accessible, and easily implementable metrics for measuring a system's explainability [20,37]. Consequently, XAI research often relies on qualitative results, insufficient user surveys (Sect. 4.2), and final predictive performance as an imperfect proxy metric for explainability. Additional resources from our work can be accessed at https://github.com/joshknize/concept-based-xai.

6 Future Work

In this section, we suggest a few future directions that may lead to progress towards truly explainable and trustworthy AI. First, we believe we must be able to quantitatively measure explainability and interpretability to properly evaluate XAI methods and improve upon them [37]. Synthetic datasets, generated programmatically rather than using generative AI, are a promising option for a step in the right direction. Synthetic datasets can enable large-scale, accurate ground truth labeling for concepts or other explainability needs if they faithfully represent real data. The synthetic dataset used in Sect. 4.2 was simple and not comprehensive of all concepts or classes, but it was still sufficient to yield insightful, quantitative results regarding our model's predictive accuracy on *concepts*. Additionally, even simpler synthetic datasets may be used to address information leakage [33] in CBMs. In our domain, images containing *only* basic features that are independent of class, such as large circles, axes, text labels, color gradients, and curved edges, might be used to train the CBL independently of the final classification layer. An alternative idea is to use state-of-the-art object detectors or semantic segmentation models to create concept-level datasets and ground truth labels. Further, prompt engineering within the current LF-CBM framework could enable improved concept generation, which may lead to more contrastive and complete explanations [23].

Finally, we believe there may be room for improvement on the CRP framework via software engineering and statistics. A lack of *contrastive* explanations and interfaces [23] is somewhat overlooked in CRP; a new, more contrastive user interface could narrow the interpretability gap. We were impressed with the flexibility and software already provided by the CRP authors, and we expect that taking novel, statistical approaches to the relevance score outputs could lead to less entangled and more interpretable concept discovery.

References

1. Achtibat, R., et al.: From attribution maps to human-understandable explanations through concept relevance propagation. Nat. Mach. Intell. **5**(9), 1006–1019 (2023)
2. Achtibat, R., et al.: AttnLRP: attention-aware layer-wise relevance propagation for transformers. In: International Conference on Machine Learning, pp. 135–168. PMLR (2024)
3. Anders, C.J., Neumann, D., Samek, W., Müller, K.R., Lapuschkin, S.: Software for dataset-wide XAI: from local explanations to global insights with Zennit, CoRelAy, and ViRelAy. PLoS ONE **21**(1), e0336683 (2026)
4. Bach, S., Binder, A., Montavon, G., Klauschen, F., Müller, K.R., Samek, W.: On pixel-wise explanations for non-linear classifier decisions by layer-wise relevance propagation. PLoS ONE **10**(7), e0130140 (2015)
5. Bostrom, N.: Superintelligence. Dunod (2024)
6. Caron, M., et al.: Emerging properties in self-supervised vision transformers. In: Proceedings of the IEEE/CVF International Conference on Computer Vision, pp. 9650–9660 (2021)
7. Chefer, H., Gur, S., Wolf, L.: Transformer interpretability beyond attention visualization. In: 2021 IEEE/CVF Conference on Computer Vision and Pattern Recognition (CVPR), pp. 782–791. IEEE (2021)
8. Christian, B.: The Alignment Problem: Machine Learning and Human Values. WW Norton & Company (2020)
9. Davila, K., et al.: CHART-Info 2024: a dataset for chart analysis and recognition. In: International Conference on Pattern Recognition, pp. 297–315. Springer (2024)
10. Davila, K., Setlur, S., Doermann, D., Kota, B.U., Govindaraju, V.: Chart mining: a survey of methods for automated chart analysis. IEEE Trans. Patt. Anal. Mach. Intell. **43**(11), 3799–3819 (2020)

11. Davila, K., Tensmeyer, C., Shekhar, S., Singh, H., Setlur, S., Govindaraju, V.: ICPR 2020 - competition on harvesting raw tables from infographics In: International Conference on Pattern Recognition, pp. 361–380. Springer (2021)
12. Davila, K., Xu, F., Ahmed, S., Mendoza, D.A., Setlur, S., Govindaraju, V.: ICPR 2022: challenge on harvesting raw tables from infographics (CHART-Infographics). In: 2022 26th International Conference on Pattern Recognition (ICPR), pp. 4995–5001. IEEE (2022)
13. Dhote, A., Javed, M., Doermann, D.S.: A survey and approach to chart classification. In: International Conference on Document Analysis and Recognition, pp. 67–82. Springer (2023)
14. Dhote, A., Javed, M., Doermann, D.S.: Swin-chart: an efficient approach for chart classification. Patt. Recogn. Lett. **185**, 203–209 (2024)
15. He, K., Zhang, X., Ren, S., Sun, J.: Deep residual learning for image recognition. In: Proceedings of the IEEE Conference on Computer Vision and Pattern Recognition, pp. 770–778 (2016)
16. Huang, K.H., et al.: From pixels to insights: a survey on automatic chart understanding in the era of large foundation models. IEEE Trans. Knowl. Data Eng. **37**(5), 2550–2568 (2024)
17. Kanroo, M.S., Kawoosa, H.S., Rana, K., Goyal, P.: C3E: a framework for chart classification and content extraction. Comput. Electr. Eng. **121**, 109861 (2025)
18. Knize, J., Davila, K.: Optimizing chart image classification: a study of data augmentation and training strategies. In: International Conference on Document Analysis and Recognition, pp. 638–655. Springer (2025)
19. Koh, P.W., et al.: Concept bottleneck models. In: International Conference on Machine Learning, pp. 5338–5348. PMLR (2020)
20. Longo, L., et al.: Explainable artificial intelligence (XAI) 2.0: a manifesto of open challenges and interdisciplinary research directions. Inf. Fusion **106**, 102301 (2024)
21. Luo, H., Sun, S., Li, P., Bu, D., Cao, H., Zhao, Y.: Comprehensive characterization of 10,571 mouse large intergenic noncoding RNAs from whole transcriptome sequencing. PLoS ONE **8**(8), e70835 (2013)
22. Margeloiu, A., Ashman, M., Bhatt, U., Chen, Y., Jamnik, M., Weller, A.: Do concept bottleneck models learn as intended? arXiv preprint. arXiv:2105.04289 (2021)
23. Miller, T.: Explanation in artificial intelligence: insights from the social sciences. Artif. Intell. **267**, 1–38 (2019)
24. Nguyen, A., Dosovitskiy, A., Yosinski, J., Brox, T., Clune, J.: Synthesizing the preferred inputs for neurons in neural networks via deep generator networks. Adv. Neural. Inf. Process. Syst. **29** (2016)
25. Niu, T., Gaff, H.D., Papelis, Y.E., Hartley, D.M.: An epidemiological model of rift valley fever with spatial dynamics. Comput. Math. Methods Med. **2012**(1), 138757 (2012)
26. Oikarinen, T., Das, S., Nguyen, L.M., Weng, T.W.: Label-free concept bottleneck models. In: The Eleventh International Conference on Learning Representations (2023)
27. Oikarinen, T., Weng, T.W.: CLIP-Dissect: automatic description of neuron representations in deep vision networks. In: ICLR 2022 Workshop on PAIR^{2}Struct: Privacy, Accountability, Interpretability, Robustness, Reasoning on Structured Data (2022)
28. Oquab, M., et al.: DINOv2: learning robust visual features without supervision. Trans. Mach. Learn. Res. J. (2024)

29. Radford, A., et al.: Learning transferable visual models from natural language supervision. In: International Conference on Machine Learning, pp. 8748–8763. PMLR (2021)
30. Ribeiro, M.T., Singh, S., Guestrin, C.: "why should i trust you?" explaining the predictions of any classifier. In: Proceedings of the 22nd ACM SIGKDD International Conference on Knowledge Discovery and Data Mining, pp. 1135–1144 (2016)
31. Rudin, C.: Stop explaining black box machine learning models for high stakes decisions and use interpretable models instead. Nat. Mach. Intell. **1**(5), 206–215 (2019)
32. Samek, W., Montavon, G., Lapuschkin, S., Anders, C.J., Müller, K.R.: Explaining deep neural networks and beyond: a review of methods and applications. Proc. IEEE **109**(3), 247–278 (2021)
33. Schoen, R., Abeloos, B., Herbin, S.: Measuring and addressing information leakage in concept bottleneck models. In: Proceedings of the IEEE/CVF International Conference on Computer Vision, pp. 624–632 (2025)
34. Sha, S., et al.: The AI index 2026 annual report. In: The AI Index 2026 Annual Report (2026)
35. Shaheen, N., Elsharnouby, T., Torki, M.: C2f-chart: a curriculum learning approach to chart classification. In: International Conference on Pattern Recognition, pp. 375–391. Springer (2024)
36. Simonyan, K., Vedaldi, A., Zisserman, A.: Deep inside convolutional networks: visualising image classification models and saliency maps. In: International Conference on Learning Representations (2014)
37. Speith, T.: How to evaluate explainability?-a case for three criteria. In: 2022 IEEE 30th International Requirements Engineering Conference Workshops (REW), pp. 92–97. IEEE (2022)
38. Speith, T.: A review of taxonomies of explainable artificial intelligence (XAI) methods. In: Proceedings of the 2022 ACM Conference on Fairness, Accountability, and Transparency, pp. 2239–2250 (2022)
39. Thiyam, J., Singh, S.R., Bora, P.K.: Chart classification: a survey and benchmarking of different state-of-the-art methods. Int. J. Docu. Anal. Recogn. (IJDAR) **27**(1), 19–44 (2024)
40. Zeiler, M.D., Fergus, R.: Visualizing and understanding convolutional networks. In: European Conference on Computer Vision, pp. 818–833. Springer (2014)

Text Pixels Are Not Enough: A Controlled Study of Canvas Sensitivity in VLM-Based OCR

Daichi Haraguchi(✉)

CyberAgent, Tokyo , Japan
haraguchi_daichi_xa@cyberagent.co.jp

Abstract. Vision-Language Models (VLMs) are increasingly used for OCR and document understanding, yet their OCR predictions can vary with canvas size even when the rendered text pixels remain unchanged. We study a simple robustness requirement for OCR: predictions should remain stable when only the surrounding canvas size changes. To enable controlled evaluation, we introduce a synthetic OCR benchmark with independently controlled text height and canvas size. We evaluate four recent VLMs under deterministic settings and find model-specific canvas sensitivity. As canvas size increases, GPT-5.2 remains nearly stable and Gemini 2.5 Flash declines gradually, whereas Qwen 3.5-27B and InternVL 3.5-8B show non-monotonic changes. For Qwen and InternVL, these changes align with public preprocessing behaviors such as resizing and tiling. Extending the protocol to four OCR systems further shows that canvas sensitivity differs across OCR pipelines. Traditional OCR systems largely remove the canvas before recognition, whereas VLM-based systems process the full image and remain canvas-sensitive. These findings suggest that OCR results should be interpreted with image-size conditions in mind and that preprocessing details should be documented.

Keywords: OCR · Vision-Language Models · OCR Benchmarking · Image Preprocessing · Canvas Sensitivity

1 Introduction

Vision-Language Models (VLMs) have rapidly become general-purpose tools for image understanding. Recent text-centric benchmarks evaluate them as OCR and document-analysis baselines [6,15]. Before recognition, however, a VLM converts the input image into visual tokens, the image units processed by its vision encoder. Common operations include downsampling, tiling, patching, and resizing [3,5,7,24]. These design choices differ substantially across models, but they are rarely treated as controlled factors in OCR evaluation. As a result, aggregate OCR benchmark scores may mix text-recognition ability with effects of model-specific image preprocessing.

F. Shafait et al. (Eds.): DAS 2026, LNCS 16851, pp. 91–107, 2027.
https://doi.org/10.1007/978-3-032-36207-0_6

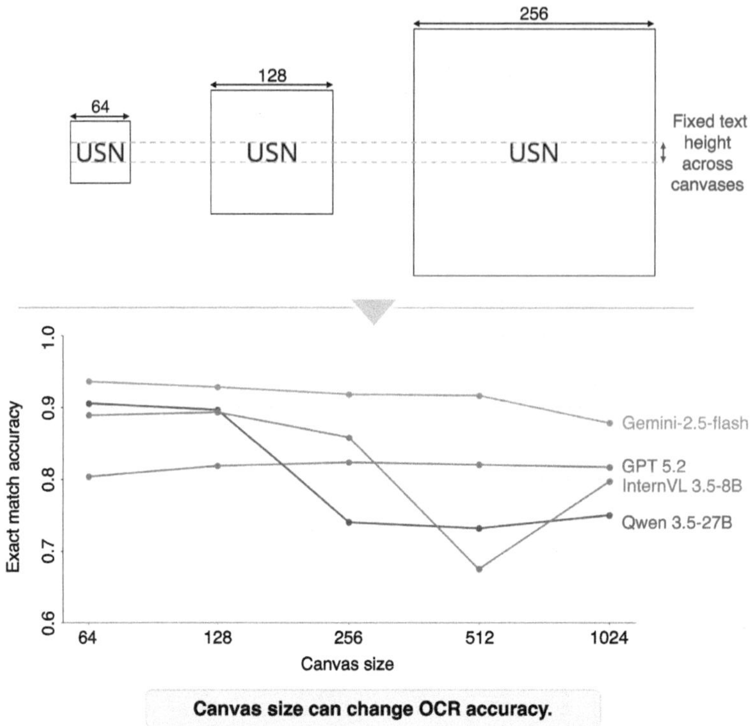

Fig. 1. Canvas-controlled OCR evaluation framework with independently controlled text height and canvas size.

This paper studies a simple robustness requirement: OCR predictions should remain stable when only the surrounding canvas changes while the rendered text remains unchanged. We refer to the surrounding input image as the *canvas*. As illustrated in Fig. 1, recent VLMs do not always meet this requirement. OCR accuracy varies substantially with canvas size even when the absolute text height is unchanged. The relation between canvas size and OCR accuracy also differs by model. This relation does not follow a simple "larger canvas ⇒ lower accuracy" trend. For some models, OCR accuracy recovers at larger canvases. For models with public preprocessing code, these accuracy changes align with identifiable resizing or tiling transitions. For API-only models, we compare the observed behavior with documented image-sizing rules. Canvas sensitivity therefore means that aggregate OCR scores should be interpreted with attention to image-size and text-height conditions.

We use synthetic renderings as controlled experimental data, not as a substitute for natural document images. Natural document images do not provide the required comparison directly, because visible text height is usually mixed with blur, compression, illumination, background texture, crop quality, and layout context. Synthetic rendering lets us generate synthetic images with independently controlled text height and canvas size. Therefore, when OCR accuracy changes across canvas conditions despite identical text rendering, the difference cannot be attributed to the text rendering itself. Such changes point to sensitivity in the image preprocessing and recognition pipeline.

Our canvas-controlled OCR evaluation protocol comprises two evaluations: single characters (Experiment 1) and multi-character strings (Experiment 2). In both, rendered text height and canvas size are independently controlled. We benchmark four representative recent VLMs whose API or weights are publicly available. The models are GPT-5.2 [18], Gemini 2.5 Flash [4], Qwen 3.5-27B [19], and InternVL 3.5-8B [25]. We use the deterministic settings available for each model. For API-only models, this describes the requested configuration rather than a guarantee that the provider will return identical outputs every time.

We also examine whether canvas sensitivity is specific to general-purpose VLMs or appears more broadly across OCR systems. For this purpose, we apply the same protocol to four OCR systems. The systems are Tesseract [23], EasyOCR [8], GOT-OCR 2.0 [27], and DeepSeek-OCR 2 [28]. Canvas sensitivity differs across these OCR pipelines, consistent with differences between traditional OCR localization pipelines and VLM-based image preprocessing.

Our contributions are:

1. We propose a canvas-controlled OCR evaluation protocol that varies canvas size and text height as independent variables.
2. Using this protocol, we show that each VLM's OCR accuracy varies with canvas size in a different way. For models with public preprocessing code, these accuracy changes align with resizing or tiling changes in the code.
3. We demonstrate that the canvas-size effect persists from single-character synthetic images to multi-character strings.
4. We extend the protocol to OCR systems with different pipeline structures. Traditional OCR pipelines largely remove canvas sensitivity before recognition. VLM-based OCR systems that process the full image remain canvas-sensitive.

These findings lead to three practical implications:

1. OCR benchmark scores remain useful. However, canvas sensitivity means that their relevance can depend on the image-size conditions of the target data. A model with strong OCR benchmark accuracy may not show the same advantage on deployment images with different canvas sizes.
2. Because canvas sensitivity is closely related to image preprocessing, documenting preprocessing rules is useful for OCR system users.
3. If different models perform well under different canvas-size conditions, input image adjustment before OCR may help match deployment images to the canvas-size conditions where the target model is more accurate.

2 Related Work

2.1 Text-Centric Benchmarks

Text reading in VLMs is commonly evaluated with OCR, scene-text, and document understanding benchmarks. OCR evaluation has long relied on benchmark datasets and shared tasks. Several OCR benchmarks have been introduced through the ICDAR competitions [9,10]. Beyond OCR benchmarks, visual question answering (VQA) datasets also require text reading. TextVQA [21] and OCR-VQA [17] focus on reading text in natural images. DocVQA [16] evaluates question answering on document images. ESTVQA [26] extends VQA to scene-text understanding.

In recent years, OCRBench [15] has emerged as a benchmark for evaluating Large Multimodal Models on text recognition, scene-text VQA, document-oriented VQA, key information extraction, and related tasks. OCRBench v2 [6] further expands this direction with additional scenarios, localization, and reasoning tasks. These benchmarks measure broad capability, but their images do not separate text height, image size, layout density, compression, and crop quality. Our protocol complements them by varying canvas size while controlling the rendered text.

2.2 OCR Pipelines and General-Purpose VLMs

OCR systems differ in how much of the input image reaches the recognition stage. Classical document OCR engines combine layout analysis, text-line segmentation, and recognition. Tesseract is a representative example of this pipeline structure [23]. Scene-text OCR commonly follows a detection-then-recognition design: detectors such as CRAFT [1] and DBNet [14] localize text regions, and recognizers such as CRNN [20] and TrOCR [12] read cropped word or line images. EasyOCR is a practical OCR toolkit in this family [8]. Because these systems localize or crop text before recognition, much of the non-text canvas can be removed before the recognizer runs.

Newer document models integrate these steps more closely. Donut [11] performs OCR-free document understanding from document images. Nougat [2] converts scientific documents into markup. GOT-OCR 2.0 [27] proposes unified end-to-end OCR, and DeepSeek-OCR 2 [28] is a VLM-based OCR system. These models often process larger image regions or full pages directly, so the amount of non-text canvas passed to the model depends on their preprocessing.

General-purpose VLMs are also increasingly evaluated on OCR and document understanding tasks. The models evaluated in this paper, GPT-5.2 [18], Gemini 2.5 Flash [4], Qwen 3.5-27B [19], and InternVL 3.5-8B [25], are not OCR-only pipelines. They usually receive the full input image and convert it into visual tokens before language decoding. This makes image preprocessing a central factor in their OCR behavior and motivates our controlled study of canvas size.

2.3 Image Resolution and Visual Tokenization in VLMs

Given the use of general-purpose VLMs for OCR, image resolution and visual tokenization become directly relevant to OCR evaluation. Modern VLMs differ substantially in how they convert variable-sized images into visual tokens. NaViT [5] packs patches from arbitrary resolutions and aspect ratios instead of forcing a fixed resize. High-resolution VLMs then developed practical variants. Monkey [13] divides images into patches. Qwen2-VL [24] uses a dynamic resolution strategy. InternVL [3] uses dynamic high-resolution tiling. LLaVA-UHD [7] uses image slices for arbitrary aspect ratios and high-resolution inputs. These designs aim to preserve fine visual details with a limited number of visual tokens. We study the OCR effect of these design choices by holding the rendered text fixed while changing only the surrounding canvas size. This manipulation can change how the input image is resized, tiled, or converted into visual tokens, making the same text easier or harder to recognize.

3 Canvas-Controlled OCR Evaluation

Overview. This section describes the OCR evaluation protocol shared by two experiments. We measure each VLM's OCR accuracy as a function of two independently controlled variables: canvas size and rendered text height. For each canvas-size comparison, the rendered text content, font, and text height are held fixed; only the surrounding canvas changes. Experiment-specific dataset details are presented in each experiment's section.

We isolate canvas sensitivity from other sources of accuracy variation through three groups of controls. *Data controls*: image generation is fully deterministic, and a single shared prompt is used across all models for each OCR experiment. *Inference controls*: we use the most deterministic inference settings available for each model. For API-only models, this refers to the requested configuration rather than a guarantee of identical outputs across repeated calls. *Evaluation pipeline*: the model receives the image and returns a response. We extract the predicted text from the response and compare it to the ground truth. We do not ask for intermediate reasoning. These controls reduce prompting, reasoning, and random decoding effects. For API-only models, image processing remains hidden, so we interpret the results in light of the provider documentation.

Models. We evaluate four representative recent VLMs whose API or weights were publicly available: GPT-5.2 [18], Gemini 2.5 Flash [4], Qwen 3.5-27B [19], and InternVL 3.5-8B [25]. We evaluate Gemini 2.5 Flash because it allows reasoning features to be disabled, whereas newer Gemini variants do not expose the same level of inference control. For each model, the input image undergoes model-specific preprocessing before being fed into the visual encoder.

Metric. We report case-sensitive exact-match accuracy, defined as the fraction of responses whose extracted letters exactly match the ground-truth string, including letter case. Because our tasks involve controlled single-character and short-string recognition, exact-match accuracy provides a direct measure of OCR correctness.

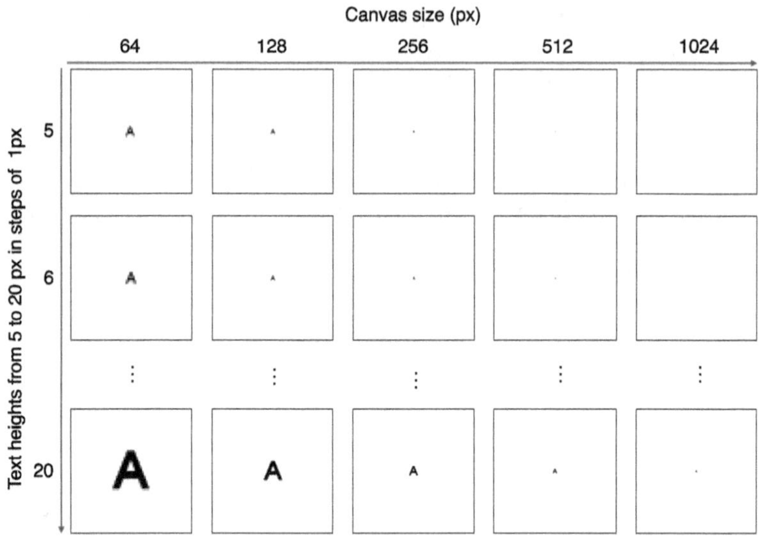

Fig. 2. Example images from the Experiment 1 dataset. Each image contains one uppercase Sloan letter centered on a square canvas. Text height and canvas size are varied independently.

4 Experiment 1: Single-Character Canvas-Controlled Analysis

Experiment 1 evaluates each VLM on the most controlled input possible: a single uppercase letter on a deterministic synthetic canvas. This setup removes variation from font diversity, layout clutter, capture conditions, and background noise. It also lets us treat canvas size and text height as two independent variables. The shared prompt is "What uppercase letter is shown in this image? Answer with only a single uppercase letter (A-Z)."

4.1 Dataset

Each image contains a single uppercase letter A–Z at the center of a square canvas. The letter is drawn in a Sloan-style font[1] inspired by visual-acuity optotypes [22]. We use this font because its uppercase glyphs have uniform stroke width and stable rendered height in our rasterization pipeline. For the letters

[1] This is available at https://github.com/denispelli/Eye-Chart-Fonts.

used here, the nominal point size has a fixed relation to rendered text height. This lets us specify the target text height directly as the font size. It also allows pixel-accurate rendering without post-hoc resampling. We use five square canvas sizes: 64, 128, 256, 512, and 1,024 px. The text heights range from 5 to 20 px in 1 px steps. The full dataset comprises $26 \times 16 \times 5 = 2{,}080$ deterministic grayscale renders. Figure 2 shows example images from this dataset.

4.2 Canvas Sensitivity Across Models

Per-model Canvas Responses. Figure 3 shows OCR accuracy against canvas size for the four VLMs. Accuracy is aggregated over all 16 text heights. The four models show different accuracy patterns as canvas size changes. GPT-5.2 is nearly insensitive to canvas size (0.940–0.947). Gemini 2.5 Flash decreases smoothly as the canvas grows (0.947 → 0.909). Qwen 3.5-27B drops sharply when the canvas size changes from 128 to 256 px (0.974 → 0.892) and then remains flat. InternVL 3.5-8B is non-monotonic, reaching its minimum at canvas 512 (0.849) and recovering at canvas 1024 (0.889).

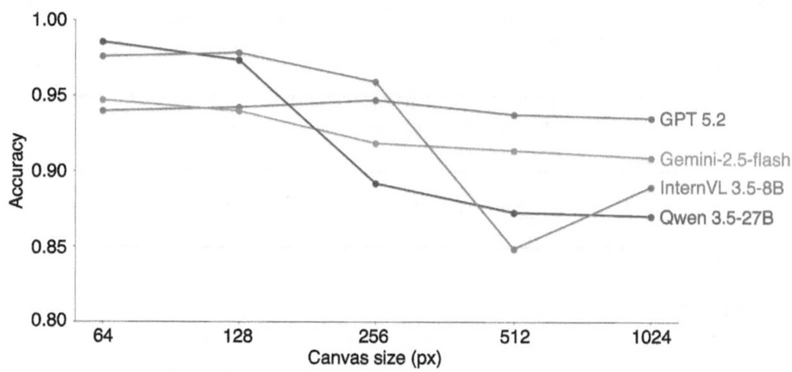

Fig. 3. Per-model canvas sensitivity in single-character recognition (Experiment 1). For each VLM, accuracy is aggregated over all 16 text heights (5–20 px) and 26 letters at each canvas size.

Correspondence with Documented Preprocessing. These responses can be interpreted against the documented image-sizing behavior and, for open-weight models, the public preprocessing code. For the open-weight models, the preprocessing behavior closely matches the observed accuracy changes.

GPT-5.2 [18] converts the input image into patches. Under our image-resolution setting, all canvases remain within the documented image-size budget.[2] This includes the largest canvas, 1,024 by 1,024 px. According to the documentation, these inputs are not expected to be forcibly downsampled. Consistent

[2] https://developers.openai.com/api/docs/guides/images-vision.

with this documented sizing behavior, GPT-5.2 shows an almost flat response across canvas sizes.

Qwen-family preprocessing uses dynamic resizing before inputs enter the vision encoder [24].[3] In our setting, both 64×64 px and 128×128 px canvas images are internally resized to 256×256 px. This effectively enlarges the rendered letter. At the 256×256 px canvas setting, however, this upscaling no longer occurs. The letter therefore appears much smaller to the model. The loss of internal upscaling produces the observed step drop in accuracy.

InternVL-family preprocessing applies dynamic high-resolution tiling [3,25]. Images are resized and split into 448 px square tiles. Under our square-canvas settings, the input is assigned to 1, 4, or 9 tiles depending on image size. Canvases up to 256 px are processed as a single upscaled tile. In contrast, a 512×512 px canvas is resized down to fit within one tile, which shrinks the rendered letters. At a 1,024×1,024 px canvas, however, the image is divided into a 3 × 3 grid of tiles. This allows the letters to be processed at a higher effective resolution again. This tile change produces the observed U-shaped accuracy response.

Gemini 2.5 Flash [4] uses a visual-token setting that the API documentation describes as "256 + Pan & Scan" in high-resolution mode.[4] Across our canvas range, OCR accuracy declines gradually as the canvas increases. Gemini does not show the large discrete transition observed in Qwen or InternVL. The gradual accuracy decline may arise from preprocessing behavior in which larger canvases reduce the effective visual resolution allocated to the text. Another possibility is that the text appears smaller in the overview representation.

5 Experiment 2: Generalization to Multi-character Strings

Experiment 2 evaluates the canvas-controlled setting on short text strings and multiple fonts. The independent variables remain the input canvas size and the rendered text height. The recognition target is now a three-character uppercase string. This setting examines whether the canvas-size effect persists when recognition requires multiple characters. The shared prompt is "Read the text in this image. Answer with only the letters exactly as shown, no explanation."

5.1 Dataset

Each image contains one of 50 randomly generated uppercase strings of length 3. Each string is rendered in one of three fonts: Sloan, Noto Sans, or Noto Serif. The same 50 strings are used for all three fonts. Thus, differences between fonts do not come from different string content. Sloan is the controlled monospaced font used in Experiment 1. Noto Sans and Noto Serif add common proportional sans-serif and serif text styles. Length 3 is used because it introduces multi-character recognition while keeping almost all strings inside the smallest canvas.

[3] https://github.com/QwenLM/Qwen3-VL/tree/main/qwen-vl-utils.
[4] https://ai.google.dev/gemini-api/docs/media-resolution.

Canvas sizes are the same as in Experiment 1: $\{64, 128, 256, 512, 1024\}$ px. Text heights are $\{5, 10, 15, 20\}$ px. Because font size is not comparable across typefaces, text height is defined by the tight bounding box of the rendered string. For each font and target height, the rendered string is scaled so that this bounding-box height matches the target within 1 px. For a fixed string, font, and text height, the same rendered text image is then centered on each canvas size.

The full grid contains 3 fonts $\times$ 50 strings $\times$ 4 text heights $\times$ 5 canvas sizes $=$ 3,000 conditions. Five Noto Serif strings do not fit at canvas 64 and text height 20, leaving 2,995 valid images. Figure 4 shows examples.

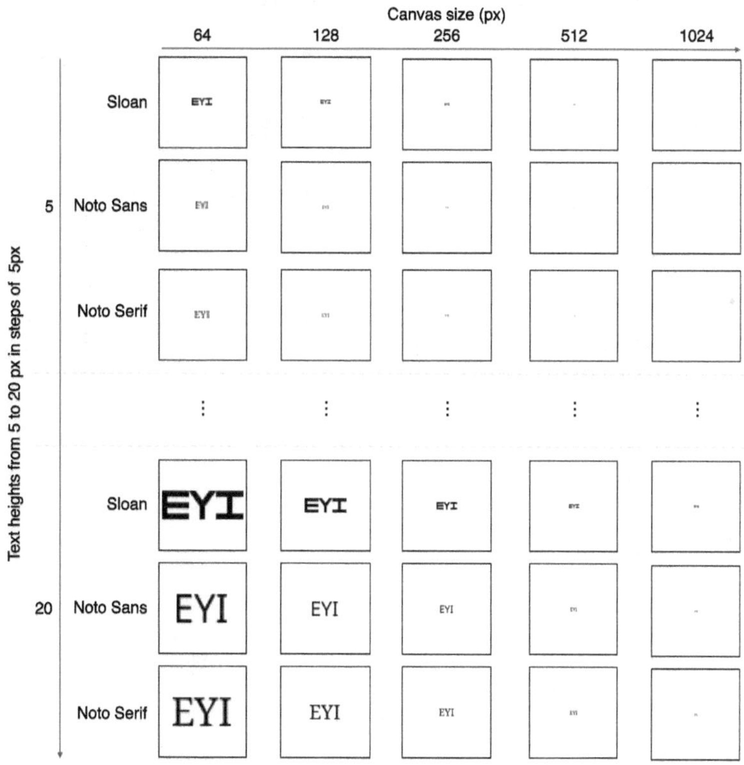

Fig. 4. Example images from Experiment 2. The same 50 three-character uppercase strings are rendered in three fonts at controlled text heights and centered on square canvases.

5.2 Canvas Sensitivity and Cross-Experiment Consistency

Accuracy as a Function of Canvas Size. Figure 5 shows exact-match accuracy against canvas size. Accuracy is aggregated over fonts, text heights, and strings.

The absolute accuracies are lower than in Experiment 1 because the output is a three-character sequence. However, OCR accuracy changes with canvas size in the same model-specific ways as in Experiment 1. GPT-5.2 remains nearly flat over canvas size. Gemini 2.5 Flash decreases gradually as canvas size increases. Qwen 3.5-27B shows a marked decrease between canvas 128 and canvas 256. InternVL 3.5-8B reaches its lowest accuracy at canvas 512 and recovers at canvas 1024. The canvas-size effect observed on isolated characters therefore persists for multi-character recognition.

Accuracy as a Function of Text Height. Figure 6 groups the same results by text height. GPT-5.2 and Gemini 2.5 Flash show relatively small differences among canvas sizes at the same text height. Qwen 3.5-27B and InternVL 3.5-8B show stronger canvas sensitivity.

For Qwen, the results for canvas sizes of 256 and larger do not improve with text height. Instead, they tend to decrease for larger text heights. This counter-intuitive trend indicates that the canvas-dependent preprocessing transition can dominate the benefit of increasing text height in this setting. Thus, increasing rendered text height is not sufficient to remove the canvas-size effect. OCR accuracy is not determined by text height alone.

InternVL 3.5-8B shows an even clearer canvas-size effect. Canvas 512 gives the lowest accuracy at every text height. At text height 5 px, its accuracy at canvas 512 is close to zero. Smaller canvases remain substantially higher. This InternVL result shows that canvas size can affect OCR performance independently of the rendered text height.

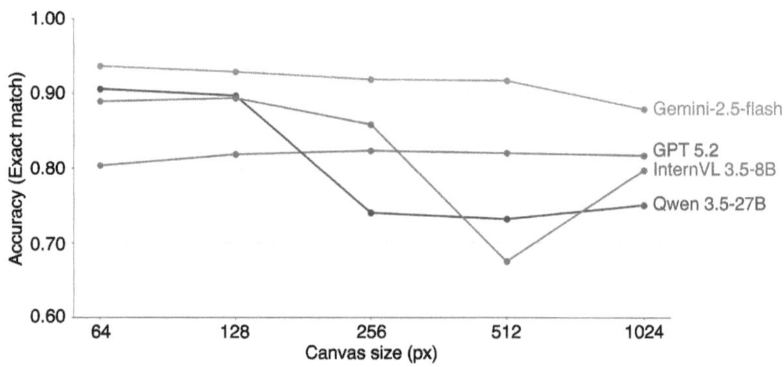

Fig. 5. Exact-match accuracy by canvas size in Experiment 2, aggregated over fonts, text heights, and strings.

Analysis by Font. Figure 7 separates the results by font. In the Sloan condition (Fig. 7(a)), all four models show the same canvas-size patterns observed in Experiment 1 and in Fig. 5. GPT-5.2 is nearly flat over canvas size. Gemini 2.5

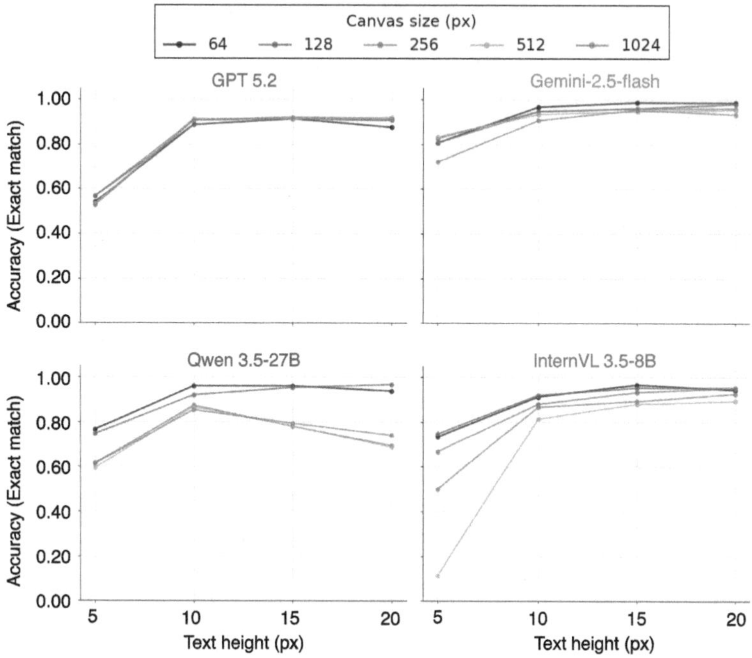

Fig. 6. Exact-match accuracy by text height in Experiment 2, grouped by canvas size.

Flash decreases gradually as canvas size increases. Qwen 3.5-27B drops around canvas 256, and InternVL 3.5-8B is lowest at canvas 512. The absolute accuracies are lowest for Sloan. This is likely because Sloan is better suited to isolated characters than to short text strings. In three-character strings, its narrow inter-character spacing can make adjacent strokes visually close or connected.

Noto Sans and Noto Serif (Fig. 7(b),(c)) give higher accuracies. The Qwen pattern may at first appear less clear. However, the finer y-axis scale shows that the same decrease around canvas 256 remains visible. Thus, although font changes the absolute recognition accuracy, the main canvas-size patterns are observed across all three fonts.

6 Canvas Sensitivity Across OCR Pipelines

Experiments 1 and 2 show that canvas sensitivity differs across VLMs. The VLM differences are related to model-specific image preprocessing behavior, especially for models with public preprocessing code. We next ask whether canvas sensitivity is specific to general-purpose VLMs. To examine this, we apply the Experiment 2 images to OCR systems with different OCR pipelines.

We use only the multi-character images. The single-character task relies on a prompt that restricts the output to uppercase A–Z. Standard OCR systems

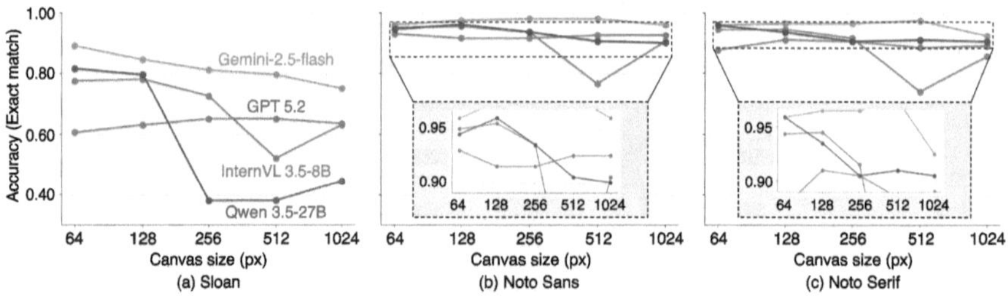

Fig. 7. Accuracy by canvas size and font in Experiment 2, aggregated over strings and text heights.

do not provide this constraint, making isolated-character predictions difficult to compare consistently.

6.1 Models Beyond General-Purpose VLMs

We evaluate two traditional OCR systems and two VLM-based OCR systems. Tesseract [23] and EasyOCR [8] localize text regions or lines before recognition. In these systems, the recognizer is not directly exposed to the full canvas. GOT-OCR 2.0 [27] and DeepSeek-OCR 2 [28] process the full input image after resizing and generate the output text with a decoder. For DeepSeek-OCR 2, we disable its automatic crop mode.

Inference is deterministic where applicable. Tesseract and EasyOCR receive only the image. GOT-OCR and DeepSeek-OCR 2 use their standard OCR prompts or tokens with fixed decoding parameters.

6.2 Canvas Sensitivity Beyond VLMs

Effect of the OCR Pipeline. Figure 8 shows two patterns in accuracy over canvas size, depending on the OCR pipeline. Tesseract and EasyOCR are nearly insensitive to canvas size. The two VLM-based OCR systems show non-monotonic responses. GOT-OCR 2.0 decreases at canvas 128 and canvas 256, then recovers from canvas 512 onward. DeepSeek-OCR 2 increases from canvas 64 to canvas 128, remains nearly flat through canvas 512, and drops at canvas 1024.

Effect of Text Height. Figure 9 shows the same results as a function of text height. Tesseract and EasyOCR fail almost completely at text height 5 px. Once the text is detectable, however, they remain nearly insensitive to canvas size. In contrast, DeepSeek-OCR 2 and GOT-OCR 2.0 show clear differences among canvas sizes. In particular, both models fail almost completely at text height 5 px at canvas 1024. Several smaller canvases still yield nonzero accuracy. This suggests that very small text can remain recognizable at some canvas sizes.

Traditional OCR Pipelines Remove Most of the Canvas. The result for Tesseract and EasyOCR follows from their OCR pipeline. Text-region detection or segmentation removes most of the surrounding blank canvas before recognition. The recognizer therefore sees a cropped text region, not the full input image. This makes the systems largely insensitive to canvas size. Text height still matters because very small text is difficult to detect and recognize.

VLM-based OCR Systems Retain the Canvas. With DeepSeek-OCR 2's crop mode disabled, both VLM-based OCR systems process the full input image. The blank canvas therefore remains part of the input to the recognizer. The non-monotonic results of GOT-OCR 2.0 and DeepSeek-OCR 2 show that canvas sensitivity is not simply a penalty for larger images. The relevant factor is how preprocessing changes the effective text resolution before recognition.

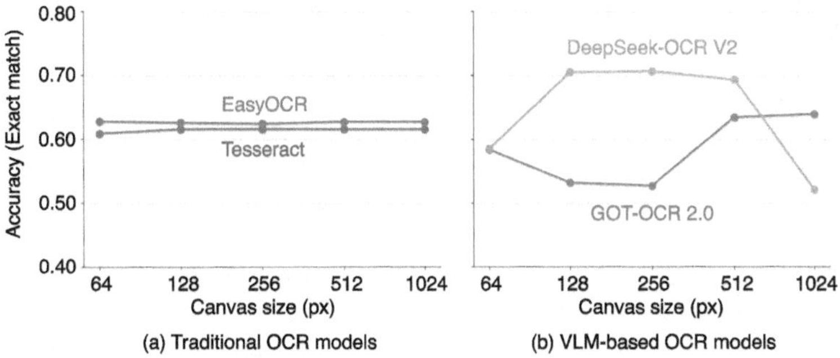

Fig. 8. Exact-match accuracy by canvas size for traditional OCR systems and VLM-based OCR systems.

6.3 Implication for Preprocessing

These results suggest that the canvas-size effect depends on where cropping or resizing occurs in the OCR pipeline. If the pipeline crops text before recognition, the canvas effect is largely removed. If the pipeline processes the full input image, the canvas remains part of the recognition problem. Canvas sensitivity therefore depends on the preprocessing steps between the input image and the recognizer, rather than on text height alone.

7 Discussion

7.1 Interpreting OCR Benchmark Scores

OCR benchmark scores can depend in part on the distribution of input image sizes and text heights in the evaluation set. A single aggregate score can hide this

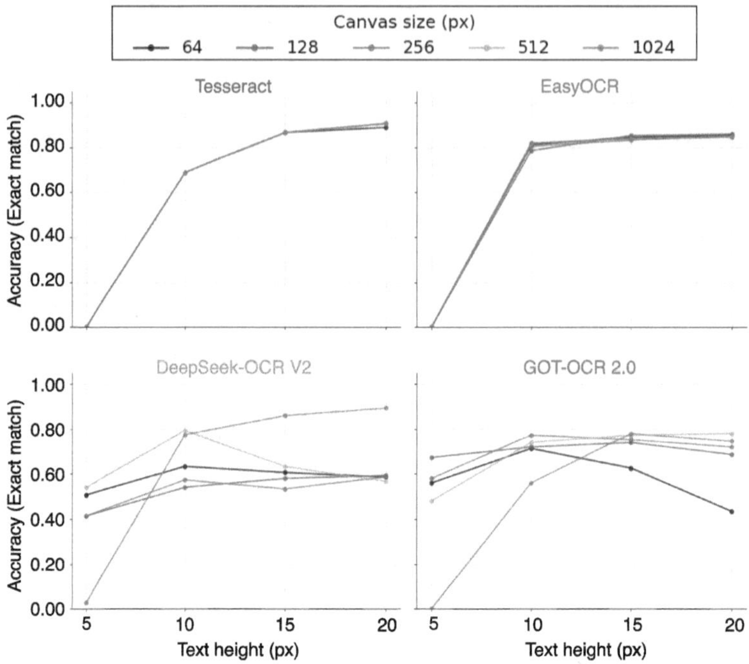

Fig. 9. Exact-match accuracy by text height for the same traditional and VLM-based OCR systems, grouped by canvas size.

dependence. A model may appear stronger or weaker depending on which canvas sizes are represented in the benchmark. For this reason, aggregate OCR scores should be interpreted together with the image-size and text-height conditions of the evaluation set. Accuracy over representative canvas sizes can make this dependence visible, but it is a tool for interpretation rather than a mandatory reporting format.

7.2 Implications for Reporting and Disclosure

For OCR system users, canvas sensitivity matters because the target data may have different image-size conditions from the benchmark data. A model with strong OCR benchmark accuracy may therefore show a smaller advantage on deployment images with different canvas sizes. Conversely, understanding canvas sensitivity may help users choose preprocessing settings that better preserve OCR performance in deployment. Cross-model OCR evaluations are easier to interpret when they make the image-size and text-height conditions clear. When canvas sensitivity is relevant, varying text height and canvas size separately can help separate text readability from canvas-dependent behavior associated with image preprocessing.

Canvas-aware OCR evaluation is easier when preprocessing is documented. Model documentation should report the main image preprocessing rules. These include resize policies, tiling behavior, maximum patch counts, and resolution settings. Without this information, OCR system users can only infer canvas sensitivity from model outputs.

7.3 Limitations and Future Work

Our experiments are intentionally synthetic. They are not designed to estimate absolute OCR accuracy on scanned or photographed documents. Instead, they isolate a basic robustness requirement for OCR systems. Predictions should not change solely because non-text canvas area changes while the rendered text pixels remain fixed. Meeting this requirement would not guarantee robustness on natural documents. However, failing it shows that OCR benchmark results can be affected by image preprocessing or recognition behavior unrelated to the text itself.

The evaluation set has several scope limits. The alphabet is restricted to uppercase A–Z. Lowercase letters, digits, and non-Latin scripts are out of scope. Experiment 2 fixes the string length at three characters, leaving longer strings to future work. We evaluate square canvases from 64 to 1,024 px. Larger canvases may expose additional model-specific resizing or tiling behavior and are left to future work. Among Gemini variants, we evaluate only Gemini 2.5 Flash, which allowed thinking to be disabled in our protocol. Other Gemini variants remain for future evaluation.

Future work should measure canvas sensitivity on natural document benchmarks and in deployment settings. A practical next step is input image adjustment before OCR. Padding, cropping, or resizing may help match deployment images to canvas-size conditions where the target model is more accurate while preserving the textual content. The best adjustment is likely model-dependent and should be validated on natural document images.

8 Conclusion

We introduced a canvas-controlled OCR evaluation protocol that varies text height and canvas size independently. Applied to four general-purpose VLMs, the protocol reveals model-specific canvas sensitivity. For OCR accuracy as a function of canvas size, GPT-5.2 is nearly flat and Gemini 2.5 Flash declines gradually as canvas size increases. Qwen 3.5-27B shows a step at the 128-to-256 canvas boundary, and InternVL 3.5-8B is lowest at canvas 512 before recovering at canvas 1024. Experiment 2 shows that these patterns remain visible for three-character strings and multiple fonts. Experiment 2 also shows that canvas size affects accuracy even when text height is fixed. Extending the same images to four OCR systems shows that canvas sensitivity is shaped by the OCR pipeline. Traditional OCR pipelines largely remove the canvas before recognition. VLM-based OCR systems retain canvas-dependent behavior. These findings point to the preprocessing path between the input image and the recognizer as a central factor in canvas sensitivity. Overall, the results show that aggregate OCR scores should be interpreted with image-size conditions in mind and that image preprocessing should be documented. Future work should extend the protocol to longer strings, additional writing systems, and natural document images. Future work should also evaluate input image adjustment before OCR.

References

1. Baek, Y., Lee, B., Han, D., Yun, S., Lee, H.: Character region awareness for text detection. In: Proceedings of the IEEE/CVF Conference on Computer Vision and Pattern Recognition (CVPR), pp. 9365–9374 (2019)
2. Blecher, L., Cucurull Preixens, G., Scialom, T., Stojnic, R.: Nougat: neural optical understanding for academic documents. In: Proceedings of the International Conference on Learning Representations (ICLR) (2024)
3. Chen, Z., et al.: How far are we to GPT-4V? closing the gap to commercial multimodal models with open-source suites. SCIENCE CHINA Inf. Sci. (2024)
4. Comanici, G., et al.: Gemini 2.5: pushing the frontier with advanced reasoning, multimodality, long context, and next generation agentic capabilities. arXiv:2507.06261 arXiv preprint (2025)
5. Dehghani, M., et al.: Patch n' pack: NaViT, a vision transformer for any aspect ratio and resolution. In: Advances in Neural Information Processing Systems (NeurIPS) (2023)
6. Fu, L., et al.: OCRBench v2: an improved benchmark for evaluating large multimodal models on visual text localization and reasoning. In: Advances in Neural Information Processing Systems (NeurIPS) (2025)
7. Guo, Z., et al.: LLaVA-UHD: an LMM perceiving any aspect ratio and high-resolution images. In: European Conference on Computer Vision (ECCV) (2024)
8. JaidedAI: EasyOCR (2020). https://github.com/JaidedAI/EasyOCR. Accessed 21 May 2026
9. Karatzas, D., et al.: Icdar 2015 competition on robust reading. In: Proceedings of the International Conference on Document Analysis and Recognition (ICDAR), pp. 1156–1160 (2015)
10. Karatzas, D., et al.: Icdar 2013 robust reading competition. In: Proceedings of the International Conference on Document Analysis and Recognition (ICDAR), pp. 1484–1493 (2013)
11. Kim, G., et al.: OCR-free document understanding transformer. In: Proceedings of the European Conference on Computer Vision (ECCV) (2022)
12. Li, M., et al.: TrOCR: transformer-based optical character recognition with pre-trained models. In: Proceedings of the AAAI Conference on Artificial Intelligence. vol. 37, pp. 13094–13102 (2023)
13. Li, Z., et al.: Monkey: image resolution and text label are important things for large multi-modal models. In: Proceedings of the IEEE/CVF Conference on Computer Vision and Pattern Recognition (CVPR), pp. 26763–26773 (2024)
14. Liao, M., Wan, Z., Yao, C., Chen, K., Bai, X.: Real-time scene text detection with differentiable binarization. In: Proceedings of the AAAI Conference on Artificial Intelligence, vol. 34, pp. 11474–11481 (2020)
15. Liu, Y., et al.: OCRBench: On the hidden mystery of OCR in large multimodal models. SCIENCE CHINA Inf. Sci. **67**, 220102 (2024)

16. Mathew, M., Karatzas, D., Jawahar, C.: DocVQA: a dataset for VQA on document images. In: Proceedings of the IEEE/CVF Winter Conference on Applications of Computer Vision (WACV), pp. 2200–2209 (2021)
17. Mishra, A., Shekhar, S., Singh, A.K., Chakraborty, A.: OCR-VQA: visual question answering by reading text in images. In: Proceedings of the International Conference on Document Analysis and Recognition (ICDAR), pp. 947–952 (2019)
18. OpenAI: Introducing GPT-5.2 (2025). https://openai.com/index/introducing-gpt-5-2/. Accessed 21 May 2026
19. Qwen Team: Qwen3.5: towards native multimodal agents (2026). https://qwen.ai/blog?id=qwen3.5. Accessed 21 May 2026
20. Shi, B., Bai, X., Yao, C.: An end-to-end trainable neural network for image-based sequence recognition and its application to scene text recognition. IEEE Trans. Pattern Anal. Mach. Intell. **39**(11), 2298–2304 (2017)
21. Singh, A., et al.: Towards VQA models that can read. In: Proceedings of the IEEE/CVF Conference on Computer Vision and Pattern Recognition (CVPR), pp. 8317–8326 (2019)
22. Sloan, L.L.: New test charts for the measurement of visual acuity at far and near distances. Am. J. Ophthalmol. **48**, 807–813 (1959)
23. Smith, R.: An overview of the Tesseract OCR engine. In: Proceedings of the International Conference on Document Analysis and Recognition (ICDAR), pp. 629–633 (2007)
24. Wang, P., et al.: Qwen2-VL: enhancing vision-language model's perception of the world at any resolution. arXiv preprint arXiv:2409.12191 (2024)
25. Wang, W., et al.: InternVL3.5: advancing open-source multimodal models in versatility, reasoning, and efficiency. arXiv:2508.18265 arXiv preprint (2025)
26. Wang, X., et al.: On the general value of evidence, and bilingual scene-text visual question answering. In: Proceedings of the IEEE/CVF Conference on Computer Vision and Pattern Recognition (CVPR), pp. 10126–10135 (2020)
27. Wei, H., et al.: General OCR theory: towards OCR-2.0 via a unified end-to-end model. arXiv:2409.01704 arXiv preprint (2024)
28. Wei, H., Sun, Y., Li, Y.: DeepSeek-OCR 2: visual causal flow. arXiv:2601.20552 arXiv preprint (2026)

BZK-LE: A Dataset for Zero-Shot and Few-Shot Archival Document Layout Classification

Marzieh Malekzadeh Mahani[1,2](✉), Genet Asefa Gesese[1,2], Mahsa Vafaie[1,2], Inger Banse-Strobel[3], Kevin Dubout[3], and Harald Sack[1,2]

[1] FIZ Karlsruhe, Leibniz Institute for Information Infrastructure, Karlsruhe, Germany
{marzieh.malekzadeh_mahani,genet-asefa.gesese,mahsa.vafaie, harald.sack}@fiz-karlsruhe.de
[2] Karlsruhe Institute of Technology, Karlsruhe, Germany
[3] Bundesarchiv Koblenz, Koblenz, Germany
{inger.banse-strobel,k.dubout}@bundesarchiv.de

Abstract. Document image classification (DIC) for historical archival documents remains challenging due to heterogeneous layouts, degraded documents, and class imbalance. Existing benchmarks mainly target contemporary documents and typically assume access to large amounts of annotated training data, which is often unrealistic in archival settings where labeled data is limited. In this work, BZK-LE is introduced as a dataset for evaluating existing DIC models under zero-shot and few-shot settings. It consists of 9,144 index cards and their corresponding layout class labels, from the *Bundeszentralkartei* (BZK), documenting postwar compensation proceedings related to Nazi persecution. The dataset comprises 22 distinct pre-printed layout types and preserves the natural class imbalance. The images exhibit typical archival artifacts such as handwritten and typewritten text, annotations, stamps, and spatial misalignment. Baseline results using state-of-the-art layout-aware DIC models are reported on both versions. The datasets are available under https://zenodo.org/records/18790631.

Keywords: Document Image Classification · Archival Dataset · Digital Cultural Heritage

1 Introduction

Historical administrative procedures have generated vast quantities of documents, including printed forms and index cards used in bureaucratic workflows. Many of these documents are preserved in physical (and sometimes digital) form in archival repositories, often without structured metadata. Although large-scale

F. Shafait et al. (Eds.): DAS 2026, LNCS 16851, pp. 108–125, 2027.
https://doi.org/10.1007/978-3-032-36207-0_7

digitization initiatives have increased their availability, the information contained in many such collections remains difficult to explore systematically. The scale of the material, combined with data protection considerations, limits direct accessibility and complicates large-scale quantitative analysis. Computational methods that support indexing and structuring of digitized archival documents are therefore essential for enabling scholarly research and improving access within memory institutions [20].

One prominent example of such a collection is the *Bundeszentralkartei (BZK)* or the German Federal Card Index [21], the central registry of compensation applications submitted especially, but not only, under the Federal Compensation Act in the Federal Republic of Germany[1]. The BZK collection documents proceedings related to material compensation granted to persecutees of National Socialism and comprises approximately 1.9 million German index cards from the State Offices for Compensation (*"Entschädigungsämter"* or *"Ämter für Wiedergutmachung"*[2] in German). As part of a broader archival corpus, the BZK represents an important historical resource for understanding post-war bureaucratic workflows on compensation [4].

Each card in the BZK records administrative details such as basic personal information on the persecutees and/or applicants, case identifiers and responsible compensation authorities, thereby serving as a reference system for compensation claims across the German federal states. Apart from persecuted natural persons, the BZK also contains a distinct subset, documenting legal entities. These records relate to proceedings in which organizations acted on behalf of persecuted individuals, as well as proceedings in which legal entities such as companies, religious communities, welfare associations, trade unions, and political parties were themselves applying for compensation as persecutees. This dimension of persecution history remains comparatively underexplored in historical research and is crucial for enabling broader historical perspectives [9].

In the BZK collection, legal entity cards present a challenging case, as no dedicated templates were created and pre-printed forms for natural persons were reused. The index cards also exhibit typical characteristics of archival documents, such as overlapping stamps, handwritten annotations, and spatially misaligned key–value regions. These challenges highlight the need for robust document image classification (DIC) as a foundational step in the digitalization of large archival collections [12]. Historical archival records consist of documents with heterogeneous structures, across different time periods, locations and institutions. Effective DIC is therefore essential, not only to manage this structural variability but also because downstream information extraction benefits substantially from prior classification: documents with similar layouts and structural

[1] The BZK is now archived material and is kept at the State Archives of North Rhine-Westphalia, Rhineland Division, in Duisburg (holding reference number BR 3015).

[2] *"Wiedergutmachung"* is a German word that translates to "making good again" or "making amends", referring to the policies of restitution and compensation provided since 1945 to survivors and victims persecuted under the Nazi regime.

properties can be processed using correspondingly adapted information extraction procedures, leading to more accurate and efficient results [19].

Despite the growing availability of DIC benchmarks, existing datasets primarily focus on born-digital or contemporary administrative documents, and exhibit limited diversity and scope in types and sources of document images [11]. This focus also means that many benchmarks are built around settings in which comparatively large amounts of labeled data are available [5] [14], whereas historical archival collections often provide only limited labeled material due to their specific provenance, access restrictions, and annotation costs [20]. To the best of our knowledge, no publicly available dataset has in its focus archival index cards documenting historical procedures, created for the task of DIC [13].

To address this gap, we introduce BZK-LE, a new dataset comprising 9,144 legal entity index cards from the *Bundeszentralkartei*. The dataset reflects the natural diversity of this archival collection and therefore, is highly imbalanced with respect to the layout types. The cards were manually annotated by four archivists from *Bundesarchiv* (The Federal Archives of Germany) with their corresponding layout type labels. To support controlled evaluation settings, additionally a balanced subset with fewer layout types is provided. All images and annotations are publicly available[3] to enable reproducibility and further research. By releasing BZK-LE, the aim is to support research in DIC and to provide a realistic benchmark for evaluating layout-aware models under archival conditions. The contributions of this work are therefore, threefold:

- **BZK-LE**: The first publicly available dataset of index cards documenting compensation proceedings involving legal entities persecuted under National Socialism, designed for evaluating zero-shot and few-shot DIC models.
- **Two dataset variants**: a naturally imbalanced version reflecting the original archival distribution and a balanced subset designed for controlled experimental evaluation.
- **Baseline results** using established DIC models are reported and analysed to provide reference benchmarks for future research.

The rest of this paper is organized as follows: Sect. 2 reviews existing DIC datasets. The BZK-LE dataset is introduced in Sect. 3. The baseline models are described in Sect. 4, followed by discussions on experimental results in Sect. 5. In Sect. 6, the intended use of the dataset is discussed. And finally, Sect. 7 concludes the paper and outlines directions for future work.

2 Related Work

DIC has been extensively studied over the past decade, supported by the release of several benchmark datasets. In this section, we focus specifically on scanned document image datasets that were explicitly created for the task of DIC, rather

[3] The dataset is accessible at: https://zenodo.org/records/18790631 under the Creative Commons Attribution 4.0 International (CC BY 4.0) license.

than born-digital collections derived from electronically generated documents. This distinction is important, as scanned documents exhibit characteristic challenges such as degradation, layout variability, and acquisition artifacts that significantly influence model design and evaluation.

RVL-CDIP Dataset. The RVL-CDIP dataset is currently the most widely adopted benchmark for large-scale DIC. It consists of 400,000 grayscale document images organized into 16 categories, such as letters, forms, invoices, and resumes [5]. The dataset was derived from the IIT-CDIP Test Collection 1.0 and provides predefined training, validation, and test splits, which has facilitated standardized evaluation across studies. Due to its scale and diversity, RVL-CDIP has been extensively used to benchmark convolutional neural networks and, more recently, transformer-based architectures for document understanding tasks. Larson et al. [11] for example have experimented with multimodal large language models (MLLMs), and the train and validation sets of RVL-CDIP. They have additionally defined a test scenario called RVL-CDIP-160 $\times$ 5, which consists of five random selections (without overlap) of 160 samples (ten per class) from the original test set, as well as the original RVL-CDIP-40K test set, which contains 40,000 samples.

Tobacco Dataset.The Tobacco-3482 dataset is a smaller benchmark comprising 3,482 document images grouped into 10 classes, including memos, emails, reports, and scientific publications [10]. Despite its limited size compared to RVL-CDIP, Tobacco-3482 has been influential in early deep learning work on document classification. It is often used for fine-tuning and transfer learning experiments, as its modest scale makes it suitable for evaluating model generalization under low-data conditions.

IMPACT Dataset. The IMPACT dataset is one of the largest publicly available corpora for document image analysis and benchmarking, particularly in the historical domain. Introduced by Papadopoulos et al. [14], the dataset contains over 600,000 high-resolution document images collected from multiple European libraries as part of the IMPACT Centre of Competence initiative, covering a wide range of layouts, scripts, languages, and degradation conditions. Due to its scale, heterogeneity, and real-world scanning artifacts (e.g., skew, bleed-through, aging effects), IMPACT presents a challenging benchmark that better reflects practical archival digitization scenarios compared to more constrained datasets such as Tobacco-3482 or RVL-CDIP.

While numerous DIC benchmarks have been proposed, they predominantly focus on contemporary administrative documents and provide limited coverage of historically complex archival materials [11]. In particular, datasets centered on archival index cards documenting structured historical procedures remain absent from the literature. To address this gap, we introduce BZK-LE, the first publicly available dataset of historical index cards documenting compensation proceedings involving legal entities persecuted under National Socialism, specifically curated for DIC.

3 BZK-LE Dataset

In this section the BZK collection and its subset of legal entity cards as the basis of BZK-LE dataset are introduced in more details, together with a description of the different labels and the annotation process.

3.1 Wiedergutmachung and the BZK Collection

The project *"Themenportal Wiedergutmachung"*[4] provides a central information system for digitized records and metadata regarding German efforts to compensate for National Socialist crimes. In the Federal Republic of Germany, *Wiedergutmachung* denotes the legal and administrative procedures established to provide compensation to victims of Nazi persecution. A cornerstone of this process was the *Bundesentschädigungsgesetz* (BEG, or the Federal Compensation Act), enacted in 1953, which regulated material claims. These workflows generated extensive archives, including application files, correspondence, and a centralized index system designed to register and track proceedings.

The *Bundeszentralkartei* (BZK) is the central registry for compensation applications in the Federal Republic of Germany. Established following the BEG, this nationwide index records whether a claim was filed, the responsible authority and the case's reference number. The collection contains approximately 1.9 million cards, each detailing essential data including names, birth/death dates, addresses, case identifiers, and the competent authority. Primarily designed to prevent duplicate filings across different regions, the BZK now serves as an archived corpus reflecting the bureaucratic practices and documentation standards of post-war compensation procedures.

3.2 BZK-LE: the Collection of Legal Entity Cards

While existing scholarly and computational studies have focused on compensation records concerning natural persons [19], documentation relating to legal entities remains largely underexplored. Companies, religious communities, welfare associations, trade unions, and political organizations were themselves formally recognized as persecuted parties and submitted compensation claims independently.

The BZK-LE[5] comprises 9,144 digitized, unrestricted index cards, documenting compensation proceedings for legal entities, paired with their corresponding German layout class labels. Archival images are provided as full-page scans without structural normalization. Originally designed for individuals, the pre-printed

[4] https://www.archivportal-d.de/themenportale/wiedergutmachung.

[5] The BZK-LE dataset is a subset of the BZK, owned by the State Archive of North Rhine-Westphalia. As these index cards contain personal and organizational names, their publication is subject to archival regulations. The use of these materials was authorized by the owning institution and complies with all legal and administrative requirements.

templates were reused for legal entities without modification, resulting in mismatches between field names and recorded values, as illustrated in the examples shown in Fig. 1. Each layout class name represents both the federal state(s) in which the layout was used and the administrative registration phase associated with it. The naming convention combines these two components into a structured label[6]. For example, `BW-Hauptphase` denotes a layout used in Baden-Württemberg during the main registration phase.

While index cards layouts vary, most follow a consistent pattern with sections for the applicant, the persecuted party, and the compensation authority. Each card contains a unique case ID, the name of the legal entity (company or association), or community, a representative and address details. German pre-printed labels are often overlaid with handwritten entries, stamps, and marginalia from the original administrative process. These elements, alongside varying print quality, create high visual diversity. Figure 1 illustrates representative examples from different layout classes.

3.3 Dataset Preparation

The BZK-LE dataset supports layout-based DIC under realistic archival conditions. This task assigns index cards to predefined layout types, which is critical because different layouts correspond to distinct archival protection periods. Identifying the correct template ensures legally compliant access management by determining publication restrictions. Furthermore, layout labels enhance information extraction, enabling context-sensitive prompting in MLLMs or layout-specific rule definition in traditional approaches [19].

A layout type is defined by recurring visual and spatial characteristics shared across index cards. These include wording and positioning of pre-printed field names, typographic properties, and formatting patterns. Layout classification is thus formulated at the structural rather than the semantic level.

Table 1. Examples of natural language descriptions for BZK-LE dataset labels in order to support prompt-based and instruction-driven approaches.

Layout label	Natural language description
BW-Hauptphase	*a scanned German index card from Baden-Württemberg during the main registration phase*
RLP-Hauptphase (abweichend 1)	*a variant layout of a scanned German index card from Rhineland-Palatinate during the main registration phase.*

To meet document classification requirements, the dataset comprises images organized into 22 distinct layout classes. Four archival experts manually verified the labels to ensure structural consistency. Each image was annotated by

[6] Coined during *Bundesarchiv* exploration (2023–2024), these names are internal working instruments rather than historical designations. They serve to "locate" layout types spatially and temporally.

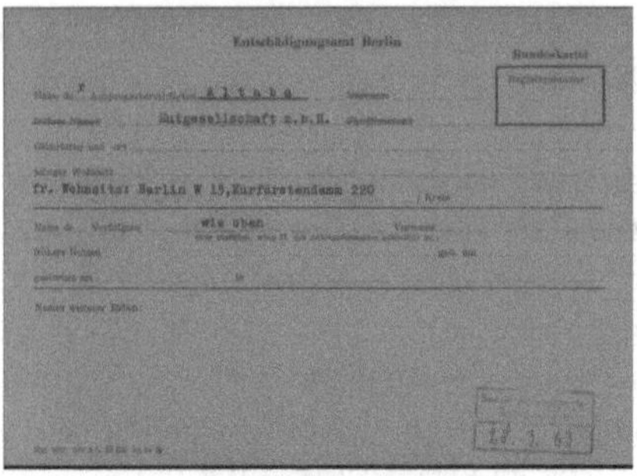

BY-BE-Hauptphase

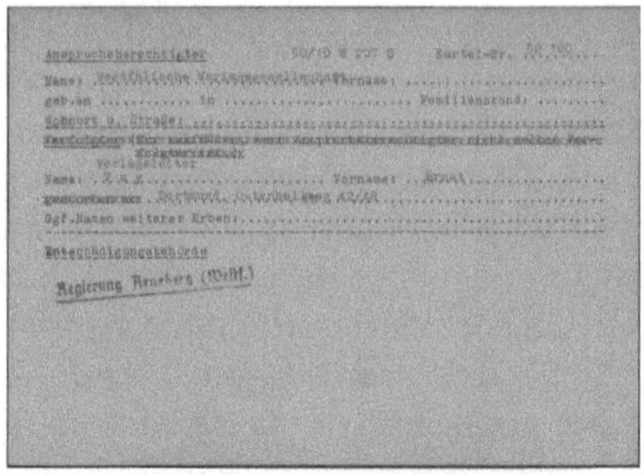

HH-NI-NRW-SH-Hauptphase

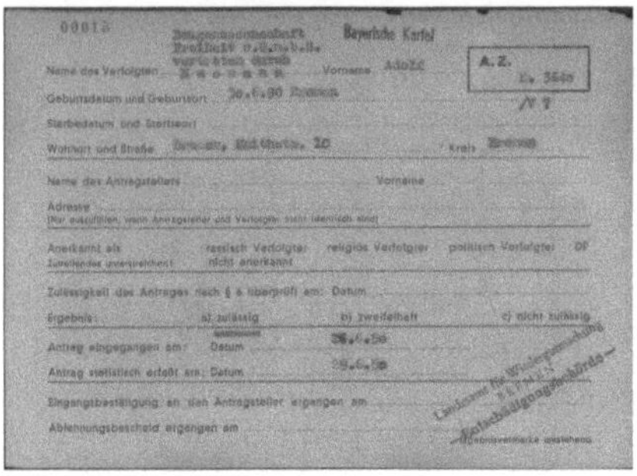

BY-HB-Frühe Phase

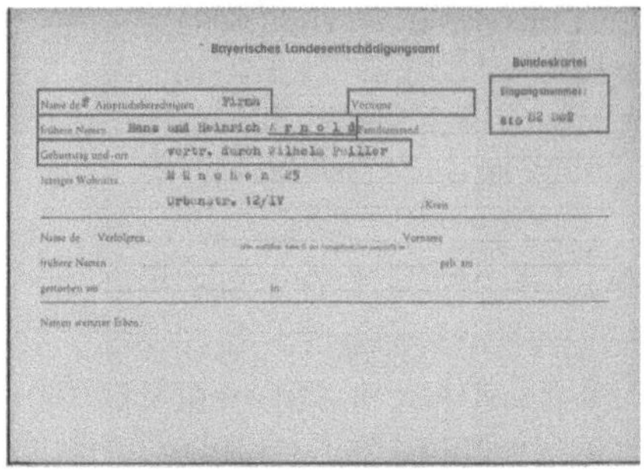

BY-BE-Hauptphase

Fig. 1. Layout class samples of BZK-LE, illustrating structural diversity, recurring sections, and the reuse of person-oriented templates for legal entity proceedings. Even within the same layout class, such as `BY-BE-Hauptphase`, subtle visual differences can be observed between individual cards, including variations in color, different stamps, and additional footer elements, further increasing complexity. The examples also show mismatches between pre-printed field names and recorded values. For instance, the bottom-right card from the layout BY-BE-Hauptphase contains the field "Name der Anspruchsberechtigten", which was originally intended to record the applicant's surname, while the recorded value instead contains "Firma", referring to the type of legal entity documented on the card.

a single expert, as layout types are defined by clearly distinguishable visual features rather than subjective judgments. This process ensures that categories correspond to historically meaningful templates, avoiding arbitrary visual groupings. Alongside categorical labels, we provide an accompanying file with natural language descriptions for each layout to support prompt-based and instruction-driven approaches. Table 1 presents a sample of these descriptions.

As shown in Fig. 2, the class distribution is heavily imbalanced: the largest class has 1,678 samples, while the smallest contains only one, yielding a median class size of 164. Notably, 11 of the 22 classes contain fewer than 100 samples, reflecting the archival collection's natural composition and introducing additional challenges for classification models. To support controlled experimental evaluation, we provide two dataset variants. The first variant preserves the natural archival distribution across all twenty-two layout classes. The second variant is a balanced subset comprising twelve layout classes with an equal number of 152 samples per class. The value of 152 was selected as a threshold that allows retaining the maximum possible number of layout classes while ensuring uniform class sizes. Classes with fewer than 152 samples were excluded from this subset, and the remaining classes were randomly down-sampled to 152 samples each without replacement. This design enables evaluation independent of class frequency effects while preserving structural diversity. For both variants, predefined training, validation, and test splits are supplied following a 70% / 15% / 15% partition to ensure reproducibility and comparability across experimental settings. All document images and corresponding layout annotations are publicly released to facilitate further research. The dataset comprises high-resolution RGB images stored in JPEG format, with widths between 2,472 and 2,560 pixels and heights between 1,648 and 3,536 pixels, indicating consistent high-resolution scanning with limited variability in document dimensions.

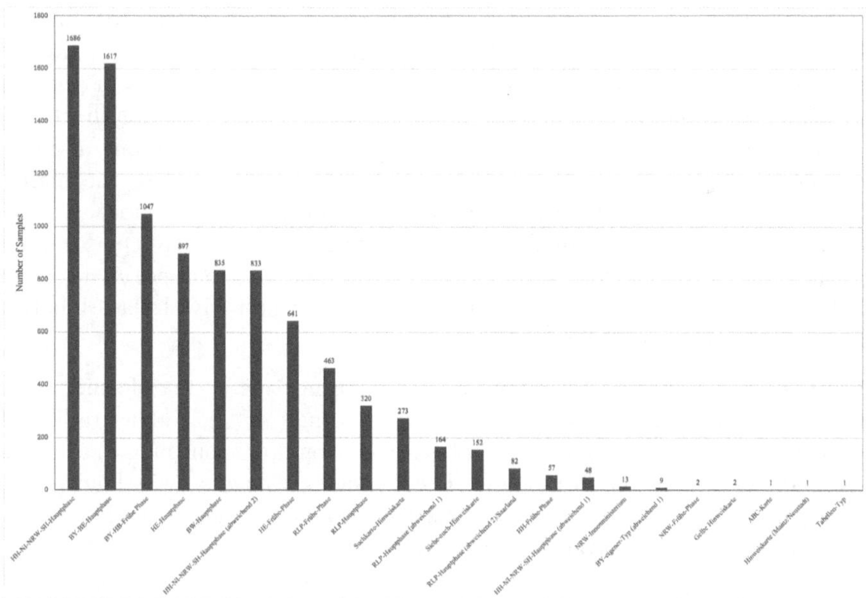

Fig. 2. Layout classes distribution in the natural BZK-LE collection.

4 Document Image Classification (DIC)

DIC is typically studied using large-scale benchmarks, where abundant labeled data facilitates training high-capacity models. Most approaches rely on extensive datasets and multimodal architectures to achieve competitive performance

[7,16,22]. However, such settings may not generalize well to specialized domains, where annotated data is scarce. In domain-specific scenarios, collecting large labeled datasets is often impractical due legal, historical, or institutional constraints. Consequently, models that require substantial supervision may fail in low-resource environments. This motivates exploring learning paradigms that remain robust under restricted data availability. Therefore, in this work, the focus is on models that have previously demonstrated good performance under limited supervision. In addition to conventional fine-tuning, results are reported for zero-shot and few-shot configurations with limited labeled data. We aim to assess whether these models remain effective when applied to our domain-specific dataset. Accordingly, we consider the following three representative models: CLIP [15], DONUT [8], and Qwen2.5-VL [1].

CLIP. (Contrastive Language–Image Pretraining) represents a major advancement in zero-shot visual recognition by learning visual concepts directly from natural language supervision [15]. It was trained on approx. 400 million internet-sourced image–text pairs. This large-scale pretraining enables the model to learn broad visual representations and their semantic alignment with natural language. A defining characteristic of CLIP is its strong zero-shot transfer capability. At inference time, no additional classifier is required. Instead, class labels are converted into textual prompts, typically following a template such as "a photo of a class". These prompts are encoded using the text encoder, and classification is performed by computing the similarity between the image embedding and each class-specific text embedding. The predicted class corresponds to the highest similarity score. This mechanism allows CLIP to generalize to previously unseen classes without further parameter updates. Given its competitive performance in prior work [18], CLIP is evaluated in a zero-shot setting (Sect. 5) to assess its ability to handle complex document images and non-trivial class labels in BZK-LE, without further fine-tuning.

DONUT. (Document Understanding Transformer) is a transformer-based model designed for Visual Document Understanding (VDU). Unlike traditional document processing systems, it operates directly on document images and does not rely on an external OCR pipeline. The model consists of a Swin Transformer encoder and a Bidirectional and Auto-Regressive Transformers (BART) decoder. The Swin Transformer extracts hierarchical visual features using shifted window attention, allowing the model to capture both local details and global layout information. This hierarchical design is particularly suitable for high-resolution document images, where structural layout and spatial relationships are essential for understanding the document. Previous studies have shown that DONUT performs well in few-shot fine-tuning settings, reporting competitive results on benchmarks such as RVL-CDIP-160×5 and RVL-CDIP-40K [17]. In this work, a fine-tuning strategy is adopted, as detailed in Sect. 5 to investigate whether similar robustness can be observed in our domain-specific classification setting with limited labeled data.

ResNet-18 as a convolutional neural network (CNN) baseline was included to complement the transformer-based DONUT model. This architecture was selected as a compact and well-established residual network for image classification [6], enabling an assessment of whether standard CNN-based visual representations are sufficient for the layout type classification task. Rather than being trained end-to-end on the target dataset, a few-shot linear probing setting was adopted, in which the pretrained backbone was kept frozen and only a lightweight classification layer was optimized. In this way, the discriminative quality of pretrained CNN features could be evaluated in a controlled manner, while the number of trainable parameters was substantially reduced and the risk of overfitting in the low-data regime was limited.

Qwen2.5-VL is a large-scale vision–language model developed by the Qwen Team that integrates advanced visual perception with a powerful language model backbone. It is designed to handle a range of multimodal tasks, including document parsing, OCR, and structured data extraction. The architecture consists of a Vision Transformer (ViT) encoder coupled with a Qwen2.5 language model decoder. Qwen2.5-VL is available in three sizes, 3B, 7B, and 72B parameters, addressing diverse computational and deployment constraints. The use of Qwen2.5-VL in this paper, detailed in Sect. 5, is inspired by the work of Macedo et al. [3], in which multiple MLLMs were evaluated for document image comparison in a zero-shot setting to assess visual similarity between document layouts. Following this protocol, Qwen2.5-VL was selected as the multimodal backbone.

5 Experiments

In order to provide a reference benchmark for future research, experiments are conducted to evaluate the performance of different models on BZK-LE dataset.

5.1 Implementation Details

This section details model-specific adaptations, preprocessing, and strategies for the BZK-LE dataset. Evaluated models include CLIP, DONUT (generative fine-tuning and linear probing), and Qwen2.5-VL-Instruct [1].

Zero-shot Inference (CLIP). A zero-shot document classification baseline using CLIP (`ViT-L/14@336px`, OpenAI pretrained weights) was evaluated. The pretrained model was used without fine-tuning on the BZK-LE dataset. For each class, a textual prompt was constructed and its corresponding embedding was computed using the CLIP text encoder. Two prompting strategies were investigated: (1) raw class labels (e.g., `BW-Hauptphase`), and (2) natural language descriptions of the document labels (e.g., *"a scanned German index card from Baden-WÃijrttemberg during the main registration phase"*). For inference, each document image was resized and center-cropped to 336 × 336 pixels following CLIP's default preprocessing pipeline [15], encoded using the visual encoder, and ℓ_2-normalized. Classification was performed by computing cosine similarities between the image embedding and all class text embeddings, scaling the similarities using the learned logit temperature parameter, and selecting the class with the highest similarity score (Top-1 prediction). No task-specific training or parameter updates were applied. Consequently, the results are directly comparable to prior zero-shot CLIP inference evaluations reported in [18].

Supervised Generative Fine-tuning (DONUT). In this setup, classification was formulated as an end-to-end generative task, following the implementation proposed
by Scius-Bertrand et al. [17]. A pretrained `VisionEncoderDecoderModel` initialized from `naver-clova-ix/donut-base-finetuned-cord-v2` was used. All document images were resized to 1280 × 960 pixels and processed using the `DonutProcessor`. Instead of predicting class labels through a standard softmax layer, the task was cast as sequence generation. For each input image, the model was trained to autoregressively generate a single class token `<label_k>`, preceded by a task-specific start token `<s_le-bzk>`. All class tokens were added to the tokenizer vocabulary, and the decoder embedding matrix was resized accordingly. The decoder start token ID was explicitly set to the task token to ensure consistent conditioning during generation. Training was performed using teacher forcing with cross-entropy loss over the generated token sequence. Fine-tuning was conducted for up to 30 epochs using the AdamW optimizer with a learning rate of 1×10^{-5} and the batch size of 8. The random seed was fixed to 2022 to ensure reproducibility. Model checkpoints were evaluated and saved at the end of each epoch, with a maximum of two checkpoints retained. No early stopping criterion was applied. In this end-to-end configuration, both the Swin-based visual encoder and the Transformer decoder were fully trainable, allowing all model parameters (approx. 200M) to be updated jointly.

Supervised Linear Probing (DONUT). A linear probing strategy [2] was adopted to evaluate the representational quality of the pretrained DONUT encoder independently of the generative decoder. The visual encoder was initialized from `naver-clova-ix/donut-base` and completely frozen, preventing any gradient updates. On top of the mean-pooled encoder hidden states, a lightweight classification head consisting of a single linear layer was added. Only the parameters of this classification layer were optimized using cross-entropy loss. The model was trained for up to 25 epochs using the AdamW optimizer with a learning rate of 5×10^{-4}. The random seed was fixed to 2022 for reproducibility. The discriminative approach removed the generative decoder, significantly reducing the number of trainable parameters. The linear probe thus evaluates how effectively pretrained DONUT visual features separate layout classes without adapting the encoder to the downstream task.

Supervised Linear Probing (ResNet-18). ResNet-18 was evaluated under the same linear probing strategy used for DONUT to assess the quality of pretrained convolutional representations for layout classification. The backbone was initialized from an ImageNet-pretrained ResNet-18 model and kept frozen. The original classification layer was replaced with a new task-specific linear head matching the number of layout classes, and only this layer was optimized. To ensure consistency across splits, a unified label mapping was applied based on the training set classes. The model was trained using cross-entropy loss and the AdamW optimizer with a learning rate of 10^{-3}, for up to 15 epochs. Input images were resized to a fixed resolution of 224×224, with light augmentation applied during training. This setup provides a lightweight discriminative baseline for evaluating how well fixed pretrained ResNet-18 features separate the layout classes.

Prototype-based Multimodal Prompting (Qwen2.5-VL-7B-Instruct). Following the visual document matching formulation of Macedo et al. [3], `Qwen/Qwen2.5-VL-7B-Instruct` was evaluated in a prototype-based layout matching setup without parameter fine-tuning. One prototype per class was randomly sampled from the training split (seed 0) and assigned a unique identifier (e.g., P1, P2). These prototypes were fixed across runs to guarantee consistent evaluation. Two labeling strategies were investigated: (i) neutral prototype identifiers and (ii) natural-language label descriptions.

All images were resized prior to inference to a fixed resolution consistent with the model's processor pipeline and encoded using the `AutoProcessor` associated with the Qwen2.5-VL architecture. During inference, all prototype images were provided jointly within a single prompt, followed by one query image. The task was explicitly framed as document layout matching, emphasizing structural cues such as header placement, text block organization, tables, and stamps, while discouraging reliance on textual semantics. The temperature was set to 0.1, and the maximum output length was limited to 128 tokens. The model was instructed to generate a single valid prototype identifier, which was then mapped to its class label for evaluation. Experiments were conducted on a system with 2× AMD EPYC 7713 64-Core CPUs, 1 TB RAM, and two 80 GB NVIDIA A100 GPUs. The environment used Ubuntu 22.04, CUDA 12.4 and PyTorch 2.5.[7]

5.2 Evaluation Metrics

Model performance was evaluated using Accuracy and Macro-F1, which provide complementary views of classification quality. Accuracy measures the overall proportion of correctly classified samples across the test set and gives an intuitive summary of performance, but it can be biased toward majority classes when the data are imbalanced. To better reflect class-wise behavior, we also consider precision, recall, and F1-score, where precision measures how many predicted instances of a class are correct, recall measures how many true instances are recovered, and F1 balances both through their harmonic mean. Macro-F1 is then computed as the unweighted average of the class-wise F1-scores, giving equal importance to all classes regardless of their frequency. This makes it particularly

[7] All codes and prompts are available at: https://anonymous.4open.science/r/BZK-LE-36BC/.

suitable for imbalanced classification settings, where performance on minority classes should contribute equally to the final evaluation.

5.3 Results

BZK-LE Challenges Zero-shot Semantic Alignment. Table 2 shows CLIP's zero-shot performance on RVL-CDIP and BZK-LE using raw class labels. For RVL-CDIP, the reported results (taken from [18]) correspond to four different splits, denoted as A, B, C, and D. Despite class variations, CLIP achieves relatively stable, competitive accuracy across splits. This suggests that when label names are semantically meaningful and aligned with pretraining data, CLIP effectively leverages its learned text–image representations. In contrast, performance on BZK-LE is substantially lower when raw labels are used, especially on the full dataset. Unlike the RVL-CDIP categories, the BZK-LE labels are domain-specific and less descriptive, which makes it harder for CLIP to associate them with visual patterns in the image. When raw labels are replaced with natural-language descriptions, performance improves noticeably on both the balanced subset and the full BZK-LE dataset. These findings indicate that richer textual context strengthens text-image alignment. However, even with descriptive prompts, performance remains limited, indicating that zero-shot CLIP faces challenges when more detailed layout understanding is required.

Table 2. Zero-shot CLIP performance with raw labels and natural-language descriptions. RVL-CDIP results (splits A–D) where each split includes 12 various document classes indicated with * are taken from [18].

Label type	RVL-CDIP* Accuracy				Balanced BZK-LE		BZK-LE	
					Accuracy	F_1	Accuracy	F_1
	A	B	C	D				
Raw-label	70.55	67.02	50.35	51.67	6.89	11.85	0.51	3.83
NL-labels	–	–	–	–	22.13	20.27	16.04	7.87

Generative DONUT Adapts Efficiently Under Few-Shot Fine-Tuning But Struggles with Domain-Shifted Classes. As shown in Table 3, the fully fine-tuned DONUT model (FF-DONUT) exhibits consistent performance gains as the number of training samples increases. Accuracy improves from 42.8% to 73.6% on RVL-CDIP-160×5 and from 44.2% to 76.4% on RVL-CDIP-40K, when scaling from 10 to 100 samples per class, demonstrating effective few-shot adaptation. The strong performance on RVL-CDIP aligns with DONUT's pretraining on the full IIT-CDIP dataset, of which RVL-CDIP is a subset, enabling efficient adaptation under limited supervision. In contrast, results on BZK-LE remain substantially lower. This can be attributed to domain shift and the generative formulation of the task, where the decoder must produce class tokens not observed during pretraining, making adaptation more challenging for unseen document types and novel label spaces.

Table 3. Performance of the fully fine-tuned DONUT model (FF-DONUT) under varying numbers of training samples per class across RVL-CDIP-160×5, RVL-CDIP-40K, and BZK-LE benchmarks. Results marked with * are taken from [17].

#Train	Model	RVL-CDIP-160×5*		RVL-CDIP-40K*		BZK-LE	
		Accuracy	Macro-F_1	Accuracy	Macro-F_1	Accuracy	Macro-F_1
10	FF-DONUT	42.8 ± 3.0	–	44.2	–	9.09	2.92
50	FF-DONUT	70.1 ± 2.6	–	71.4	–	9.09	4.15
100	FF-DONUT	73.6 ± 1.9	–	76.4	–	12.62	4.95

Table 4. Comparison of DONUT+LP and ResNet-18+LP on balanced BZK-LE under varying training sizes.

#Train	Model	Accuracy	Macro-F_1
10	DONUT+LP	74.30	71.65
50	DONUT+LP	88.14	88.06
100	DONUT+LP	90.90	90.92
100	ResNet18+LP	96.44	96.44

Linear-probing Highlights DONUT's Encoder Capabilities Compared to Generative Fine-Tuning. Table 4 presents DONUT's performance under a linear probing (LP) setting, where the pretrained visual encoder is frozen and only a linear classification head is trained. Unlike the fully fine-tuned generative formulation (FF-DONUT), this setup treats the task as a standard discriminative classification problem based on encoder embeddings. Given the weak performance of FF-DONUT on BZK-LE, it was important to examine whether the limitation stemmed from the generative decoder rather than the encoder. In the LP configuration, the decoder is removed and classification is performed on mean-pooled encoder representations, isolating the encoder's feature quality. The results show a clear improvement over the generative setup. Even with 10 training samples per class, DONUT+LP achieves 74.30% Accuracy and 71.65 Macro-F_1, and performance further increases to 90.90% Accuracy and 90.92 Macro-F_1 with 100 samples. These findings suggest that the pretrained visual encoder provides highly discriminative layout features, while the generative formulation adds difficulty when predicting unseen class tokens under limited supervision.

ResNet-18 Provides a Stronger Linear-Probing Baseline for Layout Classification. As shown in Table 4, in the linear-probing setting ResNet-18 achieves 96.44% Accuracy and 96.44 Macro-F_1 with 100 training samples per class, outperforming DONUT+LP under the same setup. This suggests that the target classes are strongly characterized by coarse visual layout properties, such as block arrangement, whitespace distribution, and overall geometric composition. Since linear probing directly tests how well fixed representations can be separated by a linear classifier, the results indicate that the pretrained ResNet-18 features are more linearly separable for this task than the frozen DONUT encoder embeddings. One possible explanation is that the classification problem is dominated by recurring structural patterns and stable layout forms, making coarse visual representations particularly effective for distinguishing document classes.

Table 5. Performance of Qwen2.5-VL-7B-Instruct with prototype-based prompting on Balanced & full BZK-LE with prototype identifiers and label descriptions.

Label type	Balanced BZK-LE		BZK-LE	
	Accuracy	F_1	Accuracy	F_1
Prototype-identifier	11.07	04.79	11.46	05.19
NL-lab-des	11.07	04.81	18.18	10.21

Qwen Benefits From Semantic Grounding When Distinguishing Visually Similar Layouts. Table 5 shows the results of Qwen2.5-VL-7B-Instruct under the prototype-based prompting setup on both the balanced and imbalanced versions of BZK-LE. Two prompting strategies were tested. Using neutral Identifiers leads to weak and unstable performance in both settings, suggesting that abstract identifiers alone do not provide enough grounding for reliable layout matching. Adding natural-language descriptions improves the results, indicating that richer semantic context helps the model better relate prototypes to query images. In contrast, incorporating natural-language descriptions improves performance on the imbalanced BZK-LE dataset. This suggests that richer semantic context helps the model better associate prototypes with query images. Moreover, the classes in the imbalanced split appear to be easier to distinguish.

5.4 Discussion

The experimental results highlight substantial differences between zero-shot inference and supervised few-shot learning for DIC. Zero-shot approaches showed limited robustness in distinguishing BZK-LE document layout classes, particularly when classes showed subtle differences. Performance improvements were mainly observed in classes where the natural-language label description corresponded directly to a distinctive visual attribute. For example, `Gelbe-Hinweiskarte` is characterized by a yellow background explicitly referenced in the label, and `Suchkarte-Hinweiskarte` contains the term "Hinweiskarte" printed on the document. In these cases, strong semantic alignment between textual label description and visual content supports classification. However, for classes defined by more complex layout differences, such as the `RLP-*` classes, performance degraded substantially, and some classes were not correctly predicted at all. This suggests that zero-shot and prompt-based strategies rely heavily on explicit semantic cues and exhibit limited sensitivity to fine-grained structural layout variations.

In contrast, few-shot experiments indicate that the pretrained visual encoder captures discriminative layout features, but that the training objective plays a decisive role. When document classification was formulated as an autoregressive sequence generation task, class-wise F1 scores remained close to zero for most classes. This instability likely stems from the large parameter space and the sequence-generation formulation, which appears difficult to optimize under limited data supervision. However, when a linear classification head was trained on top of the frozen encoder, several structurally challenging classes were successfully classified. This finding demonstrates that pretrained visual features are sufficiently expressive for layout-based DIC and that a discriminative learning objective is more suitable than generative modeling.

6 Scope and Intended Use

BZK-LE is designed as an evaluation resource for studying the robustness of pretrained document understanding models under limited supervision, rather than for large-scale supervised training or pretraining. It exhibits characteristics typical of low-resource archival document analysis, including domain shift, layout variability, and weak semantic alignment between labels and visual structures. These properties make it suitable for assessing model performance under realistic constraints in historical and administrative document processing. While BZK-LE originates from a specific archival collection, the challenges it captures, such as the lack of sufficient labeled data for large-scale experiments, are representative of a broader class of low-resource document understanding problems, including some non-archival settings [10]. Therefore, the dataset is intended to support research on zero-shot and few-shot generalization in low-resource settings, rather than to provide exhaustive coverage of all document types or to enable large-scale model training.

7 Conclusion and Future Work

This work introduced BZK-LE, a dataset of historical index cards documenting compensation proceedings involving legal entities persecuted under National Socialism. The dataset captures low-resource archival document characteristics and provides both imbalanced and balanced variants for controlled evaluation settings. Baseline results using representative DIC models, including generative, discriminative, and multimodal prompting approaches, were systematically evaluated to establish reference benchmarks. The findings highlight that, while pretrained models offer strong visual representations, layout-based classification of archival documents remains challenging under domain shift and limited-data conditions. Future work will extend evaluations on BZK-LE to additional methods, improve prototype-based evaluation strategies, and further analyze model robustness under low-resource conditions.

Acknowledgments. This work is funded by the German Federal Ministry of Finance.

References

1. Bai, S., et al.: Qwen2.5-vl technical report (2025)
2. Chen, T., Kornblith, S., Norouzi, M., Hinton, G.: A simple framework for contrastive learning of visual representations. In: Proceedings of the 37th International Conference on Machine Learning. JMLR.org. (2020)
3. De Almeida Bandeira Macedo, L., Vieira Costa, J.P., Felix De Almeida, J.P., Garcia Freitas, P., Weigang, L.: Visual document matching for zero-shot document classification. In: International Conference on Document Analysis and Recognition (2025)
4. Garay, N., Vafaie, M., Sack, H.: Named entity recognition for digitised archival documents in German. In: EKAW (Companion) (2024)
5. Harley, A.W., Ufkes, A., Derpanis, K.G.: Evaluation of deep convolutional nets for document image classification and retrieval. In: 13th International Conference on Document Analysis and Recognition (ICDAR). IEEE (2015)
6. He, K., Zhang, X., Ren, S., Sun, J.: Deep residual learning for image recognition. In: Proceedings of the IEEE Conference on Computer Vision and Pattern Recognition (2016)
7. Huang, Y., Lv, T., Cui, L., Lu, Y., Wei, F.: Layoutlmv3: pre-training for document AI with unified text and image masking. In: Proceedings of the 30th ACM International Conference on Multimedia (2022)
8. Kim, G., et al.: OCR-free document understanding transformer. In: European Conference on CV. Springer (2022)
9. Klacsmann, B., Soloshenko, V.: Lost, Looted, Unpaid: Holocaust Restitution and Historical Injustice–Editorial Introduction (2025)
10. Kumar, J., Ye, P., Doermann, D.: Structural similarity for document image classification and retrieval. Pattern Recogn. Lett. (2014)
11. Larson, S., Lim, G., Leach, K.: On evaluation of document classification with RVL-CDIP. In: Proceedings of the 17th Conference of the European Chapter of the Association for Comp. Linguistics (2023)
12. Liu, L., Wang, Z., Qiu, T., Chen, Q., Lu, Y., Suen, C.Y.: Document image classification: progress over two decades. Neurocomputing (2021)
13. Nikolaidou, K., Seuret, M., Mokayed, H., Liwicki, M.: A survey of historical document image datasets. IJDAR (2022)
14. Papadopoulos, C., Pletschacher, S., Clausner, C., Antonacopoulos, A.: The IMPACT dataset of historical document images. In: Proceedings of the 2Nd International Workshop on Historical Document Imaging and Processing (2013)
15. Radford, A., et al.: Learning transferable visual models from natural language supervision. In: International Conference on ML. PMLR (2021)

16. Saifullah, S., Agne, S., Dengel, A., Ahmed, S.: Docxclassifier: towards a robust and interpretable deep neural network for document image classification. IJDAR (2024)
17. Scius-Bertrand, A., Jungo, M., Vögtlin, L., Spat, J.M., Fischer, A.: Zero-shot prompting and few-shot fine-tuning: revisiting document image classification using large language models. In: International Conference on PR. Springer (2024)
18. Sinha, S., Khan, M.S.U., Sheikh, T.U., Stricker, D., Afzal, M.Z.: CICA: contentinjected contrastive alignment for zero-shot document image classification. In: ICDAR. Springer
19. Vafaie, M., Hertling, S., Banse-Strobel, I., Dubout, K., Sack, H.: End-to-end information extraction from archival records with multimodal large language models. In: Proceedings of the 34th CIKM (2025)
20. Wolf, F., Tüselmann, O., Matei, A., Hennies, L., Rass, C., Fink, G.A.: CM1-A dataset for evaluating few-shot information extraction with large vision language models. In: ICCKE. Springer (2025)
21. wolfthomas: Bundeszentralkartei der Entschädigungsverfahren nach dem Bundesentschädigungsgesetz (BEG) (2010). https://doi.org/10.58079/bsl5
22. Xu, Y., et al.: Layoutlmv2: multi-modal pre-training for visually-rich document understanding. In: Proceedings of the 59th Annual Meeting of the Association for Comp. Linguistics and the 11th International Joint Conference on NLPg, vol. 1 (2021)

Handwriting, Writer Identification and Music Documents

Graph-Based Aggregation for Writer Identification and Retrieval

Oliver Pilizar[1], Marco Peer[2(✉)], and Florian Kleber[1]

[1] Computer Vision Lab, TU Wien, Vienna, Austria
[2] AIBEX Group, University of Fribourg, Fribourg, Switzerland
marco.peer@unifr.ch

Abstract. Writer identification and writer retrieval aim to assign handwritten documents to their correct authors or to retrieve documents written by the same individual. While current text-independent, offline approaches achieve high performance at the page level, their accuracy degrades substantially for line- and word-level tasks, partly due to simplistic aggregation methods that fail to preserve fine-grained stylistic information. This paper proposes a graph neural network–based aggregation framework for writer identification and retrieval, representing documents or lines as graphs of local handwriting features with explicitly modeled spatial relationships. A hybrid feature extraction strategy is employed, combining different local descriptors within a unified graph representation. The approach is evaluated with respect to identification accuracy, hard Top-k, and mean average precision.

Keywords: Writer Retrieval · Writer Identification · Graph Neural Networks

1 Introduction

Writer Identification (WI) is the task of assigning a given handwritten document to its correct author by comparing it with documents of known writers, whereas Writer Retrieval (WR)aims to retrieve all documents written by the same author as a provided reference document [12].

WI and WR have applications both in historical and modern contexts. In the historical domain, they enable the identification of anonymous and potentially undated writings, such as papyri and medieval texts [5,19]. In the modern context, WI can be applied to forensic scenarios, such as linking documents to potential suspects or to clarify the authorship of disputed writings, while WR allows the retrieval of documents from an archive, for instance, within law enforcement agencies [12].

While WI/WR are active research fields [16,20,22], most approaches rely on basic aggregation methods, such as sum or mean pooling, to arrive at a representation for entire documents [18,20]. Current WR approaches achieve near-perfect Mean Average Precision (mAP) on page-level tasks (CVL: 96–97%,

F. Shafait et al. (Eds.): DAS 2026, LNCS 16851, pp. 129–145, 2027.
https://doi.org/10.1007/978-3-032-36207-0_8

IAM: 92–94%), but performance drops when less information is available, as in line-level (CVL: 47–69%, IAM: 56–75%) and word-level tasks (CVL: 3–9%, IAM: 5–14%) [20]. This performance gap may be due to basic aggregation functions discarding fine-grained details and therefore failing to capture subtle handwriting style information, which may prove crucial [8], particularly when only small amounts of handwriting are available, such as on the word- and line-level. Learned aggregation methods using transformers have, however, not been able to surpass current state-of-the-art methods using mean average pooling [16]. To address this limitation, we employ a Graph Neural Network (GNN) as an aggregation method, which is capable of capturing spatial feature relations.

The approach presented in this work represents handwritten documents, or individual lines of documents, as graphs, where each node corresponds to a combination of features extracted from a specific keypoint, and edges capture the spatial relationships between these nodes. This approach is partially inspired by sentiment analysis, in which local tokens are aggregated to predict a global meaning, a task to which GNNs, among other neural network architectures, have successfully been applied [14,28]. Similarly, in WI/WR local handwriting features are aggregated to represent an entire line or document, enabling both the identification of the writer and the retrieval of documents written by the same author. GNNs are particularly suited for this task, as they can explicitly model spatial relationships between individual features by leveraging the underlying graph structure. Lastly, the inherent ability of GNNs to handle variable input sizes makes the method applicable to both document-level and line-level representations, and allows for the use of even smaller document subsets when needed.

Thus, the following contributions are made:

- Application of GNNs as an aggregation strategy for offline, text-independent WI/WR, with a focus on line-level tasks
- Evaluation of different feature combinations to determine their impact on identification and retrieval performance
- Assessment of different graph-building strategies and comparison of different GNN architectures

The rest of this paper is structured as follows. Section 2 presents related work, Sect. 3 describes the proposed methodology, and Sect. 4 presents the experimental results.

2 Related Work

Both WI and WR encompass a variety of techniques across different aspects of the process. Kaur and Kumar [9] define these aspects as feature extraction, feature selection, classification, and distance measures. Kumar et al. [13] state that the feature extraction process of WI can be divided into three categories, namely texture, shape, and deep-learning approaches. Additionally, combined approaches leverage both learned and handcrafted features.

Graph Neural Networks Approaches in Related Handwriting Domains GNNs have largely remained unexplored in WI adjacent domains. An exception is the work by Roy et al. [23], who propose an offline signature verification system employing GNNs at the node level.

In their approach, the graph is constructed from the features representing the signature samples. A GraphSAGE architecture is used to aggregate neighboring node information to classify signatures as genuine or forged. The method is evaluated on the MCYT-75 [17] and UTSig [24] datasets.

Although the GNN is applied at the node level rather than to whole-graph representations, this work demonstrates the potential of relational modeling in handwriting analysis and suggests that GNNs can be adapted for WI and WR.

Recently, the work of Priya et al. [21] proposes a new dataset for WI, and their method relies on Delauney triangulation, whose graph is forward to a graph convolutional network to classify the writer of an word image.

Learned Aggregation Functions. Contemporary approaches to feature aggregation into global document representations in the WI and WR domains typically rely on simplistic aggregation methods such as mean or sum pooling. Mattick et al. [16] explore the potential of learned feature aggregators. Their pipeline begins with the extraction of 32×32 image patches around SIFT keypoints. A standard ResNet-18 is then used to compute their embeddings, which are aggregated into a global image representation via one of two architectures - NetVLAD [1], or a transformer [25] architecture, which applies self-attention to weight relationships between all patch embeddings, before creating a learned global representation.

Experiments are conducted on the Historical-WI dataset [6]. Performance is evaluated using mAP and precision. Although the learned aggregation methods aim to capture more complex relationships between patch embeddings, the authors find that they fall short in comparison to state-of-the-art approaches using mean pooling.

These results highlight the importance of explicitly modeling relationships between local features, a capability that GNNs naturally provide.

3 Methodology

This section presents the complete pipeline from data input to the final writer identification and retrieval results.

3.1 Datasets

As dataset, the CVL [11] dataset consisting of 310 different writers has been used. In the default CVL train/test split 27 authors are part of the training set and the other 283 are part of the test set. Since during the experiments, writer retrieval as well as writer identification is tested, the CVL train set is further split into train/validation/test (70%/15%/15%) sets, by randomly splitting each writer's documents according to those percentages, thereby keeping all 27 authors in each set. This additional split of the training data is necessary to enable the model to be used for both WI and WR. This newly created test set from the original training data is used exclusively to evaluate identification, while the validation set is used for early stopping. The CVL test set is kept unchanged, as it is already prepared for retrieval and not required for identification. The inputs used for the experiments are the individual word images as well as their ground truths.

The second dataset used is the IAM [15] dataset, consisting of 657 writers, all contributing a varying number of documents. It is not directly used in the GNN setups, but rather as an additional option for the training of the Convolutional Neural Network (CNN), which provides the image patch embeddings for the GNN.

3.2 Pre-processing and Feature Extraction

The goal of this phase is to obtain local features that are later used for WR and WI within the GNN setup. These include CNN embeddings of image patches, SIFT features, and textual data in the form of word embeddings. The usage of image patch embeddings in connection with SIFT features is inspired by the work of Christlein et al. [4]. The pre-processing pipeline, which is shown in Fig. 1a, consists of two parallel branches to extract the aforementioned data.

The upper branch begins with the extraction of SIFT features, as they are located directly at the strokes, which subsequently serve both as part of the GNN's input feature vector, but also as center points for the creation of the 32×32 pixel image patches in the following step. Figure 1b depicts coverage of the word images. These created image patches are then used to train a CNN, specifically a ResNet-20, to discriminate between the authors associated with the texts. The objective is to generate embeddings that effectively represent the visual characteristics of the handwriting. Once trained, the CNN is used to extract embeddings from the patches, which are saved as key-value pairs to allow efficient access during the graph-building process as part of the GNN setup. Simultaneously, the SIFT features are saved in the same format.

In parallel, the lower branch of the pre-processing pipeline extracts the ground truths of all words occurring in the documents. Each word is then embedded using pre-trained fastText vectors, following the approach by Bojanowski et al. [2]. To differentiate between English and German words, an additional identifier is appended to each word embedding. By providing the GNN the ground truths, the model can learn to differentiate how different authors write the same words. Additionally, by adding the language identifier, the model is able to capture how the writing style of an author may vary across languages.

Convolutional Neural Network Using Image Patches. To extract the embeddings of the image patches, a ResNet-20 architecture is employed. This selection is based on its input size compatibility with the 32×32 image patches and its output embedding dimension of 64. This dimension provides a balanced contribution to the overall feature vector, which may also incorporate SIFT

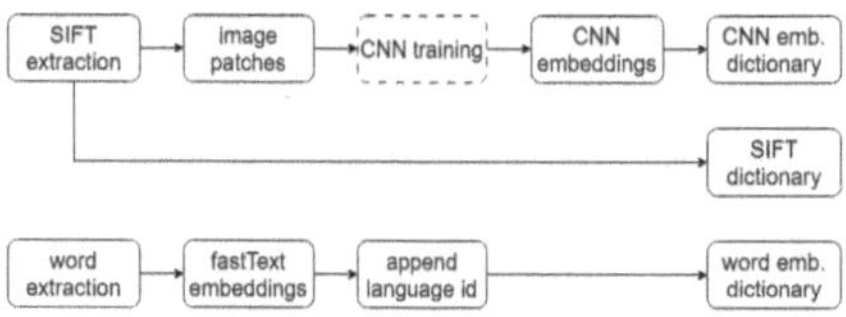

(a) Illustrated pipeline of the pre-processing phase.

(b) Contours of image patches extracted based on SIFT keypoints, overlayed on the original image to highlight coverage.

Fig. 1. Pre-processing pipeline and patch coverage visualization.

descriptors (128 dimensions), SIFT keypoint coordinates (2 dimensions), and a word embedding (101 dimensions).

3.3 Graph Building

During the graph-building phase, graphs are constructed to represent either individual lines or entire documents. The prepared data, consisting of SIFT feature information, image patch embeddings, and word embeddings, is used to create nodes. To differentiate between the representation of individual lines or entire documents, the data retrieved in this step corresponds to either the currently processed line or the document as a whole, respectively. The process is otherwise identical. Each node is a concatenated feature vector of the aforementioned data, which is always done in the same order. These nodes are then connected via one of three methods: Full, KNN or 1Line. Graphs illustrating these connection method are shown in Fig. 2.

Full Connection Method. Let $W = \{w_1, w_2, \ldots, w_N\}$ be the sequence of N words., and for each word w_i, let V_w be its set of nodes. Let E be the set of edges of the graph.

Fully connect nodes within each word:

$$E = \{\, (u, v) \mid w \in W,\ u, v \in V_w,\ u \neq v \,\} \tag{1}$$

Connect consecutive words via their outer nodes:

$$\begin{aligned} E = \{\, (u, v) \ \mid i \in \{1, \ldots, N-1\}, \\ u \in \text{rightmost}(V_{w_i}), v \in \text{leftmost}(V_{w_{i+1}}) \,\} \end{aligned} \tag{2}$$

The connection mode "Full" creates fully connected subgraphs that represent words, while sequentially connecting them.

KNN Connection Method. Let $W = \{w_1, w_2, \ldots, w_N\}$ be the sequence of N words, and for each word w_i, let V_w be its set of nodes. Let E be the set of edges of the graph. Let $\text{KNN}_k(u)$ be the set of k-nearest neighbors of node $u \in V_w$

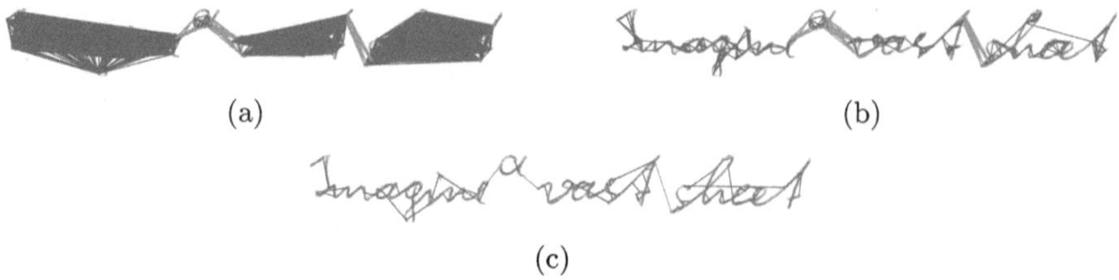

Fig. 2. Line-level depiction of the three connection modes: (a) Full, (b) KNN, (c) 1Line. Orange edges indicate connections between different words. (Color figure online)

within the same word, Components(V_w, E_w) the connected components of the word subgraph, and repr(C) a representative node from component C, chosen arbitrarily, as any node suffices to connect the component.

Connect each node to its k-nearest neighbors within the word:

$$E = \{ (u, v) \mid w \in W,\ u \in V_w,\ v \in \text{KNN}_k(u) \} \tag{3}$$

Ensure each word subgraph is connected by linking its components sequentially:

$$\begin{aligned} E = \Big\{ &(\text{repr}(C_i), \text{repr}(C_{i+1})) \ \mid w \in W, \\ &C_i, C_{i+1} \in \text{Components}(V_w, E_w) \Big\} \end{aligned} \tag{4}$$

Connect consecutive words via their outer nodes:

$$\begin{aligned} E = \{ (u, v) \mid\ & i \in \{1, \ldots, N-1\}, \\ & u \in \text{rightmost}(V_{w_i}), \\ & v \in \text{leftmost}(V_{w_{i+1}}) \} \end{aligned} \tag{5}$$

The "KNN" connection method aims to reduce the number of edges by connecting each node only to its close neighbors, utilizing the k-nearest-neighbor algorithm. The value $k = 4$ is chosen to allow the formation of natural clusters within words while preventing the creation of fully connected graphs.

1Line Connection Method. Let $W = \{w_1, w_2, \ldots, w_N\}$ be the sequence of N words, $V_w = \{v_1, v_2, \ldots, v_{n_w}\}$ the ordered set of nodes for word w_i, and E the set of edges. Let dist(u, v) denote the distance between two nodes, and first(V_w) and last(V_w) the first and last nodes of word w.

Connect nodes sequentially within a word by always linking the closest unconnected node:

Let $C_w \subseteq V_w$ be the set of connected nodes, initialized with the top-left node of the first word in the sequence $C_w = \{\text{first}(V_{w_1})\}$. Then, iteratively, connect each remaining node in order of proximity to the last added node $u_{\text{last}} \in C_w$:

$$\begin{aligned} E = \Big\{ &(u_{\text{last}}, v) \ \mid w \in W, v \in V_w \setminus C_w, \\ &v = \arg \min_{x \in V_w \setminus C_w} \text{dist}(x, u_{\text{last}}) \Big\} \end{aligned} \tag{6}$$

Repeat until all nodes in V_w are in C_w.
Connect consecutive words via the first and last nodes:

$$E = \{(\mathrm{last}(V_{w_{i-1}}),\ \mathrm{first}(V_{w_i})) \mid i \in \{2, \ldots, N\}\} \tag{7}$$

The "1Line" connection method minimizes the number of edges while maintaining a connected graph. Instead of trying to replicate the original line stroke order, this method aims to be consistent for repeated samples of the same word.

3.4 Graph Neural Networks

The general GNN architecture used for this work is shown in Fig. 3a. Figure 3b further describes the Flex Layers appearing in the general architecture, which are configurable convolutional layers. The input for the GNN is the graphs created in the previous step, which consist of the nodes representing combined feature vectors and edges denoting connections within and between words in different ways, depending on the chosen connection method as introduced in Sect. 3.3. Afterwards the Flex Layers are employed, which can be stacked any number of times. The first layer optionally changes the hidden feature dimension with subsequent Flex Layers continuing with the same size. By default, two Flex Layers are used with hidden dimensions equal to the input size. A global mean pooling layer is then employed to aggregate node features into a global representation of the graph. For WI the predicted author corresponds to the class with the highest output probability of the fully connected layer, while for WR embeddings from before the fully connected layer are compared using Euclidean distance.

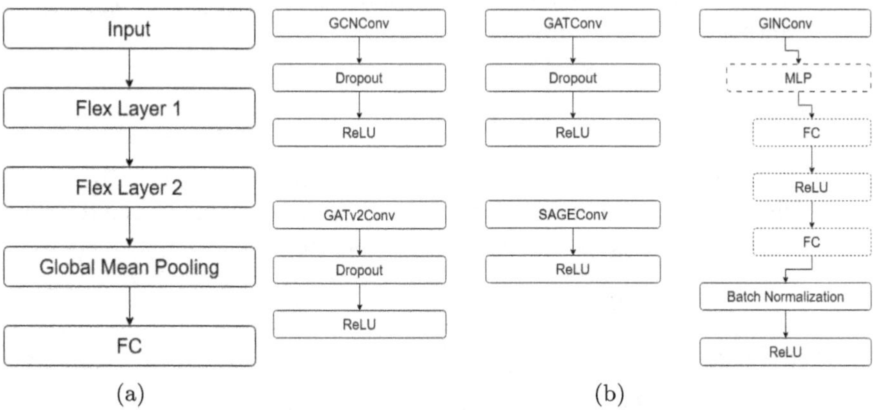

Fig. 3. Overview of the GNN architecture showing the general architecture in (a) and the detailed architecture of the Flex Layers used to test different setups in (b).

3.5 Layer Types

Graph neural network architectures commonly incorporate diverse layer types, such as convolutional, pooling, and aggregation layers. This paper investigates mainly convolutional layers. Specifically, GCNConv [10], GATConv [26], GATv2-Conv [3], SAGEConv [7] and GINConv [27] have been tested.

GCNConv is used as the standard convolutional layer during the experiments that do not specifically test the performance and viability of the different layer implementations, but, for instance, the importance of different features extracted from the documents.

In the context of writer identification and retrieval GCNConv means finding localized patterns in the provided features that characterize the writer's style, such as recurring traits in the way characters are written via SIFT and the image patch embeddings, as well as the additional contextual information about the word itself via the word embeddings.

GATConv implements an attention mechanism, which allow the network to gain a deeper understanding and highlight the most important feature nodes accordingly. For instance, anomalies in a writer's style may be assigned heavier weights, while common features or ones that do not carry as much information, for instance, the dot of a lower case"i", may get assigned lower weights. This should, in theory, allow for a more refined representation of each author's individual style. GATv2Conv is an improved version of the attention mechanism and it is analyzed if more context-dependent patterns can be found, and the overall performance of the model can thereby be improved. SAGEConv is designed to learn a function that can identify unique patterns for each writer, which can then be transferred to completely new and unseen authors. GINConv has a highly discriminative representation, which is supposed to allow the model to differentiate between authors with very similar writing styles.

4 Results

The primary focus of the evaluation lies on WR on a per-line basis, as this remains the more challenging task and provides more room for improvement compared to document-level retrieval, where state-of-the-art methods achieve over 96% mAP [20].

4.1 Retrieval

This section investigates the performance of WR, using the CVL dataset by default, unless explicitly stated otherwise. The official train/test split is used, with 70% of the training set reserved for training, 15% for validation, and 15% for testing.

Table 1 shows the standard configuration of the experiments. They stay the same through all of them unless otherwise specified, or the purpose of the experiment is to vary a certain parameter to draw conclusions on its impact. To get

Table 1. Default hyperparameters and feature dimensions used to train the GNN. Feature dimensions are shown as reference for calculating layer dimensions.

Learning Rate	0.001
Batch Size	32
Epochs	250
Patience	15
Weight Decay	0.0001
Loss	CE
Optimizer	Adam

(a) Training

Hidden dim.	Input dim
Num. of Layers	2
Layer Type	GCN
Features	SKPW
Connection	1line
Scope	line-level

(b) Model Architecture

SIFT Descriptor (S)	128
SIFT Keypoint (K)	2
Patch Embedding (P)	64
Word Embedding (W)	101

(c) Feature Dimensions

the input feature dimension of the GNN, we add up all the individual feature vector dimensions used for that GNN. The sub-table *Feature Dimensions* gives an overview of the possible features included, their notations (S - SIFT Descriptor, K - SIFT Keypoint, P - Image Patch Embedding, W - Word Embedding), and their dimensions. When combining features such as SKPW, the individual feature vectors are concatenated, resulting in a total dimensionality equal to the sum of their individual dimensions, which in the case of SKPW is 295. The default for the experiments is to use two hidden layers that both have the same dimension as the input layer.

Layer Types - GCN, GAT, GATv2, SAGE, GIN. Different layer types of the GNN are evaluated and in each experiment, the number of layers and the dimensions remain fixed, according to the overview in Table 1 . For the GAT and GATv2 layers, the number of multi-head attentions is set to one. Each GIN layer uses an underlying two-layer MLP with a ReLU activation function in between. After each GCN, GAT, and GATv2 layer, a dropout layer with a 50% probability is added, as in the original papers for the GCN [10] and GAT [26], dropout layers are used. In contrast, the original papers apply dropout before the convolutional layers, which in the experiments conducted here resulted in a decrease of up to approximately 10% mAP. The GATv2 paper [3] notes that dropout does not have a significant influence on the results, which is also the case for the results of the experiments conducted here. The results shown for GATv2 are the ones using dropout layers. SAGE and GIN do not include dropout layers for the experiments, as the results without are better, however, instead of dropout layers, a batch-norm is used after each GINConv layer. Table 2a shows those results.

The results show that all layer types, including GCN, which is used as the standard for the experiments, perform on a similar level, except for GIN, which achieves about 7% higher mAP compared to GCN. Layer types within about a percentage point can be considered effectively equivalent.

Feature Combinations - S/K/P/W. Individual and combined features used during training and inference, with the notation of S, K, P, W described in Table 1, sub-table "Feature Dimensions" are evaluated. The results are shown in Table 2c and are grouped by the number of features used.

Table 2. Retrieval performance comparison. (a) Layer types. (b) Connection methods. (c) Feature sets and combinations.

	Retrieval (%)				
	mAP	Top-1	Top-3	Top-5	Top-10
GCN	53.69	86.33	69.92	58.40	38.58
GAT	52.56	85.59	68.50	56.42	36.97
GATv2	55.93	87.36	71.87	60.69	41.67
SAGE	54.41	86.02	69.11	57.69	39.12
GIN	**60.24**	**90.46**	**77.65**	**67.53**	**49.48**

(a) Layer types

	Retrieval (%)				
	mAP	Top-1	Top-3	Top-5	Top-10
1Line	**53.77**	**86.12**	**70.00**	**58.34**	**38.55**
KNN	51.00	84.33	65.62	53.35	33.30
Full	36.18	68.50	44.01	31.51	15.92

(b) Connection methods

	Retrieval (%)				
	mAP	Top-1	Top-3	Top-5	Top-10
S	9.74	20.43	5.02	2.33	0.69
K	1.20	1.31	0.00	0.00	0.00
P	**58.79**	**89.76**	**76.06**	**65.97**	**47.29**
W	0.35	0.01	0.00	0.00	0.00
SK	6.28	15.75	2.87	1.15	0.20
SP	**60.09**	**90.55**	**77.41**	**67.70**	**48.85**
SW	9.58	19.54	4.42	2.05	0.57
KP	53.17	85.28	68.08	56.54	37.96
KW	2.02	3.55	0.05	0.01	0.00
PW	59.58	89.78	76.07	66.23	48.04
SKP	55.31	86.78	70.28	59.22	39.70
SKW	8.07	19.54	4.08	1.74	0.27
SPW	**57.96**	**89.74**	**75.65**	**64.89**	**46.21**
KPW	52.05	85.14	67.42	55.64	36.12
SKPW	52.31	84.68	67.64	55.75	36.12

(c) Feature sets and combinations

Looking at the individual features, namely S, K, P, and W, the low retrieval scores for the keypoint positions and especially the word embeddings indicate that these features alone do not provide meaningful discriminative information. SIFT descriptors, on the other hand, already show relevant results with a mAP of almost 10% and a Top-1 result of about 20% with 283 possible classes during retrieval. The image patch embeddings, however, stand out as the best-performing among the individual features with a mAP of 58.79%. Furthermore, the results indicate that both word embeddings and the SIFT feature descriptors enhance the performance of the image patches, while the inclusion of the keypoint position leads to a decrease in performance.

Graph Connection Methods - 1Line vs. KNN vs. Full. This section focuses on the results of different connection methods used during the graph-building process, which are introduced in Sect. 3.3.

The results presnted in Table 2b indicate that both 1Line and KNN achieve similar performance, with 1Line achieving slightly higher scores across all metrics. The Full connection method, which introduces a much higher number of edges compared to the other two methods, performs worse by about 15% and 17% mAP, respectively. This suggests that an edge density beyond a certain point negatively affects the model's ability to learn meaningful structural relationships.

Graph Scope - Lines vs. Document. For the document-level experiments, a graph is constructed for each document by sequentially processing each word in order and adding its corresponding nodes from all lines in a document, effectively creating one continuous line. The rest of the setup and training process follows the line-level experiments for the first part of this section and is then adapted to handle the lower sample size in the document-level task in the second part. An excerpt of the feature combinations is evaluated, selecting the top-performing combination in each category from previous experiments based on the number of features combined, and including S as the basis for the image patches. Table 3a shows the results of this experiment. The Top-5 and Top-10 results of the per-document experiments are always 0, since only 5 documents per author exist, and one of them is used as the current sample.

Table 3. Performance metrics on document-level graphs with (a) regular training and (b) adjusted, sequential training on the splits. (Top-5 and Top-10 are zero due to the number of pages per writer)

	Retrieval (%)		
	mAP	Top-1	Top-3
S	22.67	28.18	3.05
P	78.98	87.79	66.50
SP	**79.36**	**88.64**	**66.36**
SPW	78.05	87.08	64.37
SKPW	56.78	71.19	30.73

(a)

	Retrieval (%)		
	mAP	Top-1	Top-3
S	37.96	48.97	12.49
P	76.70	85.45	62.67
SP	78.33	87.44	63.73
SPW	**79.34**	**88.15**	**65.08**
SKPW	63.00	75.66	39.25

(b)

The results show that the higher-performing configurations, namely P, SP, SPW, see an increase of about 20% mAP on document-level compared to the line-level, while SKPW shows only a small increase of approximately 4%. The worst performing configuration, S, sees an improvement of over 10% mAP. Looking at the Top-1 and Top-3 metrics, per-document performances are generally lower than per-line ones, with the exception of S, which shows a higher Top-1 metric, but falls behind on the Top-3 one, following in line with the other configurations. The differences between the better-performing configurations are much less. Taking P as an example, per-document Top-1% is only about 2% worse, while the Top-3% difference increases to about 10%. For SKPW, which is in terms of mAP over 20% points lower than the best performing one, SP, the Top-1 and Top-3% decrease by ~13% and 37%, respectively. This indicates that the presence of K, the only feature in SKPW not included in any of the better-performing configurations, negatively affects the model's ability to rank documents correctly. While this effect is already observed in previous experiments in Sect. 4.1, it is amplified at the document-level. The inverse trends observed between mAP and Top-k percentages are a consequence of the limited number of samples. This sample limitation, potentially in combination with the current training method, contributes to the per-document results not reaching state-of-the-art performance.

To improve results with a fixed number of 7 samples per author during training, the data for each class is split into all possible 5-2-folds, using 5 samples for training and 2 for validation, resulting in 21 different splits. The model is trained sequentially on each split without resetting, using the updated weights from the previous split for the next iteration. The results of this training approach are shown in Table 3b.

The results demonstrate that the better performing configurations do not improve, but instead exhibit a decrease in mAP and Top-k metrics. In contrast, S and SKPW both improved, with S displaying an increase in mAP of ~15%. Since the epochs of both S and SKPW do not reach their limit during the standard training method, it suggests that the sequential training avoids local minima that these configurations are more prone to fall into. Overall, this approach does not provide a significant gain for the better-performing configurations.

Custom Layout and Comparison to State of the Art. Since the setup used for the preceding tests allows only a limited variety in terms of GNN configurations and primarily focuses on establishing a baseline of what is achievable with basic networks of different kinds, this section explores a custom architecture that build upon the learnings from the previously described results.

This experiment uses the GIN convolutional layer with the 1Line connection method, as both performed best previously. Three feature configurations are tested: P, SP, and SKPW. P and SP are included based on prior results, while SKPW is used to evaluate the effect of combining all features in a performance-focused setup rather than as a baseline.

Architectural and hyperparameter changes are explored, including larger hidden dimensions, dropout, additional GIN layers, and deeper MLPs. Only increasing the hidden dimensions leads to a notable performance improvement; all other changes reduce performance or have little effect.

Initial experiments increase hidden dimensions by roughly 30% per layer, yielding consistent gains. Larger expansions (512 and 1024 hidden units) introduce higher variability, which is mitigated by reducing the learning rate from 0.001 to 0.0001, as confirmed by five runs per feature configuration. With this adjustment, performance improves across all feature sets except one SKPW run (mAP 46.15%), while most results remain around 64%. As shown in Table 4, all configurations outperform the baseline setups from Sect. 4.1 by 3–6% points in mAP. Using only image patch embeddings achieves the best performance (mean mAP 65.14%) with the lowest variability. Adding SIFT features reduces performance, consistent with baseline results indicating that SIFT does not reliably improve outcomes. The combination of all features performs worst and shows the highest variability, though its best run slightly exceeds SP. Overall, increasing hidden dimensions combined with a lower learning rate improves performance while maintaining robustness.

Table 4. Retrieval performance (mean ± standard deviation, in %) for SKPW, SP, and P over five runs. Best mAPs achieved: SKPW 64.96%, SP 64.62%, P 66.58%. Epoch values are rounded to the nearest integer.

Setup	Epochs	mAP	Top-1	Top-3	Top-5	Top-10
SKPW	120 ± 15	60.71 ± 8.2	92.22 ± 2.4	80.93 ± 5.2	71.50 ± 6.9	53.03 ± 8.8
SP	136 ± 37	63.01 ± 1.3	93.34 ± 0.7	82.94 ± 1.5	74.14 ± 2.0	56.18 ± 2.4
P	106 ± 18	65.14 ± 1.1	94.57 ± 0.6	85.19 ± 1.1	76.91 ± 1.4	59.81 ± 1.7

Comparison with State of the Art. Figure 4 illustrates a comparison of the proposed approach against existing state-of-the-art methods presented by Peer et al. [20] on the CVL dataset. The original results include SIFT+VLAD, ResNet20+VLAD, and ResNet20+NetVLAD configurations, while the teal line indicates the performance of the best GNN-based result using only image patch embeddings as input feature, and the green line shows its mean value. As shown in the figure, the proposed method performs best for a low number of features. At around 250 feature points per line, it outperforms all state-of-the-art approaches, with ResNet20+NetVLAD surpassing the mean of the proposed approach at around 500 features, and surpassing its best run at around 1000 feature points. This demonstrates that the GNN-based aggregation method is a viable approach for WR, particularly when using a lower number of feature points.

4.2 Identification

For these experiments, the CNN used to extract image patch embeddings is trained on the IAM dataset. This prevents data leakage, as WI requires the same authors to appear in both training and testing, necessitating a split of the original CVL training set. Training the CNN on the IAM dataset, therefore, removes the need to retrain it whenever new train–test splits for the GNN are created. The WI experiments use the same standard configuration as the retrieval

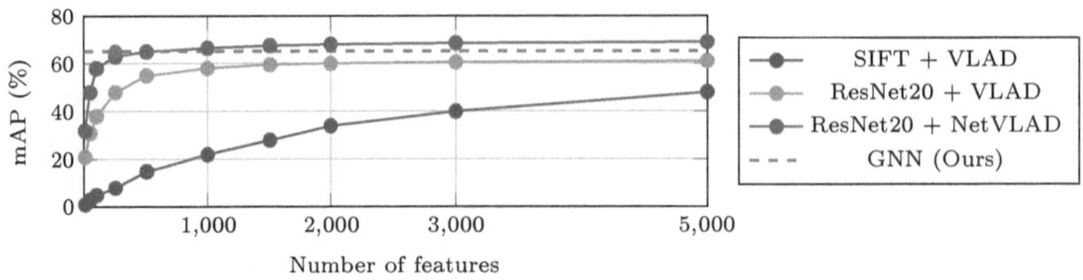

Fig. 4. Comparison of the proposed GNN-based approach, using an average of ~250 nodes per line of text, with state-of-the-art methods on the CVL dataset [20]. The dotted lines serve as an additional indicator to support comparison with state-of-the-art methods at different numbers of features.

experiments shown in Table 1. All settings remain identical unless a particular parameter is intentionally varied to evaluate its effect, such as the choice of layer type.

Layer Types - GCN, GAT, GATv2, SAGE, GIN. Similar to the tests during retrieval this experiment investigates the performance of different layer types. Table 5a shows the results.

Table 5. Test accuracy comparison. (a) Identification accuracy of different GNN layer types. (b) Identification accuracy of different feature sets and their combinations. (c) Identification accuracy of different connection methods.

	Test Accuracy (%)
GCN	97.4
GAT	97.0
GATv2	97.8
SAGE	98.1
GIN	**99.6**

(a) Layer types

	Test Accuracy (%)
S	79.18
P	**99.26**
SP	**99.26**
SPW	97.40
SKPW	97.03

(b) Feature combinations

	Test Accuracy (%)
1Line	96.28
KNN	**96.65**
Full	**96.65**

(c) Connection methods

All layer types have a test accuracy of over 97%, and exactly like the retrieval results, GCN is ahead of GAT and behind all others. Furthermore, GIN is again, just like during retrieval, the best performing layer type with a test accuracy of 99.6%.

Feature Combinations - S/K/P/W. This experiment evaluates different feature combinations during WI. As shown in the corresponding WR experiments in Sect. 4.1, the keypoint position feature K decreases performance and is therefore excluded from the subset of feature combinations used here. SKPW is tested as it serves as the standard configuration for all experiments. Table 5b shows the results.

Both P and SP achieve the highest test accuracy of 99.26%. The SIFT descriptors alone achieve the lowest accuracy, but have no negative impact on performance when combined with the image patch embeddings. On their own SIFT descriptors reach almost 80% test accuracy, which is notable since their retrieval performance, shown in Sect. 4.1, only reaches about 10% mAP. The results of SPW and SKPW indicate that the inclusion of K has no positive impact, as the observed difference is marginal, with under 1% point, which can be attributed to the individual trained models and not the method itself.

Graph Connection Methods - 1Line vs. KNN vs. Full. The results of all connection methods shown in Table 5c are nearly identical, with accuracies of almost 97% and only negligible differences. Since identification experiments that include image patch embeddings achieve results close to 100% without parameter optimization or combining the best settings of the individual experiments, it can be inferred that, based on the data used, the task of WI is easier than retrieval and that the choice between IAM and CVL trained CNNs has little impact. The results demonstrate that the use of GNNs for WI is effective.

5 Conclusion

This work investigates GNNs as learned aggregation mechanisms for offline, text-independent WI and WR, representing handwritten documents and lines as graphs whose nodes encode local handwriting features and whose edges model spatial relations. Experiments on the CVL and IAM datasets demonstrate that GNN-based aggregation achieves performance competitive with state-of-the-art pooling-based methods, particularly for WR under conditions with limited numbers of feature points, while maintaining near state-of-the-art accuracy for WI. Among the evaluated features, CNN-based image patch embeddings yield the strongest and most stable results. Sparse graph construction strategies that preserve spatial structure, such as 1Line and KNN, consistently outperform fully connected graphs for WR, whereas WI proves largely insensitive to the chosen connectivity. Overall, the results highlight the effectiveness of graph-based representations and GNN-based aggregation for capturing fine-grained handwriting characteristics, especially in low-information regimes, and establish a flexible framework for future extensions in feature design and graph construction.

Future work should investigate improved feature extraction and graph construction strategies. Dimensionality reduction methods such as PCA can be applied to high-dimensional descriptors, and the number of keypoints may be increased to better capture writer-specific characteristics; additional structural features and weighted edges are also worth-investigating. Furthermore, alternative graph construction schemes, including sparsified and multi-level graphs that integrate word-, line-, and document-level information, might be considerable to better preserve spatial relations and reduce structural bias.

Acknowledgements. This research was funded in whole or in part by the Austrian Science Fund (FWF) 10.55776/PAT4919325. Marco Peer was partly supported by the Swiss National Science Foundation (SNSF) under project grant no. 10002280, "AI-Supported Human-Computer Interaction with Large Volumes of Historical Documents".

References

1. Arandjelovic, R., Gronat, P., Torii, A., Pajdla, T., Sivic, J.: Netvlad: CNN architecture for weakly supervised place recognition. In: Proceedings of the IEEE Conference on Computer Vision and Pattern Recognition, pp. 5297–5307 (2016)
2. Bojanowski, P., Grave, E., Joulin, A., Mikolov, T.: Enriching word vectors with subword information. Trans. Assoc. Comput. Linguistics **5**, 135–146 (2017)
3. Brody, S., Alon, U., Yahav, E.: How attentive are graph attention networks? (2022). https://arxiv.org/abs/2105.14491
4. Christlein, V., Gropp, M., Fiel, S., Maier, A.: Unsupervised feature learning for writer identification and writer retrieval. In: 2017 14th IAPR International Conference on Document Analysis and Recognition (ICDAR). vol. 01, pp. 991–997 (2017). https://doi.org/10.1109/ICDAR.2017.165
5. Cilia, N.D., De Stefano, C., Fontanella, F., Marthot-Santaniello, I., Scotto di Freca, A.: Papyrow: a dataset of row images from ancient Greek papyri for writers identification. In: International Conference on Pattern Recognition, pp. 223–234. Springer (2021)
6. Fiel, S., et al.: ICDAR 2017 competition on historical document writer identification (historical-wi). In: 2017 14th IAPR International Conference on Document Analysis and Recognition (ICDAR), vol. 1, pp. 1377–1382. IEEE (2017)
7. Hamilton, W.L., Ying, R., Leskovec, J.: Inductive representation learning on large graphs (2018). https://arxiv.org/abs/1706.02216
8. He, S., Schomaker, L.: Gr-RNN: global-context residual recurrent neural networks for writer identification. Pattern Recogn. **117**, 107975 (2021)
9. Kaur, H., Kumar, M.: Signature identification and verification techniques: state-of-the-art work. J. Ambient. Intell. Humaniz. Comput. **14**(2), 1027–1045 (2023)
10. Kipf, T.N., Welling, M.: Semi-supervised classification with graph convolutional networks (2017). https://arxiv.org/abs/1609.02907
11. Kleber, F., Fiel, S., Diem, M., Sablatnig, R.: CVL-database: an off-line database for writer retrieval, writer identification and word spotting. In: 2013 12th International Conference on Document Analysis and Recognition, pp. 560–564. IEEE (2013)
12. Koepf, M., Kleber, F., Sablatnig, R.: Writer identification and writer retrieval using vision transformer for forensic documents. In: International Workshop on Document Analysis Systems, pp. 352–366. Springer (2022)
13. Kumar, V., Sundaram, S.: Offline text-independent writer identification based on word level data. arXiv:2202.10207 arXiv preprint (2022)
14. Liao, W., Zeng, B., Liu, J., Wei, P., Cheng, X., Zhang, W.: Multi-level graph neural network for text sentiment analysis. Comput. Electric. Eng. **92**, 107096 (2021)
15. Marti, U.V., Bunke, H.: A full English sentence database for off-line handwriting recognition. In: Proceedings of the Fifth International Conference on Document Analysis and Recognition. ICDAR'99 (Cat. No. PR00318), pp. 705–708. IEEE (1999)

16. Mattick, A., Mayr, M., Seuret, M., Kordon, F., Wu, F., Christlein, V.: Evaluating learned feature aggregators for writer retrieval. Int. J. Docu. Anal. Recogn. (IJDAR) **27**(3), 265–274 (2024)
17. Ortega-Garcia, J., et al.: Mcyt baseline corpus: a bimodal biometric database. IEEE Proc. Vision, Image Sign.Process. **150**(6), 395–401 (2003)
18. Peer, M., Kleber, F., Sablatnig, R.: Towards writer retrieval for historical datasets. In: Lecture Notes in Computer Science, pp. 411–427. Lecture notes in computer science, Springer Nature Switzerland, Cham (2023)
19. Peer, M., Sablatnig, R.: Feature mixing for writer retrieval and identification on papyri fragments. In: Proceedings of the 7th International Workshop on Historical Document Imaging and Processing, pp. 31–36 (2023)
20. Peer, M., Sablatnig, R., Kleber, F.: Towards the influence of text quantity on writer retrieval. In: International Conference on Document Analysis and Recognition, pp. 129–145. Springer (2025)
21. Priya, K., et al.: Graph convolutional teacher-student framework for writer inspection from intra-variable handwritten words. In: Document Analysis and Recognition - ICDAR 2025, pp. 115–129 (2026)
22. Raven, T., Matei, A., Fink, G.A.: Self-supervised vision transformers for nbsp;writer retrieval. In: Document Analysis and Recognition - ICDAR 2024: 18th International Conference, Athens, Greece, August 30–September 4, 2024, Proceedings, Part II, pp. 380–396 (2024)
23. Roy, S., Sarkar, D., Malakar, S., Sarkar, R.: Offline signature verification system: a graph neural network based approach. J. Ambient. Intell. Humaniz. Comput. **14**(7), 8219–8229 (2023)
24. Soleimani, A., Fouladi, K., Araabi, B.N.: UTSIG: a persian offline signature dataset. IET Biometrics **6**(1), 1–8 (2017)
25. Vaswani, A., et al.: Attention is all you need. Adv. Neural Inf. Process. Syst. **30** (2017)
26. Veličković, P., Cucurull, G., Casanova, A., Romero, A., Liò, P., Bengio, Y.: Graph attention networks (2018). https://arxiv.org/abs/1710.10903
27. Xu, K., Hu, W., Leskovec, J., Jegelka, S.: How powerful are graph neural networks? (2019). https://arxiv.org/abs/1810.00826
28. Zhang, C., Li, Q., Song, D.: Aspect-based sentiment classification with aspect-specific graph convolutional networks. arXiv:1909.03477 arXiv preprint (2019)

CollabScore OMR: A Hybrid System for Music Score Recognition

Aurélie Lemaitre[1,2(✉)], Bertrand Coüasnon[2,3], and Philippe Rigaux[4]

[1] IRISA, CNRS, Rennes, France
[2] Université Rennes 2, Rennes, France
aurelie.lemaitre@irisa.fr
[3] Univ. Rennes, Rennes, France
[4] CNAM, Paris, France

Abstract. We present a new optical music recognition system developed as part of the CollabScore project. It addresses the problem of recognizing images of polyphonic music scores spanning multiple pages, and enables the generation of an editable document encoded in MEI or MusicXML. The system is based on a combination of AI techniques. On the one hand, graphical elements and musical symbols are extracted from images using deep learning-based approaches. On the other hand, a syntactic rule description system allows the expression of music notation constraints. These rules enable the system to automatically detect inconsistencies in the recognition process. We provide a detailed evaluation on the new CollabScore benchmark dataset, which contains real-world printed scores from the 19th century by Camille Saint-Saëns, and show that our method outperforms a commercial OMR system.

Keywords: OMR · Deep learning · Rule based system · AI combination

1 Introduction

Optical Music Recognition (OMR) is a complex problem. Music score images remain indeed much more challenging to recognize than text pages. Unlike text, music is organized in two dimensions: the flow of monodic *events* (notes, chords, rests) is horizontally organized in *measures* that progress from left to right, whereas polyphony is vertically aligned across the stack of *staves* that compose a (music) *system*. Additionally, symbols placed at the beginning of a page or staff affect how the notation is interpreted. For example, a local accidental in a measure impacts the entire measure, a key signature at the beginning of a staff affects all the subsequent notes, and a time signature influences the number of beats in measures—potentially for dozens of pages. The reader unfamiliar with Common Western music notation is referred to [7] for an in-depth presentation.

Currently, most OMR systems can accurately recognize individual symbols but generally fail to produce a complete and reliable representation of polyphonic

F. Shafait et al. (Eds.): DAS 2026, LNCS 16851, pp. 146–162, 2027.
https://doi.org/10.1007/978-3-032-36207-0_9

musical scores spanning several pages. Our approach addresses the ambitious challenge of generating a complete, multi-pages music score. It takes as input a sequence of images featuring real-world, polyphonic scores (called *image scores* thereafter), and produces a single digital output encoded in a standard format, either MEI [14,19] or MusicXML [6,15] (called *editable scores*).

Deep-learning methods are candidate for the complete, one-shot analysis of image scores. Such techniques require however many labelled images, and the cost of labelling musical scores is very high. We propose a different approach based on a combination of several forms of AI: (i) a low level image analysis that localizes symbols and graphic elements in score images and (ii) a syntactic rule-based method that combines the detected symbols and produces a consistent editable score complying with the music notation syntax. This complete hybrid architecture is presented in Fig 1.

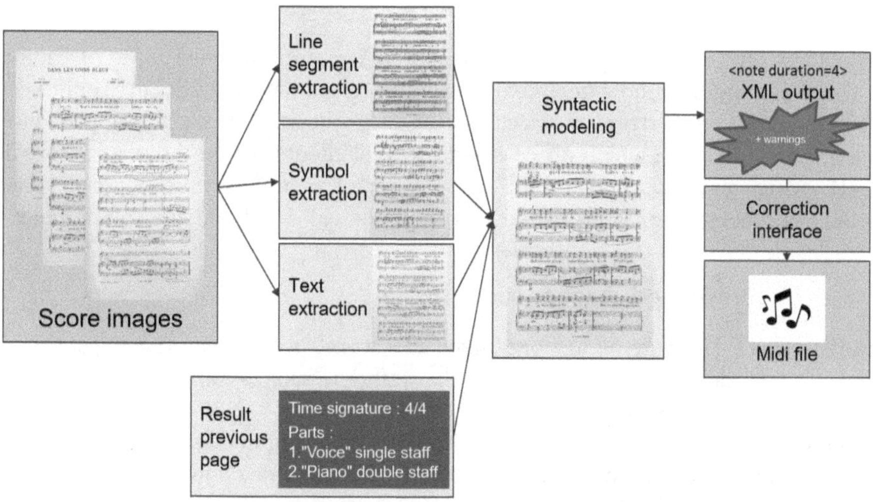

Fig. 1. Overview of CollabScore OMR: combination of extractors, applied to a sequence of pages, feeding a syntactic modeling that finally produces an editable score.

Assessing the performance of an OMR system is one of the difficulties that hinders the development and progress of OMR research. The lack of benchmark datasets and evaluation metrics has been in the past a major obstacle to characterize the quality of a specific approach and to compare systems. In the present work, we expand recent efforts to address this issue [12] and propose a complete evaluation study, including a comparison with a state-of-the-art commercial system. The study is based on the newly published CollabScore dataset, and we publicly provide all the material for the reproducibility of our results. To our knowledge, this is one of the very first complete performance study of a general-purpose OMR system, and we hope thereby to provide a solid foundations for the assessment of future OMR proposals.

This paper's contributions can therefore be summarized as follows:

1. a new OMR approach that combines deep learning techniques and syntactic analysis rules;
2. an application to real-world polyphonic multi-pages music scores;
3. and finally a quantitative and qualitative evaluation on a public real-world dataset of images, including a comparison with a state-of-the-art commercial OMR system.

The paper is organized as follows. We study the related works in Sect. 2. We then detail the two main aspects of our OMR: low level extraction of graphical elements (Sect. 3) and high level syntactic rules (Sect. 4). We finally report the results of an evaluation conducted on the CollabScore dataset in Sect. 5.

2 Related Work

In 2023, Calvo-Zaragoza et al. [1] presented a comprehensive state-of-the-art analysis, highlighting how deep-learning innovations have significantly enhanced OMR capabilities. Current research distinguishes between two main approaches: pipeline-based OMR, which processes musical elements sequentially, and holistic OMR, which employs end-to-end recognition of complete musical structures. These approaches represent the key methodological developments in the field.

2.1 Pipeline-Based Recognition

Traditional OMR systems are organized in several processing stages, each constituting an independent field of research: staff detection, symbol extraction, and production of the result.

The symbol extraction process consists in detecting and labeling pixels that correspond to a specific music symbol. This task has been widely studied in the past years, but has been improved using deep-learning [3,9]. Symbol detection can be seen as an object detection problem, but music symbol detection presents a unique challenge: the system must process very large images (approximately 4000 pixels wide) while detecting extremely small objects (as small as a few pixels for a dot indicating note duration). Several datasets were designed to train and evaluate such methods, like Muscima++ [8] and DeepScores [22].

The limit of existing approaches is the difficulty to recognize the symbols in all the contexts. There is a great variability in size, and some symbols are quite rare (eg: double sharp) [1]. Yesilkanat *et al.* [23] have shown that reducing the number of classes to focus on fixed-shape symbols, improves the recognition. We use this work in our proposal.

Regarding the evaluation aspect, most of the past works are quantified using object-detection metrics (like Mean Average Precision), but this metric does not necessarily reflect the quality of the model for an OMR. In addition, the extraction of isolated symbols does not always take into account relationships between symbols, which is required to produce the final encoding.

2.2 Holistic Recognition

The achievements of deep learning have encouraged researchers to explore end-to-end analysis of image scores. Historically, such systems were limited to monophonic scores [2]. Recent work by Li et al. [11] introduced TrOMR, an end-to-end approach trained on both monophonic and polyphonic single-staff images generated from MuseScore files. The method includes staff detection preprocessing before applying TrOMR to each individual staff.

A significant challenge remains in adapting end-to-end methods to diverse score types. Roselló et al. [20] address this as a domain adaptation problem, proposing a system capable of recognizing both mensural notation and Common Western notation. However, their work currently addresses exclusively isolated monophonic staves. A significant part of the research has been specifically devoted to piano staff recognition [13,21]. A notable contribution in this area is the GRANDSTAFF database, which contains over 50,000 polyphonic single-system piano images. While this dataset has yielded promising results for piano score recognition, it currently does not address the important challenge of maintaining continuity between systems across multiple pages.

More recently, RÃŋos-Vila et al. [18] developed SMT (Sheet Music Transformer), an end-to-end method designed for polyphonic systems. The current implementation shows significant progresses, but also some limits: it processes only isolated systems and relies on synthetic data without cross-staff voices. This highlights the need for more general and representative methods capable of handling scores extending over several pages, with complex polyphonic structures and cross-staff voice movements.

2.3 Remaining Challenges

Calvo-Zaragova [1] identify the remaining OMR challenges a: (i) learning to reconstruct music notation, (ii) recognizing multiple voices and (iii) dealing with lyrics. The CollabScore OMR addresses these three problems.

We trust in the ability of deep-learning techniques to extract musical symbols. However, we believe that it is difficult to reconstruct music notation over several pages without rules. This dual insight has led us to develop a hybrid OMR that combines deep learning approaches for robust symbol recognition with syntactic rules that maintain musical consistency across measures, systems and pages.

3 Extraction of Low Level Elements

The CollabScore OMR takes as input a complete image score, generally composed of several pages. Each page is analysed in turn. As shown in Fig. 1, the analysis of a page first consists in applying three low level graphic extractors:

- a line segment extractor, used to extract linear elements (staff lines, stems);
- a symbol extractor dedicated to the recognition of small musical symbols;
- an OCR to detect textual elements, such as lyrics, or instrument names.

Line Segment Extraction. The goal is to detect the linear elements in the image score. For this purpose, it is not necessary to apply a complex treatment based on deep learning. We apply a classical method of image analysis based on Kalman filtering [16].[1] It is very effective to detect lines, even in the presence of curvatures or crossing objects (e.g., beams and staff lines). As a result, we detect the following elements: horizontal staff lines, vertical bar lines, vertical stems, horizontal and diagonal beams. Each detected element is represented by the coordinates of its end points.

Symbol Extraction. We rely on the work proposed by Yesilkanat *et al.* [23]. The method is based on a Cascade R-CNN + Focal Net architecture. The model is first trained on the synthetic dataset "DeepScores V2 Dense" [22], which contains 1,714 images and 1.1M annotations, and then fine-tuned using a bootstrap approach on a dataset of 20 real 19th-century pages.

A key feature of this approach is that it does not process all possible classes. Unlike DeepScores, which includes 135 musical element classes, it focuses on 63 fixed-shape elements: note heads, accidentals, clefs, articulation marks, etc. (see Fig. 2). As demonstrated in [23], the extractor achieves better performance when it does not attempt to recognize variable-shape symbols such as slurs, stems, or dynamics. This is where our hybrid approach proves particularly effective: by combining a restricted (and thus more efficient) symbol detector with a rule-based system, we ensure that other objects are handled by the rule system rather than through an error-prone detection.

The output of this processing stage consists of a list of music symbols, each characterized by the coordinates of its bounding box, the associated symbol class, and a score indicating recognition confidence.

Fig. 2. In red, example of extracted objects: the detector recognizes rests, note heads and accidentals, but not linear elements like stems, staff lines and slurs. (Color figure online)

Text Detection. At this stage of analysis, we aim at localizing textual elements within the image, and particularly lyrics and instrument names. We employ a classical OCR system, namely the Pero OCR text recognizer [10]. A particular challenge in lyrics recognition arises from word segmentation in syllables separated by hyphens. For instance, in Fig. 3, the word "honte" is decomposed in two syllables, "hon" and "te". This segmentation can complicate OCR processing if

[1] Our thanks go to Jean Camillerapp, retired professor, who realised this work.

it attempts to rely on a known vocabulary. However, maintaining this syllable separation is crucial, as each syllable must be associated with its corresponding music note. We include a syllabification process that systematically divides complete words into their constituent syllables. For example, Fig. 3 demonstrates the necessary splitting of "rumeur" in "ru" and "meur". The text detection step yields isolated syllables, represented by the coordinates of their bounding boxes and the text label.

Fig. 3. Text detection: each word is segmented into syllables; syllables of a same word are then assigned to consecutive notes, see, e.g., the word "rumeur".

4 The High Level Rule-Based System

The output elements of the low-level extractors are combined using syntactic rules (Fig. 1). To this end, we leverage the existing DMOS system [4], which performs graphical and syntactic analysis.

4.1 DMOS Method for Music Notation

The DMOS method [4], proposed by CoÃijasnon *et al.*, is designed for document analysis and recognition. It adopts a syntactic approach that enables rule-based knowledge representation. As a generic framework, DMOS has been successfully applied to diverse document types. Its adaptation to new kinds of documents requires only the specification of domain-specific knowledge, using the dedicated EPF language provided with the method. A key advantage of DMOS is its ability to easily integrate multiple data sources from different extractors. Each data type represents a distinct perspective on the image and is organized within a dedicated *perceptive layer*.

In the present work, we use DMOS to describe the syntactic rules specific to music notation. This grammatical description expresses the combination, constraints and relationships that tie the extracted symbols. The syntactic modeling takes as input four *perceptive layers* (Fig. 1), containing:

- the *Line segment extraction* (staff lines, stems, beams, bar lines);
- the *Music Symbol extraction* from 63 classes (notes, clefs, accidentals...);
- the *Lyric extraction* segmented into syllables;
- and, the *Results previous pages*, whenever needed, which contains the interpretation context coming from the previous pages: number of parts, clefs, time and key signatures.

We organise the grammatical description in two steps: the *graphical* analysis followed by the *syntactic* analysis.

4.2 Graphical Analysis

The graphical analysis describes the music notation as a constrained organization of all the visual elements potentially found within a page. The rules are given by musicologists, following the Common Western notation. The system proceeds in two main stages:

Page Organization. A score page is defined as a hierarchy of systems, staves and individual measures using the *Line segments extraction* perceptive layer (Fig. 1). This ensures that subsequent analysis is confined to the relevant musical regions. A *staff* is defined as a set of five horizontal lines, while a *system* refers to a group of staves. A page may contain multiple systems. Additionally, we introduce the concept of "staff group", used for keyboard instruments (e.g., piano). In such cases, music is written on two staves but corresponds to a single part.

Measure Content. Once measures are localized, the analysis focuses on their inner graphical content. The analyser uses as input the content of the *Symbol extraction* perceptive layer (Fig. 1) that contains the music symbols, such as clefs, notes, rests and accidentals.

Figure 4 shows the rule for the content and the organisation of measure: clef, key signature, notes and rests, lyrics. The rules rely on position operators: the `AT` keyword restricts recognition to specific regions of the image. This spatial constraint ensures that only the symbols consistent with their local position are selected, filtering out potential false positives generated by the symbol recognizer.

```
measure :=
  AT(onStaff) &&
    clef C &&
  AT(rightClef C) &&
    keySignature K &&
  AT(rightKeySignat K) &&
    notesAndRests N &&
  AT(underNotes N) &&
    associatedLyrics L N &&
  AT(rightNotes N) &&
    endMeasureBar B.
```

Fig. 4. Rule for the recognition of a measure content, showing the corresponding recognized image elements (see the matching colors).

4.3 Syntactic Analysis

The syntactic analysis carries out a comprehensive verification of the recognized graphical elements consistency. It takes into account two main constraints:

- *Measure duration consistency:* the cumulative duration of music events within each measure must be consistent with the measure's metric;
- *Vertical alignment constraint:* music events intended to be played simultaneously must be vertically aligned within a system.

(a) The blue piano voice moves from one staff to the other. Two additional voices (yellow an red) appears in the middle of the 3rd measure.

(b) The piano part contains up to three voices. In measure 4, the yellow voice starts in the middle of the measure, with a note head shared with the blue voice.

Fig. 5. The syntactic analysis assigns graphical elements (notes and rests) to distinct voices, each represented by a unique color. The notation"-2.00" (in red or yellow) in the bottom-right corner of a measure indicates that the corresponding voice is incomplete. (Color figure online)

At this step of analysis, the system must take into account the metric encoded by the most recently detected time signature. Note that it can be found at the beginning of the current system, or at the beginning of the page, or at the beginning of the score, several pages before. The input perceptive layer *Results previous pages* (Fig. 1) plays here an essential role, as it provides the result of the previous page analysis to the current one.

The two constraints on duration condition the distribution of the graphical elements (notes, rests) in different voices (Fig. 5). This is an intricate part of the process, because a voice can start or stop *inside* a measure, and can move from one staff to another in a staff group.

The role of syntactic analysis is therefore to verify that the number of beats in each voice corresponds to the current metric.

5 Evaluation

An opinion frequently met in the communities of music engravers and musicologists is that OMR systems generally produce editable scores which are so faulty that it is much more effective to typeset the score from scratch using an efficient software. OMR systems will only be adopted when the cost of manually correcting the output will be less than that of a complete rewriting.

It is therefore essential to qualify new proposals with a detailed evaluation that characterizes as precisely as possible the specific cases in which the system supplies a satisfying output or, conversely, fails to reliably recognize the input image. In addition to this qualitative evaluation, a quantitative measure of the proportion of elements that must be corrected is also required.

Furthermore, the OMR behavior should be evaluated in a realistic situation, with images of actual polyphonic scores spanning several pages, and containing notation samples representative of a well-characterized genre, style or period.

To this end, we need (i) benchmark datasets, containing at least images and ground-truth scores, (ii) comparison tools, and (iii) a meaningful and well-accepted metric. Until recently, these requirements were difficult to meet. Fortunately, we are now witnessing combined efforts to converge towards a common framework for OMR validation, evaluation and comparison (see, for instance, the recent paper by Martinez-Sevilla *et al.* [12]).

We expose our evaluation methodology in what follows. It is based on the newly pusblished CollabScore dataset [17] and a metric issued from the MusicDiff method [5], consistent with the one proposed in [12]. We additionally expose and discuss a comparison with a state-of-the-art OMR commercial system, namely PhotoScore (See https://www.avid.com/products).

5.1 Reproducibility

The tests are executed following the full process described in Fig. 1. To facilitate the creation of such pipelines, the DMOS method has been recently packaged in a Python-like manner, enabling downloading the DMOS method, and integrating it with any compatible input library.

For research purposes, we provide the following resources:

- a configured DMOS environment running as a Visual Studio Code extension,
- the MUSIC grammar for syntactic modeling,
- the PyDMOS extension, which executes the full pipeline using Ray.

These resources are available at https://www-shadoc.irisa.fr/omr/.

5.2 Dataset

The CollabScore dataset [17] contains a sample of scores by Camille Saint-Saëns from the late 19th century, provided by the National Library of France (BnF). The dataset is available at https://github.com/collabscore/dataset and includes 23 scores (totaling 177 pages).

For each score, the dataset provides images of the original edition, and a ground-truth in MusicXML and MEI. All images come from the original editions, and are published on the Gallica digital library (https://gallica-bnf.fr) from the National Library of France. They are free of rights, can be directly viewed online and downloaded. The ground truth was manually edited in the CollabScore project, faithfully reflecting the sources in its editorial choices, including scrupulous respect of divisions in pages and systems. It can be considered as the target result of a "perfect" OMR.

The left part of Table 1 summarizes the dataset contents, provides links to image sources, and gives insights on the size of the score content. It presents the number of *parts*, which roughly corresponds to the number of instruments. One of the main interest of the dataset is the size of the scores. For example, the score *Quam dilecta* (see images on Gallica) consists of 5 parts on 6 staves: sopranos, contraltos, tenors, basses, and piano (two staves). The music piece spans 89 measures across 8 pages. This represents a total of 534 measures to recognize.

5.3 Metric

MusicDiff [5], available as a Python package[2], is designed to compute the edit distance between two music score documents, in the spirit of the Unix diff tool that compares two text files. It produces a list of the differences between the two scores, each "diff" being a pseudo-edition applied to a target element (a note, a rest, a symbol, etc.). The diff list is the minimal set of editions necessary to transform one score (in our case, the OMR output or "predicted" score) into the other (in our case the ground-truth or "reference" score).

This permits comparisons, but does not directly yield a normalized metric. Recently, Martinez-Sevilla *et al.* [12] proposed the OMR-NED, a global metric that divides the number of editions by the total number of elements in the notation. Intuitively, this represents the amount of the predicted score that needs to be modified to obtain the reference score. We adopt a similar approach, but distinguish three categories of element, as illustrated by Fig. 6.

Voice elements (in red, Fig. 6) cover notes, rests, and chords. The diff metric counts the number of required editions to obtain the correct notes, rests and chords. This includes modifying/inserting/deleting (i) the pitch or octave of a note, (ii) local accidentals, (iii) duration dots, (iv) ties, and (v) the number of notes inside of a chord. The correction of each element requires one edition. We define the (voice elements) *edition rate* as the number of required editions (i.e., the number of differences) divided by the total number of musical elements (note, rest, chord).

Contextual elements (in blue in Fig. 6) ensure the correct interpretation of voice elements for a performing musician. Those elements are either a key signature,

[2] See https://github.com/gregchapman-dev/musicdiff.

Fig. 6. Elements evaluated by MusicDiff: Voice elements (in red), Contextual elements (in blue), and Lyrics (in green). (Color figure online)

time signature, or a clef. They usually set a general context, at the beginning of the score, or when some changes occur. To facilitate the reading, the clef and the key signatures are usually recalled at the beginning of each new system. For instance, in Fig. 6, the second system contains a graphical reminder of the latest clef and time signature for each staff. We colored those redundancies in grey, and these elements are not considered in the total count of contextual elements. We define the (context elements) *edition rate* as the number of differences concerning contextual elements divided by the total number of context elements in the score (clef, key and time signature, in blue), excluding redundancies (in grey).

Lyrics (in green in Fig. 6) recognition is also evaluated. A lyric is correct if: (i) it is associated to the correct note, (ii) it is related to the correct verse, (iii) the syllable position in the current word (beginning, middle or end) is correctly reported and (iv) its transcription is identical (case-sensitive).

The code of this metric can be downloaded with the CollabScore dataset, and the evaluation report can be reproduced. Note that for the time being, we do not evaluate the recognition of performance indications (expressions, directions, free instructions), in yellow in Fig. 6. It will be done in a future work.

5.4 Quantitative Comparison with Photoscore

We evaluate our predicted scores with the ground-truth ones, and also include a comparison of our results with those obtained by the recognition of the dataset with a commercial OMR, Photoscore. The results are summarized in Table 1. It shows the *edition rate* for the three kinds of elements described above. This rate represents the amount of edition required to obtain the ground truth score from the predicted one (either with our OMR, or with PhotoScore). The smaller the better. Note that it can be higher than 100%, since the correction may require a

Table 1. Application to the CollabScore dataset: dataset description and comparison of our OMR vs Photoscore OMR. Evaluation of edition rate (lower is better) for the elements described in Fig. 6. (*): Photoscore failed to recognize to scores.

Dataset description					Voice elements			Context elements			Lyrics		
Original image	Score title	Nb pages	Nb parts	Nb measures	Nb elem	Edit rate Our OMR	Edit rate Photoscore	Nb elem	Edit rate Our OMR	Edit rate Photoscore	Nb elem	Edit rate Our OMR	Edit rate Photoscore
C006_0	[Dans les coins bleus]	4	2	51	801	1.37%	4.99%	56	5.36%	44.64%	182	13.74%	71.98%
C009_0	[Fière beauté]	5	2	123	1681	7.26%	23.32%	20	35.00%	95.00%	184	35.87%	50.54%
C010_0	[God save the King]	2	2	28	450	3.11%	2.89%	9	0.00%	0.00%	144	65.28%	60.42%
C013_0	[La sérénité]	3	2	51	1549	5.75%	28.15%	9	0.00%	33.33%	138	40.58%	76.09%
C024_0	[Dans ton coeur]	4	2	48	697	18.08%	33.00%	15	6.67%	20.00%	170	21.76%	63.53%
C035_0	[Avril]	5	2	66	942	8.70%	16.35%	23	30.43%	30.43%	216	33.33%	64.81%
C036_0	[L'Amour blessé]	5	2	55	1396	12.46%	79.44%	30	20.00%	46.67%	302	35.10%	81.46%
C046_0	[L'Amant malheureux]	5	2	87	1040	5.77%	2.98%	9	0.00%	33.33%	191	28.80%	60.21%
C079_0	[La coccinelle]	2	2	53	693	19.34%	10.68%	19	36.84%	36.84%	152	50.66%	110.53%
C080_0	[La feuille de peuplier]	4	2	98	1034	17.60%	19.15%	22	63.64%	81.82%	139	27.34%	44.60%
C141_0	[Guitare]	3	2	44	682	24.49%	65.54%	24	8.33%	70.83%	74	59.46%	93.24%
C307_0	[Quam dilecta]	8	5	89	1634	14.32%	17.44%	34	29.41%	32.35%	572	67.31%	50.87%
C437_0	[Calme des nuits]	5	5	58	494	4.45%	5.26%	12	8.33%	91.67%	308	45.78%	52.92%
C438_0	[Les fleurs et les arbres]	6	5	78	764	9.29%	15.71%	12	8.33%	133.33%	247	106.07%	129.55%
C449_0	[Sérénade d'hiver]	14	4	222	2751	13.85%	40.93%	28	17.86%	364.29%	1788	36.86%	66.95%
C452_0	[Les marins de Kermor]	15	5	249	3232	3.71%	66.24%	36	13.89%	33.33%	2154	49.95%	58.50%
C455_0	[Ode]	12	4	256	2909	8.25%	14.23%	48	4.17%	60.42%	2153	55.88%	83.93%
C456_0	[Les soldats de Gédéon]	38	8	357	8038	15.43%	N/A(*)	120	35.83%	N/A(*)	4715	50.92%	N/A(*)
R165_1	[Suite orchestre, Prélude]	5	10	50	1592	32.29%	29.84%	31	25.81%	219.35%	0	N/A%	N/A%
R165_2	[Suite orchestre, Sarabande]	5	9	54	1361	44.45%	52.98%	31	19.35%	135.48%	0	N/A%	N/A%
R165_3	[Suite orchestre, Gavotte]	6	7	85	1889	76.39%	27.21%	33	51.52%	206.06%	0	N/A%	N/A%
R171_0	[Danse macabre (melody)]	6	2	137	1594	9.16%	37.64%	10	0.00%	110.00%	293	23.55%	83.96%
R171_3	[Danse macabre]	15	2	471	7748	14.66%	N/A(*)	57	29.82%	N/A(*)	0	N/A%	N/A(*)
		177	88	2810	44971	**16,26%**	28,53%	688	**23,55%**	99,79%	14122	**48,78%**	74,3%

deletion followed by an insertion, resulting in a number of editions higher than the number of elements.

It is important to note that the proposed results are based on the direct output of the OMR, without considering any additional manual editing that could be driven by the warnings issued by the OMR in the case of low confidence symbol recognition.

It can be seen that, in general, our OMR performs much better than Photoscore, a notable result given the fact this latter system is distributed as a first-class component of one of the most widely used music notation software (namely, Sibelius). In many cases, the recognition quality of our OMR is very high, and the predicted score can be corrected with a much lesser effort than a complete creation. In some situation, things are getting worse, though, and these variations can be explained by some notation particularity which gives rise to a specific recognition problem. We examine and comment some of the most salient such problems in the qualitative discussion that follows.

5.5 Qualitative Results

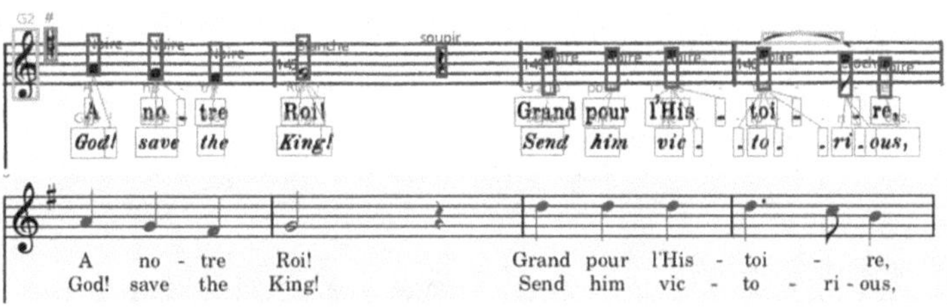

Fig. 7. Score with two-verse lyrics: association of each syllable to a note. Top part: analysis of the image score (Gallica link); bottom: final transcription.

Lyrics. The dataset contains a significant number of sung pieces. Lyric recognition is a complex task due to the syllabic division, the reconstruction of words and the possibly slightly misaligned position of the text with respect to its musical counterpart. Our OMR recognises lyrics, can split words into syllables to associate each syllable to a note. It can also handle lyrics with several verses (Fig. 7). The results in Table 1 show that lyric recognition remains a challenging problem. However, it is important to note that the metric is really strict, as it is case-sensitive, and takes into account the position of the lyric in the word.

Clef and Key Signature Changes. Music pieces by Saint-Saëns, and in particular his melodies, present very frequent key signature and time signature changes, as illustrated by Fig 8. Our OMR can detect these changes, and the detection quality is evaluated by the context edition rate in Table 1. This rate can easily become high, due to the relatively small number of contextual elements. For example, in C009_0, an edition rate of 35% only represents 7 editions.

Fig. 8. Frequent metric changes (Gallica link): original image and produced results. Note the new G Clef at the beginning of measure 2.

Cross-staff Voices. Another common feature of the dataset is the frequent crossing of voices from one staff to another in piano parts. As shown in Fig. 9, a voice may begin on a staff (the third), move to another (the second), and then return to the previous one. This makes isolated staff processing impossible, as voice continuity cannot be maintained when staves are analyzed separately. Our OMR system successfully handles these cases, whereas Photoscore struggles with such voice crossings.

Grace Note. In Table 1, the worse result appears for score R165_3. The bad edition rate is due to the presence of grace notes. In the current version, they are wrongly interpreted as first-class notes, leading a long measure. This is part of planned short-term improvements.

Multi-pages Scores. The score size in the CollabScore dataset range from from 2 to 38 pages. The challenging problem is to produce consistent notation from one system to another, and from one page to another. In addition, when an instrument does not play temporarily (typically at the beginning for a soloist that enters after an introductory part), the corresponding staff is empty and even missing. When the instrument comes back, in a new system of new page, the difficulty is to correctly insert it at the proper place in the score structure.

Let's take, for instance, images of C452_0 (follow the links). On page 7, there are 4 sung parts, and a tenor solo part is added on the top. It plays during a few systems and disappears on page 9 before coming back on page 14. The main challenge lies in transcribing each part continuously over the pages, regardless of whether the additional tenor solo is present. Our OMR use heuristics that let it correctly handle this configuration in many cases. We obtain for instance an edition rate of 3.71% on the voice elements of this C452_0 score. There

Fig. 9. Cross-staff voices: original image (top) and predicted result with our OMR (middle) and with Photoscore (bottom). Photoscore has problem organizing the voices.

exists however highly complex situation (such as large scores for orchestra and/or choirs) which lead to recognition failure, and require future work.

Implicit Tuplet. The dataset contains frequent implicit triplets notation (Fig. 10). A triplet indicates that three notes must be played within the duration of two. This is noted by the number 3, but not always repeated throughout the score. While our OMR system recognizes well explicit triplets, for the moment, it does not treat these implicit triplets correctly, and this yields high edition rates, particularly on R171_3. We will improve this in a future work, using among other indicators the constraint of the time consistency of the measure.

Fig. 10. Implicit tuplet notation: the number 3 indicates that the three notes should be played as a triplet. This information remains sometimes implicit.

6 Conclusion and Future Work

We propose a new OMR system, associated to a complete evaluation environment featuring a dataset of images, ground-truth score encoding, and metric calculation. Our systems follows a hybrid approach that combines low level extractors based on deep-learning, and high level syntactic rules. The main areas of interest for this AI combinations are: (i) to benefit from the strengths of learning-based approaches for detecting musical symbols, (ii) to avoid the need for a large amount of annotated data for a complete score, (iii) to simply express well-known rules for music notation. This hybrid methodology leverages the strengths of both paradigms to overcome the limitations of purely data-driven or purely rule-based approaches.

Proposing a new system without elements that allow to assess its property and its behavior with respect to other solutions would result in a limited contribution to the research in this area. We therefore devoted a significant part of our efforts to produce a detailed evaluation. This builds of course on previous advances in the field, and we hope that our work will participate to the ongoing movement that will lead to establish OMR to a mature and widely adopted methodology, comparable to what has been already achieved for text recognition.

Experiments show that, at least in the context of our considered dataset, our proposed OMR performs better than an existing commercial OMR, and this is a quite encouraging result. Yet, clearly, the complexity of music notation, and its countless particularities, still keep us for reaching at the moment a fully satisfying solution, and some of the remaining problems (a list of which is sketched in the discussion that precedes) require additional research. We are actively working of this challenge, with three main tracks of investigation. First, we believe that some problems (e.g., grace notes) can be solved in the current setting, by extending and improving an approach based on rules and constraints. Second, deep learning methods can give insights that could help to detect frequent patterns that correspond to highly implicit cases. And, finally, we plan to complement the fully automatic production of scores with a collaborative phase that would require a specific, limited input from users. The assignment of a staff to an instrument for instance is typically sometimes difficult to infer, and errors at this level yield catastrophic result. This is however a simple information, easy to specify, and accessible to non expert users. Such an additional, limited interactive step could therefore lead to drastic improvements of the results.

References

1. Calvo-Zaragoza, J., Martinez-Sevilla, J.C., Penarrubia, C., Rios-Vila, A.: Optical music recognition: recent advances, current challenges, and future directions. In: Document Analysis and Recognition - ICDAR 2023 Workshops, pp. 94–104 (2023)
2. Calvo-Zaragoza, J., Rizo, D.: End-to-end neural optical music recognition of monophonic scores. Appl. Sci. **8**(4), 606 (2018)
3. Choi, K.Y., Couasnon, B., Ricquebourg, Y., Zanibbi, R.: CNN-Based accidental detection in dense printed piano scores. In: ICDAR. Australia (2019)
4. Coüasnon, B., Lemaitre, A.: DMOS, It's your turnâĂŕ! In: International Workshop on Open Services and Tools for Document Analysis (ICDAR-OST), Japan (2017)
5. Foscarin, F., Jacquemard, F., Fournier-S'niehotta, R.: A diff procedure for music score files. In: 6th International Conference on Digital Libraries for Musicology, pp. 58–64 (2019)

6. Good, M.: The Virtual Score: Representation, Retrieval, Restoration, chap. MusicXML for Notation and Analysis, pp. 113–124. MIT Press (2001)
7. Gould, E.: Behind Bars, Faber Music (2011)
8. Hajic Jr, J., Pecina, P., Pacha, A., Calvo-Zaragoza, J., Vigliensoni, G., Fujinaga, I.: Muscima++: dataset and ground truth for handwritten optical music recognition. In: ICDAR, pp. 39–46 (2017)
9. Huang, Z., Jia, X., Guo, Y.: State-of-the-art model for music object recognition with deep learning. Appl. Sci. **9**(13) (2019)
10. Kišš, M., Beneš, K., Hradiš, M.: At-st: self-training adaptation strategy for OCR in domains with limited transcriptions. In: ICDAR, pp. 463–477 (2021)
11. Li, Y., Liu, H., Jin, Q., Cai, M., Li, P.: Tromr: transformer-based polyphonic optical music recognition. In: 2023 IEEE International Conference on Acoustics, Speech and Signal Processing (ICASSP), pp. 1–5 (2023)
12. Martinez-Sevilla, J.C., et al.: Sheet music benchmark: standardized optical music recognition evaluation. In: 26th International Society for Music Information Retrieval Conference (ISMIR) (2025)
13. Mayer, J., Straka, M., Hajič, J., Pecina, P.: Practical end-to-end optical music recognition for pianoform music. In: ICDAR, pp. 55–73 (2024)
14. Music Encoding Initiative (2026). http://www.music-encoding.org. last consulted Feb. 26
15. Music Notation Community Group (2026). https://www.w3.org/community/music-notation/. last consulted Feb. 26
16. Queguiner, C., Camillerapp, J., Leplumey, I.: Kalman filter contributions towards document segmentation. In: ICDAR, pp. 765–769 (1995)
17. Rigaux, P., Coüasnon, B., Guillotel-Nothmann, C., Guilloux, F., Lemaitre, A.: The collabscore dataset. towards robust and generalized OMR evaluation. In: DlFM (2026)
18. Ríos-Vila, A., Calvo-Zaragoza, J., Paquet, T.: Sheet music transformer: end-to-end optical music recognition beyond monophonic transcription. In: ICDAR (2024)
19. Rolland, P.: The music encoding initiative (MEI). In: Proceedings of the International Conference on Musical Applications Using XML, pp. 55–59 (2002)
20. Rosello, A., Fuentes-Martínez, E., Alfaro-Contreras, M., Rizo, D., Calvo-Zaragoza, J.: Source-free domain adaptation for optical music recognition. In: ICDAR (2024)
21. Ríos-Vila, A., Rizo, D., Iñesta, J.M., Calvo-Zaragoza, J.: End-to-end optical music recognition for pianoform sheet music. Int. J. Docu. Anal. Recogn. (IJDAR), 1–16 (2023). https://doi.org/10.1007/s10032-023-00432-z
22. Tuggener, L., Hajic Jr, J., Calvo-Zaragoza, J., Schmidhuber, J., Fisch, D.: Deepscores—a dataset for segmentation, detection and classification of tiny objects. In: Computer Vision and Pattern Recognition Workshops (CVPRW), pp. 53–60 (2018)
23. Yesilkanat, A., Soullard, Y., Coüasnon, B., Girard, N.: Full-page music symbols recognition: state-of-the-art deep models comparison for handwritten and printed music scores. In: DAS 2024 Workshop on Document Analysis System (2024)

Towards Zero-Shot Domain Generalization for ID Cards Presentation Attack Detection

Mario Nieto-Hidalgo[1(✉)], Juan M. Espin[1], and Juan E. Tapia[2]

[1] R and D Department, Facephi, Alicante, Spain
{marionieto,jmespin}@facephi.com
[2] da/sec-Biometrics and Internet Security Research Group, Hochschule Darmstadt, Darmstadt, Germany
juan.tapia-farias@h-da.de

Abstract. Presentation-Attack Detection (PAD) for national ID cards is limited by the lack of publicly available genuine samples, making it difficult for systems to generalize across countries. This paper introduces two main innovations: (1) a Prototypical Network head using an EfficientNet-V2-b0 backbone that requires only four genuine samples per class to create reliable prototypes; and (2) an episodic training regime that keeps PAD classes fixed while varying the card domain, allowing the network to learn universal attack cues.

Evaluated on a large multi-country dataset and the public DLC-2021 benchmark, this method achieves an average Equal Error Rate of around 9%, outperforming conventional softmax and CLIP zero-shot baselines even with data from a single source country. This approach provides accurate, privacy-preserving PAD while minimizing data collection, facilitating scalable cross-jurisdictional remote onboarding.

Keywords: PAD ID cards · Zero-shot · Few-shot

1 Introduction

Growing concerns about biometric companies in remote verification processes have increased the popularity of fully online onboarding systems. The popularity of these systems is mainly due to their user-friendly enrollment and the widespread adoption of smartphones. Users can enroll directly from home by taking a selfie or a photo of their ID card, without needing to go to a store or an office. These factors increase the number of users willing to enroll. However, this system carries a high risk of fraud, as users might attempt to exploit it. The easiest way to fool these systems is with Presentation Attacks (PAs), in which the fraudster impersonates another subject using a printed photo or a screen replay and presents it to the capture device. These systems prevent this type of fraud through PAD algorithms. Today, most solutions are based on algorithms, usually Deep Learning (DL) networks, trained on large amounts of

F. Shafait et al. (Eds.): DAS 2026, LNCS 16851, pp. 163–177, 2027.
https://doi.org/10.1007/978-3-032-36207-0_10

images to distinguish high-frequency details in PAs, such as the Moiré patterns of screen-replay attacks, printed photo textures, etc.

The State-of-The-Art (SoTA) face PAD algorithms show good performance due to the large number of open-set databases available. However, they struggle somewhat with generalization to unseen capture devices or Presentation Attack Instruments (PAI). However, PAD systems for ID cards remain in their infancy, primarily due to the difficulty of obtaining sufficient bona fide samples [12] and the lack of open-set databases. The ID cards contain sensitive information used to identify subjects; therefore, genuine bona fide samples cannot be published. This is why all publicly available datasets are used as bona fide samples of synthetic or simulated ID cards printed on Polyvinyl Chloride Plastic (PVC). The simulated bona fide samples (synthetic) are insufficient for research and for a production-ready PAD system.

The first competition on PAD on ID card was organized in the last International Joint Conference on Biometrics 2024 [12]. The results showed that the submitted algorithms lacked generalization capabilities. The 2025 edition of this challenge [13] showed improvements in the generalization of the algorithms tested, with one team surpassing the baseline. However, it remains challenging because the baseline and the best submitted algorithm differed only slightly when evaluated across four countries.

One of the main drawbacks of PAD algorithms for ID cards is their poor generalization to unseen document types or country versions [7]. Each country has its own ID card template, and multiple versions may coexist during certain periods. Each template varies significantly in its background, typography, layout, and other elements. This heterogeneity affects the performance of PAD systems, making generalization difficult. In Sanchez et al. [7], meta-learning techniques based on Few-shot learning were explored. It was suggested that at least 100 bona fide ID card users and their respective attacks were needed to obtain good performance in a new type of ID. However, even 100 bona fide users are difficult to obtain due to privacy concerns, and generating the corresponding attacks requires significant time and effort. Therefore, it is crucial to explore alternatives that could reduce this number.

Several open-set datasets are available on the SoTA. Some of the open-set datasets available are MIDV-500 [1], MIDV-2020 [2], MIDV-Holo [4], DLC-2021 [5], and SynIDPass [14], which usually contain a small number of synthetic users, printed on PVC, which is not enough to train a robust production-ready PAD system.

Motivated by the previous challenge, we proposed a new method toward Zero-shot based on Few-shot learning to predict on ID cards from a new country using pretrained features extracted from a PAD representing a different country. The main contributions of this work are:

- A Few-Shot approach is proposed to reduce the number of images needed to extend PAD to other countries.
- A comparison with a Zero-shot foundation model was explored to extend the generalization capabilities

- Our method was evaluated on private and open-set datasets using IDs from different countries.

The remainder of the article is organized as follows: Sect. 2 summarizes related works on PAD and *FSL*; Sect. 3 describes the proposed method; Sect. 4 describes the experimentation performed to test the proposed method; finally, Sect. 5 discusses the results and future work.

2 Related Work

Few-shot learning *(FSL)* has emerged as an effective learning method with considerable potential [9]. Despite recent creative efforts to tackle *FSL* tasks, rapidly learning valid information from just a few or even zero samples remains a significant challenge.

Song et al. [9] propose and analyze, in detail, a survey of more than 200 works that have raised the open challenge of *FSL* in recent years, due to the restrictions and limited access to data imposed by privacy preservation issues.

The *FSL* approaches usually work with two sets of samples: a *query set* containing samples from unknown classes and a *support set* containing samples from known classes. The *FSL* approaches use the *support set* as a source of information to infer the class of each sample of the query set. Depending on the constitution of the *support set*, the nomenclature *N-way-K-shot* is used where N is the number of classes and K the number of samples of each class in the *support set*. When $K = 0$, it is called *Zero-Shot* learning; otherwise, the term *K-Shot* or *Few-Shot* is used.

In the literature, we found two main approaches based on *FSL* theory. One of them is the Relation Network *(RL)* [10], and the other is the Prototypical Network *(PN)* [8], which stands out for its simplicity.

Both networks are based on K-Nearest Neighbors *(KNN)*, but with the advantage of being trainable end-to-end. The *KNN* can be considered as an *FSL* approach that computes the distance between the *query sample* and all *support samples*, returning the most common class among the K closest support samples. The problem with *KNN* is that its effectiveness depends on the discriminatory power of the embeddings used to compute the distance. The embeddings usually come from a previously trained backbone, and they are traditionally trained to optimize a Multi-Layer Perceptron *(MLP)* classifier, which produces an embedding space that may not be optimal for computing distances between samples of unknown class. In addition, traditional *KNN* requires lots of samples of each class to work correctly.

Both *RL* and *PN* try to solve these issues by providing an end-to-end trainable *KNN*. The *RL* replaces the *MLP* for a module that learns a distance function to optimize the discriminatory power of both the embeddings and the distance function.

On the other hand, *PN* replaces the *MLP* with a one-hot similarity score layer with a *softmax* as a loss function to optimize the distance between *query samples* and a *prototype* of each class. This prototype can be computed as the average embedding of the class. The ability of both networks to be trained end-to-end, along with training that uses a different subset of classes per batch, enables them to perform well on unseen classes.

Sanchez et al. [7] used the Prototypical Network to improve generalization with a few samples. The Prototypical Network is a metric-based *FS* learner trained to assign samples to their class prototypes. This prototype is computed by averaging the embeddings of a few samples of the class. This average, along with the training process in which the prototypes are changed in each batch, allows the network to generalize to new prototypes. In that work, additional samples of new ID card types were incrementally added to the training set to establish a benchmark for the *PN* ability to learn from a few samples. That work concludes that at least 100 distinct bona fide users and their respective attacks were required to achieve a good performance.

In addition, architectures based on foundation models such as (Contrastive Language–Image Pre-training) *CLIP* [6] have revolutionized the *Zero-Shot* scene. The *CLIP* is a family of vision-language models trained on large-scale image-text pairs to learn joint representations of visual and textual data. The *CLIP-B/16* model uses a 16×16 patch size, making it a medium-sized model that balances performance and efficiency. It is trained on 400 million image-text pairs from the internet and excels at zero-shot classification, image retrieval, and transfer learning. In contrast, *CLIP-L/14* uses a *ViT-L/14* (Vision Transformer Large, 14×14 patches) architecture, which is significantly larger and more powerful. With over 1 billion parameters, it utilizes more complex visual and semantic relationships, achieving SoTA performance.

The *CLIP* is a Vision Transformer architecture *(ViT)* [3] based network that has been trained to compare the embedding generated by an image and its text description. This enables *Zero-Shot* by computing *Cosine Similarity* (CS) between the textual class descriptions and the images of those classes.

3 Proposed Method: Few-Shot Learning Approach

In previous work, such as Sanchez et al. [7], the authors aim to expand into new countries or domains with reduced data use. In that work, the authors indicate that 100 bona fide samples and their attacks are needed to obtain competitive results. These 100 samples, in an ID card problem, aren't a small number.

Our approach is based on the premise that identifying the ID card type is easier than the PAD itself, so we assume that there is an ID card type classifier that selects the appropriate *support set*. For example, we tested an approach that is trained on different types of ID cards, driving licenses, and passports. This classifier uses the first half of an EfficientNetV2-b0 [11] backbone to generate embeddings. Those embeddings are then classified using a KNN that performs cosine distance to the embedding of the document template or specimen. This approach achieves an average accuracy of 98.5% across more than 150 different documents and 99.8% on the ID cards tested in this work. That classifier was trained with both real and synthetic samples. Although the classifier is out-of-scope for this work, it is essential to note that we could train one using only synthetic samples or specimens.

A diagram of our method is shown in Fig. 1. First of all, 1a, the baseline is shown, with a backbone (marked with an E in Figure), where an embedding is extracted by concatenating the Global Average Pooling *(GAP)* and Global Max Pooling *(GMP)* of the backbone output. At the head, a hidden layer and a Softmax Classifier Layer with four classes are present: bona fide, print, screen, and PVC, respectively.

To build our system, we modify the baseline by replacing the hidden layer with a *Prototypical Layer*. According to the *N-Way-K-Shot* nomenclature, our proposed method is a *4-Way-K-Shot* because we have 4 classes. The K is defined in preliminary experiments, which couldn't be exposed in this work due to length limitations. We defined *a support set* of 4 samples per class. Prototypes were built by averaging the embeddings of 4 samples per class of the target ID card type.

In the experiments, fewer samples (one, two, or three) showed high dependence on the selected samples, and with 4 samples, this is the lowest number at which the results are stable. We selected the lowest because we are in a scenario where bona fide samples are limited. In that case, our method is a *4-Way-4-Shot* method that uses only 16 samples: 4 bona fide samples and their respective attacks. In our method, the model never uses any samples of the target ID card type during training; *it only uses them during inference to build the prototypes.* Unlike [7], where the samples are used to fine-tune the model and to build the prototypes together.

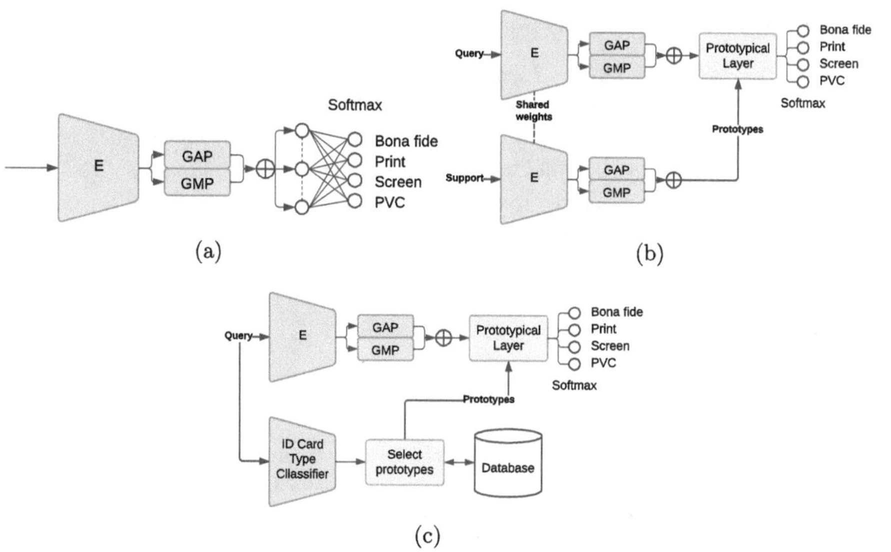

Fig. 1. Diagrams for baseline method (a), Prototypical Network train stage (b), and Prototypical Network inference stage (c).

3.1 Training Stage

Our approach proposes improving the number of images used in the previously proposed method by leveraging *Few-Shot* training. We use episodic training. An episodic task is a single, self-contained training task with its own start and end, such as a game match or a specific classification problem. Train a model to recognize these particular classes using only the support set. To test, evaluate the model on new examples (query set) of those same classes, then create new episodes with different classes and examples, and repeat the process, allowing the model to learn how to learn across many other, small-scale challenges.

In our case, each batch comprises the same number of samples per class (bona fide, print, screen, and PVC) but includes only a single ID card type. This is similar to classic Few-shot episodic training; however, *instead of changing the classes in each episode, we change the domain.* This enhances the Prototypical Network's ability to transfer across domains simply by changing the *prototypes.* The Prototypical head uses the *Euclidean distance* to compute the *logits* function. The Fig. 1b shows the diagram of the PN training process.

3.2 Inference Stage

During inference, we rely on an ID card type classifier to select the prototypes for the current sample. To work with a new ID card type, we will need samples of each class to build the prototypes: 4 bona fide and their respective attacks. These prototypes can be precomputed and stored in a database to be fetched on demand. Figure 1c shows the inference process.

3.3 Configuration and Hyperparameters

We used a backbone based on *EfficientNet-V2-b0* as a SoTA [7]. Therefore, we had an embedding of size $1 \times 2,560$, as shown in Fig. 1.

We used the unaligned card crop, resized to 224×224 as input. A batch size of 84 *query samples* (21 samples per class) and 16 *support samples* (4 samples per class). We used 4 class-output classifiers during training and validation with the *softmax* activation. However, we used the bona fide class as a binary classifier during inference. Regarding hyperparameters, we used *AdamW* as the optimizer with a learning rate of $5e-4$. We trained for up to 100 pseudo-epochs of 200 steps, with early stopping over the last 10 epochs.

3.4 Databases and Metrics

The databases used to evaluate and test the proposed methodology are described below. Two databases were used: a private database to enable extensibility to production systems and a public database to replicate the results of this work. For this purpose, to evaluate the performance of the proposed PAD method, the metrics described in ISO/IEC 30107–3[1] are to be used. The Equal Error Rate

[1] https://www.iso.org/standard/79520.html.

(EER), Bona fide Presentation Classification Error Rate (BPCER), and Attack Presentation Classification Error Rate (APCER). To further evaluate performance under fixed security constraints, the metric BPCER at fixed APCER ($BPCER_{AP}$) is employed.

The BPCER measures the proportion of bona fide presentations that are wrongly classified as attacks. The following expression defines it:

$$BPCER = \frac{1}{N_{BF}} \sum_{i=1}^{N_{BF}} RES_i, \tag{1}$$

where N_{BF} is the number of bona fide presentation samples, and $RES_i = 1$ if the sample is incorrectly flagged as an attack.

The APCER is defined as:

$$APCER_{PAIS} = 1 - \frac{1}{N_{PAIS}} \sum_{i=1}^{N_{PAIS}} RES_i, \tag{2}$$

where N_{PAIS} represents the total number of attack images for a given PAIS, and $RES_i = 1$ if the $i-th$ image is correctly identified as an attack, or 0 if it is misclassified as bona fide.

Private Dataset. A private database with different ID card types and bona fide, paper print, screen replay, and PVC print classes is used to test our approach. Table 1 shows the different ID card types and the number of users of each one. As seen there, the Chilean ID card type *CHL-2* has significantly more users than the others (Fig. 2).

(a) Example images from CHL-1, CHL-2 and Guatemala

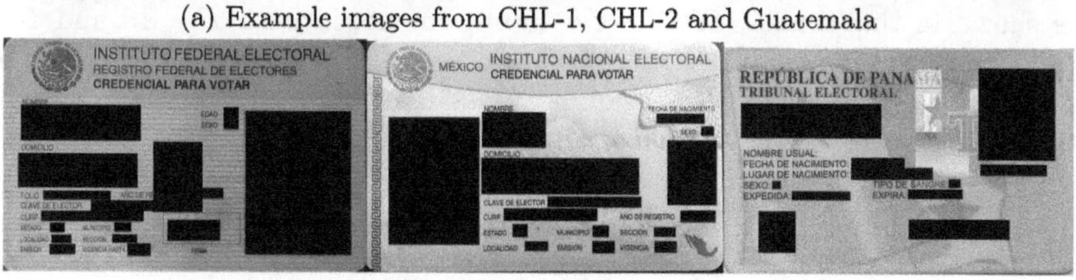

(b) Example images from Mexico-1, Mexico-2 and Panama.

Fig. 2. Example of images from a private dataset.

Public Dataset DLC-2021. To ensure reproducibility of the proposed experiments, we have replicated the experiments using a publicly available dataset. The chosen dataset is DLC-2021 [5], which contains synthetic samples of different ID card types and passports with their corresponding screen and print attacks. We used only the ID card samples from Albania *(ALB)*, Spain *(ESP)*, Estonia *(EST)*, Finland *(FIN)* and Slovakia *(SVK)*.

Table 1. Number of samples per ID card type.

	PAN	COL	CHL-1	CHL-2	MEX-1	MEX-2	GTM
Bona fide	1,006	80	696	13,000	353	660	1,001
Print	1,006	80	696	13,000	353	660	1,001
Screen	1,006	80	696	13,000	353	660	1,001
PVC	506	80	259	6,500	353	160	501

4 Experiments and Results

4.1 Experiment 0: CLIP Zero-Shot

Firstly, we evaluated *CLIP Zero-Shot* capabilities to compare them with our approach. We performed some evaluations using only the ID card for most samples, which is *CHL-2*, to adjust the text prompts and determine which *CLIP* backbone yields better results. The best performing configuration was used for comparison with our approach in both experiments.

We tested the backbone *CLIP-B16* and *CLIP-L14* from *OpenAI*, following these simple text prompts:

- *Bona fide: "a photo of an ID card"*
- *Print: "a printed photo of an ID card"*
- *Screen: "a photo of an ID card shown on a screen device",*
- *PVC: "a photo of an ID card printed on a PVC card",*

Because a PVC-printed ID card might be mistaken for the class-printed one, we also tested without that text class to see whether the results improved. Results are shown in the Table 2. We observe improvements in both $B-16$ and $L-14$ backbones when only 3 text classes are used. The backbone $B-16$ also outperforms the backbone $L-14$ on 3 classes. Therefore, we used $B-16$ with only 3 text classes as the *Zero-Shot* approach.

Table 2. Experiment 0: CLIP test with CHL-2 in private dataset.

Model/Classes	EER%	BPCER10	BPCER20	BPCER100
CLIP-B16/4C	30.53	66.66	81.63	95.24
CLIP-L14/4C	30.86	65.31	78.44	94.21
CLIP-B16/3C	**20.71**	**41.52**	**60.14**	88.06
CLIP-L14/3C	27.07	51.33	65.63	**86.33**

4.2 Experiment 1: Single Domain Training

In this experiment, we used the most common ID card type, *CHL-2*, with most samples to train both the baseline and prototypical networks. In this case, the episodic training consists of keeping the same classes and the same ID card type, but changing the samples of the support set in each episode. The goal of this experiment is to assess the generalization capabilities to unseen ID Card types when only a single ID card type is available during training. This is the most challenging case.

The results shown in Table 3 indicate that the baseline does not generalize to unseen ID card types. However, the Prototypical model shows better generalization. The best-performing ID card type for the baseline is *PAN*, which achieves an EER of 12.7% and a *BPCER10* of 17.5% while the prototypical reaches 8.44% and 7.06%, respectively. On average, the baseline achieves an EER of 26.56% and a *BPCER10* of 45.69%, while the prototypical achieves 9.07% and 10.15%, respectively, which is substantially better.

Comparing both to *Zero-Shot CLIP-B16*, we can observe that *CLIP* performs better than the baseline but worse than the prototypical approach.

Table 3. Experiment 1 on the private dataset in (%).

	EER%	BPCER10	BPCER20	BPCER100
	Baseline			
PAN	12.70	17.50	31.11	56.96
COL	38.54	66.25	75.00	92.5
CHL-1	21.35	36.64	40.80	71.84
MEX-1	56.80	100.00	100.00	100.00
MEX-2	56.80	100.00	100.00	100.00
GTM	16.73	33.87	48.55	65.83
AVG	26.56	45.69	53.79	75.96
STD	17.62	31.80	28.28	16.65
	Zero-Shot CLIP-B16			
PAN	18.31	27.53	42.94	66.30
COL	22.29	38.75	47.5	78.75
CHL-1	14.79	22.27	36.21	65.80
MEX-1	8.17	7.08	13.31	34.28
MEX-2	17.37	28.33	44.70	77.58
GTM	11.04	12.19	23.88	65.03
AVG	15.33	22.69	34.76	64.62
STD	5.13	11.55	13.48	16.08
	Prototypical Network			
PAN	8.44	7.06	14.21	36.18
COL	1.25	0.00	0.00	1.25
CHL-1	12.93	19.68	36.49	72.99
MEX-1	10.76	12.75	25.21	58.92
MEX-2	11.37	12.42	25.76	63.18
GTM	9.69	8.99	24.28	54.95
AVG	9.07	10.15	20.99	47.91
STD	4.12	6.58	12.48	25.88

4.3 Experiment 2: Multiple Domain Training

In this experiment, we performed Leave-One-Out (LOO) cross-validation on *CHL-2*, *PAN*, *COL*, and *MEX-1*, in which we trained both the baseline and the prototypical model using $N-1$ ID card types and held out the remaining ID card type for testing. In this case, episodic training keeps the same classes while varying the ID Card type. The goal of this experiment is to assess the generalization capabilities when multiple ID card types are available during training.

Observing the results in Table 4, we can see that the advantage of the prototypical is reduced. In this experiment, the best-performing ID card type for the baseline is *PAN*, which achieves an EER of 3.06% and a *BPCER10* of 0.6%, surpassing the prototypical, which achieves only 5.67% and 1.49%, respectively. However, on average, the prototypical outperforms the baseline: the baseline achieves an EER of 11.98% and a *BPCER10* of 16.95%, while the prototypical achieves 7.41% and 5.88% respectively.

Comparing with the *Zero-Shot CLIP-B16*, we can observe that both the baseline and the prototypical now perform better. The only exception is in *MEX-1*, where *CLIP* outperforms both of them.

Table 4. Experiment 2 on the private dataset in (%).

	EER%	BPCER10	BPCER20	BPCER100
	Baseline			
PAN	3.06	0.60	1.69	6.96
COL	14.17	20.00	36.25	66.25
CHL-2	11.95	14.91	29.34	63.37
MEX-1	18.74	32.29	45.04	81.87
AVG	11.98	16.95	28.08	54.61
STD	6.58	13.12	18.73	32.79
	Zero-Shot CLIP-B16			
PAN	18.31	27.53	42.94	66.30
COL	22.29	38.75	47.5	78.75
CHL2	20.71	41.52	60.14	88.06
MEX1	8.17	7.08	13.31	34.28
AVG	17.37	28.72	40.97	66.85
STD	6.35	15.64	19.82	23.47
	Prototypical Network			
PAN	5.67	1.49	6.46	25.65
COL	3.54	0.00	0.00	12.50
CHL-2	11.86	16.34	39.31	81.10
MEX-1	8.55	5.67	19.55	63.46
AVG	7.41	5.88	16.33	45.68
STD	3.61	7.38	17.34	32.00

4.4 Experiments with the Public Dataset DLC-2021

Now, the same experiments are replicated using the *DLC-2021* dataset.

Experiment 0. For *CLIP*, we tested the same backbones with *ESP* ID cards using only 3-class text prompts because, in *DLC-2021*, the bona fide is a PVC-printed ID card. We tested two sets of prompts: one with the bona fide prompt as the bona fide class, and another with the PVC prompt as the bona fide class. The results in Table 5 show that using the bona fide prompt (bf-prompt) produces the worst results compared to using the bona fide PVC-prompt, especially in *BPCER100*. The *CLIP-B16* backbone is the best in *BPCER100*; however, in all cases, performance is quite low compared to the results obtained on the private dataset. This is because the bona fide samples of *DLC-2021* are PVC-printed ID cards, which are often confused with the Print class text prompt. We chose the *CLIP-B16* backbone using the PVC prompt to compare with our approach.

Table 5. CLIP tests (%) with ESP ID cards in DLC-2021.

	EER%	BPCER10	BPCER20	BPCER100
B16-bf-Prompt	49.37	93.39	97.82	100.00
L14-bf-Prompt	38.05	89.99	95.96	99.42
B16-bf-PVC	41.81	81.33	90.64	96.92
L14-bf-PVC	42.67	89.16	96.47	99.17

Experiment 1. Replicating experiment 1 using the *DLC-2021* database, the results are similar to those obtained with our private dataset, as shown in Table 6. In Experiment 1, the baseline does not generalize to unseen ID cards, with an average EER of 27.8%, while the prototypical achieves an average EER of 10.53%.

Table 6. Experiment 1 on the DLC-2021 dataset in (%).

	EER%	BPCER10	BPCER20	BPCER100
	Baseline			
ALB	26.91	53.43	70.32	85.92
EST	21.34	35.65	44.32	68.44
FIN	27.12	53.22	64.15	82.57
SVK	35.84	71.78	83.96	94.13
AVG	27.8	53.77	65.69	82.77
STD	5.99	15.17	16.48	10.71
	Zero-Shot CLIP-B16			
ALB	38.13	76.65	88.54	98.35
EST	41.96	87.18	94.14	98.90
FIN	39.71	75.70	83.51	93.63
SVK	36.66	84.15	91.59	98.90
AVG	39.12	80.92	89.45	97.45
STD	2.27	5.63	4.57	2.56
	Prototypical Network			
ALB	13.4	18.18	33.6	77.85
EST	5.98	0.92	8.24	35.59
FIN	9.99	9.74	25.8	51.34
SVK	12.73	17.35	31.77	89.7
AVG	10.53	11.55	24.85	63.62
STD	3.37	8.04	11.56	24.62

Experiment 2. In Experiment 2 and Table 7, the advantage of the Prototypical is reduced; the average EER is 10.27% for the baseline and 9.12% for the prototypical. In this case, the advantage of the Prototypical is lost when setting the threshold to *BPCER100*, where the baseline achieves an average of 45.15% while the prototypical reaches 61.6%.

Comparing results with *CLIP-B16*, we can observe that *CLIP* does not perform well with *DLC-2021*. Both the baseline and the prototypical models significantly outperform *CLIP*, which achieves an average EER of 39.65%.

Table 7. Experiment 2 on the DLC-2021 dataset in (%).

	EER%	BPCER10	BPCER20	BPCER100
	Baseline			
ALB	7.97	6.92	12.67	47.43
ESP	5.29	1.78	5.47	29.09
EST	2.95	0.31	0.67	7.88
FIN	20.10	36.41	49.34	82.26
SVK	15.05	21.16	33.52	59.11
AVG	10.27	13.32	20.34	45.15
STD	7.13	15.32	20.5	28.39
	Zero-Shot CLIP-B16			
ALB	38.13	76.65	88.54	98.35
ESP	41.81	81.33	90.64	96.62
EST	41.96	87.18	94.14	98.90
FIN	39.71	75.70	83.51	93.63
SVK	36.66	84.15	91.59	98.90
AVG	39.65	81.00	89.68	97.34
STD	2.31	4.88	3.99	2.23
	Prototypical Network			
ALB	10.28	10.34	16.95	84.33
ESP	4.55	0.43	3.26	74.05
EST	5.37	2.93	6.04	31.99
FIN	8.87	7.18	16.43	42.91
SVK	16.54	24.84	41.76	74.72
AVG	9.12	9.15	16.89	61.6
STD	4.78	9.57	15.18	22.75

5 Conclusion and Future Work

We have shown that it is possible to generalize to unseen ID card types using only *four bona fide users and their respective attacks.* The experiments show that our approach outperforms the baseline considerably when only one ID card type is available during training. However, in Experiment 2, where more ID card types are available during training, this advantage is reduced. Even though the Prototypical still outperforms the baseline. The need for an ID card type classifier can sometimes be problematic. Therefore, we recommend using our Prototypical approach when only a few ID card samples (four) are available during training; when there are sufficient ID card types, the baseline's generalization capabilities are adequate, so there is no clear advantage to using a more complex architecture such as the Prototypical.

One of the main drawbacks of this approach is the need for bona fide and attack samples. To address this issue, we will use an anomaly detection approach in which we train the prototypical model to determine whether a sample is similar to the bona fide prototype. This could reduce the samples required to only the bona fides of the target ID card. Furthermore, we will investigate the possibility of using only specimen samples as a prototype, so we will not even need any bona fide users.

Acknowledgements. This research has been partially funded by Facephi, R&D department, and the European Union's Horizon research and innovation program under grant agreements 101121280 (EINSTEIN) and the German Federal Ministry of Education and Research, and the Hessian Ministry of Higher Education, Research, Science, and the Arts, which jointly support the National Research Center for Applied Cybersecurity ATHENE.

References

1. Arlazarov, V., Bulatov, K., Chernov, T., Arlazarov, V.: MIDV-500: a dataset for identity document analysis and recognition on mobile devices in video stream. Comput. Optics **43**(5) (2019). https://doi.org/10.18287/2412-6179-2019-43-5-818-824
2. Bulatov, K.B., et al.: MIDV-2020: a comprehensive benchmark dataset for identity document analysis. Comput. Opt. **46**(2), 252–270 (2022)
3. Dosovitskiy, A., et al.: An image is worth 16x16 Words: transformers for image recognition at scale. In: 9th International Conference on Learning Representations, ICLR 2021, Austria, 3-7 May . OpenReview.net (2021)
4. Koliaskina, L.I., et al.: MIDV-Holo: a dataset for id document hologram detection in a video stream. In: Fink, G.A., Jain, R., Kise, K., Zanibbi, R. (eds.) Document Analysis and Recognition - ICDAR 2023, pp. 486–503. Springer Nature Switzerland, Cham (2023)
5. Polevoy, D.V., et al.: Document liveness challenge dataset (dlc-2021). J. Imaging **8**(7) (2022).https://doi.org/10.3390/jimaging8070181. https://www.mdpi.com/2313-433X/8/7/181
6. Radford, A., et al.: Learning transferable visual models from natural language supervision. In: Meila, M., Zhang, T. (eds.) Proceedings of the 38th International Conference on Machine Learning. In: Proceedings of Machine Learning Research, vol. 139, pp. 8748–8763. PMLR (2021). https://proceedings.mlr.press/v139/radford21a.html
7. Sanchez, A., Espín, J.M., Tapia, J.E.: Few-shot learning: Expanding ID cards presentation attack detection to unknown ID countries. In: 2024 IEEE International Joint Conference on Biometrics (IJCB), pp. 1–9 (2024). https://doi.org/10.1109/IJCB62174.2024.10744501
8. Snell, J., Swersky, K., Zemel, R.: Prototypical networks for few-shot learning. In: Proceedings of the 31st International Conference on Neural Information Processing Systems, pp. 4080–4090. NIPS'17, Curran Associates Inc., Red Hook, NY, USA (2017)

9. Song, Y., Wang, T., Cai, P., Mondal, S.K., Sahoo, J.P.: A comprehensive survey of few-shot learning: Evolution, applications, challenges, and opportunities. ACM Comput. Surv. **55**(13s) (2023). https://doi.org/10.1145/3582688
10. Sung, F., Yang, Y., Zhang, L., Xiang, T., Torr, P.H., Hospedales, T.M.: Learning to compare: relation network for few-shot learning. In: IEEE/CVF Conference on Computer Vision and Pattern Recognition, pp. 1199–1208 (2018). https://doi.org/10.1109/CVPR.2018.00131
11. Tan, M., Le, Q.: Efficientnetv2: smaller models and faster training. In: Meila, M., Zhang, T. (eds.) Proceedings of the 38th International Conference on Machine Learning. Proceedings of Machine Learning Research, vol. 139, pp. 10096–10106. PMLR (2021). https://proceedings.mlr.press/v139/tan21a.html
12. Tapia, J.E., et al.: First competition on presentation attack detection on ID card. In: 2024 IEEE International Joint Conference on Biometrics (IJCB), pp. 1–10 (2024).https://doi.org/10.1109/IJCB62174.2024.10744475
13. Tapia, J.E., et al.: Second competition on presentation attack detection on ID card (2025). https://doi.org/10.1109/IJCB65343.2025.11411073
14. Tapia, J.E., Stockhardt, F., González-Soler, L.J., Busch, C.: Syn-idpass: passport synthetic dataset for presentation attack detection. In: 2025 IEEE International Joint Conference on Biometrics (IJCB), pp. 1–9 (2025). https://doi.org/10.1109/IJCB65343.2025.11411263

PeopleGator: A Multimodal Benchmark for Person Identification in Historical Documents

Marek Vaško(✉), Martin Kišš, Michal Hradiš, and Adam Herout

Graph@FIT, Faculty of Information Technology, Brno University of Technology, Brno, Czech Republic
{ivasko,ikiss,hradis,herout}@fit.vut.cz

Abstract. We introduce PeopleGator, a publicly available multimodal benchmark for person identification in digitized historical documents that links detected face crops with person names mentioned in nearby OCR text via spatial grounding and LLM-based name normalization. The benchmark contains 2,996 pages from 2,304 documents sourced from 16 libraries, with 3,154 named face instances covering 589 unique identities. To support systematic comparison, we define evaluation protocols for face clustering, image-to-image retrieval, and text-to-image entity retrieval under realistic archival noise. Baselines using modern face-embedding models show that hierarchical clustering can be effective, while multimodal retrieval based on page–level face–name proximity remains challenging compared to unimodal image retrieval.

Keywords: Multimodal benchmark · Person identification · Digitized historical documents · Historical document analysis · Face clustering · Text-to-image retrieval

1 Introduction

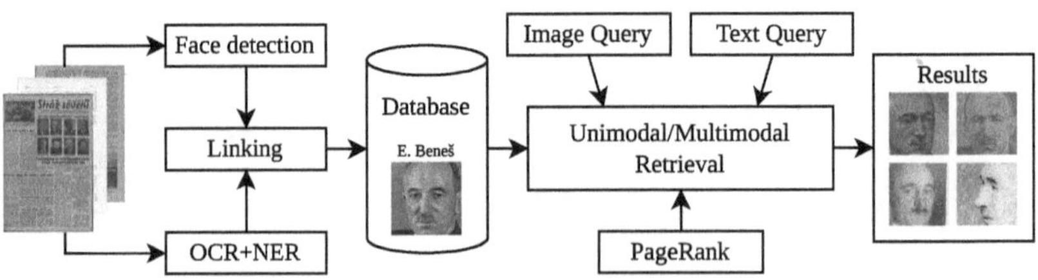

Fig. 1. Overview of the dataset construction and benchmark pipeline.

One of the goals in digital libraries and archives is to make people discoverable in historical documents – both by typing a name and by searching with

F. Shafait et al. (Eds.): DAS 2026, LNCS 16851, pp. 178–194, 2027.
https://doi.org/10.1007/978-3-032-36207-0_11

a photograph. Linking faces on scanned pages to person names in OCR text is a core requirement for person-centric search in large historical collections, but the setting is inherently noisy: the pages contain degradation artifacts (e.g., low contrast, skew, bleed-through, noise), OCR and detector failures, and only weak, static page-level grounding between a face and the nearby text.

We present PeopleGator, a publicly available multimodal benchmark built from a large corpus of historical print media with page scans, OCR transcripts with geometry, detected faces, and manually refined face–name identity clusters spanning documents (Fig. 1). Our dataset construction pipeline uses spatial association between faces and nearby text lines, LLM-based extraction and normalization of person names, and multi-annotator verification that yields a consensus-based evaluation dataset for unambiguous evaluation. On top of the dataset, we define reproducible benchmark tasks and protocols covering face clustering, image-to-image retrieval, and text-to-image entity retrieval, and we provide baseline systems using modern face-embedding models and a graph-based multimodal retrieval approach. The contributions of this work are threefold:

1. We present a new multimodal dataset in which we provide extracted named entities from historical documents and, for each named entity, a set of face images corresponding to that identity. Dataset and source code is available on Zenodo.
2. We define a suite of benchmark tasks and evaluation protocols that support both image-only and multimodal person identification scenarios.
3. We provide baseline methods and experimental results that highlight the challenges of the benchmark and establish a foundation for future work in multimodal document understanding.

2 Related Work

A survey by Ehrmann et al. [12] catalogs the challenges of NER on historical texts, namely, OCR noise, diverse scripts, and domain shift pose. Recent work shows that multimodal LLMs can jointly perform OCR, post-correction, and NER on archival sources with sub-1 % character error rate [15], and large-scale pipelines such as American Stories [10] extract structured article texts from nearly 20 M historical newspaper scans. Entity linking grounds textual mentions to knowledge bases via dense retrieval [46] or constrained generation [8,9]; multimodal NER adds visual context through noise-aware [5] and relevance-guided [2] fusion. Multimodal entity linking combines text, images, and knowledge graphs [13,23,45], with multi-grained interaction networks [31] by jointly modeling intra-modal and cross-modal feature alignment. These methods assume caption-level or article-level text–image correspondence; face–name alignment in complex historical newspaper layouts remains unaddressed. In other sources, face-name association exploits temporal continuity, such as in broadcast video [6,14,39], graph-based identity clustering from photos [1], and large-scale captioning datasets [4,41] with weakly supervised CLIP-based naming [37].

Document Understanding. Single-page document analysis benchmarks emphasize layout parsing and multimodal reasoning [7,27,30,35], retrieval [16], and OCR quality [50]. Layout-aware transformers from the LayoutLM family [20, 47,48] unify text, vision, and spatial embeddings, while self-supervised pretraining [29] and OCR-free models [22,28] bypass text extraction entirely. Existing benchmarks focus mainly on forms, key-value extraction, and QA rather than cross-modal identity disambiguation. From a graph perspective, meta-path similarity [40] provides a principled framework for typed-entity retrieval in heterogeneous networks, and the Heterogeneous Graph Transformer [19] introduces node and edge-type-dependent attention for web-scale heterogeneous graphs. PeopleGator targets person-centric identity extraction under weak face–name grounding, which can combine layout reasoning with graph-based association in a regime not addressed by current document analysis or multimodal retrieval benchmarks.

Face Recognition. Transformer backbones [25], trained with quality-adaptive margin-based loss function [24,32], on web-scale datasets [17,52], present state of the art. Yet, large-scale datasets primarily focus on contemporary photographs. ArtFace [36] shows that domain adaptation closes the gap on historical portraits, while other methods [43] can offer photo style-transfer to the contemporary domain. Archival Faces [42] introduces a face-detection benchmark for digitized historical documents. For identity clustering, embedding-graph methods such as Chinese Whispers [3] and FINCH [38] remain popular, while supervised linkage prediction with transformers [49] can further improve purity. The face clustering and cross-modal retrieval against OCR-derived person names under archival degradation remain uncovered; PeopleGator targets exactly this gap.

3 Dataset Construction

We construct a large-scale benchmark for person-centric face retrieval from historical print media. The dataset is derived from approximately 9 million digitized pages of historical newspapers and books, mainly sourced from Czech libraries. The corpus consists of scanned page images accompanied by OCR transcriptions with geometric layout information. The transcriptions were obtained using our publicly available OCR system[1] which achieves less than 0.5% character error rate (CER) on printed documents. For each page, detected faces, along with layout information, are provided. Due to the historical nature of the material, the data exhibits realistic archival artifacts such as degradation, low contrast, skew, bleed-through, noise, and detector failures.

Face Detection. We employ a two-stage pipeline: first, non-text objects are detected using YOLO11 [21] trained on AnnoPage [26], then a specialized YOLO11 trained on Archival Faces [42] extracts facial keypoints. Outputs include bounding boxes, landmarks, and aligned crops in page-coordinate space.

[1] ScribbleSense.

Face–Text Association and Name Extraction. Faces are associated with identity cues via spatial proximity: for each face, the two OCR text lines above and below it within its horizontal bounds are selected as candidates. We then use GPT-4o to extract and normalize complete person names in canonical form. This four-line window serves as a simple caption filter, favoring in-caption mentions and reducing the text required for named-entity processing, making this low recall but high precision extraction.

Identity Clustering. For each canonical name, all associated face instances and textual evidence are aggregated into an identity cluster spanning multiple pages and documents, enabling cross-document identity linking.

Manual Refinement. Four independent annotators refined the faces assigned automatically to clusters following written guidelines with three actions: *accept* (unambiguous face–name match, checked with page content), *reject* (misdetection, wrong person, or insufficient evidence), and *expand* (find additional instances via embedding-space nearest neighbors with ViT-Small, see Sect. 5). A stratified 20% subset (119 identities, covering mainly the most frequently occurring people) was annotated by all four annotators to form the test set (mean pairwise IoU: 70.1%, Fig. 2); remaining identities were annotated by at most one annotator as the development set.

Dataset Structure. The benchmark provides face bounding boxes and crops, normalized names, identity clusters with supporting OCR text, original document references with page coordinates, and a multi-annotator agreement subset.

3.1 Dataset Statistics

Our dataset contains scans from 16 libraries, covering 2,304 documents and 2,996 pages. Out of these documents 77% are periodicals and 23% are books. The distribution of pages across years is shown in Fig. 2 (bottom), with a major peak around the 1930s. We provide 3,154 unique detected faces assigned to a person's name. These assigned faces span 2,722 pages and 2,091 documents, covering 589 distinct names. On average, this corresponds to 1.30 pages per document and 1.16 named faces per annotated page. The annotation was performed by 4 annotators; 1,251 faces (39.7%) were labeled by at least two annotators. The average IoU between annotators for faces with at least two non-empty annotation sets is 70.1%. Most identities are rare: 34.3% of names appear only once, while the most frequent identity (T. G. Masaryk) has 158 faces. From a geometric perspective, the typical face crop has a median size of 266×367 pixels and covers approximately 0.98% of the page area (median). Figure 2 (right) shows the detected and annotated faces across different Czech libraries. Stacked bars distinguish between faces annotated at least once, faces annotated by multiple annotators, and not annotated faces.

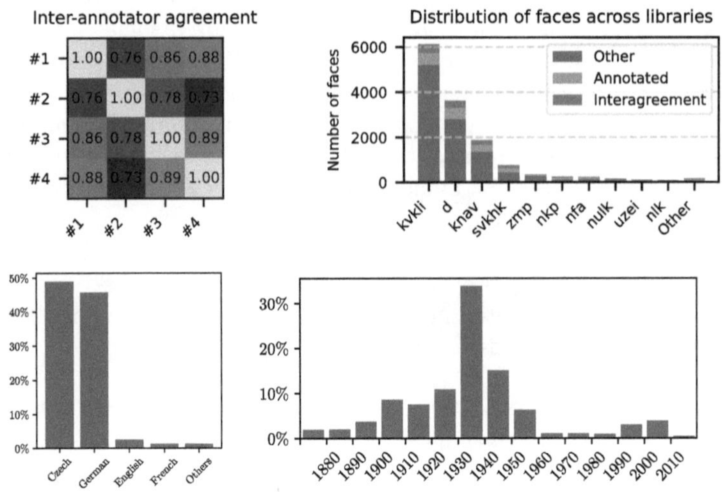

Fig. 2. Dataset statistics and annotation consistency. **Top-left:** Inter-annotator agreement as the mean (IoU) across identities. **Top-right:** Distribution of faces across libraries, also contains single-annotator and multiple-annotator annotation counts. **Bottom-left:** Distribution of pages across the years. **Bottom-right:** Portion of pages containing different languages.

3.2 Ground-Truth Representation

Each face is associated with a specific identity. Identities annotated by all four annotators form the test set; the remainder is the development set, no identity is shared across splits. Because annotators do not always agree on assignments, a single identity can have different sets of faces; we therefore take the union of all annotations per identity as ground truth.

4 Benchmark Tasks

The benchmark distinguishes between image-only tasks, which rely exclusively on visual information, and multimodal tasks, which combine face images with textual identity cues.

4.1 Face Image Clustering

Input data is provided as page scans and detected face locations, together with a unique identifier for each detection. The objective is to estimate identity clusters for each person represented in the dataset. Known identity clusters from the development set may be used in any manner, while test-set identities are used only for evaluation.

Clustering Metrics. Four standard clustering metrics are reported: Adjusted Rand Index (ARI), Homogeneity, Completeness, and V-measure. Agreement between predicted clusters and ground-truth labels is quantified by ARI while correcting for chance; given the class imbalance and the large scale of background images, ARI is used as the primary reported metric. For all metrics, evaluation is performed *only* on labeled faces; clusters formed from the unlabeled portion of the dataset are ignored.

Evaluation against human baseline. A leave-one-annotator-out protocol is adopted: identity assignments produced by one annotator are treated as predictions, while the union of labels from the other three annotators is used as the reference. Performance is evaluated on the test split using the same metrics, results are averaged across held-out annotators, and a single score per metric is reported, enabling comparison to expected human performance.

4.2 Retrieval Evaluation

We evaluate retrieval performance on two benchmark tasks: (1) image-to-image face retrieval and (2) text-to-image face retrieval. From the perspective of metric computation, both tasks are equivalent, as evaluation depends solely on the relevance of the retrieved face images to the query. Evaluated system results of each query (image or text) include a ranked list of candidate faces with scores.

Ranking Metrics. For the evaluation of retrieval systems, we calculate Normalized Discounted Cumulative Gain (nDCG@k), Mean Reciprocal Rank (MRR@k), Precision@k, Recall@k, and Mean Average Precision (mAP@k). The reported metrics are the same for image-based and text-based retrieval tasks. The main reported metric is Normalized Discounted Cumulative Gain at $k = 10$, following the standard evaluation metrics used for information retrieval [33].

Retrieval Ground-truth Construction. The ground truth for the retrieval tasks consists of queries paired with corresponding sets of relevant images. For the test split of the text-based retrieval task, we selected the test identities and unified the image annotations provided by all annotators. In contrast, for the development set, the original annotations were used without modification, as each identity had been annotated by a single annotator only. This resulted in 119 and 465 queries for the test and development sets, respectively. For the image-based retrieval task, we use the same ground-truth image sets as for the text-to-image retrieval. For each identity, we randomly selected up to 3 relevant images as query images. For a specific query, the remaining images associated with the same identity were treated as relevant targets. For identities containing only two relevant images, queries were created from both images, whereas identities with a single relevant image were treated as backgrounds. This process yielded 245 and 798 query images for the test and development sets, respectively.

4.3 Evaluation Protocol

All methods must operate under the specified conditions to allow comparison with baselines, including: using only the provided fixed set of detected faces; no use of external identity labels or external face datasets containing the same identities; and no manual post-processing or identity-specific heuristics.

For clustering tasks, evaluation is performed on the full test set without revealing the number of identities. This information should not be propagated to any evaluated method in any way.

For retrieval tasks, queries and relevant images are constructed from the test identities and provided in the dataset. The use of external same-person galleries, person-name web searches, or retrieval using images not contained in the dataset is not allowed. Each retrieval evaluation must be accompanied by 0.95 confidence intervals, computed via bootstrapping with at least 10 000 resamples of the queries.

Evaluation scripts, metric implementations, and reference baseline results are released alongside the dataset to facilitate transparent and reproducible benchmarking.

5 Baseline Methods

This section describes the baseline systems provided with the benchmark. Our goal is not to exhaustively optimize performance, but to establish reproducible reference points for multimodal person retrieval and identity clustering in historical document collections.

5.1 Feature Extraction

Face Embeddings. For visual person representation, we rely exclusively on pre-trained face recognition models. We evaluate multiple state-of-the-art (SOTA) descriptor networks trained on large-scale face datasets, including WebFace-260M [52] and MS1Mv3 [17]. Architecturally, we consider both convolutional and transformer-based backbones, including variants of ResNet and Vision Transformer (ViT) models. For each detected and aligned face crop, we extract an embedding vector. We utilize pre-trained models provided by InsightFace[2] (models Buffalo L and Antilope V2), both are trained using the ArcFace [11] objective, Adaface [24] iResnet101 model pre-trained on WebFace14M[3] and ViT Small pre-trained on MS1Mv3 using CosFace [44] objective[4]. These models sample diverse training objectives and different architectures.

---

[2] InsightFace Python Package.

[3] CVLFace AdaFace model on HuggingFace.

[4] ViT Small model on HuggingFace.

Named Entity Extraction. Named entities are extracted from the page transcriptions using GPT 5.2. The model is prompted to return normalized, non-truncated person names whenever possible. The input to the model is a CSV file containing transcription lines and corresponding line indices. For each unique person name, the model returns the list of line indices where that name is mentioned.

Name Embeddings. For each extracted name, we compute a dense text embedding using Gemma 3-based KaLM Embedding [18,51] model, which demonstrates strong multilingual retrieval performance on the Massive Text Embedding Benchmark [33]. All name embeddings are ℓ_2-normalized before similarity computation.

Identity Clustering Baselines. Clustering is performed over face embeddings only, reflecting the common unsupervised identity discovery setting. We evaluate the following clustering algorithms as implemented in scikit-learn, k-means clustering, Agglomerative clustering, and DBSCAN. For each method, hyperparameters are tuned on the development split.

5.2 Image-Only Retrieval Baseline

Our baseline method for the image-only retrieval task is implemented as nearest-neighbor search in embedding space. Given a query image, we calculate the cosine similarity between the embedding of that image and the rest of the images in the dataset. The computed similarities are then used as the final scores for the query.

5.3 Multimodal Retrieval Baseline

We propose a graph-based multimodal retrieval baseline integrating visual and textual signals.

Graph Construction. We construct three similarity matrices face–face similarity matrix (cosine similarity of face embeddings) $A_{f\leftrightarrow f}$, name–name similarity matrix (cosine similarity of name embeddings) $A_{n\leftrightarrow n}$, and face–name adjacency matrix $A_{f\leftrightarrow n}$. For the face–name adjacency matrix, we consider three construction strategies: (1) normalized spatial closeness on the page (denoted as *distance* strategy), (2) equal assignment of all names to all faces (denoted as *occurrence* strategy), and (3) per-face uniform weights inversely proportional to the number of names detected on the page (denoted as *uniform* strategy). In the first strategy, if multiple instances of the same name appear on a page, only the closest instance to a given face is considered. Each of the three matrices is sparsified, and together they are merged into a single adjacency matrix

$$A = \begin{pmatrix} A_{f\leftrightarrow f} & A_{f\leftrightarrow n} \\ A_{f\leftrightarrow n}^T & A_{n\leftrightarrow n} \end{pmatrix}. \tag{1}$$

Retrieval via PageRank. Our multimodal retrieval baseline is built upon the Personalized PageRank [34] algorithm[5], which enables relevance propagation over a heterogeneous graph structure. Given a graph represented by an adjacency matrix A and an initial distribution over nodes, which is derived from the query, PageRank returns a relevance ranking for all nodes of the graph. In our setting, w is initialized as a one-hot vector whose non-zero entry corresponds to the query node. Retrieval results are then obtained by selecting the scores associated with face nodes, which are interpreted as the similarity between the query and candidate images.

Although the graph is primarily constructed to support multimodal text-to-image retrieval, the same framework naturally enables unimodal image-to-image retrieval. For the text-to-image retrieval, the non-zero value is set to the corresponding text node, while in the image-to-image retrieval, the entry corresponds to an image node. Otherwise, the process in both cases remains identical.

Table 1. Overview of best hyperparameters for clustering algorithms.

	Agglomerative		**KMeans**			**DBScan**	
	linkage	**#clusters**		**#init**	**init**	ϵ	**min samples**
Antelope V2	complete	1000	2000	10	random	16	1
Buffalo	complete	2000	1000	50	random	16	1
ViT Small	ward	1000	2000	20	random	1	1
iResNet101	ward	2000	1000	50	random	16	1

6 Experimental Results

To evaluate the performance of baseline approaches, we use the evaluation protocol from Sect. 4 and the baseline solutions described in Sect. 5.

6.1 Image Clustering Results

Clustering experiments were conducted using embeddings for all detected faces, computed with the models described in Sect. 5. Off-the-shelf clustering algorithms were applied to these feature vectors, and hyperparameters were tuned via a grid search with approximately log-spaced numerical values. For DBSCAN, the neighborhood radius (epsilon) was swept from 0.1 to 32 (10 values) and the minimum number of samples from 1 to 20 (6 values). For agglomerative clustering and k-means, the number of clusters was varied from 100 to 5000 (8 values). For agglomerative clustering, the linkage criterion was evaluated using average,

[5] We use implementation from scikit-network.

Table 2. Evaluation of clustering models against human baseline.

		ARI	Homogeneity	Completeness	V-measure
Agglo.	Antelope V2	**0.549**	0.936	**0.761**	**0.840**
	Buffalo L	**0.492**	**0.975**	**0.749**	**0.847**
	iResNet101	0.156	0.908	0.623	0.739
	ViT Small	0.201	0.888	0.636	0.742
KMeans	Antelope V2	0.219	0.924	0.661	0.771
	Buffalo L	0.401	0.898	0.738	0.810
	iResNet101	0.259	0.824	0.640	0.721
	ViT Small	0.151	0.925	0.613	0.737
DBScan	Antelope V2	0.067	**0.975**	0.571	0.720
	Buffalo L	0.063	**0.993**	0.573	0.727
	iResNet101	0.048	0.827	0.565	0.671
	ViT Small	0.108	0.782	0.609	0.685
Human baseline		*0.373*	*0.729*	*0.875*	*0.792*

complete, and Ward linkage; for k-means, the initialization method (k-means++ vs. random) and the number of initializations (10–50; 3 values) were varied. For each configuration, metrics were computed on the development set, converted to ranks, and averaged to select the best setting. The best configurations for each embedding model and clustering algorithm are reported in Table 1.

Using these hyperparameters, performance was evaluated on the test set and compared to the human baseline (Table 2). Under identical evaluation conditions, agglomerative clustering with Antelope V2 and Buffalo L achieved the strongest overall results, exceeding the human baseline in ARI and V-measure. DBSCAN produced near-perfect homogeneity but low completeness (identity splitting), whereas the human baseline achieved the highest completeness.

6.2 Image-Only Retrieval Results

We performed evaluation of our image-only retrieval method, which calculates cosine similarity between embeddings, with all the models on the image-based retrieval task. The results are visualized in Fig. 3. The results show that the differences between the evaluated models are negligible.

6.3 Text-Based Multimodal Retrieval Results

In the text-based multimodal task, we used our multimodal retrieval method which is based on the Personalized PageRank. First, we performed a grid search on the development set over a set of hyperparameters that modify the creation of the adjacency matrix A: sharpening factors for all three sub-matrices, sparsity

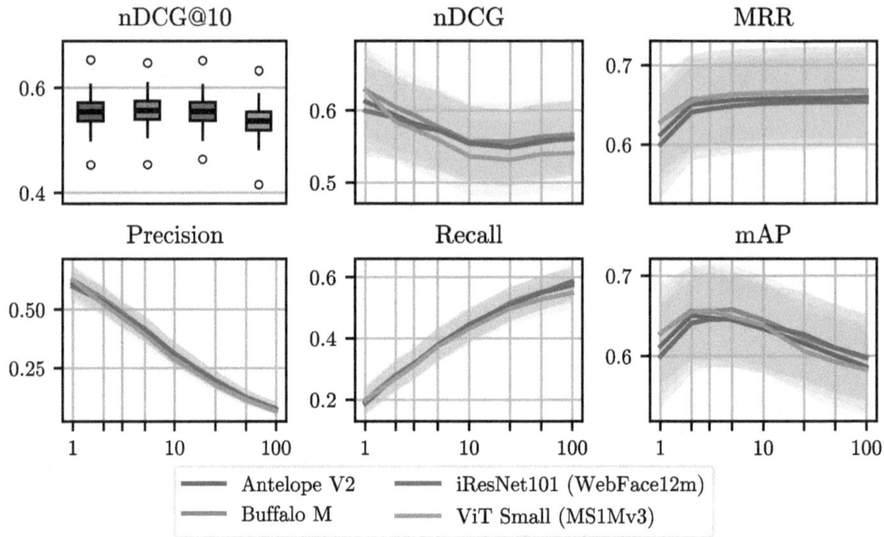

Fig. 3. Results for the image-only retrieval task. In all plots, the vertical axis represents the evaluation metric indicated in the plot title. The horizontal axis shows the values of k used for computing metric@k, except for the upper-left plot. In the upper-left plot, the horizontal axis corresponds to the evaluated models, and each boxplot summarizes the bootstrap distribution of nDCG@10 scores: the box indicates the interquartile range (Q1–Q3), the central line denotes the median, the whiskers show the confidence interval, and the circles represent the minimum and maximum values.

parameter, and the three strategies for the face–name matrix construction. We calculate the sharpened matrix $\hat{A}_\mu$, for modality $\mu \in \{f \leftrightarrow f, n \leftrightarrow n, f \leftrightarrow n\}$, using the original matrix A_μ and the sharpening factor τ_μ in the following way:

$$\hat{A}_\mu = \exp(\tau_\mu \cdot A_\mu). \tag{2}$$

Since the face–name matrix $A_{f\leftrightarrow n}$ is sparse, we perform its sharpening only on non-zero elements to maintain its sparsity. After sharpening, we use the $\hat{A}_\mu$ instead of the A_μ in the adjacency matrix A. We denote the sharpening factors for the three sub-matrices as $\tau_{f\leftrightarrow f}$, $\tau_{n\leftrightarrow n}$, and $\tau_{f\leftrightarrow n}$. For the face–face similarity matrix $A_{f\leftrightarrow f}$ and the name–name similarity matrix $A_{n\leftrightarrow n}$, we define the sparsity parameter as the ratio of non-zero elements to the number of all elements in the matrices.

We ran the evaluations on the development set with all combinations of the hyperparameters. All the considered values for each hyperparameter and the best-performing combination on the main metric nDCG@10 are summarized in Table 3. It is worth mentioning that the top-performing combinations differed in the sparsity parameters, while the sharpening factors and the face–name matrix strategy remained constant. Also note that the sparsity parameters were chosen so that the smaller values cover the elements on the diagonal and only a very small portion of the non-diagonal ones. In fact, this value was derived by rounding up the number of diagonal elements. Even so, the baseline method tends to work slightly better with this lower value rather than with the higher one. We evaluated the best-performing combination on the test set, and the results are shown in Fig. 4a. Again, the results show that the differences between the evaluated models are negligible.

Table 3. Hyperparameter overview. In the last column we report the best-performing combinations of parameters for the text-based and image-based multimodal baselines.

Hyperparameter	Considered values	Best-performing combination	
		Text-based ret.	Image-based ret.
$\tau_{f\leftrightarrow f}$	1, 2, 4, 8	8	4
$\tau_{n\leftrightarrow n}$	1, 2, 4, 8	1	4
$\tau_{f\leftrightarrow n}$	1, 2, 4, 8	2	4
$A_{f\leftrightarrow f}$ sparsity	1×10^{-4}, 1.5×10^{-4}	1×10^{-4}	1×10^{-4}
$A_{n\leftrightarrow n}$ sparsity	5×10^{-5}, 1×10^{-4}	1×10^{-4}	5×10^{-5}
$A_{f\leftrightarrow n}$ strategy	distance, occurrence, uniform	uniform	uniform

6.4 Image-Based Multimodal Retrieval Results

In the image-based multimodal baseline evaluation, we performed the same hyperparameter grid search as in the text-based evaluation. The best-performing combination of parameters is also reported in Table 3. We evaluated our baseline method on the development set with the best-performing combination of parameters and the results are in Fig. 4b. Also here, the differences between the models are negligible.

6.5 Evaluation Conclusions

The image clustering results show that agglomerative clustering combined with identity-optimized face embeddings achieves the best overall performance, outperforming the human baseline across most clustering metrics. Automated methods therefore enable more precise identity separation, whereas human annotations tend to merge identities more conservatively, leading to higher completeness.

From the retrieval evaluations, the results indicate broadly comparable performance across evaluated models, with image-only retrieval achieving the best overall results. Hyperparameter analysis suggests that spatial proximity between faces and names has a limited impact, while the models instead favor sparser associations and modality-dependent sharpening. Overall, however, multimodal image-based retrieval performs notably worse than its unimodal counterpart.

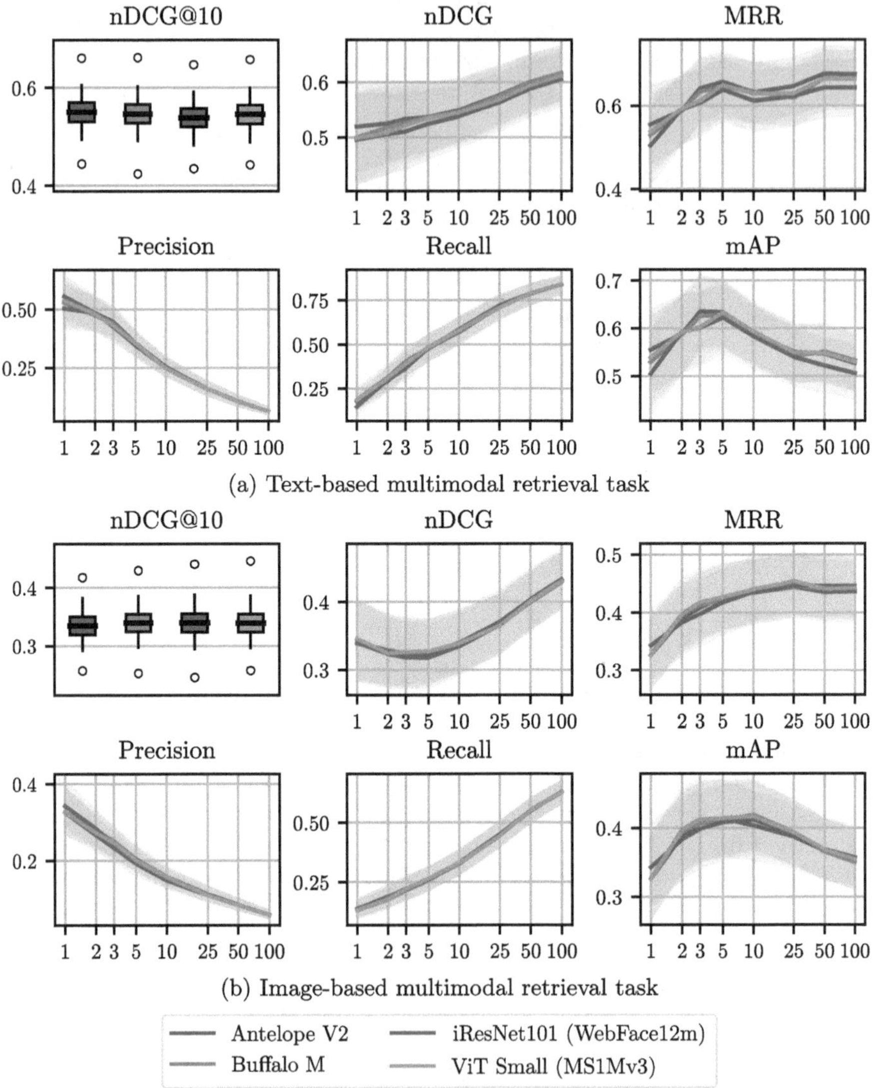

Fig. 4. Results for the text-based and image-based multimodal retrieval tasks. In all plots, the vertical axis represents the evaluation metric indicated in the plot title. The horizontal axis shows the values of k used for computing metric@k, except for the upper-left plot in both subfigures. In the upper-left plots, the horizontal axis corresponds to the evaluated models, and each boxplot summarizes the bootstrap distribution of nDCG@10 scores: the box indicates the interquartile range (Q1–Q3), the central line denotes the median, the whiskers show the confidence interval, and the circles represent the minimum and maximum values.

7 Conclusion

In this paper, we introduced PeopleGator–a multimodal benchmark for person identification in historical documents. The dataset was created by linking detected face crops to nearby OCR text and normalized person names, and by validating cross-document identity clusters through multi-annotator refinement. By packaging this resource with standardized protocols for face clustering, image-to-image retrieval, and text-to-image entity retrieval, the benchmark enables reproducible comparison of methods under realistic archival conditions.

Baseline experiments show that strong face embeddings paired with hierarchical clustering can separate identities effectively–often exceeding the annotator-derived baseline on separation-focused metrics. For retrieval, simple embedding-based image search remains the strongest reference point, and the provided multimodal approach highlights that naïve use of page-level face–name proximity is not sufficient on its own. Future gains will likely require more reliable cross-modal grounding and layout-aware reasoning.

Acknowledgments. This work has been supported by the Ministry of Culture Czech Republic in NAKI III project Orbis Pictus–book revival for cultural and creative sectors (DH23P03OVV033). This work was supported by the Ministry of Education, Youth and Sports of the Czech Republic through the e-INFRA CZ (ID:90254). We would like to especially thank our annotators for their commitment and perseverance, which made this dataset possible.

Disclosure of Interests. The authors have no competing interests to declare that are relevant to the content of this article.

References

1. Interesting faces: a graph-based approach for finding people in news. Patt. Recogn. **43**(5), 1717–1735 (2010)
2. Improving multimodal named entity recognition via text-image relevance prediction with large language models. Neurocomputing **651**, 130982 (2025)
3. Biemann, C.: Chinese whispers: an efficient graph clustering algorithm and its application to natural language processing problems. In: Proceedings of TextGraphs (2006)
4. Biten, A.F., Gómez, L., Rusiñol, M., Karatzas, D.: Good news, everyone! context driven entity-aware captioning for news images. In: Proceedings of the IEEE/CVF Conference on Computer Vision and Pattern Recognition (CVPR), pp. 12466–12475 (2019)
5. Chen, S., Aguilar, G., Neves, L., Solorio, T.: Can images help recognize entities? a study of the role of images for multimodal NER. In: Xu, W., Ritter, A., Baldwin, T., Rahimi, A. (eds.) Proceedings of the Seventh Workshop on Noisy User-generated Text (W-NUT 2021), pp. 87–96. Association for Computational Linguistics, Online (2021)
6. Chen, Z.N., Ngo, C.W., Zhang, W., Cao, J., Jiang, Y.G.: Name-face association in web videos: a large-scale dataset, baselines, and open issues. J. Comput. Sci. Technol. **29**(5), 785–798 (2014)

7. Da, C., Luo, C., Zheng, Q., Yao, C.: Vision grid transformer for document layout analysis. In: Proceedings of the IEEE/CVF International Conference on Computer Vision (ICCV), pp. 19462–19472 (2023)
8. De Cao, N., Izacard, G., Riedel, S., Petroni, F.: Autoregressive entity retrieval. In: 9th International Conference on Learning Representations (ICLR) (2021)
9. De Cao, N., et al.: Multilingual autoregressive entity linking. Trans. Assoc. Comput. Linguistics **10**, 274–290 (2022)
10. Dell, M., et al.: American stories: a large-scale structured text dataset of historical U.S. newspapers. In: Advances in Neural Information Processing Systems (NeurIPS), Datasets and Benchmarks Track (2023)
11. Deng, J., Guo, J., Xue, N., Zafeiriou, S.: Arcface: additive angular margin loss for deep face recognition. In: 2019 IEEE/CVF Conference on Computer Vision and Pattern Recognition (CVPR), pp. 4685–4694 (2019)
12. Ehrmann, M., Hamdi, A., Linhares Pontes, E., Romanello, M., Doucet, A.: Named entity recognition and classification in historical documents: a survey. ACM Comput. Surv. **56**(2), 27:1–27:47 (2023)
13. Galanopoulos, D., Goulas, A., Mezaris, V.: Cross-modal image recommendation for news articles by multimodal foundation models-based retrieval-reranking. In: Proceedings of the MediaEval 2025 Workshop (2025)
14. Gay, P., Meignier, S., Deléglise, P., Odobez, J.M.: Crf-based context modeling for person identification in broadcast videos. Frontiers in ICT **Volume 3 - 2016** (2016)
15. Greif, G., Griesshaber, N., Greif, R.: Multimodal LLMs for OCR, OCR post-correction, and named entity recognition in historical documents (2025)
16. Guo, H., et al.: Towards natural language-based document image retrieval: new dataset and benchmark. In: Proceedings of the Computer Vision and Pattern Recognition Conference (CVPR), pp. 29722–29732 (2025)
17. Guo, Y., Zhang, L., Hu, Y., He, X., Gao, J.: MS-Celeb-1M: a dataset and benchmark for large-scale face recognition. In: Leibe, B., Matas, J., Sebe, N., Welling, M. (eds.) ECCV 2016. LNCS, vol. 9907, pp. 87–102. Springer, Cham (2016). https://doi.org/10.1007/978-3-319-46487-9_6
18. Hu, X., et al.: Kalm-embedding: superior training data brings a stronger embedding model (2025)
19. Hu, Z., Dong, Y., Wang, K., Sun, Y.: Heterogeneous graph transformer. In: Proceedings of The Web Conference 2020, pp. 2704–2710. WWW '20, Association for Computing Machinery, New York, NY, USA (2020)
20. Huang, Y., Lv, T., Cui, L., Lu, Y., Wei, F.: LayoutLMv3: pre-training for document AI with unified text and image masking. In: Proceedings of the 30th ACM International Conference on Multimedia (MM), pp. 4083–4091. ACM (2022)
21. Jocher, G., Qiu, J., Chaurasia, A.: Ultralytics yolo (2023)
22. Kim, G., et al.: OCR-free document understanding transformer. In: Computer Vision – ECCV 2022. Lecture Notes in Computer Science, vol. 13688, pp. 498–517. Springer (2022)
23. Kim, J., Lee, G., Kim, T., Shin, K.: KGMEL: knowledge graph-enhanced multimodal entity linking. In: Proceedings of the 48th International ACM SIGIR Conference on Research and Development in Information Retrieval, pp. 3015–3019. SIGIR '25, Association for Computing Machinery, New York, NY, USA (2025)
24. Kim, M., Jain, A.K., Liu, X.: Adaface: quality adaptive margin for face recognition. In: Proceedings of the IEEE/CVF Conference on Computer Vision and Pattern Recognition (CVPR), pp. 18750–18759(2022)

25. Kim, M., Su, Y., Liu, F., Jain, A., Liu, X.: Keypoint relative position encoding for face recognition. In: Proceedings of the IEEE/CVF Conference on Computer Vision and Pattern Recognition (CVPR), pp. 244–255 (2024)
26. Kišš, M., Hradiš, M., Dvořáková, M., Jiroušek, V., Kersch, F.: Annopage dataset: dataset of non-textual elements in documents with fine-grained categorization. In: Jin, L., Zanibbi, R., Eglin, V. (eds.) Document Analysis and Recognition - ICDAR 2025 Workshops, pp. 50–66. Springer Nature Switzerland, Cham (2026)
27. Landeghem, J.V., et al.: Document understanding dataset and evaluation (dude). In: 2023 IEEE/CVF International Conference on Computer Vision (ICCV), pp. 19471–19483 (2023)
28. Lee, K., et al.: Pix2Struct: screenshot parsing as pretraining for visual language understanding. In: Proceedings of the 40th International Conference on Machine Learning (ICML). In: Proceedings of Machine Learning Research, vol. 202, pp. 18893–18912. PMLR (2023)
29. Li, J., Xu, Y., Lv, T., Cui, L., Zhang, C., Wei, F.: DiT: self-supervised pre-training for document image transformer. In: Proceedings of the 30th ACM International Conference on Multimedia (MM), pp. 3530–3539. ACM (2022)
30. Li, M., et al.: DocBank: a benchmark dataset for document layout analysis. In: Scott, D., Bel, N., Zong, C. (eds.) Proceedings of the 28th International Conference on Computational Linguistics, pp. 949–960. International Committee on Computational Linguistics, Barcelona, Spain (Online) (2020)
31. Luo, P., Xu, T., Wu, S., Zhu, C., Xu, L., Chen, E.: Multi-grained multimodal interaction network for entity linking. In: Proceedings of the 29th ACM SIGKDD Conference on Knowledge Discovery and Data Mining, pp. 1583–1594. ACM (2023)
32. Meng, Q., Zhao, S., Huang, Z., Zhou, F.: Magface: a universal representation for face recognition and quality assessment. In: Proceedings of the IEEE/CVF Conference on Computer Vision and Pattern Recognition (CVPR), pp. 14225–14234 (2021)
33. Muennighoff, N., Tazi, N., Magne, L., Reimers, N.: MTEB: massive text embedding benchmark. In: Vlachos, A., Augenstein, I. (eds.) Proceedings of the 17th Conference of the European Chapter of the Association for Computational Linguistics, pp. 2014–2037. Association for Computational Linguistics, Dubrovnik, Croatia (2023)
34. Page, L., Brin, S., Motwani, R., Winograd, T.: The pagerank citation ranking: Bringing order to the web. Tech. rep., Stanford infolab (1999)
35. Pfitzmann, B., Auer, C., Dolfi, M., Nassar, A.S., Staar, P.: Doclaynet: a large human-annotated dataset for document-layout segmentation. In: Proceedings of the 28th ACM SIGKDD Conference on Knowledge Discovery and Data Mining, pp. 3743–3751. KDD '22, Association for Computing Machinery, New York, NY, USA (2022)
36. Poh, F., George, A., Marcel, S.: ArtFace: towards historical portrait face identification via model adaptation. In: Proceedings of the ICCV 2025 Workshop on Biometrics for Arts (ArtMetrics) (2025)
37. Qu, T., Tuytelaars, T., Moens, M.F.: Visually-aware context modeling for news image captioning. In: Proceedings of the 2024 Conference of the North American Chapter of the Association for Computational Linguistics: Human Language Technologies (Volume 1: Long Papers), pp. 2927–2943. Association for Computational Linguistics (2024)
38. Sarfraz, S., Sharma, V., Stiefelhagen, R.: Efficient parameter-free clustering using first neighbor relations. In: CVPR (2019)
39. Satoh, S., Nakamura, Y., Kanade, T.: Name-it: naming and detecting faces in news videos. IEEE Multimedia **6**(1), 22–35 (1999)

40. Sun, Y., Han, J., Yan, X., Yu, P.S., Wu, T.: PathSim: meta path-based top-K similarity search in heterogeneous information networks. Proc. VLDB Endowment **4**(11), 992–1003 (2011)
41. Tran, A., Mathews, A., Xie, L.: Transform and tell: entity-aware news image captioning. In: Proceedings of the IEEE/CVF Conference on Computer Vision and Pattern Recognition (CVPR), pp. 13035–13045 (2020)
42. Vaško, M., Herout, A., Hradiš, M.: Archival faces: detection of faces in digitized historical documents. In: Document Analysis and Recognition – ICDAR 2025 Workshops: Wuhan. China, 20–21 September 2025, Proceedings, Part II, pp. 17–34. Springer-Verlag, Berlin, Heidelberg (2026)
43. Wan, Z., et al.: Bringing old photos back to life. In: Proceedings of the IEEE/CVF Conference on Computer Vision and Pattern Recognition (CVPR), pp. 2747–2757 (2020)
44. Wang, H., et al.: Cosface: large margin cosine loss for deep face recognition. In: CVPR (2018)
45. Wang, X., et al.: WikiDiverse: a multimodal entity linking dataset with diversified contextual topics and entity types. In: Muresan, S., Nakov, P., Villavicencio, A. (eds.) Proceedings of the 60th Annual Meeting of the Association for Computational Linguistics (Volume 1: Long Papers), pp. 4785–4797. Association for Computational Linguistics, Dublin, Ireland (2022)
46. Wu, L., Petroni, F., Josifoski, M., Riedel, S., Zettlemoyer, L.: Scalable zero-shot entity linking with dense entity retrieval. In: Proceedings of the 2020 Conference on Empirical Methods in Natural Language Processing (EMNLP), pp. 6397–6407. Association for Computational Linguistics (2020)
47. Xu, Y., Li, M., Cui, L., Huang, S., Wei, F., Zhou, M.: Layoutlm: pre-training of text and layout for document image understanding. In: KDD (2020)
48. Xu, Y., Xu, Y., Lv, T., Cui, L., Wei, F., Zhou, M.: Layoutlmv2: multi-modal pre-training for visually-rich document understanding. In: ACL (2021)
49. Ye, J., Peng, X., Sun, B., Wang, K., Sun, X., Li, H., Wu, H.: Learning to cluster faces via transformer (2021)
50. Zhang, Y.: OCR-quality: a human-annotated dataset for OCR quality assessment (2025)
51. Zhao, X., et al.: KaLM-embedding-v2: superior training techniques and data inspire a versatile embedding model. In: The Fourteenth International Conference on Learning Representations (2026)
52. Zhu, Z., et al.: Webface260m: a benchmark unveiling the power of million-scale deep face recognition. In: Proceedings of the IEEE/CVF Conference on Computer Vision and Pattern Recognition (CVPR), pp. 10492–10502 (2021)

Scale-Aware Feature-Level Encoding for Historical Document Stamp Detection

Mohamed Chelali[1](✉), Florence Cloppet[2], Camille Kurtz[2], Isabelle Bloch[3], and Daniel Foliard[1]

[1] Université Paris Cité, ECHELLES, CNRS, Paris F-75013, France
{mohamed.chelali,daniel.foliard}@u-paris.fr,
{mohamed.chelali,daniel.foliard}@sorbonne-universite.fr
[2] LIPADE, Université Paris Cité, Paris F-75006, France
{florence.cloppet,camille.kurtz}@u-paris.fr,
{florence.cloppet,camille.kurtz}@sorbonne-universite.fr
[3] Sorbonne Université, CNRS, LIP6, Paris, France
isabelle.bloch@u-paris.fr, isabelle.bloch@sorbonne-universite.fr

Abstract. Stamps are generally used to authenticate, validate, and officially identify a document. Their analysis is particularly complex, especially for historical documents, which can be damaged. This article focuses on the particular task of stamp detection from old photographs. These stamps allow historians to trace the provenance and distribution of photographs within historical press agencies and, therefore, convey important information. Their varied, sometimes complex shapes, their non-standardized spatial location on the document, and the presence of textual information at different scales make their detection very challenging. In addition, the print quality of the stamp on the paper also has an impact, as some stamps may be faded or even completely invisible. To cope with the multi-scale nature of the features to be detected and with the potential degradation of scanned documents, we introduce a Scale-Aware Positional Encoding (SAPE) module designed for vision Transformer-based architectures that rely on multi-scale feature maps. Our module encodes the normalized logarithmic stride as a continuous variable using two complementary components: (1) an interpolation within a learnable table of scale anchors, and (2) a sinusoidal projection capturing fine-grained variations. The resulting embedding provides a smooth and adaptive modulation of multi-resolution features, leading to better spatial coherence across feature-pyramid levels. Experimental results demonstrate that SAPE improves the DeepSolo baseline and achieves competitive performance against representative detection approaches. These results confirm the effectiveness of SAPE in handling high scale variability and degradation in archival scans. The code is available at: https://github.com/mchelali/SA-DETR.

Keywords: Document analysis · Stamp detection · Historical documents · Multi-scale features · Vision Transformers · Digital humanities

F. Shafait et al. (Eds.): DAS 2026, LNCS 16851, pp. 195–212, 2027.
https://doi.org/10.1007/978-3-032-36207-0_12

1 Introduction

Stamps have long served as the primary mechanism to authenticate, validate, and identify documents in administrative and legal contexts [8,9]. In archival preservation, they provide a metadata layer recording a document history [19]. However, their analysis is challenging because stamps are often faded, partially occluded, or nearly invisible. This is particularly true in early news picture agency collections such as Forbin [5], where stamps document the provenance, circulation, and editorial lifecycle of photographs. These markings, accumulated across stages such as archiving, dating, and editorial use, are essential for tracing the provenance and circulation of photographs. In this context, automatic stamp detection supports cultural heritage preservation by enabling the reconstruction of circulation patterns from stamped markings. This task remains difficult due to the heterogeneity of stamp patterns and the presence of overlapping marks and degradation (see Fig. 1 (a)), which introduce geometric and scale variability.

The analysis of document layout has progressively shifted from binarization-based heuristics [3] to end-to-end detectors such as R-CNN [10,11,25], YOLO [20], and DETR-based models [4,29]. However, axis-aligned detectors remain limited for irregular, rotated, and overlapping stamps. Recent text-spotting Transformers such as TESTR [27] and DeepSolo [26] address arbitrary shapes using ordered point sequences and Bézier curves.

In this paper, we build upon DeepSolo, which encodes semantic information and precise geometry in a single pass. However, its reliance on discrete level embeddings for multi-scale feature pyramids remains sub-optimal for historical documents. To address this, we introduce the *Scale-Aware Positional Encoding* (SAPE) module that treats feature levels as a continuous variable. SAPE provides a smooth and adaptive feature-pyramid modulation enabling the model to bridge the gap between fine-grained textual details and global stamp geometries.

This paper is organized as follows. Section 2 provides an overview of state-of-the-art methods and models for object detection. Section 3 details the proposed module integrated into the DeepSolo model. The experimental protocols and results are presented in Sect. 4. Finally, Sect. 5 concludes the paper and outlines future perspectives.

2 Related Works

These documents present significant challenges, including physical degradation, layout variability, and the coexistence of multiple information layers such as handwritten text, marginalia, and official stamps. Stamps are especially difficult to analyse dur to their variable sppearance, degradation, and overlap with surrounding content. While earlier approaches relied on color-based processing, seal extraction, and document-layer segmentation [6,18,21], recent methods increasingly use deep neural networks for robust detection in complex document collections. This section reviews deep learning methods for document analysis, focusing on multi-scale modeling.

2.1 Deep Neural Networks in Document Analysis

Deep Neural Networks moved the field away from hand-crafted features toward end-to-end representation learning. Early two-stage architectures, such as the R-CNN family [10,11,25] introduced Region Proposal Networks and anchor-based refinement, enabling high-precision localization. However, historical documents

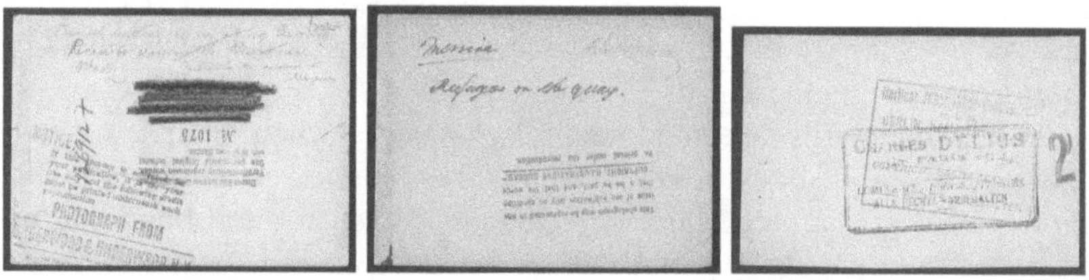

(a) Samples from the Forbin dataset illustrating stamp quality under different configurations, such as varying sizes or overlapping instances.

(b) Samples from the Historical Postcards dataset illustrating postal markings with variable contrast and visibility.

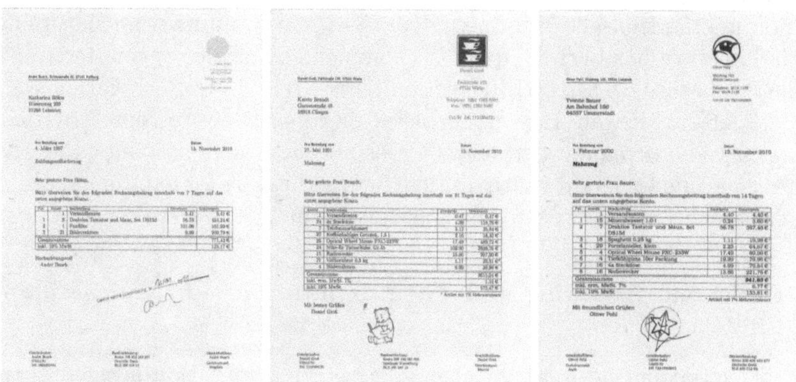

(c) Samples from StaVer showing structured administrative seals with diverse shapes and appearances.

Fig. 1. Representative samples from the evaluation datasets. (a) The Forbin images illustrate degraded and overlapping stamps in historical press-agency archives; (b) the Historical Postcards samples show postal markings with varying visibility; and (c) the StaVer samples contain more structured administrative seals. Together, these datasets cover a broad range of stamp appearances, degradation levels, and geometric configurations, motivating the need for robust multi-scale detection.

often exhibit unpredictable object scales and non-rectangular geometries, such as circular administrative seals or handwritten notes, that are difficult to capture with fixed-ratio anchors. Furthermore, the reliance on hand-designed components like Non-Maximum Suppression (NMS) often leads to instability in crowded document pages where text lines and stamps frequently overlap. To mitigate this, single-stage models like EAST (Efficient and Accurate Scene Text detector) model [28] and YOLO [20] family emerged to predict shapes directly from feature maps, often utilizing Feature Pyramid Networks (FPN) [13] to manage extreme scale variance. While FPN-like structures in models such as YOLOv8 have improved the detection of overlapping official seals [2], these grid-based regression methods still face fundamental challenges when dealing with highly irregular or curved agency stamps. Such complex instances require a more flexible representation than standard axis-aligned or even rotated bounding boxes.

2.2 Transformers for Vision and Multi-scale Approaches

Transformers were adopted in computer vision through the Vision Transformer (ViT) [7], demonstrating that global self-attention can effectively replace convolutions but it lacks the inductive biases for spatial continuity and multi-scale reasoning essential for historical document details. Hierarchical models like Swin Transformer [17] and the Pyramid Vision Transformer (PVT) [24] introduced multi-resolution processing, but the representation of scale remains largely implicit and lacks a learnable representation of pyramid levels. Meanwhile, the DEtection TRansformer (DETR) [4] shifted detection toward set-prediction, eliminating anchors and NMS, while Deformable DETR [29] introduced a sparse attention mechanism, allowing the model to attend to multi-scale feature maps. This architecture has been adapted for complex geometries encountered in text spotting such as TESTR [27], Detection-based Text Line Recognition [1], and DeepSolo [26], which leverage point-based or contour-aware representations to capture curved or irregular text regions. Despite these advancements, these architectures continue to operate with discrete level indicators for handling multi-scale features, which fails to capture the continuous character of scale variations.

2.3 Positional Encoding and Their Limitations for Multi-scale Models

Since the self-attention mechanism is inherently permutation-invariant, Transformers require explicit positional information to preserve spatial ordering. Original sinusoidal encodings [23] provide fixed, translation-consistent signals but are fundamentally scale-agnostic and fail to encode variations in feature-map resolution. By contrast, learned positional encodings [7] are limited by fixed grid sizes, which hinders generalization to unseen resolutions. Two-dimensional extensions and relative biases [17] improve spatial modeling but are confined to local token interactions. In multi-scale transformers such as Deformable DETR and its text-recognition variants, the level embeddings are simply parameters $\ell_k \in \mathbb{R}^d$, one per feature level, which introduces several limitations:

- **No continuity**: nothing enforces a meaningful relationship between embeddings, as neighboring scales are learned in isolation;
- **Weak cross-level generalization**: underrepresented feature levels cannot be smoothly related to neighboring levels, limiting robustness across heterogeneous document layouts;
- **Limited expressiveness**: a single vector per level cannot capture nonlinear scale interactions, e.g. one vector per level is too restrictive to capture nonlinear scale effects, such as the way curved handwriting varies with scan quality.

These constraints motivate the proposed *Scale-Aware Positional Encoding* module, which treats the feature pyramid as a continuous space. By combining sinusoidal scale encodings with and interpolated learnable anchors, SAPE captures feature-level scale information and augments spatial positional encodings with continuous pyramid-level features.

3 SA-DETR: Scale-Aware DEtection TRansformer

Modern DETR-like architectures for text detection, such as DeepSolo [26], rely on multi-scale feature pyramids to handle the vast range of object sizes present in complex layouts. However, their discrete level embeddings provide limited scale awareness and do not fully capture the continuous scale drift found in historical archives. To address this, we introduce the **Scale-Aware Positional Encoding** module, which provides Transformer tokens with explicit continuous scale information and promotes smoother interactions across pyramid levels. Figure 2 illustrates the overall framework, whose components are detailed in the following sections.

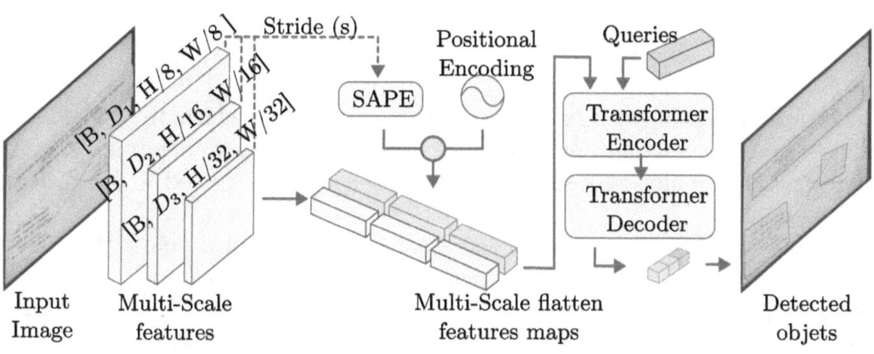

Fig. 2. Overview of the adapted DeepSolo pipeline for stamp detection. Multi-scale feature maps are extracted from the backbone and enriched with 2D Positional encodings for spatial information and with the proposed Scale-Aware Positional Encoding (SAPE) for feature-level scale conditioning. The resulting tokens are then processed by the Transformer encoder-decoder, which uses point-based queries to predict the precise geometry of historical stamps, as highlighted in the output results.

3.1 DeepSolo Model Overview

Our framework builds upon the DeepSolo model [26], originally proposed for scene text spotting and distinguished by its explicit point-based representation. By modeling instances using sets of control points rather than axis-aligned bounding boxes, DeepSolo effectively captures elongated geometries. Inspired by ABCNet [16], it represents instances using *Bézier* curves, making it particularly well suited for detecting curved, circular, elliptical, and other irregular shapes.

Feature Extraction. Given an input image $I \in \mathbb{R}^{H\times W\times 3}$, where H and W represent respectively the height and the width of the image, a CNN Backbone (e.g., ResNet-50) is used to extract a multi-scale feature pyramid $\{F^{(\ell)}\}_{\ell=1}^{L}$, where each feature map $F^{(\ell)} \in \mathbb{R}^{H_\ell\times W_\ell\times D_\ell}$ corresponds to the pyramid level ℓ. To ensure architectural consistency, each feature map is projected to a unified embedding dimension D via a 1×1 convolution, yielding $\tilde{F}^{(\ell)} \in \mathbb{R}^{H_\ell\times W_\ell\times D}$.

Spatial–Scale Encoder. In the original DeepSolo architecture, a learnable level embedding $e^{(\ell)}$ is added to the 2D sinusoidal positional encoding $PE^{(\ell)}$ to distinguish between pyramid levels. By contrast, we introduce a *Scale-Aware Positional Encoding* (SAPE), which explicitly encodes scale information using the stride s_ℓ of each feature map (see Sect. 3.2). Combining with spatial positional encoding $PE^{(\ell)}$, it jointly models spatial location and feature-map resolution. The resulting spatial–scale encoder (SSE) is defined as

$$\mathrm{SSE}^{(\ell)} = PE^{(\ell)} + \mathrm{SAPE}(s_\ell). \tag{1}$$

Following the Deformable DETR framework used in DeepSolo, multi-scale feature maps are flattened and concatenated into a unified token sequence:

$$\mathcal{Z} = Concat^{\ell\in\{1,L\}}\left(\mathrm{flatten}(\tilde{F}^{(\ell)} + \mathrm{SSE}^{(\ell)})\right). \tag{2}$$

The Transformer encoder processes these tokens, after which the top-K candidates are selected and refined by the decoder through deformable cross-attention [29]. Finally, task-specific prediction heads are applied to produce the final outputs.

For detection, each instance is represented by N sampled points $\mathbf{P} = \{\mathbf{p}_n \in \mathbb{R}^2, n = 1...N\}$ describing the geometric *extrema* (the corners and outermost boundary points). In addition, each instance is represented by M control points defining a *Bézier* curve $\mathbf{B} \in \mathbb{R}^{M\times 2}$ which captures its internal structure and curvature. These points serve as supervision for point-based detection queries.

Bipartite Matching. Following the DETR paradigm, the training objective is formulated as a set-prediction problem. A bipartite matching between the K predicted queries and the G ground truth instances ($G \leq K$) is computed using the Hungarian algorithm [12]. This process finds an injective mapping

$\hat{\sigma} \in \mathcal{S}_K$, where $\mathcal{S}_K$ is the set of all possible permutation, that minimizes the total matching cost:

$$\hat{\sigma} = \arg\min_{\sigma \in \mathcal{S}_K} \sum_{g=1}^{G} \mathcal{C}\left(Y^g, \hat{Y}^{\sigma(g)}\right). \tag{3}$$

For each pair of a ground-truth instance Y^g and a prediction $\hat{Y}^{\sigma(g)}$, the matching cost $\mathcal{C}$ is a weighted combination of classification and geometric terms:

$$\mathcal{C}\left(Y^g, \hat{Y}^{\sigma(g)}\right) = \lambda_{cls}\mathcal{L}_{cls}\left(\hat{y}^{\sigma(g)}\right) + \lambda_{bez}\mathcal{L}_{bez}\left(P^g, \hat{P}^{\sigma(g)}\right) \tag{4}$$

where $\mathcal{L}_{cls}$ is the classification cost, derived from the focal loss [14], as proposed in [27]. Let $\hat{y}$ be the predicted probability for the target class $\hat{Y}^{\sigma(g)}$. To handle the foreground-background imbalance during matching, the derived loss function computes the difference between the positive and negative focal terms:

$$\mathcal{L}_{cls}\left(\hat{Y}^{\sigma(g)}\right) = -\mathbb{1}_{\{g \in \mathrm{Im}(\sigma)\}}\alpha(1-\hat{y})^{\gamma}\log(\hat{y}) - \mathbb{1}_{\{g \notin \mathrm{Im}(\sigma)\}}(1-\alpha)\hat{y}^{\gamma}\log(1-\hat{y}) \tag{5}$$

where $\mathbb{1}$ is the indicator function and $\mathrm{Im}(\sigma)$ is the image of the mapping σ. α and γ are two hyper-parameters that control respectively class balancing and the importance of hard examples.

The term $\mathcal{L}_{bez}$ is the L1 distance between the ground truth P^g and predicted $\hat{P}^{\sigma(g)}$ *Bézier* curves. The distance is computed on M points sampled on the curve as follows:

$$\mathcal{L}_{bez}(P^g, \hat{P}^{\sigma(g)}) = \mathbb{1}_{\{g \in \mathrm{Im}(\sigma)\}} \sum_{n=1}^{N} \left\| p_n^g - \hat{p}_n^{\sigma(g)} \right\|. \tag{6}$$

Overall Loss Function. After assignment, the model is optimized using a loss function composed of a classification $\mathcal{L}_{cls}$, which is the focal loss used in the cost function, and regression loss functions defined over both *Bézier* control points $\mathcal{L}_{bez}$ as in Eq. 6, and over boundary points $\mathcal{L}_{bd}$ according to:

$$\mathcal{L}_{bd}(P^g, \hat{P}^{\sigma(g)}) = \mathbb{1}_{\{g \in \mathrm{Im}(\sigma)\}} \sum_{n=1}^{G} \left(\left\| top_n^g - \hat{top}_n^{\sigma(g)} \right\| + \left\| bot_n^g - \hat{bot}_n^{\sigma(g)} \right\| \right) \tag{7}$$

where top_n and bot_n denote the n-th pair of sampled coordinates on the upper and lower boundaries of the instance, respectively. The final loss function is defined as a weighted sum:

$$\mathcal{L} = \lambda_{cls}\mathcal{L}_{cls} + \lambda_{bez}\mathcal{L}_{bez} + \lambda_{bd}\mathcal{L}_{bd} \tag{8}$$

where the coefficients λ_{cls}, λ_{bez}, and λ_{bd} act as balancing factors to normalize the gradient contributions of each task. This formulation enables end-to-end training without heuristic post-processing such as Non-Maximum Suppression (NMS).

3.2 Scale-Aware Positional Encoding (SAPE)

In standard multi-scale Transformer detectors like DeepSolo [26], feature levels are distinguished using discrete learnable embeddings $e^{(\ell)}$. To address this limitation, we introduce the **Scale-Aware Positional Encoding** module that generates a continuous scale-conditioned signal. Figure 3 illustrates the architecture of SAPE, which is composed of two parallel branches: a *Discrete Interpolation Path* and a *Continuous Sinusoidal Path.* Rather than discarding discrete scale representations, SAPE augments them with a continuous formulation. The discrete component preserves the structural prior of the feature pyramid, while the continuous component enables smooth modeling of cross-scale relationships.

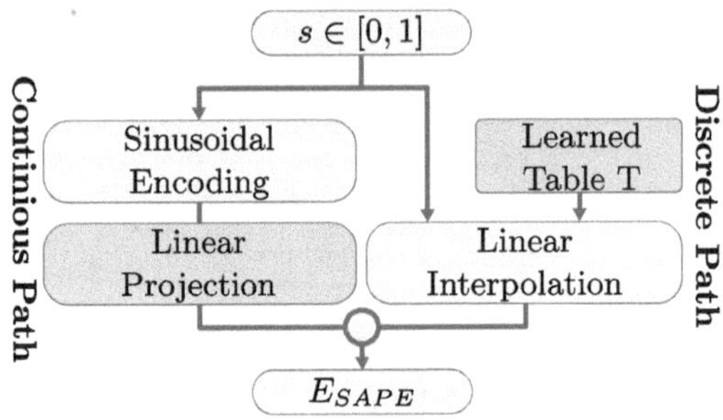

Fig. 3. Architecture of the Scale-Aware Positional Encoding (SAPE) module. It maps the log-normalized scale coordinate s into two parallel paths: (1) a Discrete Interpolation Path that approximates the scale manifold through piecewise-linear interpolation of learnable anchors, and (2) a Continuous Sinusoidal Path that induces a geometric frequency prior. Their fusion bridges the granularity gap between feature levels, enabling a feature-level scale-aware representation for historical document analysis.

SAPE operates on the stride information of each feature level. To account for the exponential nature of the feature pyramid's receptive fields, we map each level to a continuous scale coordinate $s \in [0, 1]$ using a log-normalized mapping. Given a stride d_i, its scale coordinate is computed as:

$$s_i = \frac{\log_2(d_i) - \log_2(d_{\min})}{\log_2(d_{\max}) - \log_2(d_{\min})} \tag{9}$$

where $d_{\min}$ and $d_{\max}$ represent the minimum and maximum strides of the feature pyramid, respectively. Based on this normalized coordinate, the SAPE module processes the scale information through two complementary mechanisms, capturing both specialized learnable features and global geometric continuity.

The Discrete Interpolation Path: Instead of treating each resolution level as an isolated category, this branch models the feature-level scale space through a learnable scale table $\mathcal{T} \in \mathbb{R}^{K \times D}$. $\mathcal{T}$ is a learnable matrix where each of the K rows constitutes a scale anchor. Rather than acting as a simple look-up table, $\mathcal{T}$ provides a piecewise-linear approximation of the scale space.

For any normalized scale coordinate $s \in [0, 1]$, we project it into the table's index space to obtain a virtual index $idx = s \cdot (K - 1)$. The core innovation lies in the use of linear interpolation between the adjacent anchors $\lfloor idx \rfloor$ and $\lceil idx \rceil$:

$$E_{interp}(s) = (1 - \delta)\mathcal{T}\lfloor idx \rfloor + \delta\mathcal{T}\lceil idx \rceil \tag{10}$$

where $\delta = idx - \lfloor idx \rfloor$. Unlike a discrete look-up table, this formulation is piecewise-linear and differentiable with respect to s. This allows the model to compute representative features for intermediate scales not present in the backbone features, enabling the model to synthesize representative features for feature levels that are not explicitly represented in the original pyramid hierarchy. This design preserves the anchoring to the original feature hierarchy while relaxing the rigidity of purely discrete embeddings.

The Continuous Sinusoidal Path: While the previous branch learns features from the data distribution, the sinusoidal branch imposes a geometric-continuity inductive bias. We apply a sinusoidal encoding function $\mathrm{PE}(s_i)$ following the standard Transformer formulation [23], providing a stable and feature-level consistent signal. The resulting encoding is then projected into the model embedding space using an affine transformation:

$$E_{sinus}(s_i) = W_p \, \mathrm{PE}(s_i) + b_p, \tag{11}$$

where $W_p \in \mathbb{R}^{d_{\sin} \times D}$ and b_p are learnable parameters. This branch preserves a global notion of scale continuity, ensuring that the model maintains a consistent relative ordering of scales across the entire pyramid. This branch provides a global and smooth representation of scale, which complements to the locally adaptive interpolation mechanism.

Feature Fusion: The two branches play complementary roles. The interpolation path ensures adaptation to the feature pyramid structure, while the sinusoidal path enforces a global and smooth scale prior. The final scale-aware embedding is obtained by fusing both branches. By combining E_{interp} with E_{sinus}, the SAPE module generates a hybrid scale encoding:

$$E_{SAPE} = \frac{1}{2}\left(E_{interp} + E_{sinus}\right) + \mathcal{B}, \tag{12}$$

where $\mathcal{B}$ denotes an optional learnable level bias. The resulting vector is broadcast across the spatial dimensions $(H \times W)$ of the corresponding feature map and added to the spatial positional encodings. Integrating SAPE into the DeepSolo framework explicitly informs each token of its feature-level scale context, enabling robust detection across the heterogeneous resolutions commonly encountered in historical document images.

4 Experimental Study

The proposed SA-DETR is experimented on three datasets described in Sect. 4.1. We then report the implementation details, evaluation metrics, comparisons with competitive methods, qualitative results, and ablation study in Secs. 4.2–4.5. The objective is to assess both the detection performance and the robustness of SA-DETR under heterogeneous historical document conditions.

4.1 Evaluation Datasets

Three stamp detection datasets with different document types and annotation protocols are used. **The Forbin dataset**[1] [5] contains early 20th-century photographs collected by Victor Forbin (1864–1947). The collection includes 120,106 recto–verso image pairs, where the verso side often contains captions, editorial notes, and stamps. In this work, we focus on stamp detection from the verso pages. A subset of images from different geographical regions was annotated using Arkindex, resulting in 5,752 stamp instances across 3,475 images. The dataset is divided into training, validation, and testing subsets with ratios of 70%, 15%, and 15%, respectively, and representative samples are shown in Fig. 1(a). **The Historical Postcards dataset** [22] contains 4,293 postcard images from the Image'Est archive, covering the Grand Est region of France between 1899 and 1930. It provides annotations for text regions and postal markings. In this work, we retain only images containing stamps, resulting in 353 images with 372 annotated instances. This subset is used to evaluate stamp detection under visual configurations that differ from Forbin, particularly in terms of stamp size, contrast, and layout structure, as illustrated in Fig. 1(b). Finally, we evaluate the method on a combined **StaVer/DDI-100** benchmark. StaVer [18], originally introduced for the ICDAR 2011 stamp detection challenge, contains 400 images with bounding-box annotations. It is complemented with DDI-100 [2], which adds 366 images and 1,311 stamp instances and is publicly available on Roboflow[2]. In the following experiments, **StaVer** denotes this combined benchmark. Example samples are shown in Fig. 1(c).

4.2 Implementation Details

We integrated the SAPE module into the DeepSolo [26] architecture by replacing the standard discrete level embeddings. Multi-scale features are extracted from a ResNet-50 backbone at strides $\{8, 16, 32, 64\}$. The transformer hidden dimension is set to 256, with six layers in both the encoder and decoder.

We conduct ablation studies to evaluate the two paths of the SAPE module. For the discrete path, we vary the number of anchors in the Scale Table $\mathcal{T}$ with $K \in \{4, 8, 16\}$. This configuration controls the granularity of the continuous feature-level representation between the standard pyramid levels. The

[1] https://huggingface.co/datasets/mchelali/forbin_dataset.

[2] https://universe.roboflow.com/marcos-7aslt/stampdet/dataset/10.

Table 1. Overall detection performance comparison across the three datasets. The proposed SA-DETR achieves competitive mAP scores and improves localization precision, particularly under the stricter AP_{75} metric. Best, second-best, and third-best results in each column are indicated in bold, underlined, and gray bold, respectively.

	Forbin			HistPostCard			StaVer		
Method	mAP	AP50	AP75	mAP	AP50	AP75	mAP	AP50	AP75
Faster R-CNN [25]	**34.06**	**75.73**	25.93	82.60	**99.50**	94.10	70.84	**96.00**	**78.56**
RetinaNet [15]	30.83	72.57	22.61	**86.80**	**100.00**	94.10	**76.36**	95.53	**84.20**
YOLOv8-s [20]	32.73	73.25	24.83	77.70	94.30	89.20	57.70	81.70	62.20
YOLOv11-s [20]	33.73	**74.82**	**26.02**	**78.50**	94.10	91.80	61.27	81.99	67.99
Deform. DETR [29]	27.79	68.20	18.47	52.20	83.10	64.40	55.34	89.77	61.55
DeepSolo [26]	44.14	68.73	50.31	56.11	92.01	63.53	67.46	95.01	74.02
SA-DETR	**46.19**	74.91	**50.35**	76.82	**100.00**	**95.66**	**70.78**	**95.48**	80.10

sinusoidal branch follows the standard Transformer formulation with base frequency 10,000 [23].

The model is optimized using AdamW with an initial learning rate of 10^{-4} and a weight decay of 10^{-4}, together with a linear learning-rate scheduler and a warm-up. We employ an early-stopping strategy with a patience of 30 epochs, monitored using the validation mean Average Precision (mAP).

The training uses a multi-task loss function (Eq. 8) combining classification and geometric regression terms. The weights are set to $\lambda_{cls} = 1.0$, $\lambda_{bez} = 1.0$, and $\lambda_{bd} = 0.5$. The lower weight assigned to the boundary term acts as a regularizer, encouraging boundary refinement without overfitting. The classification is optimized using focal loss with $\alpha = 0.25$ and $\gamma = 2.0$ [14,29].

Finally, we apply standard data augmentations including random resizing, rotation, and cropping. These augmentations are used only to improve robustness to document layout variability and degraded scans.

4.3 Comparison with Competitive Methods

We compare the proposed SA-DETR framework with representative object detectors, including axis-aligned two-stage detectors (Faster R-CNN), one-stage detectors (RetinaNet, YOLOv8-s, YOLOv11-s), and transformer-based approaches (Deformable DETR). The small configurations YOLOv8-s and YOLOv11-s provide a balanced trade-off between accuracy and computational efficiency while serving as strong real-time baselines. For each dataset, models are trained and evaluated independently, since the stamp appearance differs substantially across collections.

Table 1 reports the quantitative comparison across datasets using mAP, AP_{50}, and AP_{75}. These metrics allow us to evaluate both overall detection performance and localization precision under moderate and strict IoU thresholds.

On the Forbin dataset, bounding-box detectors achieve mAP scores below 35%, highlighting the difficulty of modeling rotated and irregular historical stamps using axis-aligned representations. Deformable DETR also underperforms, suggesting that global attention alone is insufficient without an appropriate geometric representation. By contrast, the DeepSolo baseline significantly improves detection performance thanks to its point-based curve representation, reaching 44.14% mAP. SA-DETR further improves performance to 46.19% mAP while also achieving the best AP_{50} and AP_{75} scores. These results confirm the benefit of explicit scale-aware modeling for historical documents.

On the Historical Postcards dataset, RetinaNet achieves the highest mAP score due to the predominance of rectangular annotations and visually structured layouts. However, SA-DETR achieves the best AP_{50} and AP_{75} values, reaching 100.00% and 95.66%, respectively. This indicates more accurate localization under stricter overlap constraints and confirms the effectiveness of the proposed polygonal representation for rotated and irregular postal markings.

On the StaVer dataset, RetinaNet achieves the highest mAP score due to the use of axis-aligned binding box annotations, while Faster R-CNN remains competitive. However, SA-DETR significantly enhances localization quality, with AP_{75} attaining 80.10%, which indicates better geometric alignment despite the evaluation bias toward rectangular predictions.

Overall, these results show that combining point-based geometric modeling with explicit scale-aware feature encoding leads to more robust and precise stamp detection across heterogeneous historical document collections.

We use Critical Difference (CD) diagrams to compare the average ranks of all methods across datasets. The Friedman test is first used to assess whether statistically significant differences exist among the evaluated methods. When significant differences are detected, pairwise Wilcoxon signed-rank tests are applied as a post-hoc analysis to identify which methods differ.

We generated separate CD diagrams for mAP, AP_{50}, and AP_{75}, as well as a global CD diagram combining all dataset-metric pairs. The diagrams show that SA-DETR ranks third for mAP, second for AP_{50}, and first for AP_{75}. This indicates that the proposed method is particularly effective under stricter localization criteria. When all dataset-metric pairs are jointly considered, SA-DETR obtains the best global average rank, confirming its balanced performance across datasets and metrics. Figure 4 shows the global CD diagram.

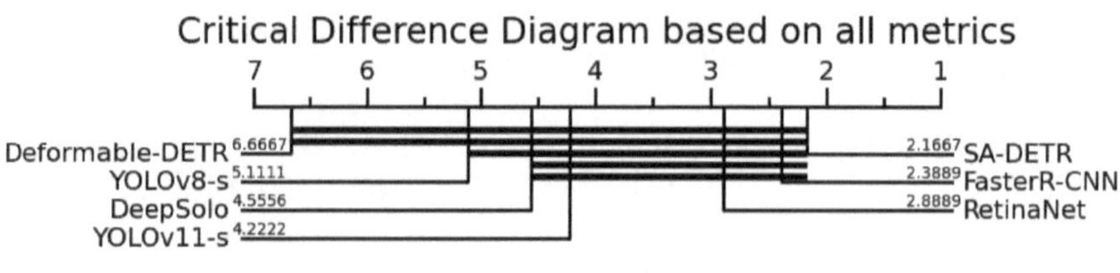

Fig. 4. Critical Difference diagram computed by jointly considering all dataset-metric pairs, i.e., mAP, AP_{50}, and AP_{75} over the Forbin, Historical Postcards, and StaVer datasets. Lower ranks indicate better performance.

4.4 Qualitative Results

To complement the quantitative evaluation, Fig. 5 reports qualitative comparisons of stamp detection results across the evaluated models on the three datasets. Only predictions with a confidence score higher than 30% are displayed.

On the Forbin sample, the presence of dense editorial notes, overlapping markings, and degraded visual content makes stamp localization particularly challenging. Faster R-CNN, RetinaNet, YOLOv11-s and Deformable DETR detect stamp regions, but their predictions are often poorly aligned with the actual stamp boundaries. DeepSolo misses stamp instances, showing the difficulty of modeling degraded and partially visible markings. By contrast, SA-DETR provides a more accurate localization of the visible stamp regions. Some ambiguities remain, especially when date stamps or blue editorial marks are visually close to the target stamp appearance. It is worth noting that one detected date stamp is not included in the ground-truth annotations, which may lead to an apparent false positive despite being visually meaningful.

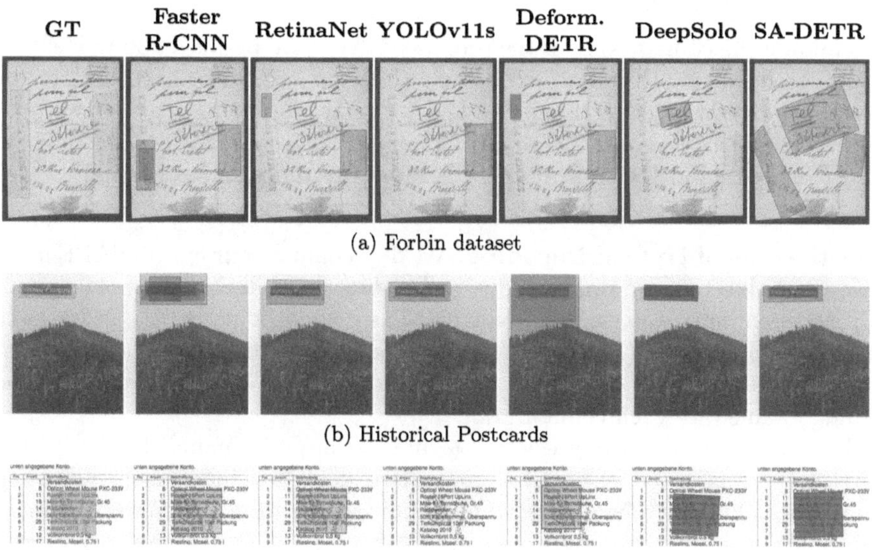

(a) Forbin dataset

(b) Historical Postcards

(b) StaVer dataset

Fig. 5. Visualization of stamp ground truth and detected results across different models on the three datasets. The detected stamps by all models have score equal or greater than 30%.

For the Historical Postcards sample, Faster R-CNN and Deformable DETR tend to produce multiple detections, some of which only partially cover the stamp or extend to a larger surrounding area. RetinaNet, YOLOv11-s, DeepSolo and SA-DETR provide more precise predictions on this example, with bounding regions that better match the annotated stamp. This suggests that visible stamps can be correctly localized by several detectors, while region-based and Deformable DETR baselines may still over-cover the target.

On the StaVer sample, the stamp overlaps with textual content, making it difficult to separate the target instance from the background structure. Faster R-CNN and RetinaNet successfully localize the stamp, whereas YOLOv11-s fails to detect it. Deformable DETR produces a detection but with incomplete spatial coverage. Both DeepSolo and SA-DETR correctly detect the overlapping stamp, highlighting the benefit of sequence-based localization for complex layouts. Compared with DeepSolo, SA-DETR maintains competitive localization while benefiting from the proposed scale-aware encoding.

Overall, these qualitative observations support the quantitative results. SA-DETR provides robust detections under degradation, overlap, and scale variability. This suggests that the proposed scale-aware positional encoding improves the model's ability to capture both fine-grained degraded details and global stamp geometry.

4.5 Ablation Studies

To evaluate the contribution of the proposed SAPE, we perform ablation experiments on a subset of the Forbin dataset. These experiments aim to isolate the contributions of the hybrid branches and to assess the model sensitivity to the number of anchor points K. We report the mAP and standard deviation of the model, as each model is trained and evaluated five times across the splits.

Effectiveness of Hybrid Encoding. We first compare our hybrid SAPE module against the standard discrete-level embeddings used in the DeepSolo. To this end, we train and evaluate four configurations: (1) without SAPE, (2) only the Discrete Interpolation (DI) path, (3) only the Continuous Sinusoidal path, and finally (4) the full hybrid SAPE combining both paths. Table 2 summarizes the results obtained for each configuration.

Removing SAPE entirely leads to a substantial performance drop, with the mAP decreasing to 50.24% and the standard deviation increasing, indicating reduced stability. Introducing the DI path improves performance to 58.69% mAP, confirming its ability to capture scale information. The CS path provides an even larger gain, reaching 63.90% mAP. It outperforms the DI path by nearly 5% and significantly improves both of AP_{50} and AP_{75}, showing that continuous scale encoding plays a crucial role in modeling fine-grained scale variations. Combining both paths in the full SAPE module achieves the best overall performance with 64.24% mAP and reduces variance across runs.

Table 2. Ablation of the SAPE components on the Forbin dataset. "DI" refers to the Discrete Interpolation table, and "CS" refers to the sinusoidal projection.

DI	CS	mAP	AP50	AP75
–	–	50.24±5.53	76.38±5.46	61.31±7.42
✓	–	58.69±4.28	76.71±5.16	67.80±5.88
–	✓	63.90±2.16	83.21±2.50	73.95±2.41
✓	✓	**64.24±0.8**	**83.80±2.0**	**73.90±1.9**

Overall, these results indicate that discrete interpolation provides strong anchor-level scale priors and the continuous sinusoidal encoding is essential for handling intermediate scale drift. This leads to improve strict localization accuracy, as reflected by the increase in AP_{75}.

Impact of Scale Table Size (K). We further analyze the sensitivity of the SAPE module to the number of anchor points K used in the scale table $\mathcal{T}$. Given that the feature pyramid comprises four levels, we evaluate $K \in \{4, 8, 16\}$ to study the effect of interpolation granularity on detection performance. Table 3 reports the corresponding results.

Table 3. Impact of the number of anchor points K on detection performance.

Anchor Points (K)	mAP	AP50	AP75
$K = 4$	63.74±2.04	82.59±2.65	73.17±3.28
$K = 8$	**64.24±0.80**	**83.80±2.00**	**73.90±1.90**
$K = 16$	61.76±3.64	83.21±4.36	71.96±4.28

Using $K = 4$ provides strong performance, indicating that a coarse scale discretization is sufficient to capture the main pyramid structure. Increasing the number of anchor points to $K = 8$ consistently improves performance across all metrics, yielding the highest mAP (64.24%) and the lowest variance. However, increasing the table size to $K = 16$ leads to a performance drop and higher variance. This shows that a moderate increase ($K = 8$) in scale resolution enables the model to better account for intermediate scale variations, whereas an excessive increase ($K = 16$) negatively affects performance and leads to potential over-parameterization.

The ablation studies show that the full SAPE configuration, combining the discrete and continuous paths using a scale-table size of $K = 8$, achieves the best performance. This setting provides the best trade-off between detection accuracy, stability, and model complexity, consistently delivering the strongest performance across all evaluation metrics.

5 Conclusion

In this work, we introduced the Scale-Aware Positional Encoding (SAPE) module to explicitly encode scale information across the multi-scale feature pyramid. SAPE combines a continuous branch that encodes scale using the stride of each feature map through sinusoidal encoding, and a discrete branch relying on a learnable anchor table to model scale as a continuous manifold. Their fusion provides a richer representation of feature-map. Integrated into the DeepSolo, the resulting SA-DETR improves stamp detection performance across three datasets and achieves the best overall statistical ranking.

As future work, we will evaluate the runtime behavior of SA-DETR to better assess its practical deployment in archival workflows. We will also investigate the integration of SAPE into recent document understanding architectures, including text spotting frameworks and emerging vision-language models.

We also plan to extend the framework toward the joint detection of text and stamps for historical-document analysis. We aim to bridge the gap between model architecture and dataset granularity. While DeepSolo targets word-level recognition, the Forbin dataset provides annotations at the title or legend level. To address this mismatch, we will investigate hierarchical architectures capable of decomposing large text regions into words, enabling end-to-end coarse localization and fine-grained recognition.

Acknowledgment. This work was supported by the ANR-24-CE38-4079 project and by GENCI-IDRIS under grant 2025-AD011016637 on the Jean Zay supercomputer.

References

1. Baena, R., Kalleli, S., Aubry, M.: General detection-based text line recognition. In: NeurIPS (2024). http://papers.nips.cc/paper_files/paper/2024/hash/4acb23d5d9b4bea566799afac0ee3125-Abstract-Conference.html
2. Bento, J., Paixão, T., Alvarez, A.B.: Performance evaluation of YOLOv8, YOLOv9, YOLOv10, and YOLOv11 for stamp detection in scanned documents. Appl. Sci. **15**(6) (2025). https://www.mdpi.com/2076-3417/15/6/3154
3. Binmakhashen, G.M., Mahmoud, S.A.: Document layout analysis: a comprehensive survey. ACM Comput. Surv. **52**(6) (2019). https://doi.org/10.1145/3355610
4. Carion, N., Massa, F., Synnaeve, G., Usunier, N., Kirillov, A., Zagoruyko, S.: End-to-end object detection with transformers. In: Vedaldi, A., Bischof, H., Brox, T., Frahm, J.-M. (eds.) ECCV 2020. LNCS, vol. 12346, pp. 213–229. Springer, Cham (2020). https://doi.org/10.1007/978-3-030-58452-8_13
5. Chelali, M., Gosselet, S.K., Cloppet, F., Kurtz, C., Bloch, I., Foliard, D.: The FORBIN Dataset: a collection of historical photographs with archival metadata. JOHD (2026). https://doi.org/10.5334/johd.487
6. Dey, S., Mukherjee, J., Sural, S.: Consensus-based clustering for document image segmentation. Int. J. Docu. Anal. Recogn. (IJDAR) **19**(4), 351–368 (2016). https://doi.org/10.1007/s10032-016-0275-1

7. Dosovitskiy, A., et al.: An image is worth 16x16 words: transformers for image recognition at scale. In: ICLR (2021). https://openreview.net/forum?id=YicbFdNTTy
8. Forczmański, P.: Stamp detection in scanned documents. Ann. UMCS, Inf. **10**(1), 61–68 (2010). https://doi.org/10.2478/v10065-010-0036-6
9. Gayer, A., Ershova, D., Arlazarov, V.: Fast and accurate deep learning model for stamps detection for embedded devices. Pattern Recognit. Image Anal. **32**(4), 772–779 (2022). https://doi.org/10.1134/S1054661822040046
10. Girshick, R.B.: Fast R-CNN. In: ICCV, pp. 1440–1448 (2015). https://doi.org/10.1109/ICCV.2015.169
11. Girshick, R.B., Donahue, J., Darrell, T., Malik, J.: Rich feature hierarchies for accurate object detection and semantic segmentation. In: CVPR, pp. 580–587. IEEE Computer Society (2014). https://doi.org/10.1109/CVPR.2014.81
12. Kuhn, H.W.: The Hungarian method for the assignment problem. Naval Res. Logistics Quarterly **2**(1-2), 83–97 (1955). https://doi.org/10.1002/nav.3800020109
13. Lin, T.Y., Dollar, P., Girshick, R., He, K., Hariharan, B., Belongie, S.: Feature pyramid networks for object detection. In: CVPR, pp. 936–944 (2017). https://doi.org/10.1109/CVPR.2017.106
14. Lin, T.Y., Goyal, P., Girshick, R., He, K., Dollár, P.: Focal loss for dense object detection. IEEE Trans. Patt. Anal. Mach. Intell. **42**(2), 318–327 (2020). https://doi.org/10.1109/TPAMI.2018.2858826
15. Lin, T., Goyal, P., Girshick, R.B., He, K., Dollár, P.: Focal loss for dense object detection. In: ICCV, pp. 2999–3007 (2017). https://doi.org/10.1109/ICCV.2017.324
16. Liu, Y., Chen, H., Shen, C., He, T., Jin, L., Wang, L.: ABCNet: real-time scene text spotting with adaptive Bezier-curve network. In: CVPR, pp. 9806–9815 (2020). https://doi.org/10.1109/CVPR42600.2020.00983
17. Liu, Z., et al.: Swin transformer: hierarchical vision transformer using shifted windows. In: ICCV, pp. 9992–10002 (2021). https://doi.org/10.1109/ICCV48922.2021.00986
18. Micenková, B., van Beusekom, J.: Stamp detection in color document images. In: ICDAR, pp. 1125–1129 (2011). https://doi.org/10.1109/ICDAR.2011.227
19. Nikolaidou, K., Seuret, M., Mokayed, H., Liwicki, M.: A survey of historical document image datasets. Int. J. Document Anal. Recognit. **25**(4), 305–338 (2022). https://doi.org/10.1007/S10032-022-00405-8
20. Redmon, J., Divvala, S., Girshick, R., Farhadi, A.: You only look once: unified, real-time object detection. In: CVPR, pp. 779–788 (2016). https://doi.org/10.1109/CVPR.2016.91
21. Sun, D.: Seal extraction and incomplete restoration in complex document image, pp. 200–205. ICBDC '21, Association for Computing Machinery (2021). https://doi.org/10.1145/3469968.3470001
22. Tabbone, S., Pelingre, M.: Benchmarking OCR tools for historical postcards: a dataset and evaluation. In: 7th SUMAC @ ACM Multimedia, pp. 61–68. SUMAC '25 (2025). https://doi.org/10.1145/3746273.3760201
23. Vaswani, A., et al.: Attention is all you need. In: NeurIPS, pp. 5998–6008 (2017). https://proceedings.neurips.cc/paper/2017/hash/3f5ee243547dee91fbd053c1c4a845aa-Abstract.html
24. Wang, W., et al.: Pyramid vision transformer: a versatile backbone for dense prediction without convolutions. In: ICCV, pp. 548–558 (2021). https://doi.org/10.1109/ICCV48922.2021.00061

25. Ye, M., Zhang, J., Zhao, S., Liu, J., Du, B., Tao, D.: DPText-DETR: towards better scene text detection with dynamic points in transformer. In: AAAI, pp. 3241–3249 (2023). https://doi.org/10.1609/AAAI.V37I3.25430
26. Ye, M., et al.: DeepSolo: Let transformer decoder with explicit points solo for text spotting. In: CVPR, pp. 19348–19357 (2023). https://doi.org/10.1109/CVPR52729.2023.01854
27. Zhang, X., Su, Y., Tripathi, S., Tu, Z.: Text spotting transformers. In: CVPR, pp. 9509–9518 (2022). https://doi.org/10.1109/CVPR52688.2022.00930
28. Zhou, X., et al.: EAST: an efficient and accurate scene text detector. In: CVPR, pp. 2642–2651 (2017). https://doi.org/10.1109/CVPR.2017.283
29. Zhu, X., Su, W., Lu, L., Li, B., Wang, X., Dai, J.: Deformable DETR: deformable transformers for end-to-end object detection. In: ICLR (2021). https://openreview.net/forum?id=gZ9hCDWe6ke

OCR, Layout and Recognition Systems

TrOCR for Medieval HTR: A Systematic Ablation Study with Cross-Dataset Validation

Sachin Sharma(✉), Michele Flammini, and Federico Simonetta

GSSI – Gran Sasso Science Institute, L'Aquila, Italy
{sachin.sharma,michele.flammini,federico.simonetta}@gssi.it

Abstract. Fine-tuning transformer-based handwritten text recognition (HTR) models on medieval manuscripts is challenging because these models are pre-trained on modern text and must adapt to a very different visual domain. This paper studies how three controllable fine-tuning choices (contrast normalization, data augmentation, and layer freezing) affect recognition accuracy when adapting TrOCR to small historical datasets. We run controlled experiments on a 13th-century Italian manuscript (I-CT 91 "Cortonese") and replicate the same experimental grid on the public READ-16 benchmark as robustness evidence. On Cortonese, our best configuration achieves 8.03% character error rate (CER). Statistical comparisons across 13 configurations show that freezing up to three encoder layers or six decoder layers does not significantly harm accuracy, while deeper freezing becomes progressively detrimental. Removing contrast normalization (CLAHE) yields 7.84% CER, comparable to a domain-specialized baseline, suggesting strong optimization can reduce reliance on image preprocessing. Cross-dataset validation on READ-16 shows that decoder freezing thresholds transfer more robustly than encoder thresholds, and combined freezing strategies require dataset-specific re-validation. Finally, we use Grad-CAM gradient attributions and decoder cross-attention maps to diagnose error patterns and failure modes revealed by the ablations. Source code is available at https://github.com/LaudareProject/TrOCR-analysis.

Keywords: TrOCR · Ablation Study · Medieval Manuscripts · Transfer Learning · Digital Humanities

1 Introduction

Handwritten text recognition (HTR) is a prerequisite for turning medieval manuscripts into searchable, analyzable corpora, yet performance remains fragile. Relative to modern printed documents, manuscripts combine shifting letterforms, pervasive abbreviations, degraded supports (faded ink, bleed-through, staining), and irregular layouts [22]. Systems tuned to clean print therefore generalize poorly in this domain [15,16].

F. Shafait et al. (Eds.): DAS 2026, LNCS 16851, pp. 215–228, 2027.
https://doi.org/10.1007/978-3-032-36207-0_13

Recent transformer-based HTR models have improved modern handwriting recognition. TrOCR [14], for instance, performs strongly in contemporary domains, but medieval material lies far from its pre-training distribution in handwriting style, orthographic conventions, and linguistic structure [1,2]. We select TrOCR as the baseline because it is a strong and widely used encoder–decoder HTR model with public checkpoints, making its fine-tuning behavior practically relevant. At the same time, historical collections often provide only limited supervision (typically a few thousand labeled lines), so the details of fine-tuning can dominate the final outcome [17,23].

The historical HTR literature includes a wide range of application-focused studies [6,25,26], and recent work confirms that transformer-based models can be adapted to historical documents [7,11,24,27]. However, reported gains are often hard to interpret because preprocessing, augmentation, and optimization are frequently adjusted together; controlled ablations remain relatively uncommon [17]. It is also unclear how far a single hyperparameter configuration transfers across scripts, languages, and periods.

Interpretability is similarly underexplored. Attention visualizations are a common starting point; Donut, for example, reports decoder cross-attention maps with emergent localization behavior [10]. Yet attention weights are not explanations by default [3]. Gradient-based attributions such as Grad-CAM [20] are more tightly coupled to the training objective, but attention- and gradient-derived signals are still rarely used as quantitative diagnostics in OCR/HTR [10,12].

Within Document Analysis Systems, this work addresses the line-level HTR component of a broader transcription pipeline: manuscript image → segmentation → line-level HTR → correction/review → document-level transcription use.

We study this question with controlled ablations that isolate preprocessing, data augmentation, and encoder/decoder layer freezing. Overall, this work contributes practical recommendations for TrOCR users:

1. We provide controlled ablations on a 13th-century Italian manuscript, isolating preprocessing, augmentation, and encoder–decoder freezing effects on CER and generalization, with accompanying statistical significance tests.
2. We provide a full 13-configuration ablation replicated on READ-16 with line-level statistical validation, quantifying which findings transfer across scripts, languages, and centuries and which require dataset-specific re-validation.[1]
3. a qualitative diagnostic analysis using Grad-CAM and cross-attention to interpret ablation-linked error patterns and failure modes.

In order to test whether fine-tuning conclusions are specific to a single corpus or transfer across typologically distant collections, we perform a cross-dataset

[1] The source code used for the experiments is available at https://github.com/LaudareProject/TrOCR-analysis. The dataset is currently under review and will be released upon completion of the review.

validation using two datasets that differ along several complementary axes: script, language, time period, and document type (I-CT 91 and READ-16).

We do not claim to outperform all existing HTR systems. Instead, we provide a controlled ablation study and actionable fine-tuning guidance for practitioners working with small medieval HTR datasets.

2 Dataset and Experimental Setup

2.1 Dataset

Fig. 1. Example of a page from the I-CT 91 "Cortonese" manuscript.

All experiments were performed on the I-CT 91 "Cortonese" manuscript (Fig. 1) [21], an Italian laude source dating from the 13th century. The dataset comprises 259 manuscript pages, 2,517 annotated text lines, 14,214 words, and 75,667 characters. Each line is provided with verified ground truth transcriptions. The average image resolution is 2082×2717 with 300 DPI and average text line size of 1352.85×159.64. The manuscript presents multiple paleographic challenges, including the presence of two different scribal hands, the coexistence of textual and neumatic notation, and substantial physical degradation such as faded ink and parchment staining. These characteristics make the collection low-resource medieval test case for line-level HTR adaptation under real-world manuscript conditions.

To ensure reproducibility and fair evaluation, we adopted one of the standardized splits provided by our project. The partition preserves page-level separation to prevent data leakage between training and testing subsets. The dataset includes 1,666 text lines for training (66.2%), 397 for validation (15.8%), and 454 for testing (18%). Unlike random splitting, this configuration guarantees consistent comparisons across studies and future reproducibility for subsequent research on the same material. The exact split indices and all experimental source code are publicly available in the code repository to facilitate replication once the dataset is released.

3 Methodology

In order to define a rigorous protocol, we define the main set the following research question: *when adapting a pre-trained TrOCR model to a small medieval HTR dataset, which architectural choices impact the CER metric the most, and to which extent the impact is measurable?*

To investigate this question, we examine three controllable fine-tuning levers: *a)* preprocessing, *b)* data augmentation, and *c)* parameter-efficient fine-tuning through layer freezing.

These axes were selected because they are common, controllable design decisions in low-resource HTR pipelines and directly trade off accuracy, compute, and manual pipeline effort.

All other training factors are held constant throughout the experiments. The experimental design modifies one lever at a time relative to a fixed full fine-tuning baseline, thereby enabling a clean attribution of performance differences to individual methodological choices.

Preprocessing Pipeline: The preprocessing pipeline was designed to enhance image readability and improve OCR robustness under varying manuscript conditions. Contrast enhancement was performed using Contrast Limited Adaptive Histogram Equalization (CLAHE) with a clip limit of 2.0 and an 8×8 tile grid [19,28]. For comparison, standard histogram equalization was also considered during first experimentation. Deskewing was applied using a projection profile method constrained within $\pm5°$, ensuring consistent text line orientation. Finally, images were normalized to a fixed height of 64 pixels and standardized according to ImageNet statistics.

To improve robustness, several augmentation techniques were applied during training. Geometric transformations included random rotations within $\pm5°$ (probability 0.4) and elastic deformations ($\alpha = 20$, $\sigma = 5$, $p = 0.25$). Photometric augmentations involved Gaussian blur ($p = 0.35$), brightness adjustment ($p = 0.40$), and contrast variation ($p = 0.30$). Additionally, manuscript degradation effects were simulated through speckle noise ($p = 0.25$), morphological operations ($p = 0.15$), and random shadow or staining overlays ($p = 0.15$). These augmentations emulate common challenges in medieval texts such as uneven illumination, ink bleed, and parchment defects.

Ablation Experiments: We run controlled ablations over three levers: preprocessing (CLAHE), data augmentation, and parameter-efficient fine-tuning via layer freezing. We keep the dataset split, evaluation protocol, and TrOCR-base initialization fixed, and change one lever at a time. For preprocessing and augmentation, we compare four variants under full fine-tuning (`enc_0_dec_0`): (i) CLAHE + augmentation (baseline), (ii) no CLAHE, (iii) no augmentation, and (iv) no CLAHE + no augmentation. For layer freezing, we sweep $K \in \{0, 3, 6, 9, 12\}$ frozen layers in the encoder or decoder while keeping the other module fully trainable. Because both modules have 12 layers, this corresponds to 25% steps (0%–100% frozen). We then test `enc_3_dec_6` to verify whether the individually non-significant freezing thresholds remain non-significant when applied jointly.

Following Following Hüttner et al., we train with AdamW and a One-Cycle schedule for 50 epochs (no early stopping). The learning rate starts at 1×10^{-9}, rises to 5.5×10^{-6} over the first five epochs (10% of the run), and then anneals to a value lower than the peak by a factor of 2.2×10^4. Momentum follows the same One-Cycle schedule (0.85–0.95). We regularize with weight decay (0.01) and label smoothing (0.1). We train with 8 samples per device and accumulate gradients for two steps, for an effective batch size of 16.

Kraken Baseline: In order to obtain a domain-specific baseline, we fine-tuned a CNN model using Kraken [9]. We used the pre-trained CATMuS Medieval model (version 1.6.0) [18]. This model was trained using the default Kraken architecture on a large dataset comprising 160 000 text lines and 5 millions characters from documents from the 8th to the 16th centuries [4]. Importantly, this dataset included various European languages, including Italian. We fine-tuned the CATMuS model on our I-CT 91 using early stopping with patience of 15 epochs and the augmentation strategy implemented by Kraken.[2]

Evaluation Metrics: Model performance was evaluated using the standard Character Error Rate (CER) and Word Error Rate (WER) metrics. The Character Error Rate is defined as:

$$CER = \frac{S + D + I}{N},$$

where S, D, and I represent the number of substitutions, deletions, and insertions respectively, and N is the length of the reference transcription. The Word Error Rate follows the same formulation but is computed at the word level. Both metrics were used to provide a comprehensive assessment of transcription accuracy at fine-grained and semantic levels.

[2] For more info about the augmentation pipeline, see the Python module `kraken/lib/dataset/recognition.py` as implemented in version 5.3.0.

4 Results

Table 1. I-CT 91 Ablation Study

Configuration	Frozen (enc/dec)	Trainable Params	CER	WER	mean ΔCER	Sign.
enc_0_dec_0 ⋆	0 / 0	333.9M (100%)	8.0	31.9	—	—
No Augm	0 / 0	333.9M (100%)	8.4	32.1	+0.45	ns
No CLAHE No Aug	0 / 0	333.9M (100%)	8.2	32.3	+0.18	ns
No CLAHE	0 / 0	333.9M (100%)	**7.8**	**31.1**	-0.19	ns
enc_3_dec_0	3 / 0	~312.7M (93.6%)	8.2	32.4	+0.20	ns
enc_6_dec_0	6 / 0	~291.4M (87.3%)	9.7	34.8	+1.66	***
enc_9_dec_0	9 / 0	~270.2M (80.9%)	13.8	42.5	+5.79	***
enc_12_dec_0	12 / 0	248.9M (74.5%)	16.5	47.5	+8.43	***
enc_0_dec_3	0 / 3	~285.1M (85.4%)	8.2	32.0	+0.18	ns
enc_0_dec_6	0 / 6	~236.3M (70.8%)	8.4	32.0	+0.34	ns
enc_0_dec_9	0 / 9	~187.5M (56.1%)	8.9	33.4	+0.84	*
enc_0_dec_12	0 / 12	138.7M (41.5%)	10.3	37.4	+2.23	***
enc_3_dec_6	3 / 6	~215M (64.4%)	8.6	32.6	+0.52	ns
Kraken-CATMuS (Baseline)	—	—	7.9	31.2	−0.1	—

Notes: Sign. = Wilcoxon signed-rank test vs baseline, Holm-corrected, computed between line-wise CER ($N = 454$ lines). Friedman omnibus test: $\chi^2 = 1001.19$, $p < 10^{-151}$. ⋆ marks the full fine-tuning. *** $p < 0.001$, ** $p < 0.01$, * $p < 0.05$, ns = not significant ($p_{\text{holm}} \geq 0.05$). Kraken–CATMuS is an external domain-specialized baseline included for context; it is not a TrOCR variant and is not part of the statistical comparisons.

4.1 Ablation Analysis on I-CT 91

We assess two common pipeline components—contrast normalization (CLAHE) and data augmentation—while keeping the model, split, and optimization fixed (full fine-tuning with `enc_0_dec_0` under the same One-Cycle schedule). This allows us to attribute differences to the pipeline rather than to the training setup. Our augmentation pipeline combines geometric, photometric, and degradation-style transforms. Rather than isolate each transform in a factorial design, we evaluate the full pipeline as a single intervention and compare it against a no-augmentation baseline. On Cortonese, neither CLAHE nor the augmentation pipeline moves accuracy in a reliable direction once per-line variability is taken into account (Table 1). In the same setting, TrOCR is broadly comparable to the Kraken–CATMuS reference.

To study parameter-efficient adaptation, we progressively freeze larger prefixes of the encoder or decoder and compare configurations using line-level statistical testing (Table 1). The outcome is markedly asymmetric. Encoder freezing is quickly detrimental once it extends beyond the first few layers. In contrast, the decoder is more tolerant: freezing a moderate fraction of decoder layers remains comparable to full fine-tuning, with clear degradation emerging only when most of the decoder is held fixed. On Cortonese, the combined setting `enc_3_dec_6`

Table 2. READ-16 Ablation Study

Configuration	Frozen (enc/dec)	Trainable Params	CER	WER	mean ΔCER	Sign.
enc_0_dec_0 ⋆	0/0	333.9M (100%)	**5.16**	**21.18**	–	–
No Augm	0/0	333.9M (100%)	5.92	23.97	+0.75	**
No CLAHE No Augm	0/0	333.9M (100%)	5.62	23.24	+0.45	ns
No CLAHE	0/0	333.9M (100%)	5.30	21.91	+0.14	ns
enc_3_dec_0	3/0	~312.7M (93.6%)	5.94	23.86	+0.78	*
enc_6_dec_0	6/0	~291.4M (87.3%)	7.37	25.86	+2.20	***
enc_9_dec_0	9/0	~270.2M (80.9%)	11.80	33.25	+6.63	***
enc_12_dec_0	12/0	248.9M (74.5%)	15.71	39.82	+10.54	***
enc_0_dec_3	0/3	~285.1M (85.4%)	5.60	22.45	+0.44	ns
enc_0_dec_6	0/6	~236.3M (70.8%)	5.69	22.14	+0.53	ns
enc_0_dec_9	0/9	~187.5M (56.1%)	5.96	24.00	+0.80	**
enc_0_dec_12	0/12	138.7M (41.5%)	8.14	31.78	+2.97	***
enc_3_dec_6	3/6	~215M (64.4%)	6.49	24.37	+1.33	***
VAN [5]	–	–	4.10	16.29	–	–
GatedLexiconNet [13]	–	–	2.13	6.52	–	–
TrOCR base [8]	–	–	4.81	23.22	–	–

Sign. = Wilcoxon signed-rank test vs enc_0_dec_0 baseline, Holm-corrected (N = 1,138 lines).
Friedman omnibus test: $\chi^2 = 1978.68$, $p < 10^{-151}$.
⋆ = full fine-tuning baseline.
*** $p < 0.001$, ** $p < 0.01$, * $p < 0.05$, ns = not significant ($p_{\text{Holm}} \geq 0.05$). VAN, GatedLexiconNet, and TrOCR base are external third-party systems included as reference points only; they are not TrOCR ablation variants and are not subject to statistical comparison.

remains comparable to full fine-tuning while reducing the number of trainable parameters.

4.2 Cross-Dataset Validation on READ-16

We replicate the full ablation grid on READ-16 to test transfer beyond a single manuscript (Table 2). READ-16 is used here as robustness evidence rather than as proof of universal generalization.

The CLAHE conclusion is stable: under the present optimization and model, contrast normalization does not emerge as a required ingredient. Augmentation, however, is more clearly beneficial on READ-16 than on Cortonese. The encoder–decoder asymmetry in freezing also persists, yet encoder freezing becomes less forgiving under this dataset shift, whereas decoder freezing remains comparatively safe up to a moderate depth.

The combined configuration `enc_3_dec_6` is non-significant on Cortonese but significantly worse than full fine-tuning on READ-16.

5 Discussion

5.1 Ablation and Baseline Results

The freezing ablation yields a clear encoder–decoder asymmetry (Table 1). The encoder is the component that needs to learn the most: freezing beyond the first few layers is consistently harmful, which matches the intuition that the visual stack must adapt to manuscript-specific letterforms and degradation. The decoder, by contrast, is more transferable, and moderate decoder freezing can reduce training cost without a measurable loss on Cortonese.

The preprocessing ablation is equally instructive. On Cortonese, neither CLAHE nor the full augmentation bundle produces a reliable gain (Table 1). This suggests that, under One-Cycle training, TrOCR can learn robust features from minimally processed inputs, and that aggressive contrast enhancement is not a prerequisite for competitive accuracy.

The READ-16 replication (Table 2) sharpens the main cautions. The encoder remains the brittle part under domain shift, whereas the decoder continues to tolerate freezing up to a moderate depth. Augmentation, however, behaves differently and becomes clearly beneficial on READ-16, which is consistent with a dataset-dependent regularization effect. Finally, a combined freezing recipe that is acceptable on one corpus can become significantly worse on another, so freezing strategies should be re-validated per dataset rather than reused wholesale.

With respect to baselines, Kraken–CATMuS remains a strong reference point because it benefits from domain-relevant pre-training. The fact that TrOCR can reach a similar regime with careful fine-tuning and minimal preprocessing suggests that optimization choices can partially compensate for the lack of specialized pre-training, but it does not remove the motivation for domain-aware pre-training when such data are available.

Practical Implications. For TrOCR fine-tuning on low-resource historical line images, our results suggest four practical recommendations: 1) treat CLAHE as optional rather than default; 2) validate augmentation per dataset rather than assume a universal gain; 3) avoid aggressive encoder freezing; and 4) prefer full fine-tuning unless compute constraints justify moderate decoder freezing.

The ablations identify which fine-tuning choices change accuracy; the qualitative analysis is included to help interpret where these choices fail, especially in ambiguous or high-loss tokens. We therefore use Grad-CAM and cross-attention as diagnostic tools for understanding observed error modes.

5.2 Diagnostic Interpretation of Ablation Results

Normalized entropy is defined as:

$$H' = \frac{-\sum_i p_i \log p_i}{\log(n)}$$

where p_i denotes the normalized activation at patch i, and n is the total number of patches. The normalized entropy satisfies $H' \in [0, 1]$ and measures

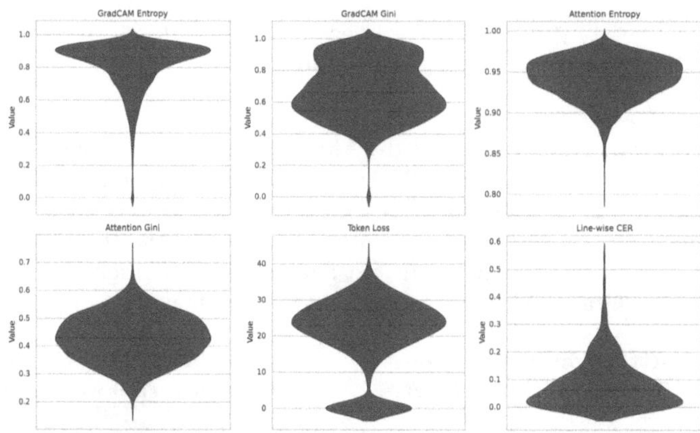

Fig. 2. Distribution of metric values related to losses and Grad-CAM/attention maps. Violin plots and box plots are computed on the test set of the I-CT 91 dataset.

how diffusely the attention map distributes its activation. A value of $H' = 1$ corresponds to a uniform attention map with no spatial preference, whereas values of H' close to 0 indicate that the activation is concentrated on only a few patches.

In the context of HTR, a high-entropy map associated with an erroneous token suggests that the model lacks a confident spatial hypothesis for that character, making the metric diagnostically informative. The Gini coefficient is defined as:

$$G = \frac{2\sum_{i=1}^{n} i\,p_i}{n\sum_{i=1}^{n} p_i} - \frac{n+1}{n}$$

where the activations p_i are sorted in ascending order before computation, and n denotes the total number of patches. If the activations are normalized such that $\sum_{i=1}^{n} p_i = 1$, the expression simplifies accordingly. The coefficient satisfies $G \in [0, 1]$, where values close to 1 indicate highly concentrated activation on a small number of patches, while values close to 0 correspond to a nearly uniform attention distribution. Figure 2 summarizes Token Loss and two map statistics (normalized entropy and Gini) for both Grad-CAM and decoder cross-attention. Because such aggregates can hide qualitative structure, we also inspected token-level overlays. Table 3 shows four representative tokens spanning both low- and high-error lines, and highlights the recurring patterns and failure modes we observed.

Both maps are computed token-by-token from the final decoder block under teacher forcing. For *attention*, we average the cross-attention weights over heads, reshape them to the patch grid, and normalize. For *Grad-CAM*, we backpropagate the ground-truth logit for the same step, form a gradient-weighted sum over heads, apply ReLU, and normalize. In both cases we drop the vision `[CLS]` patch. Token Loss is computed on the same forward pass, and we summarize each map with normalized entropy and Gini.

Table 3. Comparative analysis of Grad-CAM and attention summaries, Token Loss, and CER in four prediction scenarios. The first case pairs a high-error line with broadly distributed maps (high entropy, low Gini). In the second case, Grad-CAM is sharply localized while attention remains comparatively diffuse. In the third case, the Grad-CAM map collapses to (near-)zero. In the fourth case, attention is relatively more concentrated than Grad-CAM on a correctly recognized line.

Grad-CAM Entr/Gini	Attention Entr/Gini	Token Loss	Line CER	Grad-CAM (above) and Attention (below) maps
0.97/0.31	0.98/0.26	24.5	39%	 Token "ran" in "et don' alegrança." TrOCR predicted "et don alegrança a mer"
0.35/0.99	0.91/0.51	19.3	0.0%	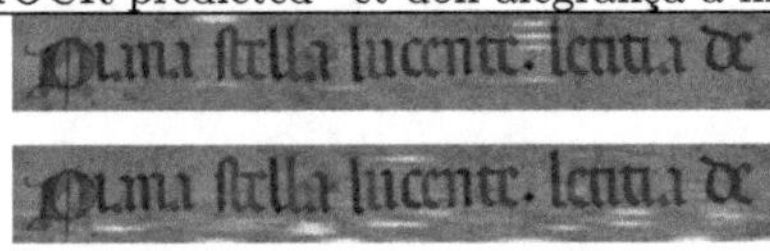 Token "l" in "Diana stella lucente letitia de" TrOCR predicted without errors.
0.0/0.0	0.93/0.45	35.3	0.0%	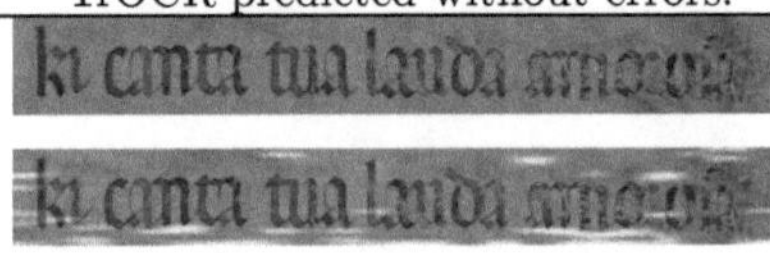 Token "t" in "ki canta tua lauda amorosa" TrOCR predicted without errors.
0.77/0.71	0.85/0.63	12.2	0.0%	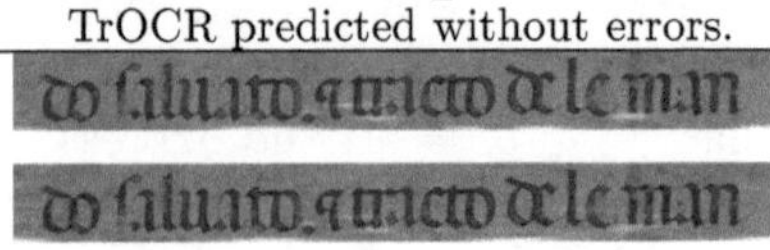 Token "et" in "do salvato et tracto de le man" TrOCR predicted without errors.

In the first two examples, Grad-CAM mostly follows token contours, emphasizing stroke boundaries more than filled interiors. This "edge-seeking" behavior is consistent with the model relying on high-frequency ink–background transitions, and with Grad-CAM highlighting precisely the regions to which the token-level loss is most sensitive. Cross-attention is often less crisp in the overlays: even after averaging across heads, it can remain spread over nearby patches, including in correctly decoded cases, which suggests that the decoder may maintain multiple plausible alignments rather than committing to a single peak.

The metrics in Table 3 align with these visual regimes. When the line is difficult and the token loss is high, both maps tend to have high entropy and low Gini, indicating broad, weakly concentrated support. Conversely, a line can be transcribed correctly while still containing tokens with elevated loss; in such cases, Grad-CAM may remain sharply localized even when attention looks comparatively diffuse. This is a useful reminder that "correct" is not the same as "confident", and that the two map types capture different aspects of the decoding process.

The third row highlights a practical caveat: Grad-CAM can collapse to an all-zero map, making entropy and Gini uninformative by construction. We treat this as a failure mode of the signal (e.g., negative contributions removed by ReLU or vanishing gradients), not as evidence of extreme certainty. In these cases, attention remains well-defined, so attention-derived diagnostics can still provide a fallback.

The fourth row shows the complementary situation, where attention is relatively more concentrated than Grad-CAM on a correctly recognized line.

Taken together, the examples reinforce that attention and Grad-CAM answer different questions. For downstream diagnostics, this motivates a simple hybrid rule: use Grad-CAM-based measures when the map is non-degenerate, and fall back to attention-based measures (or explicitly flag the token) when Grad-CAM energy is near zero.

6 Conclusion and Future Work

This paper presents a controlled ablation study of practical TrOCR fine-tuning choices for low-resource medieval HTR. Across Cortonese and READ-16, the most stable findings are that encoder freezing is more fragile than decoder freezing, CLAHE is not required under the present recipe, and augmentation and combined freezing strategies remain dataset-dependent.

On Cortonese, CLAHE and our augmentation bundle are optional under this recipe: neither produces a statistically reliable gain (Table 1). Freezing behaves differently. The encoder is the brittle part: freezing past `enc_3` quickly becomes harmful, whereas the decoder remains tolerant up to `dec_6`. Under these constraints, `enc_3_dec_6` remains non-significant with respect to full fine-tuning.

READ-16 preserves the broad asymmetry (decoder more transferable than encoder) and again does not elevate CLAHE to a required component (Table 2). What changes is that augmentation becomes beneficial, and the combined `enc_3_dec_6` recipe turns significantly worse than full fine-tuning. In other words, encoder freezing thresholds and, especially, combined freezes do not automatically carry over across corpora.

Beyond accuracy, Grad-CAM and decoder cross-attention are useful as diagnostic tools for interpreting ablation-linked errors, but they are best treated as complementary evidence alongside sequence-level signals such as token loss.

These conclusions are strongest for TrOCR-like encoder–decoder HTR models on low-resource historical line images. They should not be overgeneralized to all HTR architectures or datasets, and READ-16 should be interpreted as robustness evidence rather than proof of universal transfer. Even so, the study can help future researchers avoid redundant ablation work when adapting TrOCR in similar settings. These conclusions should be validated on additional scripts, periods, datasets, and HTR architectures.

Future work is largely methodological. The most direct next step is domain-aware pre-training on paleographic corpora (e.g., CATMuS-like collections), followed by the same cross-dataset evaluation protocol, to separate "better initialization" from "better fine-tuning." In parallel, uncertainty estimation should be made robust to edge cases by masking structural tokens (`<s>`, `</s>`, `[PAD]`), detecting degenerate Grad-CAM maps, and combining map-based summaries with calibrated sequence-level signals. These signals are useful only if they hold up downstream, so we consider active learning and semi-supervised experiments—measured as annotation savings at a fixed target quality—to be the next required validation. Finally, freezing is not the only parameter-efficient option; comparing it against alternatives such as low-rank adapters, and replicating across scripts, languages, and material conditions, would make the results more generalizable.

Funding Statement

This work has been funded by the European Union (Horizon Programme for Research and Innovation 2021–2027, ERC Advanced Grant "The Italian Lauda: Disseminating Poetry and Concepts Through Melody (12th-16th century)", acronym LAUDARE, project no. 101054750). The views and opinions expressed are, however, only those of the author and do not necessarily reflect those of the European Union or the European Research Council. Neither the European Union nor the awarding authority can be held responsible for such matters.

References

1. Aguilar, S.T.: TRIDIS: a comprehensive medieval and early modern corpus for HTR and NER (2025). https://doi.org/10.48550/arXiv.2503.22714
2. Belay, B.H., et al.: A historical handwritten dataset for ethiopic OCR with baseline models and human-level performance. In: Barney Smith, E.H., Liwicki, M., Peng, L. (eds.) IEEE International Conference on Document Analysis and Recognition (2024)
3. Chefer, H., Gur, S., Wolf, L.: Transformer interpretability beyond attention visualization. In: Proceedings of the IEEE/CVF Conference on Computer Vision and Pattern Recognition (CVPR), pp. 782–791 (2021). https://doi.org/10.1109/CVPR46437.2021.00083
4. Clérice, T., et al.: CATMuS medieval: a multilingual large-scale cross-century dataset in latin script for handwritten text recognition and beyond. In: Barney Smith, E.H., Liwicki, M., Peng, L. (eds.) Document Analysis and Recognition - ICDAR 2024, pp. 174–194, Springer Nature Switzerland, Cham (2024). https://doi.org/10.1007/978-3-031-70543-4_11

5. Coquenet, D., Chatelain, C., Paquet, T.: End-to-end handwritten paragraph text recognition using a vertical attention network. IEEE Trans. Patt. Anal. Mach. Intell. **45**(1), 508–524 (2023). https://doi.org/10.1109/TPAMI.2022.3144899. ISSN 1939-35
6. Fischer, A., Frinken, V., Fornés, A., Bunke, H.: Transcription alignment of latin manuscripts using hidden markov models. In: Proceedings of the 2011 Workshop on Historical Document Imaging and Processing, pp. 29–36, HIP '11, Association for Computing Machinery (2011). https://doi.org/10.1145/2037342.2037348
7. Fischer, N., Hartelt, A., Puppe, F.: Line-level layout recognition of historical documents with background knowledge. Algorithms **16**(3), 136 (2023). https://doi.org/10.3390/a16030136
8. Hüttner, L., et al.: Low-rank adaptation vs. fine-tuning for handwritten text recognition. In: 2025 IEEE/CVF Winter Conference on Applications of Computer Vision Workshops (WACVW), pp. 1233–1242 (2025). https://doi.org/10.1109/WACVW65960.2025.00146. ISSN 2690-621
9. Kiessling, B.: Version 5 of the kraken ATR engine for the humanities. In: Yin, X.C., Karatzas, D., Lopresti, D. (eds.) Document Analysis and Recognition - ICDAR 2025, pp. 443–458. Springer Nature Switzerland, Cham (2026). https://doi.org/10.1007/978-3-032-04624-6_26
10. Kim, G., et al.: OCR-free document understanding transformer. In: Avidan, S., Brostow, G., Cissé, M., Farinella, G.M., Hassner, T. (eds.) European Conference on Computer Vision (2021a)
11. Kim, G., et al.: Donut: Document understanding transformer without OCR **7**(15), 2. arXiv preprint arXiv:2111.15664 (2021). https://arxiv.org/abs/2111.15664
12. Kim, G., Yokoo, S., Seo, S., Osanai, A., Okamoto, Y., Baek, Y.: On text localization in end-to-end OCR-free document understanding transformer without text localization supervision. In: Coustaty, M., Fornés, A. (eds.) ICDAR Workshops (2023)
13. Kumari, L., Singh, S., Rathore, V.V.S., Sharma, A.: GatedLexiconNet: a comprehensive end-to-end handwritten paragraph text recognition system (2024). https://doi.org/10.48550/arXiv.2404.14062
14. Li, M., et al.: TrOCR: transformer-based optical character recognition with pre-trained models. In: Proceedings of the AAAI Conference on Artificial Intelligence, vol. 37, pp. 13094–13102 (2023). https://doi.org/10.1609/aaai.v37i11.26538
15. Loss, E., Guernaccini, F., Carassai, M.: From manuscript to metadata: experiments on handwritten text recognition, tagging and importation for the memoriali series (1265-1452). Jlis.it **16**(2), 59–85 (2025). ISSN 2038-102. https://doi.org/10.36253/jlis.it-641
16. Matos, A., Almeida, P., Correia, P., Pacheco, O.: iForal: automated handwritten text transcription for historical medieval manuscripts. J. Imaging **11**(2), 36 (2025). ISSN 2313-433. https://doi.org/10.3390/jimaging11020036
17. Meoded, E.: Handwritten text recognition of historical manuscripts using transformer-based models (2025). https://doi.org/10.48550/arXiv.2508.11499
18. Pinche, A., et al.: Catmus medieval (2024). https://doi.org/10.5281/zenodo.12743230
19. Pizer, S.M., et al.: Adaptive histogram equalization and its variations. Comput. Vision Graph. Image Process. **39**, 355–368 (1987). https://api.semanticscholar.org/CorpusID:62771950
20. Selvaraju, R.R., Cogswell, M., Das, A., Vedantam, R., Parikh, D., Batra, D.: Grad-cam: visual explanations from deep networks via gradient-based localization. In:

Proceedings IEEE International Conference Computer Vision (ICCV), pp. 618–626 (2017). ISSN 2380-750. https://doi.org/10.1109/ICCV.2017.74
21. Sharma, S., Simonetta, F., Flammini, M.: Experimenting active and sequential learning in a medieval music manuscript. In: IEEE International Workshop on Machine Learning for Signal Processing, arXiv, Istanbul (2025). https://doi.org/10.48550/arXiv.2507.15633
22. Simonetta, F., Mondal, R., Ludovico, L.A., Ntalampiras, S.: Optical music recognition in manuscripts from the ricordi archive. In: AudioMostly '24, ACM, Milan, Italy (2024). https://doi.org/10.1145/3678299.3678324
23. Ströbel, P.B., Hodel, T., Boente, W., Volk, M.: The adaptability of a transformer-based OCR model for historical documents. In: Coustaty, M., Fornés, A. (eds.) ICDAR Workshops, (2023)
24. Ströbel, P.B., Clematide, S., Volk, M., Hodel, T.: Transformer-based HTR for historical documents (2022). https://arxiv.org/abs/2203.11008
25. Toselli, A.H., Romero, V., Villegas, M., Vidal, E., Sánchez, J.A.: HTR dataset ICFHR 2016 [data set]. Zenodo (2018). https://doi.org/10.5281/zenodo.1164045
26. Toselli, A.H., Vidal, E.: Handwritten text recognition results on the bentham collection with improved classical n-gram-hmm methods. In: Proceedings of the 3rd International Workshop on Historical Document Imaging and Processing, pp. 15–22. HIP '15, Association for Computing Machinery (2015). https://doi.org/10.1145/2809544.2809551
27. Xu, Y., Li, M., Cui, L., Huang, S., Wei, F., Zhou, M.: Layoutlm: pre-training of text and layout for document image understanding. In: Proceedings of the 26th ACM SIGKDD International Conference on Knowledge Discovery & Data Mining, pp. 1192–1200, KDD '20, ACM (2020). https://doi.org/10.1145/3394486.3403172
28. Zuiderveld, K.: Contrast limited adaptive histogram equalization, pp. 474–485. Academic Press Professional, Inc., USA (1994). ISBN 0123361559

DocScribe: Handwriting-Aware Document Templating with Compact OCR-Free Models

Marcel Lamott[1(✉)], Nauman Riaz[2,3], Sheraz Ahmed[2,3], Andreas Dengel[2], Ulrich Schwanecke[1], and Adrian Ulges[1]

[1] Hochschule RheinMain–University of Applied Sciences and Arts, Wiesbaden, Germany
marcel.lamott@hs-rm.de
[2] German Research Center for Artificial Intelligence, Kaiserslautern, Germany
[3] DeepReader GmbH, Kaiserslautern, Germany
https://lavis.cs.hs-rm.de/

Abstract. We study whether compact OCR-free models can recover visually faithful structured markup, i.e. HTML that reproduces a document's layout and text, directly from rendered document images. Training such models is held back by the scarcity of document images paired with faithful structural ground truth at scale. We address this with synthetic supervision: vision–language-model-based document generation pipelines produce HTML and render it into paired images, yielding image–HTML training data without manual annotation, and can additionally mark handwriting regions explicitly via a class attribute, an annotation rarely present in real corpora. We fine-tune Pix2Struct$_{Base}$ (282M parameters) on such data generated by DocDjinn, with Claude producing the HTML targets, and evaluate structural and visual fidelity against frontier and open-source vision–language models and WebCoder-1.3B, the closest architectural precedent. Our compact model is competitive with Qwen2.5-VL-7B on structural fidelity (TreeBLEU 0.42 vs. 0.43) at roughly 25× fewer parameters, and preserves handwriting markup with high precision where larger general-purpose models flatten it. Despite never having seen web-page training data, it reaches visual-similarity parity with WebCoder-1.3B on the longer subsets of WebCode2M using 17× less fine-tuning data, suggesting that, in our setting, the training domain can matter more than model scale for image-to-HTML generation.

Keywords: Document Understanding · OCR-Free Models · Handwriting Recognition · Image-to-Markup Generation · Synthetic Data

1 Introduction

Recovering structured markup from rendered document images is a core capability for downstream document understanding, search, and re-rendering tasks.

F. Shafait et al. (Eds.): DAS 2026, LNCS 16851, pp. 229–245, 2027.
https://doi.org/10.1007/978-3-032-36207-0_14

A key challenge is the lack of annotated datasets pairing document images with visually faithful HTML: manual annotation is slow and expensive, and privacy constraints frequently prevent sharing real document collections, motivating synthetic document generation as a practical alternative. While prior work has used rendered web pages as a source of such image–HTML pairs [7,18], that domain lacks characteristics specific to documents, such as dense multi-column layouts, form and table structures, and handwritten content.

We explore a different source: recent Vision Language Model (VLM)-based document generation pipelines [17,21], which generate HTML for synthetic documents that, when rendered, yields document images paired with their underlying, visually faithful HTML. This provides a supervision signal for image-to-markup models analogous to that used in restricted domains such as academic papers [3] and web pages [7,18]. Beyond visible content, the HTML can carry an annotation real corpora rarely expose: handwriting regions marked explicitly by a class attribute [17], filled at rendering time by a diffusion handwriting model, so that each image is paired with markup distinguishing typeset from handwritten regions.

A templating engine that preserves this distinction would address a gap in existing image-to-markup systems. Pipeline-based converters [1,27] and compact VLM-based systems [13,25,30] transcribe handwritten text into typeset markup, flattening a structurally meaningful distinction. Recovering it supports downstream operations on document structure, such as re-rendering with substituted content for privacy or augmentation, or feeding templated HTML back into synthesis pipelines as seed material.

This raises two questions. First, can HTML from synthetic document generators supervise the reverse direction, training a model to recover structured markup from a rendered document image? Second, can this supervision additionally teach the model to preserve handwriting rather than flattening it? We investigate both with a compact Optical Character Recognition (OCR)-free model that produces a document image's underlying HTML, with handwriting regions marked by a `handwritten` class. We approach this as training on synthetic, teacher-generated supervision: a large VLM teacher (Claude, embedded in the DocDjinn pipeline) generates HTML targets at corpus scale, and a compact student (Pix2Struct$_{Base}$, 282M parameters) is trained to reproduce them. Unlike classical response-based knowledge distillation, the student is not trained to match the teacher's output distributions on shared inputs; the teacher-generated HTML is treated as ground-truth supervision. Pix2Struct [19], pre-trained on screenshot-to-simplified-HTML pairs, provides a natural backbone; fine-tuning adapts it to DocDjinn's richer, document-specific HTML. The closest prior work is WebCoder-1.3B [7], which fine-tunes Pix2Struct$_{Large}$ (1.3B) on 2.56 million web pages for image-to-HTML generation; our setting differs in three respects: the visual domain is documents rather than web pages, the supervision targets markup distinguishing handwritten from typeset content, and the student is roughly 4.6× smaller.

We evaluate visual fidelity, structural fidelity, and handwriting reconstruction against Claude as a frontier-VLM reference point, the open-source VLM Qwen2.5-VL-7B, and image-to-HTML baselines, on an in-distribution test split and on cross-domain transfer to web pages (WebCode2M).

Our contributions are: (i) we introduce *handwriting-aware document templating*, producing HTML from a rendered document image that distinguishes typeset and handwritten content, unlike existing image-to-markup systems that flatten this distinction; (ii) we show that VLM-based synthetic document pipelines such as DocDjinn [17] can supply supervision for this task, including handwriting-region annotations unavailable in real corpora; (iii) we show that a 282M-parameter Pix2Struct student trained on Claude-generated DocDjinn supervision is competitive with Qwen2.5-VL-7B on structural fidelity at roughly 25× fewer parameters and preserves handwriting content with high precision; and (iv) our document-trained model transfers partially to web-page templating: on the longer WebCode2M subsets it reaches visual-similarity parity with WebCoder-1.3B despite never seeing web-page data and using 17× less fine-tuning data, while the same web-trained model yields much lower structural fidelity on documents (TreeBLEU 0.12 vs. 0.42), suggesting that document-domain supervision transfers across visual domains more readily than the reverse.

2 Related Work

Synthetic Document Generation. Approaches towards generating synthetic annotated documents fall into four paradigms [17]. Template- and sampling-based methods such as SynthDoG [15] compose documents from predefined layouts, producing structured annotations but constrained to the layouts encoded in their templates. Generative Adversarial Network (GAN)- and Diffusion Probabilistic Model (DPM)-based methods [4,36] achieve strong visual fidelity but cannot generate contextually grounded content or task-specific annotations. VLM-based approaches address this by leveraging large pre-trained models for content generation: DocGenie [21] conditions a VLM on real document images, and DocDjinn [17] additionally generates task-specific annotations for downstream document understanding tasks. A distinguishing feature of DocDjinn for the present work is that the VLM-generated HTML carries explicit annotations of handwriting regions via a class attribute, which are subsequently filled by a diffusion handwriting model at rendering time. These annotations are intrinsic to the synthesis pipeline and unavailable in any real document corpus. We adopt DocDjinn as the supervision source for this work.

OCR-Free Document Models. OCR-free document models read pixels directly through a visual encoder and produce structured output through an autoregressive decoder, eliminating the error propagation associated with separate OCR stages. Donut [15] pairs a Swin Transformer encoder with a BART-based decoder for structured information extraction. Pix2Struct [19] adopts a Vision Transformer (ViT) encoder and is pre-trained on screenshot-to-simplified-HTML pairs, an objective that develops strong representations for visually rich documents. These models operate at moderate scale (143M parameters for Donut, up to 1.3B for Pix2Struct$_{LARGE}$) and provide the OCR-free backbones used in the image-to-markup studies surveyed below.

Image-to-Markup Generation. A separate but closely related line of work studies the problem of producing structured markup directly from rendered images. The task originates with Im2Tex [5] and has since been applied at varying granularities: tables (PubTabNet [40], TableFormer [24]), forms (Sarkar et al. [32]), academic pages (Nougat [3]), and web pages (WebSight [18], Design2Code [33], WebCode2M [7]). For general-document conversion, pipeline-based systems such as Docling [1] and Marker [27] chain together specialised models for layout analysis, OCR, and table-structure recognition. VLM-based systems such as olmOCR [30] and LightOnOCR [20] fine-tune compact vision–language models for PDF-to-Markdown conversion through distillation from larger teachers. Most relevant to this work, SmolDocling [25] and its successor Granite-Docling [13] achieve single-pass document conversion in compact 256–258M parameter models comparable in size to Donut and Pix2Struct, producing DocTags, a structural markup format that separates content from layout.

Across this body of work, image-to-markup is treated as a *content-recovery* problem: textual content is transcribed and structural elements are identified. The *origin* of textual content (whether a given region was typeset on the page or filled in by hand) is not preserved in the output markup. Handwritten text is OCR'd and emitted as if it were typeset, collapsing a structurally meaningful distinction. A further contrast is visual fidelity: most general-document converters emit Markdown or DocTags that capture content and logical structure but discard visual appearance, whereas templating to renderable HTML targets a representation from which the original document can be reconstructed. Recovering markup that preserves both visual fidelity and the typeset-versus-handwritten distinction has not been systematically investigated for general documents, despite being a natural fit when training supervision comes from a synthesis pipeline that explicitly annotates handwriting regions.

Synthetic Supervision from Large Models. Training compact models on data generated by large VLMs is related to response-based knowledge distillation [10,38], though it is closer to supervised training on synthetic, teacher-generated data: the student learns from generated targets rather than matching the teacher's output distributions. Prior work has applied this paradigm in adjacent settings, notably LightOnOCR [20], which trains a compact student on PDF-to-Markdown targets generated by a large VLM teacher. We adopt the same strategy for handwriting-aware document templating, with supervision generated by Claude through the DocDjinn synthesis pipeline.

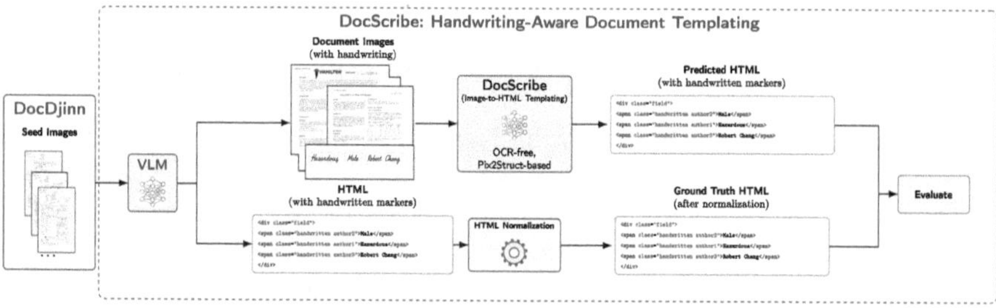

Fig. 1. Our DocScribe pipeline for handwriting-aware document templating. The VLM-based DocDjinn generator produces HTML / document image pairs, on which the DocScribe model is fine-tuned to convert document images to structured HTML. HTML normalization aligns predictions with ground truth for evaluation.

3 Approach

We fine-tune Pix2Struct$_{Base}$ [19] on (image, HTML) pairs produced by DocDjinn [17], as illustrated in Fig. 1. The HTML, generated by the Claude VLM as part of the DocDjinn pipeline contains two classes of synthesis-specific markers: handwriting regions carry a `handwritten` class with author subclasses `authorN`, and visual-element regions (charts, logos, barcodes, photos, stamps) carry `data-placeholder` and `data-content` attributes. Handwriting markers are filled at rendering time by a latent diffusion synthesiser trained on the IAM [22] corpus. We use the rendering variant in which visual elements are not rasterised, so each training image contains the typeset content and the injected handwriting but leaves the visual-element regions blank.

Our model is trained to recover handwriting markup from the rendered image: marking handwritten regions with the `handwritten` class and attributing each to an author via the `authorN` label. Visual elements are out of scope; we treat their placeholders as opaque layout slots in the HTML target. The method comprises two components: HTML normalization and tokenizer vocabulary extension.

3.1 HTML Normalization

Each HTML target passes through a multi-stage canonicalization pipeline. The first stage strips markup that does not affect the rendered image: non-visual tags (`<script>`, `<meta>`, `<link>`, `<title>`, and comments), non-rendering attributes (event handlers, the `aria-*` and `data-*` families, `id`, `tabindex`, `role`, `xmlns`), no-op CSS properties such as `cursor` and `transition`, CSS rules referencing classes absent from the DOM, and class names referenced by no selector. The second stage standardizes what remains. We apply NFKC normalization to text nodes, which prevents SentencePiece encode–decode artifacts. We rewrite CSS values into a uniform form: numeric units are trimmed, colors become lowercase hex (shortened where possible), and font names are lowercased and unquoted.

Within <style> blocks, we drop @page, @import, @charset, and @font-face at-rules and sort the remaining declarations alphabetically. The output is serialized as a single dense HTML line.

DocDjinn introduces three synthesis-specific markers, each handled differently by the pipeline. Visual-element placeholders (data-placeholder, data-content) are emptied of children and stripped of decorative styling, but the elements themselves remain in the DOM so their layout slots survive. The internal signature class is removed outright, both from elements and from any CSS rule that mentions it. Handwriting markers receive the inverse treatment: every rule targeting .handwritten or .authorN is replaced by a single canonical rule applied uniformly to all documents, .handwritten{font-family:cursive;font-size:26px}, while the marker classes themselves stay on their elements and are exempt from the unstyled-class strip.

The model is therefore left with two compositional tasks on handwriting: marking handwritten regions with the handwritten class and grouping them by author through consistent authorN labels within each document. The styling itself is fixed across the corpus and need not be learned.

3.2 Tokenizer Vocabulary Extension

Pix2Struct uses a T5 SentencePiece tokenizer. Encoding normalized HTML with this tokenizer produces long sequences with substantial redundancy from HTML markup that recurs across documents. We extend the tokenizer with domain-specific tokens covering frequent HTML constructs in three categories: opening and closing tags, CSS property names, and attribute anchors.

Two design choices matter. First, opening tag tokens are added in *prefix form* without the closing > (e.g., <div, not <div>), so that a single token matches both the bare and attributed forms under SentencePiece's literal substring matching. Closing tags are added in full form. Second, the new tokens are added as regular tokens rather than special tokens, so they appear in decoded model outputs at evaluation; only framing tokens (BOS, EOS, PAD) are stripped.

3.3 Training Objective

Given a rendered document image X and its normalized target HTML token sequence $Y = (y_1, \ldots, y_T)$, we fine-tune the model to minimize the token-level negative log-likelihood

$$\mathcal{L} = -\sum_{t=1}^{T} \log p_\theta(y_t \mid y_{<t}, X),$$

i.e. standard autoregressive cross-entropy over the HTML token sequence, starting from the pre-trained Pix2Struct$_{Base}$ checkpoint. Normalization (Sect. 3.1) is applied to all targets before training, so the model is supervised on the canonical HTML form rather than raw DocDjinn output.

4 Experiments

Given our compact 282M-parameter model DocScribe, trained on synthetic supervision generated by large-scale VLMs via the DocDjinn pipeline, our evaluation addresses two questions: (1) how closely can DocScribe approach its frontier-VLM teacher on visually faithful handwriting-aware document templating, and (2) how does DocScribe compare to existing image-to-HTML systems transferred to the document domain? To answer these, we compare DocScribe against Claude Sonnet 4.5 (a frontier-VLM reference point), Qwen2.5-VL-7B-Instruct (an open-source general-purpose VLM), and WebCoder-1.3B (Pix2Struct$_{LARGE}$ fine-tuned on web pages) on the in-distribution test split.

4.1 Experimental Setup

Dataset. The corpus consists of 155083 synthetic document samples produced by DocDjinn from 40 synthesis variants over 11 benchmark datasets (Table 1). Benchmark datasets supplied seed images for synthesis; a *variant* denotes one synthesis configuration (generation hyperparameter setting), with per-benchmark counts inherited from the DocDjinn corpus. We construct a stratified 94/3/3 split per variant with random seed 42, with a floor of one sample per variant in each split to ensure all variants appear in validation and test. Final split sizes are 145775 training, 4654 validation, and 4654 test samples.

Table 1. Base benchmarks and per-split sample counts in the training corpus. Variants per benchmark reflect different synthesis configurations inherited from DocDjinn.

Benchmark	Variants	Train	Val	Test	Total
DocVQA [23]	6	46946	1498	1498	49942
PubLayNet [39]	12	44496	1421	1421	47338
RVL-CDIP [8]	6	21884	700	700	23284
DocLayNet [29]	4	14677	468	468	15613
CORD [26]	6	6684	214	214	7112
Tobacco3482 [16]	1	3846	123	123	4092
Kleister [34]	1	3235	103	103	3441
ICDAR2019 [6]	1	1425	45	45	1515
WTQ [28]	1	1391	44	44	1479
SROIE [12]	1	948	30	30	1008
FUNSD [14]	1	243	8	8	259
Total		**145775**	**4654**	**4654**	**155083**

Implementation Details. *Tokenizer Extension.* We select tokens at a document-frequency threshold of $\geq 1\%$ on the normalized training corpus, yielding 163 additions: 66 tag tokens (e.g., `<div`, `</td>`), 95 CSS property tokens (e.g., `font-size:`, `margin-bottom:`), and 2 attribute anchors (`class=''`, `style=''`). Table 2 reports the resulting token-length distribution; the extension reduces mean encoded length by 10.5% and nearly halves the number of documents exceeding the decoder limit.

Table 2. Token-length on the normalized HTML corpus under the base Pix2Struct T5 tokenizer and the extended tokenizer with 163 added domain tokens. p99: 99th percentile of per-document token length. Pix2Struct's decoder maximum length is 4096 tokens.

	Base tokenizer	Extended tokenizer
mean	1700	1522
p99	4250	3992
#docs > 4096	2128	1222

Training Data Preparation. We use DocDjinn PDFs in the variant without rasterised visual elements. Each PDF is converted to PNG at native resolution and rescaled (aspect-ratio preserved) to a maximum longest side of 1024 pixels. At training time, Pix2Struct's image processor re-tiles each image into at most 2048 16×16 patches.

Handwriting Synthesis (Background). Handwriting is produced upstream by a latent diffusion model over IAM writer styles [22]; DocDjinn samples 8 styles, all of which appear in both training and evaluation. The model's word-level outputs are composited onto elements bearing the `handwritten` class.

Evaluation Rendering. For metric computation, predicted and reference HTML are rendered to PDF and image using DocDjinn's rendering pipeline (Playwright with headless Chromium 140; single-page PDFs sized to content via custom dimensions measurement). Documents that render to multiple pages have their first page extracted, while documents that fail to render are omitted.

Training and Inference. Fine-tuning runs on $8\times$A100 40 GB GPUs with BF16 mixed precision and gradient checkpointing. We start from `google/pix2struct-base`. Per-GPU batch size is 1; gradient accumulation of 7 steps yields an effective batch size of 56. We use AdamW with learning rate 1×10^{-4}, weight decay 0.01, linear warmup for 10% of steps and cosine decay thereafter; other optimizer parameters use PyTorch defaults.[1] Training runs for

[1] The learning rate was selected from a sweep conducted with a Donut-based model in preliminary experiments and reused here without a Pix2Struct-specific sweep.

50 epochs (130200 steps, approximately 6.3 days). For stability, we apply gradient clipping at norm 1.0 and skip optimizer steps when pre-clip gradient norm exceeds 100. The encoder consumes up to 2048 patches, while the decoder consumes up to 4096 tokens; both use eager attention. We select the checkpoint that minimizes validation cross-entropy, which corresponds to the final epoch. Generation at both validation and test time uses greedy decoding (`num_beams` = 1, `no_repeat_ngram_size` = 4, `repetition_penalty` = 1.2).

Prompting. The prompt shown in Fig. 2 was used to elicit HTML output from the control run VLMs (Claude Sonnet 4.5 and Qwen2.5-VL-7B-Instruct), paired with the document image as input. Prompting Claude Sonnet 4.5 on the 4654-document test split in batch mode consumed 9.66M input and 8.98M output tokens (2076 / 1930 per document), corresponding to 82\$ ($\approx$ 0.0176\$ per 1000 documents).

```
Exactly reproduce the attached document image as a single self-contained HTML document. When rendered in a browser, the HTML
must match the input image: same text (verbatim), same layout, same typography, same visual structure. Aim for an HTML
rendering that is visually indistinguishable from the input.

Rules:
- Output ONLY the HTML, starting with <html> and ending with </html>. No commentary, no markdown fences.
- Use one <style> block in <head> for shared CSS; inline style="..." for per-element positioning. No <script>, <meta>, <link
>, <title>, or HTML comments.
- No id, event handlers, tabindex, role, xmlns, aria-*, or data-* attributes.
- Reproduce visible text verbatim: same wording, case, punctuation, reading order. Do not paraphrase, translate, correct
typos, or invent content. Omit illegible or cropped text rather than guessing.
- Reproduce page layout via HTML structure (headings, paragraphs, divs, tables, lists, spans) and CSS (position, font-size,
font-weight, text-align, margin, padding, color, background).

Handwriting:
- Wrap each handwritten element (signatures, filled-in fields, margin notes, annotations) in class="handwritten authorN".
- N is 1-based: author1 for the first handwriting style in reading order (top-to-bottom, left-to-right), author2 for the next
 visually distinct hand, etc. Same writer (judged by slant, letter shapes, stroke style) keeps the same N across the page.
Always include an authorN class, even when only one writer is present.
- Printed text must NOT carry .handwritten.
- If handwriting is present, the <style> block must contain exactly this rule once, with no other styling for .handwritten or
 .authorN:
    .handwritten{font-family:cursive;font-size:26px}

Example: <span class="handwritten author1">John Smith</span> <span class="handwritten author2">12/05/2024</span>
```

Fig. 2. Prompt used to convert document images to HTML with Claude Sonnet 4.5 and Qwen2.5-VL-7B-Instruct.

Baselines. We compare DocScribe against the following baselines.

Claude Sonnet 4.5. Anthropic's frontier VLM, prompted zero-shot to produce HTML in the DocDjinn schema. Since Claude generated the training targets within DocDjinn, this serves as a frontier reference point for the same model under zero-shot conditions.

Qwen2.5-VL-7B-Instruct [37]. A strong open-source general-purpose VLM with documented document-parsing capabilities, prompted analogously to Claude. This baseline tests how a moderately-sized open VLM performs the task zero-shot.

WebCoder-1.3B [7]. Pix2Struct$_{Large}$ fine-tuned on WebCode2M for image-to-HTML generation on web pages, run zero-shot on our test set. This is the closest architectural and task precedent to our work and isolates the role of training domain: WebCoder-1.3B shares the Pix2Struct image-to-HTML architecture with our model and differs primarily in the fine-tuning corpus (web pages vs. documents). We follow the authors' published inference instructions.[2]

Pix2Struct$_{Base}$ (Intuned). The base `google/pix2struct-base` checkpoint run zero-shot on our test set, without fine-tuning on DocDjinn. This ablation isolates the contribution of fine-tuning from the pre-trained visual encoder and decoder.

Metrics. To assess how faithfully a model reconstructs the source document as HTML, in both rendered appearance and underlying structure, we evaluate along four axes: structural fidelity, visible-text fidelity, visual similarity, and handwriting reconstruction.

Structural Fidelity. We report two metrics on 1-height subtrees of the parsed DOM, following WebCode2M [7]. For each element with element children, a 1-height subtree consists of the parent tag and its ordered list of child tags. *Tree-BLEU* is the multiplicity-weighted fraction of candidate subtrees occurring in the reference (candidate-anchored precision); *TreeRouge* is the set-recall variant $|S(\text{cand}) \cap S(\text{ref})|/|S(\text{ref})|$.[3]

Visible-text Fidelity. Text F1 is the word-level F_1 between the lowercased, whitespace-tokenized visible text of prediction and reference, each treated as a bag of words. *Text ANLS* is the Average Normalized Levenshtein Similarity (ANLS) [2] over the joined visible-text string (order-sensitive, character-granular). Text F1 forgives reordering but punishes spelling drift; Text ANLS the reverse.

Visual Similarity. Following WebCode2M, we render both predicted and reference HTML and report *CLIP similarity* [31] (cosine similarity of CLIP image embeddings) and *Visual Score* [33], a block-level metric: text regions are detected on both renders via EasyOCR, prediction blocks greedily matched to reference blocks by text similarity, and four sub-similarities (text, position, mean RGB colour, CLIP) averaged over matched pairs and weighted by the block-match ratio. We additionally report two distribution-level Fréchet-distance metrics: *FID* [9], over Inception-V3 [35] embeddings of the rendered images, and *Layout-*

[2] The WebCode2M paper [7] refers to this model as "WebCoder"; the official repository releases it as `xcodemind/UICoder`. We follow the repository's inference code: https://github.com/CGCL-codes/naturalcc/tree/main/examples/webcode2m.

[3] WebCode2M's published "TreeBLEU" column matches our TreeBLEU (their released code, https://huggingface.co/datasets/xcodemind/webcode2m/discussions/3); the formula in their paper text corresponds to our TreeRouge.

FID [21], over LayoutLMv3 [11] `[CLS]`-token embeddings combining the page's words, word-level bounding boxes, and rendered image.[4]

Handwriting Reconstruction. For each document, we extract HTML elements bearing the `handwritten` class from both reference and prediction, yielding unordered sets R and P; each element carries visible text and an author label (`authorN`). We align R and P via minimum-cost bipartite matching (Hungarian algorithm) with text-only pairwise cost $1 - \text{ANLS}(r_i, p_j)$. On the matching π, with $S = \sum_{(i,j)\in\pi} \text{ANLS}(r_i, p_j)$, we report *Soft Precision* $S/|P|$, *Soft Recall* $S/|R|$, their harmonic mean *Soft F1* (all counting unmatched elements as zero), and *Mean Pair ANLS* $S/|\pi|$, which isolates content quality among matched pairs. We additionally report *Pairwise Authorship Agreement*, the fraction of $\binom{|\pi|}{2}$ matched-element pairs for which reference and prediction agree on whether the two elements share an author (scoring document-local partition agreement, not label equality). Missing `authorN` labels form their own cluster; documents with fewer than two matched elements are excluded from authorship aggregates. Corpus aggregates use micro-pooling.

Marker-less Handwriting Recall. Baselines that do not emit `handwritten` markup cannot be evaluated with the metrics above. For these we report *Handwriting Recall*: for each reference handwritten element, the maximum normalized Levenshtein similarity over predicted-document substring windows, averaged across reference elements. This measures only whether each handwritten span appears in the prediction, not whether it is correctly identified or grouped, and is strictly weaker than Soft Recall.

4.2 Results

We report quantitative metrics in Tables 3 and 4, with qualitative examples shown in Fig. 3.

Document Templating. Table 3 reports document templating metrics across different baselines and control runs. Claude achieves the strongest overall results, as expected for a frontier model with substantial inference cost. DocScribe produces meaningfully better outputs than the untuned Pix2Struct-Base ablation and the WebCoder-1.3B web-domain baseline, and is competitive with Qwen2.5-VL-7B on structural fidelity (TreeBLEU 0.42 vs. 0.43) despite being approximately 25× smaller.

Domain Shift and Metric Trade-offs. Our compact model achieves the highest structural fidelity of any non-frontier baseline (TreeBLEU 0.42), competitive with Qwen2.5-VL-7B at 25× fewer parameters. WebCoder-1.3B, despite being 4.6× larger, reaches only TreeBLEU 0.12: it reproduces document text reasonably well (F1 0.70) and produces visually plausible renders (Fig. 3d), yielding lower FID and Layout-FID, but generates HTML in a structure that does

[4] Visual Score uses the Design2Code v1 formulation following WebCode2M. Layout-FID's bounding boxes are extracted from the rendered PDF via `pdfplumber`.

not match the DocDjinn schema. This shows that image-to-HTML training on web-page screenshots does not transfer cleanly to document structure despite the architectural and task similarity, and that distributional visual metrics measure general plausibility rather than templating-specific correctness; TreeBLEU and Text F1 are more direct measures of the task. Combined with the untuned Pix2Struct-Base ablation (near-zero structural and text fidelity), this indicates that, in our setting, document-domain training rather than architecture is the main driver of performance. Claude outperforms our model across all metrics, trading orders-of-magnitude higher inference cost for absolute performance.

Table 3. Document templating metrics on the in-distribution test split, grouped by evaluation axis: structural fidelity (TreeBLEU, TreeRouge), text fidelity (Text F1, Text ANLS), and visual similarity (CLIP, Visual Score, FID, Layout-FID). Arrows denote metric direction. Sample count n reflects documents with parseable output. Best per column in bold.

Model	n	Structural Fidelity		Text Fidelity		Visual Similarity			
		TreeBLEU (↑)	TreeRouge (↑)	F1 (↑)	ANLS (↑)	CLIP (↑)	Visual (↑)	FID (↓)	LFID (↓)
Claude	4632	**0.60**	**0.50**	**0.98**	**0.97**	**0.86**	**0.61**	**4.94**	**0.97**
Qwen	4581	0.43	0.37	0.97	0.94	0.79	0.43	15.95	3.34
Pix2Struct$_{Base}$	4654	0.00	0.00	0.00	0.00	0.64	0.01	209.07	73.13
WebCoder-1.3B	4654	0.12	0.18	0.70	0.61	0.72	0.29	31.13	6.66
DocScribe (Ours)	4654	0.42	0.39	0.52	0.48	0.72	0.38	52.65	11.98

Handwriting Recovery. Table 4 shows results of the handwriting evaluation. Among the models that emit `handwritten` markup, DocScribe and Qwen2.5-VL-7B exhibit complementary strengths. DocScribe achieves higher Soft Precision (0.77 vs. 0.52) and Mean Pair ANLS (0.81 vs. 0.72), indicating that when our model labels a region as handwritten, the prediction is generally correct and the matched content is well-reproduced. Qwen achieves higher Soft Recall (0.41 vs. our 0.31) and Pairwise Authorship Agreement (0.75 vs. our 0.38); the two models tie on Soft F1 (0.46 vs. 0.45).

The gap between our Soft Recall and marker-less Handwriting Recall (0.31 vs. 0.61) indicates that the missing content is largely present in our predictions but lacks the `handwritten` class attribute: the model is conservative in committing to handwriting markup, a pattern also reflected in the low Pairwise Authorship Agreement (0.38). Claude achieves the strongest handwriting metrics overall (Soft F1 0.87, Authorship Agreement 0.90). Our 282M-parameter model preserves handwriting content at substantially smaller scale, trading recall for precision relative to the 7B Qwen baseline.

4.3 Cross-Domain Transfer to Web Pages

We evaluate our document-trained model on the WebCode2M test set [7] without any web-page training, comparing against WebCoder-1.3B, the released Pix2Struct$_{LARGE}$ checkpoint fine-tuned on 2.56 million web pages (Table 5). The

(a) Target (b) DocScribe

(c) Qwen2.5-VL-7B-Instruct (d) WebCoder-1.3B (e) Claude Sonnet 4.5

Fig. 3. Example renderings comparing DocScribe and baselines on a test-split document. (a) Target: the reference HTML rendered. (b) DocScribe preserves the overall layout, headings, and body-text blocks, with its main errors in fine typographic detail rather than structure. (c) Qwen2.5-VL transcribes the text correctly but reformats the layout. (d) WebCoder renders a structurally different document with a reformatted header and denser paragraphs, despite its web-page training. (e) Claude reproduces structure and content closely. The untuned Pix2Struct-Base baseline produced near-empty output and is omitted.

comparison is asymmetric in architecture (4.6× fewer parameters) and in training data: WebCoder saw roughly 17× more fine-tuning samples than DocScribe. WebCoder outperforms us on WebCode2M-Short, as expected for an in-domain specialist on shorter web pages, and retains an advantage on structural fidelity (TreeBLEU) across all subsets. On WebCode2M-Mid and WebCode2M-Long, however, our document-trained model reaches parity with WebCoder on visual similarity (Visual Score and CLIP) despite being smaller and trained on far fewer samples. The contrast with WebCoder's much lower structural fidelity on documents (TreeBLEU 0.12, Table 3) points to an asymmetry: document-domain training transfers partially to web pages, while web-domain training yields much lower structural fidelity on documents. We attribute the visual-similarity transfer to two plausible factors: the structural complexity of the DocDjinn corpus (tables, forms, multi-column layouts) shares more primitives with longer web pages than short web pages do, and our decode length cap (4096 tokens) sub-

Table 4. Handwriting reconstruction metrics on the in-distribution test split. PAA: Pairwise Authorship Agreement. HW Recall: marker-less Handwriting Recall, applicable to all models. Models that do not emit **`handwritten`** markup (Pix2Struct$_{Base}$, WebCoder-1.3B) are evaluated only on HW Recall. Sample count n reflects documents with parseable output. Best per column in bold.

Model	n	Mean Pair ANLS (↑)	Soft Precision (↑)	Soft Recall (↑)	Soft F1 (↑)	PAA (↑)	HW Recall (↑)
Claude	4632	**0.90**	**0.85**	**0.89**	**0.87**	**0.90**	**0.93**
Qwen	4581	0.72	0.52	0.41	0.46	0.75	0.85
Pix2Struct$_{Base}$	4654	–	–	–	–	–	0.04
WebCoder-1.3B	4654	–	–	–	–	–	0.50
DocScribe (Ours)	4654	0.81	0.77	0.31	0.45	0.38	0.61

stantially exceeds WebCoder's (2048), better accommodating long-form HTML targets. We do not claim DocScribe as a web-page templating model, but the cross-domain result suggests that document-domain supervision yields image-to-HTML representations that transfer partially to web pages, particularly on longer, denser layouts.

Table 5. DocScribe (trained on documents only) vs. WebCoder-1.3B (trained on web pages) on the WebCode2M test set [7], split into Short, Mid, and Long subsets by code length. Mean ± standard deviation; best per column in bold. Metrics computed using the WebCode2M evaluation protocol.

Model	WebCode2M-Short			WebCode2M-Mid			WebCode2M-Long		
	Visual (↑)	CLIP (↑)	TreeBLEU (↑)	Visual (↑)	CLIP (↑)	TreeBLEU (↑)	Visual (↑)	CLIP (↑)	TreeBLEU (↑)
WebCoder-1.3B [7]	$.78 \pm .25$	$.73 \pm .13$	$.35 \pm .17$	$.69 \pm .19$	$.71 \pm .10$	$.22 \pm .11$	$.65 \pm .21$	$.69 \pm .12$	$.15 \pm .07$
WebCoder-1.3B†	$.75 \pm .19$	$\mathbf{.72 \pm .12}$	$\mathbf{.36 \pm .25}$	$.72 \pm .16$	$.71 \pm .11$	$\mathbf{.39 \pm .20}$	$.67 \pm .18$	$\mathbf{.75 \pm .19}$	$\mathbf{.39 \pm .18}$
DocScribe (Ours)‡	$\mathbf{.76 \pm .14}$	$.71 \pm .11$	$.13 \pm .08$	$\mathbf{.75 \pm .11}$	$\mathbf{.72 \pm .11}$	$.18 \pm .10$	$\mathbf{.69 \pm .16}$	$.68 \pm .11$	$.22 \pm .11$

† We re-run WebCoder-1.3B evaluation following the official repository implementation. Our reproduction diverges from the published numbers on the Mid and Long subsets, particularly for TreeBLEU; we report our reproduction as the apples-to-apples comparison against DocScribe, evaluated under identical conditions, but were unable to fully reconcile the difference with the published results. Excluding 8/2 of 256 samples on Short/Mid that failed evaluation.
‡ Excluding 43/37/20 of 256 samples on Short/Mid/Long that failed evaluation.

5 Conclusion

We introduced *handwriting-aware document templating* and showed that a 282M-parameter Pix2Struct student trained on Claude-generated DocDjinn supervision performs it competitively against larger general-purpose VLMs: it nearly matches Qwen2.5-VL-7B on structural fidelity at 25× fewer parameters and preserves handwriting content with high precision. Cross-domain evaluation on WebCode2M further shows that document-domain supervision transfers across visual domains: on the longer subsets, our model reaches visual-similarity parity with WebCoder-1.3B despite never having seen web-page data and using 17× less fine-tuning data, while the converse transfer yields much lower structural fidelity. Overall, our findings suggest that, in these experiments, the training domain can matter more than model scale for image-to-HTML generation.

5.1 Limitations

Scope of Evaluation. Our in-distribution evaluation uses a held-out test split drawn from the same 40 synthesis variants and 8 writer styles seen in training, demonstrating that the model learns the task on the DocDjinn distribution. A real-world counterpart is difficult to construct: we are not aware of a public document corpus pairing scanned images with faithful structural HTML and explicit handwriting-region annotations, precisely the supervision the synthesis pipeline provides. We do, however, evaluate out-of-distribution generalization on real web pages (Sect. 4.3), where our model is competitive on visual similarity. Holding writers out would have cost a substantial fraction of the handwriting training signal at the current corpus size; scaling to a larger writer pool and broader source distributions would enable principled held-out evaluation.

Handwriting Recall. DocScribe commits to the `handwritten` class conservatively, achieving high precision but low Soft Recall (0.31). The higher marker-less Handwriting Recall (0.61) indicates the missing content is largely present but unmarked: the model transcribes handwriting yet underuses the markup. Loss reweighting or sampling toward the rare positive class are natural remedies.

5.2 Future Work

Validation cross-entropy was still decreasing at termination, suggesting extended training may help. The cross-domain result further motivates joint training on documents and web pages to study how training-domain mixing affects image-to-HTML capability, and ablating which aspects of document-domain supervision drive the partial transfer.

Acknowledgments. This work was partially funded by the German Federal Ministry of Research, Technology and Space (BMFTR), Project 'SLIMDOC' (FKZ: 13HAW15PX4).

Disclosure of Interests.. The authors have no competing interests to declare that are relevant to the content of this article.

References

1. Auer, C., et al.: Docling technical report. arXiv preprint arXiv:2408.09869 (2024)
2. Biten, A.F., et al.: Scene text visual question answering. In: ICCV, pp. 4290–4300 (2019)

3. Blecher, L., et al.: Nougat: neural optical understanding for academic documents. In: The Twelfth International Conference on Learning Representations (2024)
4. Bui, Q.A., et al.: Automatic synthetic document image generation using generative adversarial networks: application in mobile-captured document analysis. In: 2019 International Conference on Document Analysis and Recognition (ICDAR), pp. 393–400 (2019). https://doi.org/10.1109/ICDAR.2019.00070
5. Deng, Y.: Image-to-markup generation with coarse-to-fine attention. In: International Conference on Machine Learning, pp. 980–989 (2017)
6. Gao, L., et al.: ICDAR 2019 competition on table detection and recognition (CTDAR). In: ICDAR, pp. 1510–1515 (2019)
7. Gui, Y., et al.: Webcode2m: a real-world dataset for code generation from webpage designs. In: Proceedings of the ACM on Web Conference 2025, pp. 1834–1845 (2025)
8. Harley, A.W., et al.: Evaluation of deep convolutional nets for document image classification and retrieval. In: ICDAR, pp. 991–995 (2015)
9. Heusel, M., et al.: Gans trained by a two time-scale update rule converge to a local nash equilibrium. Adv. Neural Inf. Process. Syst. **30** (2017)
10. Hinton, G., et al.: Distilling the knowledge in a neural network. In: NIPS Deep Learning and Representation Learning Workshop (2015)
11. Huang, Y., et al.: Layoutlmv3: pre-training for document AI with unified text and image masking. In: Proceedings of the 30th ACM International Conference on Multimedia, pp. 4083–4091 (2022)
12. Huang, Z., et al.: ICDAR2019 competition on scanned receipt OCR and information extraction. In: ICDAR, pp. 1516–1520 (2019)
13. IBM Research: Granite-docling: End-to-end document conversion (2025). https://www.ibm.com/new/announcements/granite-docling-end-to-end-document-conversion. Accessed 18 May 2026
14. Jaume, G., et al.: FUNSD: a dataset for form understanding in noisy scanned documents. In: 2nd International Workshop on Open Services and Tools for Document Analysis, OST@ICDAR 2019, pp. 1–6 (2019)
15. Kim, G., et al.: Ocr-free document understanding transformer. In: Computer Vision – ECCV 2022: 17th European Conference, Tel Aviv, Israel, 23–27 October 2022, Proceedings, Part XXVIII, pp. 498–517. Springer-Verlag (2022). https://doi.org/10.1007/978-3-031-19815-1_29
16. Kumar, J., et al.: Structural similarity for document image classification and retrieval. PRL **43**, 119–126 (2014)
17. Lamott, M., et al.: Docdjinn: controllable synthetic document generation with VLMs and handwriting diffusion. Int. J. Docu. Anal. Recogn. (2026). under review
18. Laurencon, H.: Unlocking the conversion of web screenshots into html code with the websight dataset. arXiv:2403.09029 arXiv preprint (2024)
19. Lee, K., et al.: Pix2struct: screenshot parsing as pretraining for visual language understanding. In: International Conference on Machine Learning, pp. 18893–18912 (2023)
20. LightOn AI: Making knowledge machine-readable: introducing LightOnOCR-1B (2025). https://lighton.ai/lighton-blogs/making-knowledge-machine-readable. Accessed 18 May 2026
21. M., H.P.: Docgenie: a framework for high-fidelity synthetic document generation via seed-guided multimodal LLM and document-aware evaluation. In: Synthetic Data for Computer Vision Workshop @ CVPR 2025 (2025)
22. Marti, U.V., et al.: The IAM-database: an English sentence database for offline handwriting recognition. Int. J. Doc. Anal. Recogn. **5**(1), 39–46 (2002)

23. Mathew, M., et al.: DOCVQA: a dataset for VQA on document images. In: WACV (2021)
24. Nassar, A., et al.: Tableformer: table structure understanding with transformers. In: Proceedings of the IEEE/CVF Conference on Computer Vision and Pattern Recognition, pp. 4614–4623 (2022)
25. Nassar, A., et al.: Smoldocling: an ultra-compact vision-language model for end-to-end multi-modal document conversion. In: Proceedings of the IEEE/CVF International Conference on Computer Vision, pp. 21972–21983 (2025)
26. Park, S., et al.: CORD: a consolidated receipt dataset for post-OCR parsing. In: Document Intelligence Workshop at NeurIPS (2019)
27. Paruchuri, V., et al.: Marker: convert PDF to markdown + JSON quickly with high accuracy (2024). https://github.com/datalab-to/marker. Accessed 18 May 2026
28. Pasupat, P., et al.: Compositional semantic parsing on semi-structured tables. In: ACL, pp. 1470–1480 (2015)
29. Pfitzmann, B., et al.: DocLayNet: a large human-annotated dataset for document-layout segmentation. In: Proceedings of 28th ACM SIGKDD Conference on Knowledge Discovery and Data Mining, pp. 3743–3751 (2022)
30. Poznanski, J., et al.: OLMOCR: unlocking trillions of tokens in pdfs with vision language models. arXiv:2502.18443 arXiv preprint (2025)
31. Radford, A., et al.: Learning transferable visual models from natural language supervision. In: International Conference on Machine Learning, pp. 8748–8763 (2021)
32. Sarkar, M., et al.: Document structure extraction using prior based high resolution hierarchical semantic segmentation. In: European Conference on Computer Vision, pp. 649–666 (2020)
33. Si, C., et al.: Design2code: benchmarking multimodal code generation for automated front-end engineering. In: Proceedings of the 2025 Conference of the Nations of the Americas Chapter of the Association for Computational Linguistics: Human Language Technologies (Volume 1: Long Papers), pp. 3956–3974 (2025)
34. Stanisławek, T., et al.: Kleister: key information extraction datasets involving long documents with complex layouts. In: Lladós, J., Lopresti, D., Uchida, S. (eds.) ICDAR 2021. LNCS, vol. 12821, pp. 564–579. Springer, Cham (2021). https://doi.org/10.1007/978-3-030-86549-8_36
35. Szegedy, C., et al.: Rethinking the inception architecture for computer vision. In: Proceedings of the IEEE Conference on Computer Vision and Pattern Recognition, pp. 2818–2826 (2016)
36. Tanveer, N., et al.: Diffusion models for document image generation. In: Document Analysis and Recognition - ICDAR 2023, pp. 438–453. Springer Nature Switzerland (2023)
37. Wang, P., et al.: Qwen2-vl: enhancing vision-language model's perception of the world at any resolution (2024)
38. Xu, X., et al.: A survey on knowledge distillation of large language models (2024)
39. Zhong, X., et al.: PubLayNet: largest dataset ever for document layout analysis. In: ICDAR, pp. 1015–1022 (2019)
40. Zhong, X., et al.: Image-based table recognition: data, model, and evaluation. In: European Conference on Computer Vision, pp. 564–580 (2020)

Evaluating Multimodal Large Language Models for End-to-End Diagram Recognition

Moritz Burmester, Jonas Scharfenberger(✉), and Burkhardt Funk

Institute of Information Systems, Leuphana University Lüneburg, Lüneburg, Germany
moritz.burmester@stud.leuphana.de,
{jonas.scharfenberger,burkhardt.funk}@leuphana.de

Abstract. Graph-structured diagrams (GSD) are frequently used in documents to visualize processes or relationships in an easily understandable way for humans. Extracting the underlying information from the diagram images in a structured format is important for various downstream applications. In contrast to prior work, which either focuses on visual question answering (VQA) for diagram understanding or relies on computer vision pipelines for full diagram recognition, we systematically test three open-weights multimodal large language models (MLLM) for end-to-end graph parsing under different experimental settings and compare to an object-detection baseline. We curate a publicly available dataset of 843 hand-drawn and machine-generated diagrams from related work covering flowcharts, business process modeling diagrams, and others. We evaluate the extractions with string matching against ground truth. Our findings show that MLLMs can compete with traditional computer vision approaches on simple diagrams but still struggle with complex layouts and relational extraction.

Keywords: diagram recognition · multimodal large language models · flowcharts · graph-structured diagrams · graph parsing

1 Introduction

Diagrams can be an effective way to communicate complex ideas or relationships. An important subset is graph-structured diagrams (GSD). By this term, we mean diagrams consisting of texts, nodes, and edges between nodes that are used to visualize relations or process flows. GSDs are frequently used by researchers and practitioners in the form of flowcharts (FC), business process modeling notation (BPMN) diagrams, and finite automata (FA), among others.

At their core, diagrams are designed to effectively communicate, for example, complex processes or relations to other humans. To that end, diagrams use shapes, text labels, and spatial arrangement to represent information in a two-dimensional layout. While creating a clear and effective diagram can already be

F. Shafait et al. (Eds.): DAS 2026, LNCS 16851, pp. 246–262, 2027.
https://doi.org/10.1007/978-3-032-36207-0_15

a challenging task, so is extracting its underlying information in a structured and machine-readable format. This task is usually referred to as diagram recognition or graph parsing in related work. It involves recognizing the structure of the diagram, extracting text information, and matching structural elements and texts. The desired output of diagram recognition systems is a structured text representation (e.g., JSON-formatted) of the GSD.

There is growing interest in making GSDs machine-readable, as they enable a broad range of downstream tasks. These include, among others, code generation and knowledge extraction from flowcharts [5,14], extracting structured data from organizational charts or ownership diagrams in financial documents [19], extracting models from UML diagrams [7,9], and model verification in safety-critical systems [32]. Diagram recognition systems can also facilitate the transfer of hand-drawn diagrams to a suitable digital format, which can enable further analysis, search and editing [17].

Graph parsing systems typically consist of three components: (1) a detection or segmentation model, (2) an optical character recognition (OCR) engine, and (3) a matching heuristic to reconstruct the diagram structure. Traditional approaches, such as "Tahuti" [13], purely rely on rule-based algorithms. Recent approaches use fine-tuned off-the-shelf object detection models, such as YOLO, or custom architectures such as Arrow R-CNN [23]. The aforementioned models only detect and classify the diagram's structural elements. Matching the recognized objects and extracting the text elements is handled separately. In contrast, multimodal large language models (MLLMs) that can simultaneously handle text and image inputs and generate text outputs have the potential to provide a unified solution to the diagram parsing problem. However, the current capabilities of MLLMs for diagram recognition are mostly limited to visual question answering (VQA) benchmarks [15,29,31,38] or evaluated on small datasets [7].

In our study, we seek to contribute to the ongoing efforts in diagram recognition and graph parsing by systematically evaluating the capabilities of MLLMs. Our contribution is twofold: (1) we curate a publicly available dataset[1] with a test set of 843 diagram Image-JSON pairs from related work across multiple diagram domains, and (2) we systematically evaluate three open-weight MLLMs (Gemma4, Qwen3.5, and InternVL-3.5) on this dataset under different settings to assess their ability to extract textual, relational, and positional information from diagrams in a structured format.

2 Related Work

2.1 Multimodal Large Language Models

Multimodal large language models (MLLMs) integrate various data types such as text and images to enhance understanding of spatial and conceptual information [34]. Recent open-weights MLLMs include InternVL3.5 [33], Gemma4 [12] and

[1] https://github.com/diagram-parsing/mllm-diagram-recognition.

the Qwen3.5 series [21]. An important measure for the performance of MLLMs are the MMMU/MMMUPro benchmarks, which evaluate multimodal reasoning across diverse academic subjects using diagrams, charts, and figures [35,36]. Qwen3.5-27B and Gemma4-31B achieve a score of 86.1 [20] and 85.2 [11] on MMMUPro respectively. To the best of our knowledge, no benchmark scores are reported for InternVL3.5-14B on either of the benchmarks, but its larger variant InternVL3.5-30B-A3B reaches 77.7 on MMMU [18]. For reference, a recent commercial model (GPT-5.4) achieves a score of 82.1 according to the MMMUPro leaderboard [35]. While these scores may be optimized to beat the benchmarks, we think that it also highlights the growing capabilities of open-source MLLMs.

2.2 Evaluation of MLLMs in the Diagram Domain

Systematic evaluations of multimodal large language models (MLLMs) in the diagram domain are primarily based on large-scale visual question answering (VQA) benchmarks. Datasets such as FlowchartQA [31] and FlowVQA [15] focus on geometric and topological reasoning over diagrams, typically in a multiple-choice setting. For instance, models are evaluated on questions about node connectivity (e.g., "Does A connect to B?") or spatial properties (e.g., "Which is the leftmost node?"), rather than requiring reconstruction of an underlying structured representation.

On the other hand, existing benchmarks for structured information extraction from diagrams often focus on extracting the content of specific nodes or relationships between nodes based on natural language descriptions. FlowCE employs manually curated real-world flowchart images paired with human-authored QA annotations spanning multiple evaluation dimensions, including information extraction. In particular, some questions target subsets of the diagram, such as "What is the text content in all the green boxes in this flowchart?" [38]. In related work, Hou et al. [15] also evaluate MLLMs using multiple-choice QA tasks designed to assess visual-language understanding and reasoning. Their evaluation includes questions such as "How many text labels are there in the diagram?" and "How many entities are connected to cow?", which primarily measure the ability to identify entities (nodes) and reason about their relationships rather than to reconstruct the full diagram structure. Other evaluations are further limited in scale. For instance, only four images are used to evaluate an MLLM's ability to convert UML diagrams to PlantUML [7]. Overall, prior work primarily evaluates MLLMs on diagram understanding through localized question answering or partial extraction tasks. In contrast, we focus on the task of diagram recognition and evaluate the capabilities of MLLMs to faithfully extract and reconstruct the underlying graph of a diagram image in a structured data format.

2.3 Computer Vision Approaches to Diagram Recognition

In diagram recognition, approaches typically vastly differ between online and offline diagram recognition. In online recognition of hand-written diagrams, we assume to have access to the individual strokes of the diagram and oftentimes also to the temporal information of the strokes. Offline recognition refers to the assumption that we only have an image of the hand-written or machine-generated diagram. In the online setting, related work usually involves grouping strokes, classifying strokes to symbols or arrows, and identifying related symbols based on the vicinity to arrow tails or heads. The classification can be template-based (on geometric features) or based on machine learning models. More recent approaches such as SpaceGTN [16] use graph neural networks to tackle this task. This work can also be applied to the offline scenario through a set of pre-processing steps to divide the diagram into strokes. However, the majority of approaches in the offline setting use object detection or segmentation models to recognize the structural elements of the diagrams. In post-processing steps, the detected arrows are usually matched to detected nodes based on distance-based matching heuristics. Notable works in this field are Arrow R-CNN [23] and DiagramNet [25]. Other approaches rely on simply fine-tuning off-the-shelf object detection models (e.g., YOLO) on small datasets of manually annotated diagrams or large datasets of synthetic datasets which are automatically annotated (e.g., [28]).

2.4 Diagram Recognition Datasets

To evaluate the aforementioned diagram recognition models, various datasets have been proposed in related work. These datasets cover flowcharts, finite automata (FA), business process modeling notation (BPMN) diagrams, UMLs, block diagrams, structural equation models, and iStar diagrams and vary in their complexity.

In 2011, the CLEF-IP dataset was introduced, which consists of machine-generated flowcharts extracted from patent documents. Around the same time, early hand-drawn diagram datasets like the OHFCD (FC_A) dataset were proposed, which contains 419 (train: 248, test: 171) hand-drawn flowchart images [1]. Bresler et al. [3] later introduced an online hand-drawn flowchart dataset containing 672 (train: 280, val: 196, test: 196) images collected from multiple participants. In a separate work, they also introduced an online dataset of 300 (train: 132, val: 84, test: 84) hand-drawn finite automata collected under a similar setting [4]. Due to the lack of a large dataset of hand-drawn diagrams, Gervais et al. [10] introduce the DIDI dataset which consists of 22,287 (train: 16,717, validation: 2,785, test: 2,785) hand-drawn flowcharts. The authors let 364 participants re-draw randomly generated GraphViz graphs. Schäfer et al. [26] extend the existing corpus datasets by an offline hand-drawn BPMN diagram dataset. It consists of 502 (train: 308, validation: 102, test: 92) photographs of hand-drawn BPMN models from 107 participants. This dataset is more complex due to the higher number of different classes, document background noise, and number of symbols and edges per diagram. Bhushan and Lee [2] present a dataset of 502 (train: 300, validation: 106, test: 96) block diagram images obtained through web crawling and manually annotating the images. Another dataset that is not specifically created for diagram recognition purposes but can be used for such is

SAP-SAM [30]. This dataset consists of more than a million modeling files for diagrams, which can be reconstructed and used for diagram recognition tasks. Other large datasets of machine-generated diagrams are synthetically generated using GraphViz and use random configurations of nodes, edges, and text labels to create a diverse set of images [22]. Datasets that are not publicly available cover structural equation models [22] and other hand-written flowchart diagrams [37].

3 Methodology

Our goal is to obtain a structured text representation of a given graph-structured diagram. Given a GSD image from a certain domain (FA, BPMN, etc.), we task MLLMs to parse the GSD into a structured data format. To this end, we curate a publicly available test dataset of 843 image-JSON pairs consisting of a variety of diagram domains from related work, which encompass hand-drawn and machine-generated diagrams, as well as synthetic diagram images. The text representation of the graph structure consists of a set of node objects as well as edge objects which represent relations between nodes. A node object consists of a text label and a class which is usually depicted through the node shape. An edge object is represented by a 4-tuple of a head node, a tail node, a class, and a text label. We note that certain diagram domains may be extended to also depict node clusters (e.g., pools and lanes in BPMNs), and a more nested representation for node objects in the case of UMLs through a list of attributes and methods, however, we do not include these in our evaluation. Illustrative examples of our proposed data format can be found in our GitHub repository

We evaluate the performance of the MLLMs across a range of experiments and settings, including (1) a comparison to an object-detection (OD) baseline, (2) end-to-end recognition, which evaluates the MLLMs capacity to reconstruct the entire graph structure given a (few-shot) prompt and an input image, (3) isolated edge recognition, where we provide ground-truth node names to the MLLM to focus on the task of extracting relations between nodes, and (4) positional recognition, where we prompt the model to generate bounding boxes for nodes and edges.

3.1 Evaluation Strategy

Diagram recognition models are traditionally evaluated based on node and edge recognition. Evaluations of the computer vision approaches are based on measuring the intersection over union (IoU) between the predicted bounding boxes and ground truth bounding boxes. To evaluate the positional recognition, we adopt the node and edge recognition metrics used by Schäfer et al. [23] where we consider a lower IoU threshold (0.5 vs. 0.8):

Node$_{bbox}$: A detected node is considered correct if (a) its predicted bounding box exceeds an IoU threshold of 0.5 with the ground truth, and (b) its predicted class matches the ground truth class.

Edge$_{bbox}$: An extracted edge is considered correct if (a) its predicted bounding box exceeds an IoU threshold of 0.5 with the ground truth, and (b) the edge is matched to the correct head and tail node, also evaluated at an IoU threshold of 0.5.

It is important to note that the evaluation strategy above ignores the correctness of textual information. Since our evaluation goes beyond traditional computer vision approaches and also considers the correctness of textual information, we modify the evaluation criteria. To evaluate the correctness of extracted elements, we first normalize all relevant strings (node text labels, edge text labels, classes) by removing whitespace, underscores, and converting text to lowercase. Node matching is solved as a minimum-cost assignment problem using the Hungarian algorithm, with normalized Levenshtein distance as the cost function between node text labels and a threshold of τ. We evaluate at a threshold of $\tau = 0$ (exact string matching) and $\tau = 0.1$ (accounting for small deviations in the text labels). For matching ground truth and predicted edges, the predicted endpoint node IDs are remapped to their matched ground-truth IDs. Each edge is then represented as a tuple consisting of its tail node, its head node, and optionally its class and label. We pose the following conditions for node and edge recognition scores:

Node$_{text}$: An extracted node is considered correct if it matches the ground truth node in the (a) text label and (b) class label.
Edge$_{text_{part}}$: An extracted edge is considered correct if it matches the ground truth edge in the (a) head node and (b) tail node.
Edge$_{text_{full}}$: An extracted edge is considered correct if it matches the ground truth edge in the (a) head node, (b) tail node, (c) class label, and (d) edge label.

In order to evaluate the model's accuracy of extracted diagram components, we compute precision, recall, and report the micro F1-score.

3.2 Prompting Strategy

We use a "training" set of 100 images to develop an n-shot ($n \in \{0, 2, 5\}$) prompt and to tune the temperature parameter (model default vs. 0.1), where a lower temperature of 0.1 yields better results.

We use three different variations of the few-shot prompt, each targeting a distinct aspect of the diagram recognition evaluation. The *end-to-end extraction* prompt is our base prompt which specifies a JSON schema, lists of valid node and edge classes, the diagram type, and n image-JSON examples. The *bounding box detection* variant extends the base prompt by additionally requesting bounding boxes for every recognized node and edge. This information is also added to the image-JSON examples for this prompt. In the *isolated edge recognition* variant, ground truth node annotations are provided for each of the n examples as well as for the query image. Thus, we seek to decouple edge recognition from node recognition by eliminating node-level extraction errors that are not caught by the relaxed Levenshtein distance evaluation.

We modify some important MLLM configuration parameters for our task of diagram recognition. First, we set a maximum number of 8192 output tokens for all models. Since diagram recognition is sensitive to the resolution of the input images, we further raise each model's per-image visual token budget. The exact parameter differs across the MLLMs. For Gemma4 we increase `max_soft_tokens` from the default of 280 to 560 for end-to-end extraction and 1120 for bounding box extraction. For Qwen3.5 we increase `max_pixels` to $2048 \cdot 28 \cdot 28$, corresponding to an upper limit of roughly 2048 tokens per image. For InternVL-3.5 we increase the parameter `max_dynamic_patch` from 4 to 8, doubling the number of tiles into which each image may be partitioned.

3.3 Evaluation Datasets

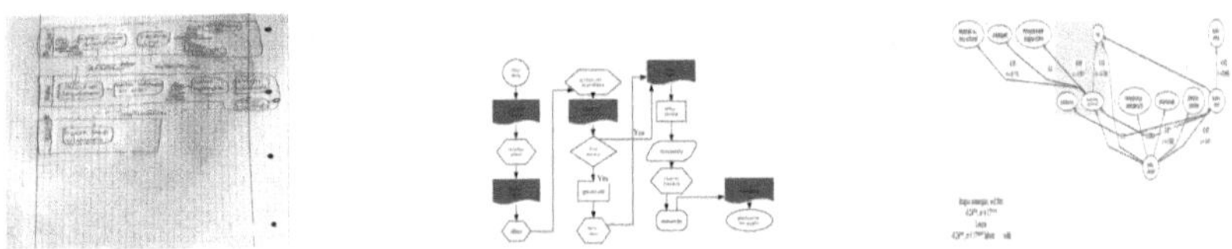

(a) hdBPMN [26] (b) SYN-FC [22] (c) SYN-SEM [22]

Fig. 1. Exemplary images from the diagram datasets used for our evaluation.

We use a mixture of publicly available datasets from related work that contain text and bounding box annotations, as well as novel datasets covering flowcharts, finite automata, BPMN, UML, and SEM diagrams (see Fig. 1 and Table 1). The annotations are publicly available on the GitHub repository.

The three hand-drawn datasets consist of 146 (FC_A), 83 (FA), and 80 (hdBPMN) images. We excluded 25 images from the test set of the FC_A data and 12 images from the hdBPMN test set, since these images do not contain any text information and thus do not fit our extraction task. Lastly, we excluded one image from the FA test set, since we found errors in the annotation. Among the hand-drawn datasets, the hdBPMN dataset is the most challenging, containing on average 14.5 nodes and 18.1 edges. The Sketch2Process system [27] achieves F1 scores of 0.958 for recognizing shapes and 0.93 for edges. While their system also extracts texts, they only report character error rates (mean of 5.5%). The FC_A and FA datasets are comparatively easy, where F1 scores of 0.998 for nodes and 0.967 for edges for the FC_A dataset [25] and an averaged F1 over all classes of 0.993 for the FA dataset [23] are reported.

We also include four datasets that are machine-generated and cover structural equation models (SEM), flowcharts (FC), UMLs, and BPMNs. The SEM dataset consists of images of structural equation models from information systems research papers which contain a comparatively large number of edge labels (8.2 per image) that have to be assigned to the correct edge. The CBD [2] dataset also covers diagrams with interrupted edges as well as interlinked and regular edges. The remaining datasets (SAM-UML, SAM-BPMN) are generated from the underlying JSON files published by Sola et al. [30] and have not been used for the purpose of diagram recognition in related work. The BPMN data contains a high average number of nodes and edges per image. For the SAP-SAM datasets and the SEM dataset, there are no benchmark scores reported in related work.

Lastly, since MLLMs are trained on large web-crawled corpora we cannot rule out that the aforementioned datasets have been previously seen during training. To assess generalization beyond potentially memorized examples, we additionally evaluate on two synthetic datasets of machine-generated diagrams that resemble SEMs and flowcharts (SYN-SEM and SYN-FC). These diagrams are generated using the GraphViz engine and contain random words as node text labels.

Table 1. Statistics of our test datasets. N: Number of images; Nodes/img, Edges/img: Avg. number of nodes/ edges per image; NC, EC: Number of node/ edge/ classes.

Dataset	N	Nodes/img	Edges/img	NC	EC
FC_A	146	7.12	7.2	5	1
FA	83	4.92	9.48	2	1
hdBPMN	80	14.53	18.05	10	3
CBD	96	7.02	6.89	1	1
SEM	132	9.3	10.39	1	2
SAM-UML	50	8.28	7.94	5	8
SAM-BPMN	50	12.4	12.48	22	4
SYN-FC	100	17.3	19.5	8	1
SYN-SEM	106	11.5	12.8	1	2

4 Results

In this section, we present the main findings of our experiments. For a more detailed breakdown of our results, we refer the reader to our GitHub repository, which provides fine-grained scores for all experiments as well as the code necessary to reproduce our results.

4.1 Object Detection Baseline

As mentioned in the related work section, computer-vision based diagram recognition achieves near-perfect recognition scores on easy datasets (e.g., recognition scores of 0.998 for nodes and 0.967 for edges on FC_A [25]). However, to the best of our knowledge, these evaluations typically ignore the textual content of nodes and edges during their evaluation.

We address this evaluation gap by re-running approaches from related work [2,23,26] and including a text extraction step. For object detection, we re-implement Arrow R-CNN [23] on datasets where node and edge bounding boxes as well as arrow keypoint annotations are available (FA, FC_A, hdBPMN, and SEM) and Faster R-CNN on datasets where keypoint annotation is unavailable (CBD). For reconstructing the diagram structure, we follow the methods outlined in related work for BPMNs [26], hand-drawn finite automata and flowcharts [23] and machine-generated flowcharts [2]. For extracting the textual content of nodes and edges we investigate two different methods. First, we use the traditional OCR engine PaddleOCR [8]. Second, we prompt the Qwen3.5-4B model to extract the text from a region indicated by a colored bounding box.

Without considering texts, our node and edge recognition F1 scores (evaluated at an IoU threshold of 0.5) mostly reproduce scores reported in related work [23,26]: node recognition F1 reaches 0.9976 (FC_A), 1.0 (FA), and 0.9022 (hdBPMN), and edge recognition F1 reaches 0.9244 (FC_A), 0.9728 (FA), and 0.8337 (hdBPMN). However, the edge recognition scores from related work are higher, especially on the hdBPMN dataset (0.918) [26].

When requiring the textual content of nodes and edges to match instead of their bounding boxes to sufficiently overlap, the node and edge recognition scores drastically drop across all datasets. This indicates that text extractions can be a major bottleneck in the diagram reconstruction. The performance gap is most notable on the hand-drawn datasets, where PaddleOCR yields node recognition scores of only 0.4085 (FC_A), 0.7083 (FA), and 0.2049 (hdBPMN), with edge recognition scores of 0.1077, 0.4927, and 0.011, respectively. In contrast, node and edge recognition scores using Qwen3.5-4B are superior on these datasets: node F1 scores of 0.7948 (FC_A), 0.9632 (FA), and 0.5865 (hdBPMN), and edge F1 scores of 0.5435, 0.8271, and 0.2164, respectively. On the machine-generated datasets the results are mixed: PaddleOCR performs better on CBD (node: 0.8891 vs. 0.7151; edge: 0.6184 vs. 0.3676) whereas Qwen3.5-4B yields better results on SEM (node: 0.8058 vs. 0.6761; edge: 0.5785 vs. 0.4678).

4.2 End-to-End Diagram Recognition

Table 2 presents node and edge recognition F1 scores for five-shot prompting for InternVL-3.5 (14B), Gemma4 (31B), and Qwen3.5 (27B) as well as an object detection (OD) baseline across our test datasets. For the MLLMs, we report mean F1 score and standard deviation across 5 runs, with temperature set to 0.1.

Gemma4 produces valid JSON syntax in every run, and InternVL-3.5 fails on only 0.8 files on average for the hdBPMN dataset. Qwen3.5 is the only model that regularly produces invalid outputs, ranging from 4.4 files on CBD to 37.8 on hdBPMN and 54.8 on SYN-SEM. We manually investigate Qwen's extractions and find that these errors partially stem from repetitive reasoning that exhausts the 8192 token output limit before the JSON closes. This issue is most severe on complex datasets with a high number of relations per image and often worsens as the number of input-output pairs is increased from 0 to 5.

Across all datasets except for the SAM-BPMN and SEM dataset, Gemma4 outperforms the object detection baseline and the other MLLMs. On the SAM-BPMN dataset, Gemma4 achieves very low scores compared to the other models. We investigated the extractions and found that instead of extracting the information from the provided example, it generated a default output for a BPMN diagram not found in our test set. The object detection baseline surpasses InternVL-3.5 and Qwen3.5 on the FA, hdBPMN, and SEM datasets but fails to reach the node recognition scores of Gemma4 on any dataset except SEM. The node recognition score is limited by the weakest-performing individual metric (node classification and node text extraction), since a mismatch in any of the individual metrics causes the combined metric to fail. Dropping the class requirement raises node F1 scores only modestly: at most 0.06 for InternVL-3.5 on the FA and SAM-UML datasets, 0.0367 for Gemma4 and 0.02 for Qwen3.5 on the SAM-UML dataset.

Gemma4 also performs better in edge recognition on all datasets except for SAM-BPMN. The object detection baseline outperforms InternVL-3.5 and marginally surpasses Qwen3.5 on the FA and SEM dataset. On the SEM (up to -0.0638 for Gemma4), SAM-UML (up to -0.3276 for Gemma4), and SAM-BPMN (up to -0.0474 for Gemma4) datasets we notice a considerable performance drop in edge recognition when additionally requiring edge classes and text labels to match ($\text{Edge}_{text_{full}}$). On the SEM datasets, this drop is caused by the high number of edges carrying a path coefficient as a text label. On the SAM-UML dataset, the mean drop in performance when only including classes is -0.0832 for Gemma4, -0.1578 for InternVL3.5 and -0.0584 for Qwen3.5. On the SAM-BPMN dataset, the performance drop is primarily caused by including edge labels as well.

The results described above required exact string matching between ground truth texts and model output texts. We note that this may penalize the MLLM unfairly, as minor deviations in strings (accents, typos) may still be semantically correct and do not affect the quality of the extraction. To account for this possibility, we also assess the node and edge recognition scores when relaxing the string matching by allowing a normalized Levenshtein distance of 0.1 between ground truth and extracted text label. On average, the node recognition increases by 0.0179 for Gemma4, by 0.0249 for InternVL-3.5, and by 0.0225 for Qwen3.5. Edge recognition increases by 0.0213 for Gemma4, by 0.0243 for InternVL-3.5, and by 0.0338 for Qwen3.5. The datasets most affected by this increase are CBD, hdBPMN, and SEM.

4.3 Isolated Edge Recognition

Since edge recognition results in the previous subsection may be influenced by extracted node names, we evaluate the extraction performance conditioned on ground-truth node names to better isolate edge recognition capabilities. Where

Table 2. F1 scores (mean ± std across 5 runs) with five-shot prompting and a temperature of 0.1. For the object detection baseline (OD) best scores are stated.

Dataset	InternVL-3.5	Gemma4	Qwen3.5	**OD**
$Node_{text}$: Node recognition				
FC_A	0.8287 ± 0.0019	**0.8365** ± 0.0017	0.8098 ± 0.0119	0.7948
FA	0.8596 ± 0.0036	**0.9865** ± 0.0000	0.9228 ± 0.0281	0.9632
hdBPMN	0.4767 ± 0.0033	**0.7716** ± 0.0024	0.5216 ± 0.0263	0.5865
CBD	0.9367 ± 0.0004	**0.9657** ± 0.0000	0.9354 ± 0.0051	0.8891
SEM	0.6373 ± 0.0039	**0.7291** ± 0.0026	0.6006 ± 0.0184	0.8058
SAM-UML	0.7500 ± 0.0185	**0.9461** ± 0.0052	0.7327 ± 0.0378	—
SAM-BPMN	**0.6227** ± 0.0022	0.3967 ± 0.0052	0.5788 ± 0.0569	—
SYN-FC	0.6342 ± 0.0023	**0.7458** ± 0.0046	0.3611 ± 0.0306	—
SYN-SEM	0.5821 ± 0.0054	**0.7455** ± 0.0054	0.3720 ± 0.0192	—
$Edge_{text_{part}}$: Edge recognition (source and target match only)				
FC_A	0.6199 ± 0.0026	**0.7012** ± 0.0021	0.6760 ± 0.0140	0.5435
FA	0.7362 ± 0.0022	**0.8833** ± 0.0005	0.8167 ± 0.0256	0.8271
hdBPMN	0.1348 ± 0.0022	**0.4041** ± 0.0027	0.3223 ± 0.0153	0.2164
CBD	0.7857 ± 0.0033	**0.8919** ± 0.0016	0.8579 ± 0.0067	0.6184
SEM	0.4245 ± 0.0079	**0.7553** ± 0.0045	0.5663 ± 0.0151	0.5785
SAM-UML	0.4091 ± 0.0119	**0.5310** ± 0.0105	0.3883 ± 0.0232	—
SAM-BPMN	0.2515 ± 0.0071	0.2616 ± 0.0039	**0.4036** ± 0.0516	—
SYN-FC	0.3211 ± 0.0058	**0.6549** ± 0.0038	0.2642 ± 0.0159	—
SYN-SEM	0.1976 ± 0.0047	**0.4148** ± 0.0043	0.2071 ± 0.0108	—

available, we compare the isolated edge recognition score to object detection benchmark scores not considering text labels, since this should be eliminated through providing the ground truth node names.

To verify that the provided node names were correctly used by the model, we first evaluate the node recognition scores of all models. Gemma4 only marginally fails to achieve a perfect node recognition score on SYN-SEM (0.967), CBD (0.9973), hdBPMN (0.9935), and SEM (0.9992). Similarly, InternVL-3.5 achieves scores of 0.9985 (CBD), 0.9757 (hdBPMN), and 0.9877 (SEM), where the deviation on hdBPMN is caused by invalid or missing JSON outputs. In contrast, Qwen3.5 fails to reach near-perfect node recognition on all datasets due to erroneous or missing JSON outputs with scores ranging from 0.5962 (hdBPMN) and 0.7373 (SAM-UML) to 0.9722 (FC_A). While this causes a low node recognition recall score, the precision remains close to 1 for Qwen3.5.

In the previous subsection, severe performance drops were observed from node to edge recognition, even on datasets (FC_A, FA, and hdBPMN) where the object detection baseline exhibited relatively small declines. The datasets affected most by the performance drop from node to edge recognition were hdBPMN, SEM, SAM-UML, and SAM-BPMN. Except for the SAM-UML data, the edge recognition scores in the isolated setting (see Table 3) increase notably on these datasets as well as on the FC_A and synthetic datasets. On the FC_A dataset, Gemma4 surpasses the object detection benchmark from related work (0.967) [26] and Qwen3.5 nearly matches the performance. On the hdBPMN dataset, Gemma4's partial edge recognition scores increases to 0.7797, which still falls short of our object detection benchmark of 0.8336 for edge recognition without considering texts. Similarly, on the FA dataset the best performing model Gemma4 achieves an F1-score of 0.8929 which cannot match the benchmark score of 0.984 from related work [23]. On the SEM dataset, all three models' partial edge recognition scores improve, where Gemma4 now achieves a solid score of 0.9426. For the SEM and CBD dataset, a direct comparison is not possible, as its ground truth does not annotate which bounding boxes are connected by an edge. The most substantial performance gains can be observed on the SAM-BPMN dataset, where partial edge recognition scores increase to 0.9133 (Gemma4), 0.6718 (InternVL-3.5), and 0.6653 (Qwen3.5). However, on the SAM-UML and FA datasets the isolated setting only marginally increases the edge recognition scores which remain comparatively low.

Table 3. F1 scores (mean ± std across 5 runs) for isolated edge recognition ($Edge_{text_{part}}$). We used five-shot prompting and a temperature of 0.1.

Dataset	InternVL-3.5	Gemma4	Qwen3.5
FC_A	0.9326 ± 0.0047	**0.9896** ± 0.0000	0.9594 ± 0.0088
FA	0.7667 ± 0.0028	**0.8929** ± 0.0007	0.8415 ± 0.0153
hdBPMN	0.3529 ± 0.0029	**0.7797** ± 0.0019	0.5069 ± 0.0208
CBD	0.8963 ± 0.0026	**0.9605** ± 0.0004	0.8950 ± 0.0136
SEM	0.7216 ± 0.0064	**0.9426** ± 0.0003	0.8085 ± 0.0230
SAM-UML	0.4476 ± 0.0193	**0.5337** ± 0.0023	0.3949 ± 0.0403
SAM-BPMN	0.6718 ± 0.0102	**0.9133** ± 0.0001	0.6653 ± 0.0250
SYN-FC	0.7767 ± 0.0037	**0.8103** ± 0.0012	0.4157 ± 0.0534
SYN-SEM	0.5163 ± 0.0078	**0.6507** ± 0.0027	0.2078 ± 0.0486

4.4 Positional Recognition

To assess the ability of the best performing model, Gemma4, to generate bounding boxes for nodes and edges, we evaluate its node and edge recognition performance at an IoU threshold of 0.5. Where available, we also provide benchmark scores from object detection models.

On the FC_A and FA datasets, Gemma4 achieves solid mean node recognition scores of 0.8562 ± 0.003 and 0.9477 ± 0.0025, respectively, which fall short of our Arrow R-CNN benchmarks (0.9976 and 1.0). On the remaining datasets,

Gemma4's node scores drop to 0.5126 ± 0.0038 (hdBPMN), 0.6048 ± 0.0052 (SEM), and 0.7427 ± 0.0036 (SYN-FC). The edge recognition scores are substantially lower across all datasets. On FA, where Gemma4 reaches its highest edge recognition score of 0.621 ± 0.0111, Arrow R-CNN scores 0.9728. This performance gap widens even further on the FC_A (0.2091 ± 0.0036 vs. 0.9244) and hdBPMN (0.0257 ± 0.0013 vs. 0.8337) datasets.

This discrepancy is partly an artifact of the IoU based evaluation. Gemma4 tends to produce bounding boxes for vertical and horizontal edges with a width (resp. height) of 0 that approximately connect the edge's head and tail (example images can be found in the Github repository). While these detections often correctly localize the underlying node relations, they are marked as incorrect predictions at the IoU threshold of 0.5. This phenomenon is observed more often on datasets containing small or elongated edges (e.g., hdBPMN or SEM).

4.5 Influence of Number of Few-Shot Examples

The influence of the number of samples in few-shot prompting on the node and edge recognition performance varies considerably across datasets and MLLMs. In the following, we compare zero-shot to five-shot prompting where the first value represents zero-shot performance and the second five-shot performance.

Gemma4 shows improvements in node recognition on the FC_A dataset (0.7594 to 0.8365) and in edge recognition on the FC_A (0.6112 to 0.7012) and SEM dataset (0.6099 to 0.7553) when providing five image-JSON pairs in the prompt. However, performance declines substantially on SAM-BPMN (node: 0.7294 to 0.3967, edge: 0.4818 to 0.2616). This decline on the SAM-BPMN dataset is unique to Gemma4. On the remaining datasets the effects of using five-shot prompting are minor. InternVL-3.5 benefits from five-shot prompting with respect to node recognition on FA (0.7826 to 0.8596), FC_A (0.5303 to 0.8287), and SAM-UML (0.4890 to 0.7500) as well as in edge recognition on FC_A (0.5033 to 0.6199). In contrast to Gemma4, the edge recognition decreases for InternVL-3.5 on the SEM dataset (0.5011 to 0.4245). For Qwen3.5, five-shot prompting typically increases the number of missing or erroneous JSON outputs. However, node recognition improves on the FC_A (0.7610 to 0.8098) and SAM-UML (0.2595 to 0.7327) datasets and edge recognition improves on FC_A (0.6157 to 0.6760), SAM-UML (0.3038 to 0.3883), and SAM-BPMN (0.3044 to 0.4036).

5 Discussion

Overall, our results suggest that open-weight, general purpose MLLMs are becoming competitive for diagram recognition, especially on simpler diagrams and for node recognition. In aggregate, Gemma4 achieves the best *end-to-end* diagram recognition performance across datasets and substantially outperforms the other MLLMs in both node and edge recognition. It also outperforms the object detection baseline on several datasets, despite relying on a non-specialized approach in comparison to traditional diagram recognition pipelines. However, isolating the edge recognition performance reveals that this superior end-to-end performance is mostly due to text extraction rather than superior diagram structure recognition. Once OCR errors are removed, MLLMs no longer surpass the object detection baseline in edge recognition on the hdBPMN and FA datasets. This is consistent with previous work in the VQA domain, where it has been shown that MLLMs excel at identifying nodes, but struggle at identifying relations and reasoning about them [15]. It is worth noting, however, that in the aforementioned isolated edge recognition setting, the edge recognition F1 scores substantially improve. Our evaluation further shows that node and edge recognition scores vary across datasets. More complex datasets like hdBPMN, SAM-UML, SAM-BPMN, and the synthetic datasets still pose a significant challenge for MLLMs. In positional recognition of nodes and edges, Gemma4 achieves solid scores but still lags behind specialized diagram recognition computer vision architectures like Arrow R-CNN [24]. Our evaluation, however, also reveals that commonly used evaluation metrics like IoU do not fully capture the MLLMs' spatial reasoning ability.

Text extraction is a major bottleneck in current diagram recognition pipelines relying on object detection models. While computer-vision based architectures achieve excellent structural detection scores, there is a disconnect once textual correctness is required. This is especially apparent on noisy diagram images from hand-drawn datasets, where OCR struggles. At the same time, the small-weight Qwen3.5 model (4B) outperformed PaddleOCR. Recent work integrated OCR via tool calling with MLLMs, which allows the model to leverage the capabilities of specialized systems for extraction tasks [6]. We argue that joint pipelines of MLLMs and specialized systems are a promising direction for further research on end-to-end diagram recognition.

There are several limitations that we identify in our evaluation. First, we represent GSDs through nodes and relations. While this captures the essence of diagram recognition in all domains, it overlooks features specific to more complex diagram types such as BPMNs and UMLs. Here, additional elements such as node groups as well as nested node structures containing attributes and methods have to be considered. Additionally, edges may be extended to also include cardinalities for UML diagrams. Second, the varying model sizes, as well as different context sizes and output limits, make comparing the MLLMs with each other difficult. Third, it would be interesting to also vary the output format in the evaluation. MLLMs might be biased towards certain data formats, and additionally, not all diagram domains are represented by JSON. For instance, BPMNs are native to XML formats.

6 Conclusion

We evaluated the performance of three different medium-sized open-weights MLLMs (InternVL3.5-14B, Gemma4-31B, Qwen3.5-27B) on the diagram recognition task of extracting information from GSDs. We created a test set of 843 diagram-JSON pairs based on different diagram databases from related work. Unlike prior approaches from the VQA domain, which focused mostly on simple subtasks (e.g., text labels of specific nodes) with multiple-choice questions or used very small, narrowly focused datasets, our work investigates graph-structured diagram parsing, where the focus is put explicitly on overall graph extraction and reconstruction. Our results show that MLLMs achieve strong performance on node recognition but struggle with edge recognition. This gap persists for certain datasets, even when the models are provided with ground truth nodes, which shows that relational extraction remains a core limitation. Furthermore, our findings suggest that OCR is a weak point in existing computer vision-based diagram recognition approaches, especially for hand-written diagrams. Combining MLLMs with these methods may improve accuracy. Future work could explore integrating MLLMs in modular diagram recognition pipelines where different tasks of the process are distributed among specialized components. Another direction of research is to investigate fine-tuning MLLMs on our provided dataset. As MLLMs advance, benchmarking should also extend output formats to capture the structural complexity of diagrams such as BPMN and UML.

References

1. Awal, A.M., Feng, G., Mouchère, H., Viard-Gaudin, C.: First experiments on a new online handwritten flowchart database. In: Agam, G., Viard-Gaudin, C. (eds.) Document Recognition and Retrieval XVIII, vol. 7874, p. 78740A. International Society for Optics and Photonics, SPIE (2011). https://doi.org/10.1117/12.876624
2. Bhushan, S., Lee, M.: Block diagram-to-text: Understanding block diagram images by generating natural language descriptors. In: He, Y., Ji, H., Li, S., Liu, Y., Chang, C.H. (eds.) Findings of the Association for Computational Linguistics: AACL-IJCNLP 2022, pp. 153–168. Association for Computational Linguistics, Online only (2022). https://doi.org/10.18653/v1/2022.findings-aacl.15
3. Bresler, M., Prusa, D., Hlaváč, V.: Online recognition of sketched arrow-connected diagrams. Int. J. Doc. Anal. Recogn. **19**(3), 253–267 (2016). https://doi.org/10.1007/s10032-016-0269-z
4. Bresler, M., Phan, T.V., Prusa, D., Nakagawa, M., Hlavac, V.: Recognition system for on-line sketched diagrams. In: 2014 14th International Conference on Frontiers in Handwriting Recognition, pp. 563–568. IEEE, Greece (2014). https://doi.org/10.1109/ICFHR.2014.100
5. Chakraborty, S., Paul, S., Ahsan, S.M.M.: A novel approach to rapidly generate document from hand drawn flowcharts. In: 2020 IEEE Region 10 Symposium (TENSYMP), pp. 702–705 (2020). https://doi.org/10.1109/TENSYMP50017.2020.9231033
6. Chen, Q., Zhang, X., Guo, L., Chen, F., Zhang, C.: Dianjin-ocr-r1: Enhancing ocr capabilities via a reasoning-and-tool interleaved vision-language model (2026). https://arxiv.org/abs/2508.13238
7. Conrardy, A., Cabot, J.: From image to uml: First results of image based uml diagram generation using llms (2024). https://arxiv.org/abs/2404.11376
8. Cui, C., et al.: Paddleocr 3.0 technical report (2025). https://arxiv.org/abs/2507.05595

9. Di Martino, B., Esposito, A.: Automatic recognition of design patterns from uml-based software documentation. In: Proceedings of International Conference on Information Integration and Web-based Applications & Services, pp. 280–289. ACM, Vienna Austria (2013). https://doi.org/10.1145/2539150.2539173
10. Gervais, P., Deselaers, T., Aksan, E., Hilliges, O.: The didi dataset: Digital ink diagram data (2020). https://arxiv.org/abs/2002.09303
11. Google: Gemma 4 31b (2026). https://huggingface.co/google/gemma-4-31B, Accessed 19 May 2026
12. Google DeepMind: Gemma 4 (2026). https://deepmind.google/models/gemma/gemma-4/, Accessed 24 June 2026
13. Hammond, T., Davis, R.: Tahuti: a geometrical sketch recognition system for UML class diagrams. In: ACM SIGGRAPH 2006 Courses on - SIGGRAPH '06, p. 25. ACM Press, Boston, Massachusetts (2006). https://doi.org/10.1145/1185657.1185786
14. Herrera-Camara, J.I., Hammond, T.: Flow2Code: from hand-drawn flowcharts to code execution. In: Proceedings of the Symposium on Sketch-Based Interfaces and Modeling, pp. 1–13. ACM, Los Angeles California (2017). https://doi.org/10.1145/3092907.3092909
15. Hou, Y., Giledereli, B., Tu, Y., Sachan, M.: Do vision-language models really understand visual language? (2025). https://arxiv.org/abs/2410.00193
16. Hu, H., et al.: SpaceGTN: A Time-Agnostic Graph Transformer Network for Handwritten Diagram Recognition and Segmentation. In: Proceedings of the AAAI Conference on Artificial Intelligence, vol. 38, no. (3), pp. 2211–2219 (2024). https://doi.org/10.1609/aaai.v38i3.27994
17. Liu, H., et al.: Flowmind2digital: The first comprehensive flowmind recognition and conversion approach (2024). https://arxiv.org/abs/2401.03742
18. OpenGVLab: Internvl3.5-14b (2026). https://huggingface.co/OpenGVLab/InternVL3_5-14B, Accessed 19 May 2026
19. Qiao, M., Wang, J., Xiang, J., Hou, Q., Li, R.: Structure diagram recognition in financial announcements. In: Document Analysis and Recognition - ICDAR 2023 (2023). https://doi.org/10.1007/978-3-031-41676-7_2
20. Qwen Team: Qwen3.5-27b (2026). https://huggingface.co/Qwen/Qwen3.5-27B, Accessed 19 May 2026
21. Qwen Team: Qwen3.5 blog: Towards native multimodal agents (2026). https://qwen.ai/blog?id=qwen3.5, Accessed 24 June 2026
22. Scharfenberger, J., Funk, B.: Parsing causal models – an instance segmentation approach. In: Intelligent Information Systems, vol. 477, pp. 43–51. Springer, Cham (2023). https://doi.org/10.1007/978-3-031-34674-3_6
23. Schäfer, B., Keuper, M., Stuckenschmidt, H.: Arrow R-CNN for handwritten diagram recognition. In: International Journal on Document Analysis and Recognition (IJDAR) (2021). https://doi.org/10.1007/s10032-020-00361-1
24. Schäfer, B., Stuckenschmidt, H.: Arrow R-CNN for flowchart recognition. **1**, 7–13 (2019). https://doi.org/10.1109/ICDARW.2019.00007
25. Schäfer, B., Stuckenschmidt, H.: DiagramNet: Hand-Drawn Diagram Recognition Using Visual Arrow-Relation Detection. In: Document Analysis and Recognition – ICDAR 2021 (2021). https://doi.org/10.1007/978-3-030-86549-8_39
26. Schäfer, B., van der Aa, H., Leopold, H., Stuckenschmidt, H.: Sketch2BPMN: Automatic Recognition of Hand-Drawn BPMN Models. In: La Rosa, M., Sadiq, S., Teniente, E. (eds.) CAiSE 2021. LNCS, vol. 12751, pp. 344–360. Springer, Cham (2021). https://doi.org/10.1007/978-3-030-79382-1_21

27. Schäfer, B., Van Der Aa, H., Leopold, H., Stuckenschmidt, H.: Sketch2Process: end-to-end BPMN sketch recognition based on neural networks. IEEE Trans. Softw. Eng. **49**(4), 2621–2641 (2023). https://doi.org/10.1109/TSE.2022.3228308
28. Schölch, L., et al.: Towards automatic parsing of structured visual content through the use of synthetic data. In: 2022 26th International Conference on Pattern Recognition (ICPR), pp. 1607–1613 (2022)
29. Singh, S., et al.: FlowVQA: Mapping multimodal logic in visual question answering with flowcharts. In: Ku, L.W., Martins, A., Srikumar, V. (eds.) Findings of the Association for Computational Linguistics: ACL 2024, pp. 1330–1350. Association for Computational Linguistics, Bangkok, Thailand (2024). https://doi.org/10.18653/v1/2024.findings-acl.78
30. Sola, D., Warmuth, C., Schäfer, B., Badakhshan, P., Rehse, J.R., Kampik, T.: Sap signavio academic models: A large process model dataset (2022). https://arxiv.org/abs/2208.12223
31. Tannert, S., et al.: FlowchartQA: The first large-scale benchmark for reasoning over flowcharts. In: Proceedings of the 1st Workshop on Linguistic Insights from and for Multimodal Language Processing, pp. 34–46. Association for Computational Lingustics (2023). https://aclanthology.org/2023.limo-1.5/
32. Waldvogel, A., Annighoefer, B.: Model-based block diagram recognition for model visualization verification. In: Proceedings of the ACM/IEEE 27th International Conference on Model Driven Engineering Languages and Systems, pp. 1131–1140. MODELS Companion '24, Association for Computing Machinery, New York, NY, USA (2024). https://doi.org/10.1145/3652620.3687825
33. Wang, W., et al.: Internvl3.5: Advancing open-source multimodal models in versatility, reasoning, and efficiency (2025)
34. Wu, J., Gan, W., Chen, Z., Wan, S., Yu, P.S.: Multimodal large language models: A survey (2023). https://arxiv.org/abs/2311.13165
35. Yue, X., et al.: Mmmu: A massive multi-discipline multimodal understanding and reasoning benchmark for expert agi (2024). https://arxiv.org/abs/2311.16502
36. Yue, X., et al.: Mmmu-pro: A more robust multi-discipline multimodal understanding benchmark (2025). https://arxiv.org/abs/2409.02813
37. Yun, X.L., Zhang, Y.M., Yin, F., Liu, C.L.: Instance GNN: a learning framework for joint symbol segmentation and recognition in online handwritten diagrams. IEEE Trans. Multimedia **24**, 2580–2594 (2022). https://doi.org/10.1109/TMM.2021.3087000
38. Zhang, E., Yao, R., Liu, H., Yu, J., Wang, J.: First multi-dimensional evaluation of flowchart comprehension for multimodal large language models (2024). https://arxiv.org/abs/2406.10057

Lect-Former: Localized Self-attention Skip Connections for Whiteboard Content Extraction

Kenny Davila(✉), Fei Xu, Sunyoung Ro, and Min Song

School of Computing, DePaul University, Chicago, IL 60604, USA
{kdavilac,fxu14,SR02,msong13}@depaul.edu

Abstract. Accurate extraction of handwritten lecture video content facilitates downstream applications such as lecture video summarization, indexing, and search. This task requires both local foreground modeling and contextual reasoning to distinguish handwriting from complex background artifacts. In this work, we propose the Lect-Former model for detection and segmentation of handwritten content in lecture videos. The proposed architecture follows a U-Net-style fully convolutional design, where the skip connections are replaced by transformer-based modules employing spatial attention within local neighborhoods. The model uses separate branches for region-level content detection, pixel-level background estimation, and pixel-level binarization. We further investigate multiple pretraining objectives and attention backends for low-data handwritten content extraction. Experimental results on lecture-video and general text segmentation datasets show that masked contextual reconstruction substantially improves downstream adaptation, while scalable localized attention mechanisms provide favorable performance-efficiency tradeoffs under limited fine-tuning data.

Keywords: handwritten text segmentation · foreground extraction · lecture video analysis

1 Introduction

Accurate extraction of handwritten content from lecture videos is the first step in many downstream applications such as lecture video summarization, indexing, and search. Correctly extracting the content at the pixel level requires both local foreground modeling and contextual reasoning to distinguish handwriting from complex background artifacts. Fully convolutional architectures can capture contextual information within local receptive fields, but recent transformer-based architectures [25] generally provide stronger performance by explicitly modeling interactions between distant spatial regions through attention mechanisms.

In this work, we propose the Lect-Former model for detection and segmentation of handwritten content in lecture videos. The proposed architecture follows a U-Net-style fully convolutional design [21], where the skip connections

F. Shafait et al. (Eds.): DAS 2026, LNCS 16851, pp. 263–281, 2027.
https://doi.org/10.1007/978-3-032-36207-0_16

are replaced by transformer-based modules inspired by UCTransNet [26]. These modules employ localized spatial attention mechanisms similar to Neighborhood Attention Transformers (NAT) [9]. This allows the skip connections to enrich encoder representations with localized contextual information before they are fused into the decoder. Following our prior work on handwritten lecture content segmentation [6], the proposed architecture employs a primary branch for pixel-level text segmentation (e.g., binarization), together with lightweight auxiliary heads for region-level text detection and pixel-level background estimation.

The resulting model is relatively large, while fully annotated lecture-video data remains scarce. To address this limitation, we propose a three-stage training protocol to improve representation learning under limited supervision. The first stage performs self-supervised reconstruction pretraining using masked feature reconstruction. Unlike conventional masked-image reconstruction approaches [10], masking is applied to encoder features rather than the input image, allowing both the convolutional decoder and transformer-based skip connections to be trained on stable inputs. The second stage is inspired by recent text-removal approaches for scene understanding [19], and uses weak supervision generated through image inpainting to train the primary text segmentation branch and the auxiliary text detection and background estimation heads. The third stage fine-tunes the complete model for handwritten text segmentation using a multi-stage optimization strategy including warm-up and progressively harder sample selection.

We experimented using an extended version of the LectureMath dataset [6], where the proposed Lect-Former architecture and training protocol consistently improve handwritten foreground extraction from lecture video frames. Additional ablation studies analyze different reconstruction objectives, localized attention mechanisms, and fusion strategies under limited training data. Our code is availble at: https://github.com/kdavila/LectFormer_DAS2026.

2 Related Works

Our approach combines prior work on handwritten lecture-content extraction, localized attention for dense prediction, transformer-based skip connections, and representation learning under limited supervision.

2.1 Handwritten Content Extraction in Lecture Videos

Extracting content from lecture videos can be considered a form of scene text segmentation. These videos mainly contain handwritten content on whiteboards and/or chalkboards, but some occasional typeset content is still expected in edited videos. Some common challenges include low-contrast handwriting, video-compression artifacts, relatively low-resolution text, non-uniform illumination, shadows and occlusions, traces with inconsistent strength, left-over noise on the writing surface, among others. Our previous work, FCN-LectureNet [6], has explored the effectiveness of using deep segmentation networks based on U-Net with reasonable success, but showing space for improvement where the model is unable to differentiate between high-contrasting edges and real handwriting due to the lack of explicit context modeling. Nevertheless, even imperfect content extraction can be useful for automatic lecture video summarization and downstream retrieval tasks [7].

2.2 Localized Attention for Dense Prediction

Convolutional architectures for dense prediction often face a trade-off between semantic abstraction and spatial precision [17]. Encoder-decoder architectures such as U-Net [21] mitigate this through multi-scale skip fusion, but feature transformation and interaction remain primarily convolutional. These limitations become more severe for handwritten text extraction, where challenging visual conditions require stronger contextual reasoning to preserve coherent structures and suppress false positives.

Following the success of attention mechanisms in sequence modeling [25], transformer architectures have been increasingly adopted in computer vision through Vision Transformers (ViT) [8]. By modeling interactions across spatial regions, transformers provide richer contextual reasoning than local convolutional filters alone, which is beneficial for coherent handwriting reconstruction. However, global self-attention introduces substantial computational cost for dense prediction tasks. Neighborhood Attention Transformer (NAT) [9] addresses this limitation by restricting self-attention to local neighborhoods while preserving contextual feature interaction, making localized attention particularly attractive for dense image prediction tasks.

2.3 Transformer-Based Skip Connections

U-Net-based architectures perform well for segmentation due to their encoder-decoder design and skip connections that preserve spatial detail [21]. However, direct cross-scale feature copying can introduce semantic mismatch between low-level encoder features and high-level decoder representations. To address this issue, UCTransNet [26] replaces conventional skip connections with transformer-based channel attention that aligns encoder and decoder representations before feature fusion. By contrast, many transformer-based segmentation architectures place transformer-based spatial attention mainly in the encoder or broader feature-extraction pipeline [2,4].

Our transformer-based skip connections are similarly inspired by these works. However, unlike approaches like UCTransNet [26] that aggregate features jointly across scales, each skip connection in our model applies local attention independently at its corresponding feature resolution.

2.4 Auxiliary Supervision for Scene Text Segmentation

Prior scene text segmentation methods have explored auxiliary supervision and text-aware modeling to improve segmentation quality and suppress background

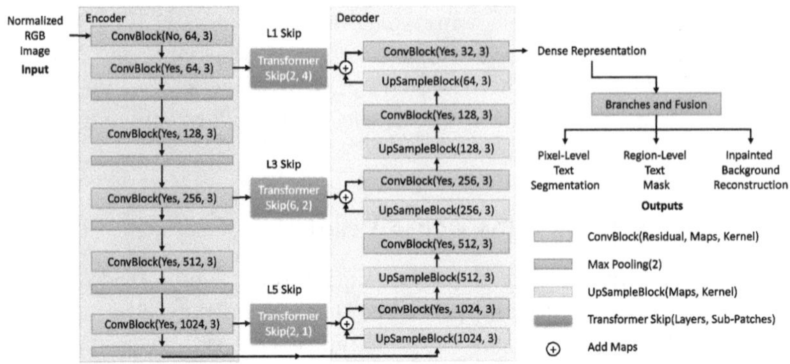

Fig. 1. Lect-Former Architecture. The proposed network follows a U-Net style design, but replaces the skip connections with windowed, spatial self-attention modules.

artifacts. Recent transformer-based approaches jointly optimize text localization with downstream segmentation objectives. SwinTextSpotter [12] combines text detection and recognition within a unified transformer framework, while Yu et al. [29] argue that accurate text localization is more important than explicit character recognition, jointly optimizing text detection and segmentation.

Related ideas have also been explored in text removal and weakly supervised reconstruction settings. ViT-Eraser [19] integrates text localization and image inpainting using segmentation-guided pretraining and masked image modeling. Our prior work, FCN-LectureNet [6], used similar auxiliary supervision for handwritten content extraction, jointly optimizing text-mask prediction and background estimation objectives with the final binarization task. Our multi-stage training protocol builds upon these ideas.

3 Methodology

In this section, we detail our Lect-Former architecture (Sect. 3.1), and the proposed training protocol (Sect. 3.2).

3.1 The Lect-Former Architecture

The proposed Lect-Former architecture builds on FCN-LectureNet [6], but introduces improvements to the base design, emphasizing attention-based modules in the skip connections. Figure 1 illustrates the overall architecture.

Convolutional Encoder-Decoder. The architecture follows a conventional U-Net [21] encoder-decoder design (Fig. 1). The first stage uses a standard convolutional block, followed by residual blocks based on the ResNet-18 design [11]). All blocks except the output layers use Group Normalization [27] with a single group, approximating Layer Normalization [1] on convolutional feature maps.

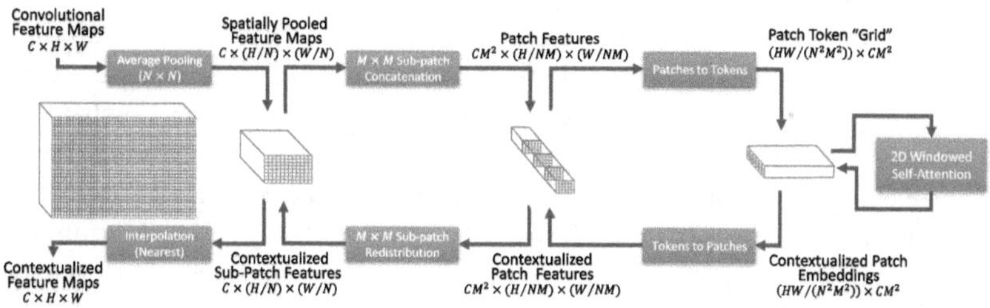

Fig. 2. Transformer-based Skip Connections. Skip connections are replaced by 2D windowed self-attention modules that produced contextualized feature maps.

Unlike our prior FCN-LectureNet [6], which concatenated feature maps from lateral connections, Lect-former treats them as residuals by adding them to the up-sampled feature maps from the up-sampling blocks. This design encourages some degree of feature alignment between the encoder and the decoder. Finally, our design does not use a bottleneck in between the encoder and decoder.

Transformer Skip Modules. Following UCTransNet [26], the proposed architecture replaces standard skip connections with transformer-based modules. However, unlike approaches that aggregate features jointly across multiple scales, our model applies localized attention independently at different encoder resolutions. Specifically, transformer-based Skip modules are used for low-, mid-, and high-level feature representations (L1, L3, and L5), as illustrated in Fig. 1.

Each skip module operates on convolutional features extracted immediately before max-pooling and enriches them with localized contextual information prior to decoder fusion. Since feature maps at different encoder levels have different spatial resolutions (e.g., 256×256, 64×64, and 16×16 for L1, L3, and L5), each level is divided into spatially aligned logical patches with progressively smaller spatial sizes (16×16, 4×4, and 1×1, respectively). To maintain comparable token dimensionality across scales, each patch is further divided into sub-patches, whose features are average pooled and concatenated into a single patch-level representation. This produces fixed-length 1024-dimensional representations for all scales (e.g., $64 \times 4 \times 4$, $256 \times 2 \times 2$, and $1024 \times 1 \times 1$). These patch-level representations become the tokens used by the localized self-attention blocks, as illustrated in Fig. 2.

After constructing the token-level representation, a 2D sinusoidal positional encoding is added. Similar to DETR [3], it is constructed from 1D sinusoidal embeddings [25]. The embedding is divided equally between the horizontal and vertical coordinates, each encoded using sine and cosine functions at multiple frequencies. Coordinates are normalized to relative image positions in the range 0 to 2π. To support arbitrary image sizes, stochastic positional scaling is introduced during training. Each crop is treated as if it were located at a random position within a larger virtual image of random size. At inference time, the true relative coordinates derived from the actual image size are used.

Since our downstream task is pixel-level content extraction, we adopt localized 2D windowed attention to capture local contextual interactions while preserving spatial locality and computational efficiency. The localized self-attention modules were implemented using transformer encoders, where encoder depth strongly affects both model capacity and computational cost. Preliminary experiments showed that the optimal depth depends on the corresponding feature scale, with mid-level representations (L3) benefiting most from deeper attention modeling. The final configuration uses 2 transformer layers at L1, 6 at L3, and 2 at L5. The attention window size is another key hyper-parameter. The final configuration uses a fixed 7×7 attention window across all feature levels, as larger or scale-dependent windows consistently performed worse during pretraining. We evaluated multiple implementations of localized self-attention including additive attention masks, explicit windowed attention using SPDA, shifted-window attention (inspired by Swin Transformers [16]), Neighborhood Attention Transformer (NATTEN [9]), and FlexAttention. Their segmentation accuracy, runtime, and memory trade-offs are analyzed in Sect. 4.4.

After localized self-attention, the contextualized patch embeddings are reshaped back into spatial feature maps and redistributed into C contextualized sub-patch feature maps. Nearest-neighbor interpolation is then used to restore the original spatial resolution without modifying the learned sub-patch representations (Fig. 2). Finally, the resulting contextualized feature maps are fused with the corresponding up-sampled decoder features.

Task-Specific Branches. Like FCN-LectureNet [6], the proposed network uses three task-specific branches: a main binarization branch together with auxiliary text-detection and background-estimation heads. The auxiliary heads are intentionally lightweight, encouraging the shared encoder-decoder backbone to learn representations useful across all tasks.

The text-detection and background-estimation heads share a similar shallow design composed of a small residual bottleneck followed by task-specific prediction layers. The background-estimation head predicts a normalized RGB reconstruction of the text-free background, while the text-detection head predicts region-level text-presence logits.

The main binarization branch receives both the decoder features and an enhanced image derived from the auxiliary predictions. Specifically, the estimated background is subtracted from the normalized input image and multiplied by the soft text mask:

$$I_{enh} = (I - B) \cdot \sigma(M) \qquad (1)$$

where I is the normalized input image, B is the estimated background, and M corresponds to the text-detection logits. Ideally, this enhanced representation suppresses non-text regions while amplifying foreground handwriting structures, simplifying the downstream binarization task.

The binarization branch follows the FCN-LectureNet [6] design, replacing Batch Normalization with Group Normalization. In FCN-LectureNet, the enhanced image was progressively fused with decoder representations through repeated multi-stage concatenation. In this work, we additionally investigate alternative fusion strategies, including early fusion, late additive refinement, and a hybrid of both. The effect of these fusion mechanisms is analyzed experimentally in Sect. 4.5.

3.2 Training Protocols

The final goal of the proposed Lect-Former architecture is content extraction, mainly handwriting, from lecture video frames. However, fully annotated datasets for this task remain extremely limited relative to the capacity of transformer-based segmentation models. To address this limitation, we propose a three-stage training protocol that progressively pretrains the network using data from multiple domains before fine-tuning on the target lecture-video dataset. This follows the general motivation of our prior work [6], while extending the pretraining strategy to stronger supervision objectives.

Stage 1: Reconstruction Pretraining. The goal of the first training stage is to initialize the encoder, decoder, and Transformer-based Skip modules using a reconstruction objective before downstream fine-tuning. Since encoder-decoder architectures with skip connections can easily learn near-identity mappings, we adopt masked reconstruction pretraining to encourage the network to learn non-trivial contextual representations. Our reconstruction strategy differs from earlier pretraining schemes such as the median-filter reconstruction used in FCN-LectureNet [6], extending reconstruction pretraining toward contextual masked modeling and multi-scale supervision objectives.

Unlike approaches that mask the input image [10], we mask at the encoder feature level. During training, 75% of the encoder patches are replaced by zero-valued feature vectors, forcing the decoder and Transformer-based Skip modules to reconstruct missing content from incomplete contextual information. Unlike masked self-attention schemes that explicitly suppress attention to masked tokens, our Transformer-based Skip modules still process the complete token sequence, allowing masked tokens to participate in contextual attention despite containing no semantic feature information. Thus, both the localized attention layers and convolutional decoder operate under the same masking constraints.

We additionally explored multi-scale reconstruction targets inspired by Laplacian pyramids, where the reconstructed image is decomposed into multiple frequency bands represented at a common spatial resolution. We also evaluated combinations of masked reconstruction and Laplacian-based targets. In all cases, the reconstruction objective combines a pixel-level L1 loss with a perceptual loss [22] computed from intermediate VGG-19 feature activations [13]. The effects of different reconstruction objectives are analyzed experimentally in Sect. 4.3.

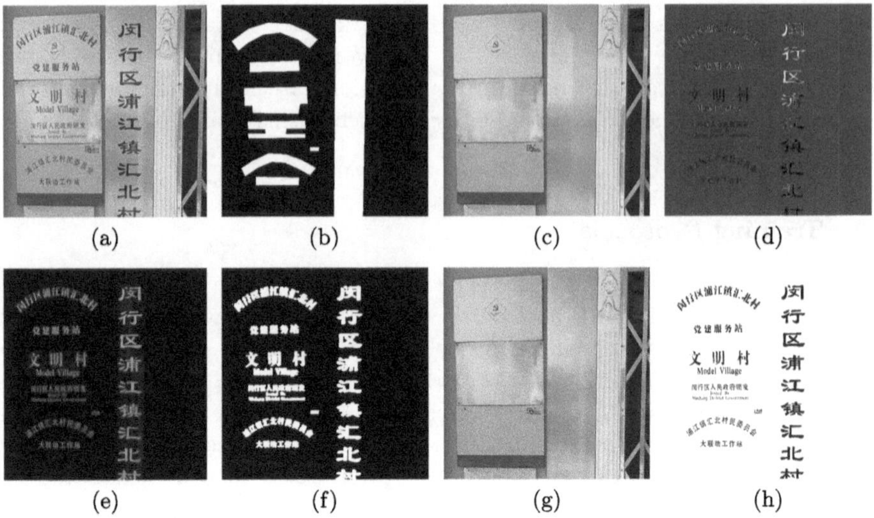

Fig. 3. Example of the generation of weakly labeled targets for text deletion and binarization. (a) input image; (b) text mask ground truth; (c) initial inpainting; (d) result of subtracting (c) from (a); (e) channel-wise max of (d); (f) Applying Otsu and then dilation to (e); (g) final background estimation target; and (h) final binarization target.

Stage 2: Text Deletion Pretraining. The goal of the second stage is to train the auxiliary background-estimation and text-detection heads used to construct the enhanced image representation defined in Eq. (1). More importantly, this stage also generates weak character-level pseudo-labels for the final binarization branch using only coarse text-region annotations available in scene-text datasets.

Given an input image containing text and its corresponding text-region mask, we remove the text using classical image inpainting [24], producing an initial text-less image (Fig. 3c). We then compute the absolute difference between the original and inpainted images, where character pixels typically produce high-magnitude responses (Fig. 3d–e). The RGB difference image is collapsed into a single-channel response map using the per-pixel maximum absolute difference across channels (Fig. 3e). Otsu thresholding yields character-level pseudo-labels for the binarization branch (Fig. 3h). The resulting character mask is further expanded using morphology-based refinement (Fig. 3f) to identify likely background pixels within text regions, allowing portions of the original image to be restored into the inpainted background target before boundary smoothing.

The generated text-region masks are used to supervise the text-detection head using BCE loss, while the pseudo character-level masks supervise the final binarization branch. The refined inpainted images are used to train the auxiliary background-estimation head using a foreground-weighted perceptual reconstruction loss, where text-region pixels receive weight 1.0 and background pixels receive weight 0.155 based on the average foreground/background ratio observed in the training data. Unlike Stage 1, this stage explicitly biases the model toward accurate reconstruction and localization of handwritten foreground content. Compared with the median-filter supervision used in FCN-LectureNet [6], the proposed procedure preserves non-text high-frequency image details while simultaneously generating weak supervision for downstream binarization.

Stage 3: Text Segmentation Training. The final stage fine-tunes the network for handwritten content extraction using pixel-level binarization ground truth. With relatively small lecture-video datasets and training on random crops extracted from high-resolution images, optimization can become unstable and highly sensitive to sampling difficulty. Our prior handwritten content extraction pipeline, FCN-LectureNet [6], explored foreground-focused sampling, where training crops are biased toward regions with handwriting foreground pixels rather than sampled uniformly at random. However, transitioning abruptly from reconstruction-based pretraining to aggressive foreground-focused sampling may destabilize learned representations and induce catastrophic forgetting [14].

To address this, we adopt a progressive curriculum-based fine-tuning strategy. Training begins with a warm-up stage where the shared backbone modules (encoder, decoder, and Transformer-based Skip modules) remain frozen while task-specific branches adapt to the downstream objectives. This stage uses standard random crop sampling. After warm-up, full-network optimization is enabled before progressively introducing foreground-focused sampling.

Foreground-focused sampling is then introduced progressively. This strategy biases sampling toward image regions whose foreground-pixel ratio exceeds a given threshold. The threshold for these difficult samples is gradually increased from 0% to 10% in increments of 1%. At each stage, only 50% of the sampled crops are constrained by the foreground ratio requirement, while the remaining crops are sampled uniformly at random. This preserves exposure to easy/background regions while still emphasizing difficult handwritten regions.

Optimization uses an exponential moving average (EMA, $\alpha = 0.9$) of the validation loss. A patience hyper-parameter controls both learning-rate scheduling and curriculum progression. Whenever the EMA validation loss fails to improve for a fixed number of epochs, the learning rate is reduced by a factor of 10, up to two reductions. Training terminates after a third plateau event.

The same patience parameter also controls the progression of foreground-focused sampling difficulty. Specifically, the foreground-ratio threshold increases by 1% whenever the EMA validation loss fails to improve for $0.25\times$ patience epochs, until reaching the maximum threshold of 10%. Learning-rate scheduling and curriculum progression remain independently controlled, allowing foreground difficulty to continue increasing even after learning-rate reductions occur.

During fine-tuning, the same branch-specific objectives from Stage 2 are preserved. The text-detection and binarization branches use BCE losses, while the background-estimation branch continues using the foreground-weighted perceptual reconstruction loss. Since the background reconstruction objective produces substantially larger loss magnitudes, we apply loss weights of $W_{bg} = 1.0$, $W_{txt} = 1.0$, and $W_{bin} = 10.0$ to prioritize the final binarization objective.

4 Experiments

In this work, we proposed the Lect-Former architecture together with a multi-stage training protocol for handwritten foreground extraction from lecture-video frames. This section first describes the datasets and shared experimental setup, followed by ablation studies analyzing different pretraining stages and architectural design choices. Finally, we present a detailed evaluation on the target domain together with a comparison against the closest baseline model.

4.1 Datasets

Each pipeline stage used different datasets for training and validation. Stage 1 pretraining used the MS-COCO 2017 [15], with the standard training (118K) and validation (5K) splits. Since this stage only requires image reconstruction, no additional annotations were needed. Stage 2 text-deletion pretraining used the annotated LSVT [23] portion (30K images). LSVT contains Chinese and English scene-text images with polygon-based text annotations, which were converted into binary masks. After removing 10 narrow images (minimum side < 256 pixels), the remaining 29,990 images were divided into training (26,991) and validation (2,999) splits. For Stage 3 foreground extraction, we used the TextSeg [28] and Total-Text [5] scene-text segmentation datasets for ablation studies, since they provide pixel-level text annotations while remaining substantially different from the final handwritten lecture-video domain.

Final domain-specific evaluation used an extended version of the Lecture-Math dataset [6]. 16 new hand-only lecture videos were added, for 50 total. The dataset includes manually annotated key-frames with pixel-level binary masks of handwritten content. The videos were split into 32 training, 8 validation, and 10 testing videos. Unlike prior work [6], the splits are writer-disjoint between training and testing to reduce handwriting-style bias during evaluation. The resulting image-level splits contain 353 training, 85 validation, and 100 testing images.

4.2 Experimental Setup

Across all experiments, we used AdamW [18] with weight decay 1×10^{-4} and $(\beta_1, \beta_2) = (0.9, 0.99)$. The base learning rate is 1×10^{-4} for all tasks. Data augmentation included horizontal and vertical flipping, rotation, zooming, color jittering, illumination perturbations, and additive Gaussian noise.

For pretraining stages 1 and 2, optimization followed the same validation-driven protocol. Training continued until the validation loss failed to improve for 5 consecutive epochs. The first two plateaus trigger a $\times 0.1$ learning-rate reduction, while the third triggers early stopping. We set a maximum of 200 training epochs, though early stopping always ends it earlier. Stage 3 fine-tuning

Table 1. Ablation study on different Stage-1 pretraining objectives. "Mask." indicates masked feature reconstruction, "Lap." indicates Laplacian reconstruction targets, "S2" indicates text-deletion pretraining, and "FT" indicates the number of downstream binarization fine-tuning epochs. "ZS" corresponds to zero-shot binarization after Stage 2. Results are reported on TextSeg and Total-Text using fgIoU and F1 metrics. Each value represents the mean ± standard deviation across five independently trained models.

ID	Pre-T			Train	TextSeg		Total-text	
	Mask.	Lap.	S2	FT	FgIoU	F1	FgIoU	F1
1	–	–	✓	ZS	57.30 ± 0.59	67.77 ± 0.56	47.62 ± 0.95	60.09 ± 1.00
2	–	–	✓	500	76.26 ± 0.56	84.37 ± 0.46	61.95 ± 1.26	73.18 ± 1.11
3	–	–	–	1000	71.42 ± 1.93	80.51 ± 1.65	57.23 ± 0.80	68.85 ± 0.79
4	✓	–	✓	ZS	59.04 ± 0.88	69.48 ± 0.76	51.44 ± 0.69	63.91 ± 0.60
5	✓	–	✓	500	78.26 ± 0.80	85.85 ± 0.64	65.84 ± 1.49	76.47 ± 1.20
6	✓	–	–	1000	78.99 ± 0.64	86.41 ± 0.48	65.80 ± 1.28	76.50 ± 1.10
7	–	✓	✓	ZS	60.22 ± 0.55	70.48 ±0.58	49.30 ± 0.35	61.71 ± 0.34
8	–	✓	✓	500	77.54 ± 0.63	85.31 ± 0.52	64.51 ± 0.48	75.32 ± 0.43
9	–	✓	-	1000	76.33 ± 0.56	84.43 ± 0.45	58.32 ± 1.33	69.89 ± 1.23
10	✓	✓	✓	ZS	59.35 ± 1.55	69.78 ± 1.47	51.51 ± 0.94	63.92 ± 0.94
11	✓	✓	✓	500	78.25 ± 1.34	85.81 ± 1.03	66.31 ± 2.41	76.85 ± 2.06
12	✓	✓	–	1000	78.66 ± 0.68	86.21 ± 0.54	65.76 ± 0.99	76.45 ± 0.84

used experiment-specific optimization schedules. Fixed schedules were used for ablation studies, while the final lecture-video evaluation employed the Sect. 3.2 adaptive optimization protocol, with both patience and warm-up equal to 50 epochs.

For high-resolution inputs, inference was performed using tiled prediction to control memory usage. Images whose diagonal exceeded 2400 pixels were divided into 1024 × 1024 tiles with an overlap of 256 pixels. Predictions from overlapping regions were merged using spatially weighted interpolation to reduce boundary artifacts. This procedure was used for large images in Total-Text and, more frequently, TextSeg. In contrast, the 1080p lecture-video frames in the target whiteboard/chalkboard domain fall below this threshold and were therefore processed in a single forward pass.

Experiments ran on multiple NVIDIA GPU systems with at least 24 GB VRAM. Unless otherwise stated, all experiments used an effective batch size of 16 images. Some larger attention configurations required gradient accumulation to preserve this batch size under memory constraints. Runtime and memory comparisons in Sect. 4.4 were measured on the same hardware platform.

For ablation studies, we followed the scene text segmentation evaluation protocol used in TextSeg [28], reporting foreground Intersection over Union (FgIoU) and foreground F1-score. Both metrics focus exclusively on foreground text pixels. Final evaluation on lecture-video images reports standard document-image binarization metrics from the H-DIBCO [20] competition.

Stage 1 pretraining is performed on a sufficiently large dataset and generally produces stable results. In contrast, Stage 2 and especially Stage 3 exhibit more noticeable fluctuations due to the smaller training datasets and stochastic optimization effects. To reduce this variance, all experiments involving Stage 2 or Stage 3 training were repeated multiple times using different random seeds. Reported values correspond to the average performance across runs, while the number of repetitions for each condition is indicated in the corresponding tables.

4.3 Ablation Study on Stage 1 Pretraining Objectives

Our goal in this ablation study was to measure the effectiveness of four pretraining objectives for Stage 1: simple image reconstruction, reconstruction with masked features, reconstruction with multi-scale reconstruction targets (Laplacian targets), and a hybrid that combines the last two. For each Stage 1 pretraining objective, we evaluated three different additional training configurations: zero-shot after Stage 2, 500-epoch fine-tuning after Stage 2, and 1000-epoch fine-tuning after Stage 1. These conditions further allowed us to study how different Stage 1 pretraining objectives lead to better or worse downstream adaptations. The corresponding metrics can be found in Table 1.

These results show that pretraining with masked features generally leads to the best downstream performance across most experimental conditions. The Laplacian-inspired multi-scale reconstruction objective consistently outperforms simple image reconstruction, although it generally remains slightly below masked feature reconstruction. Combining both strategies leads to results that are generally comparable to, or slightly worse than, masked feature reconstruction alone, while also exhibiting higher variances across runs. This suggests that the benefits of both objectives are not trivially additive in the current training configuration.

We can also observe that the benefits of using Stage 2 pretraining are more obvious for the weaker pretraining objective of simple image reconstruction, where fine-tuning after Stage 2 substantially outperforms directly fine-tuning the Stage 1 model. Meanwhile, when stronger pretraining objectives are used, this benefit seems to shrink and in some cases disappear. However, these trends are not necessarily consistent with our final experiments on our target domain. In all of our following experiments, when Stage 1 pretraining is used, we refer to the model trained for reconstruction with masked features only.

4.4 Ablation Study on Localized Self-Attention Implementations

This ablation study measures the effectiveness and feasibility of using different implementations for the localized self-attention skip connection modules: Masked Attention, Windowed Attention, Shifted, Flex Attention and Neighborhood Attention Transformers. Windowed Attention did not use fused kernels and therefore required explicit window materialization, leading to substantially higher memory consumption during training. we used gradient accumulation

Table 2. Ablation study on different backend for transformer-based skip-connection attention. "Type" refers to the backend used. All networks were pretrained using masked features only, and then they are fine-tuned for exactly 1000 epochs. Results are reported on TextSeg and Total-Text using fgIoU and F1 metrics. Each value represents the mean ± standard deviation across five independently fine-tuned models.

ID	Type.	TextSeg		Total-text	
		FgIoU	F1	FgIoU	F1
1	No Attention	72.15 ± 1.16	81.08 ± 1.01	60.64 ± 1.45	71.89 ± 1.28
2	Masked Attention	78.99 ± 0.64	86.41 ± 0.48	65.80 ± 1.28	76.50 ± 1.10
3	Windowed Attention	77.12 ± 1.01	84.96 ± 0.77	65.49 ± 1.26	76.29 ± 1.06
4	Shifted (Interleaved)	78.58 ± 0.65	86.31 ± 0.51	65.40 ± 1.58	76.24 ± 1.28
5	Shifted (Hybrid)	78.19 ± 0.45	85.82 ± 0.37	66.29 ± 1.22	76.93 ± 1.01
6	Flex Attention	78.13 ± 0.51	85.85 ± 0.42	67.69 ± 0.38	78.07 ± 0.28
7	Neighborhood Attention	77.87 ± 0.66	85.61 ± 0.48	66.47 ± 0.93	77.10 ± 0.73

Table 3. Ablation study on different backend for transformer-based skip-connection attention. Results are reported on the resources used to execute inference on the entire test split of the TextSeg dataset. Most images are processed at their original resolution, while extremely large images are tiled.

ID	Type.	Peak Allocated Memory (MB)		Inference Time (s)	
		Mean	Max	Mean	Max
1	No Attention	1823.58	4647.02	0.45	15.75
2	Masked Attention	7287.24	36717.13	1.78	24.55
3	Windowed Attention	5224.13	11163.02	0.60	19.71
4	Shifted (Interleaved)	2658.47	6058.48	0.51	17.80
5	Shifted (Hybrid)	2658.45	6058.48	0.53	18.58
6	Flex Attention	2684.72	6680.06	0.53	18.17
7	Neighborhood Attention	5953.59	28642.75	1.04	43.24

with two microbatches of 8 images to preserve an effective batch size of 16. For Shifted Attention, we considered two variations, one that used interleaved regular and shifted attention windows, while the second one used a hybrid combination that re-used the same attention weights to compute both regular and shifted window attention within each layer fusing both outputs before the next layer. As a baseline, we also considered the use of standard skip connections (No Attention). All conditions start from Stage 1 pretraining followed by a 1000-epoch fine-tuning on the target dataset. In addition to segmentation performance, we also consider resource usage metrics such as peak allocated memory and inference time. These resources were measured by running one version of each condition on the entire test set of TextSeg, and we report the average and maximum values for all predictions. The corresponding segmentation performance metrics can be found in Table 2, while resource usage metrics can be found in Table 3.

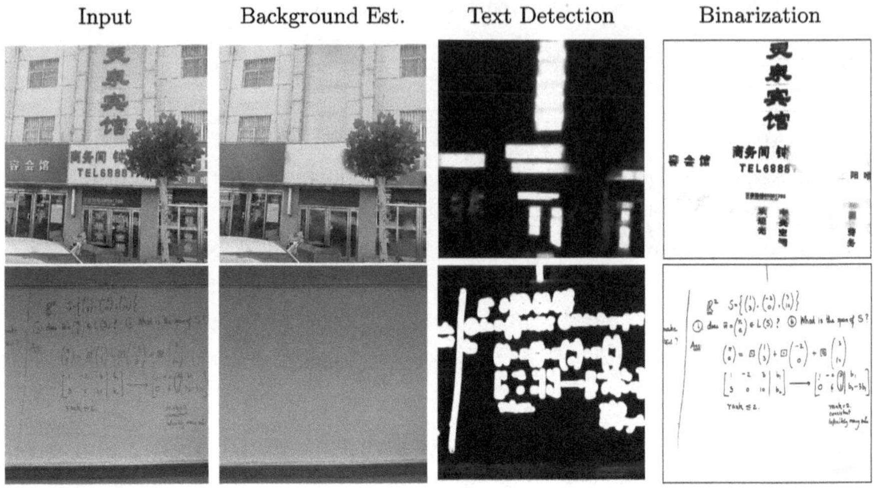

Fig. 4. Examples of intermediate and final outputs from the proposed multi-stage training pipeline. The top row corresponds to Stage 2 pretraining, while the bottom row shows the final binarization model after Stage 3 training.

It can be seen that all localized self-attention variants outperform the baseline without attention. Meanwhile the simplest Masked Attention tends to outperform other implementations in TextSeg, while Flex Attention achieves the strongest performance on Total-Text. Windowed Attention is generally outperformed by other attention implementations despite its higher memory requirements. Meanwhile Shifted attention (both interleaved and hybrid) produces competitive results in all cases. Based on these numbers alone, no single implementation consistently dominates across both datasets.

However, once that we consider the resource numbers in Table 3, some clear patterns emerge. Masked Attention achieves strong segmentation performance, but its memory requirements are substantially higher than the more efficient alternatives. Neighborhood Attention also exhibits high memory consumption, limiting its practicality on lower-memory GPUs. On the other hand, Shifted Attention and Flex Attention are very memory efficient, faster than other conditions that use attention, and each one achieves the best performance on different datasets. These results indicate that both Shifted Attention and Flex Attention provide strong tradeoffs between segmentation performance and computational efficiency. In the following experiments, we use the Shifted Interleaved implementation in the following experiments due to its competitive segmentation performance, low peak memory usage, and slightly faster inference.

Table 4. Ablation study on different fusion mechanisms of multiple branch outputs. "Type" refers to the fusion mechanism used. "FT" indicates the number of downstream binarization fine-tuning epochs. "ZS" corresponds to zero-shot binarization after Stage 2. All networks were pretrained using masked features only. Results are reported on TextSeg and Total-Text using fgIoU and F1 metrics. Each value represents the mean $\pm$ standard deviation across three independently fine-tuned models.

ID	Type.	FT	TextSeg		Total-text	
			FgIoU	F1	FgIoU	F1
1	FCN-LectureNet [6]	ZS	60.82 ± 1.10	71.43 ± 1.01	51.92 ± 1.59	64.23 ± 1.58
2	FCN-LectureNet [6]	500	78.95 ± 0.49	86.57 ± 0.38	67.90 ± 0.65	78.05 ± 0.54
3	Early Concat.	ZS	61.10 ± 0.53	71.74 ± 0.40	52.36 ± 0.46	64.82 ± 0.67
4	Early Concat.	500	78.73 ± 0.53	86.34 ± 0.48	67.43 ± 0.49	77.67 ± 0.56
5	Aux. Only	ZS	61.40 ± 0.44	71.87 ± 0.44	51.48 ± 0.50	63.70 ± 0.52
6	Aux. Only	500	74.90 ± 0.33	83.93 ± 0.33	66.43 ± 0.31	77.22 ± 0.32
7	Late Add.	ZS	61.67 ± 0.59	71.98 ± 0.68	52.75 ± 0.90	65.19 ± 0.95
8	Late Add.	500	79.04 ± 0.22	86.62 ± 0.17	66.52 ± 0.15	77.26 ± 0.05
9	Hybrid (Early + Late)	ZS	61.01 ± 0.29	71.67 ± 0.32	52.27 ± 0.79	64.74 ± 0.87
10	Hybrid (Early + Late)	500	78.83 ± 0.48	86.50 ± 0.36	67.77 ± 0.35	78.02 ± 0.33

4.5 Ablation Study on Auxiliary Branch Fusion

This ablation study investigates the effect of different auxiliary branch fusion mechanisms. We study how the binarization branch integrates the enhanced image I_{enh} produced by the auxiliary heads. We consider multiple fusion strategies, including Early Concatenation, Auxiliary Only, Late Additive fusion, and a Hybrid configuration that combines Early Concatenation and Late Additive fusion. Early concatenation directly concatenates I_{enh} with the decoder features at the input of the binarization branch. Auxiliary only uses I_{enh} as the input to a shallower binarization head. Late Additive fusion reuses this shallow auxiliary head and performs a late additive fusion between its prediction and the output of the main branch. As a baseline, we follow the aggressive fusion strategy from our prior FCN-LectureNet [6], which concatenates I_{enh} with the inputs of every block in the binarization branch. All Conditions were pretrained using Stage 1 and 2. We report both zero-shot prediction results and results after a 500-epoch fine-tuning stage on the target dataset. Metrics are reported in Table 4. The top row of Fig. 4 presents representative outputs after Stage 2 on LSVT.

Overall, the different fusion strategies achieve comparable performance, especially after fine-tuning. Interestingly, the Auxiliary Only configuration still produces competitive results, suggesting that the enhanced image I_{enh} already contains meaningful segmentation information. However, this simpler configuration underperforms on the more challenging TextSeg dataset after fine-tuning, indicating that integrating decoder features remains beneficial. The Hybrid configuration does not consistently improve over Early Concatenation or Late Additive

Table 5. Evaluation of Text Binarization performance using the extended LectureMath dataset. We use the binarization metrics from H-DIBCO. [20] to evaluate and compare the performance of the proposed Lect-Former (Ours) and the baseline model FCN-LectureNet [6] using different combinations of Pretraining (Pre-T) stages and Fine-tuning (Train). Each number represents the average of 5 independently trained models.

	Pre-T		Train				Standard			Pseudo		
ID	Rec	T.Del	Bin	Model	PSNR ↑	DRD ↓	Rec ↑	Prec ↑	F1 ↑	Rec ↑	Prec ↑	F1 ↑
1	–	–	✓	[6]	19.83	14.34	87.94	72.11	76.90	93.08	70.04	77.87
2	–	–	✓	Ours	19.39	18.6	81.13	70.24	73.36	88.41	68.12	75.27
3	–	✓	–	[6]	20.43	6.86	66.38	93.98	75.62	76.97	93.47	82.89
4	–	✓	–	Ours	20.31	6.34	64.58	93.58	74.30	75.14	92.74	81.58
5	–	✓	✓	[6]	21.28	10.51	88.28	79.07	81.73	93.00	76.97	82.68
6	–	✓	✓	Ours	22.02	8.94	86.73	82.00	83.13	93.39	80.18	85.31
7	✓	–	✓	[6]	20.64	12.51	87.78	75.37	79.33	92.52	72.78	79.95
8	✓	–	✓	Ours	21.65	10.23	87.10	81.36	82.28	92.92	79.08	83.82
9	✓	✓	–	[6]	20.14	7.72	65.52	92.19	74.05	75.86	91.65	81.13
10	✓	✓	–	Ours	20.75	5.79	68.84	93.41	77.71	79.18	92.35	84.18
11	✓	✓	✓	[6]	21.84	9.11	86.83	82.14	82.94	92.31	79.93	84.37
12	✓	✓	✓	Ours	22.96	6.69	88.13	85.41	85.91	94.24	83.50	87.84

fusion, while the aggressive FCN-LectureNet fusion strategy [6] also remains competitive without clearly outperforming the simpler alternatives. Overall, no fusion mechanism consistently dominates across all conditions. For the final experiments, we select Late Additive fusion due to its competitive performance and architectural simplicity.

4.6 Final Performance on Lecture Videos

In our final experiment, we compare the performance of the proposed Lect-Former architecture against our prior work, FCN-LectureNet [6]. Our training protocol is overall stronger, including important changes such as AdamW instead of SGD optimization. To make a fairer comparison, the baseline model is directly train under our full optimization framework. This should minimize differences related to training protocols, and highlight performance differences caused by the architecture design itself. We evaluate the effect of using different combinations of the proposed training stages. The goal is to measure how much does each stage contribute to the final performance of the network on the target domain. Our final results based the H-DIBCO metrics can be found in Table 5. The bottom row of Fig. 4 presents representative outputs after fine-tuning on real data.

It can be seen that the baseline model still performs better under low-data training such as only using Stage 3. The gap reduces on zero-shot predictions after Stage 2 pretraining, and the new model starts achieving better performance once that these two stages are combined. The gaps increase once that

Stage 1 is further added showing that the proposed Lect-Former can benefit even further from the initial masked reconstruction pretraining. Another important observation is that unlike ablation studies on TextSeg and Total-Text, the Stage 2 pretraining still plays an important role in helping both consistently achieve better results than when it is not used. The proposed model final configuration reported here has over 160M. Our results show that this network can still be effectively trained on just 350 domain images given appropriate training objectives on larger datasets from different domains.

5 Conclusion

In this work, we have proposed the Lect-Former architecture for accurate extraction of foreground from lecture-video frames. Our experiments show that the proposed transformer-based skip connections with localized self-attention are effective in helping the model to achieve better performance on general text segmentation tasks. We conducted different ablation studies that show how different design choices and pre-training stages contribute to the final performance of the model. In the future, we would like to explore how variations of this architecture can be used for higher level tasks such as handwriting segmentation, differentiating between mathematical expressions, text and hand-drawings. The proposed framework is also capable of producing text-less outputs (see Fig. 4), which can also be useful for other applications.

References

1. Ba, J.L., Kiros, J.R., Hinton, G.E.: Layer normalization. arXiv preprint arXiv:1607.06450 (2016)
2. Cao, H., et al.: Swin-UNet: UNet-like pure transformer for medical image segmentation. In: Karlinsky, L., Michaeli, T., Nishino, K. (eds.) ECCV 2022. LNCS, vol. 13803, pp. 205–218. Springer, Cham (2022). https://doi.org/10.1007/978-3-031-25066-8_9
3. Carion, N., Massa, F., Synnaeve, G., Usunier, N., Kirillov, A., Zagoruyko, S.: End-to-end object detection with transformers. In: Vedaldi, A., Bischof, H., Brox, T., Frahm, J.-M. (eds.) ECCV 2020. LNCS, vol. 12346, pp. 213–229. Springer, Cham (2020). https://doi.org/10.1007/978-3-030-58452-8_13
4. Chen, J., et al.: TransUNet: transformers make strong encoders for medical image segmentation. arXiv preprint arXiv:2102.04306 (2021)
5. Ch'Ng, C.K., Chan, C.S.: Total-text: a comprehensive dataset for scene text detection and recognition. In: 2017 14th IAPR International Conference on Document Analysis and Recognition (ICDAR), vol. 1, pp. 935–942. IEEE (2017)
6. Davila, K., Xu, F., Setlur, S., Govindaraju, V.: FCN-LectureNet: extractive summarization of whiteboard and chalkboard lecture videos. IEEE Access **9**, 104469–104484 (2021)
7. Davila, K., Zanibbi, R.: Visual search engine for handwritten and typeset math in lecture videos and latex notes. In: 2018 16th International Conference on Frontiers in Handwriting Recognition (ICFHR), pp. 50–55. IEEE (2018)

8. Dosovitskiy, A., et al.: An image is worth 16×16 words: transformers for image recognition at scale. arXiv preprint arXiv:2010.11929 (2020)
9. Hassani, A., Walton, S., Li, J., Li, S., Shi, H.: Neighborhood attention transformer. In: Proceedings of the IEEE/CVF Conference on Computer Vision and Pattern Recognition, pp. 6185–6194 (2023)
10. He, K., Chen, X., Xie, S., Li, Y., Dollár, P., Girshick, R.: Masked autoencoders are scalable vision learners. In: Proceedings of the IEEE/CVF Conference on Computer Vision and Pattern Recognition, pp. 16000–16009 (2022)
11. He, K., Zhang, X., Ren, S., Sun, J.: Deep residual learning for image recognition. In: Proceedings of the IEEE Conference on Computer Vision and Pattern Recognition, pp. 770–778 (2016)
12. Huang, M., et al.: SwinTextSpotter: scene text spotting via better synergy between text detection and text recognition. In: Proceedings of the IEEE/CVF Conference on Computer Vision and Pattern Recognition, pp. 4593–4603 (2022)
13. Johnson, J., Alahi, A., Fei-Fei, L.: Perceptual losses for real-time style transfer and super-resolution. In: Leibe, B., Matas, J., Sebe, N., Welling, M. (eds.) ECCV 2016. LNCS, vol. 9906, pp. 694–711. Springer, Cham (2016). https://doi.org/10.1007/978-3-319-46475-6_43
14. Kirkpatrick, J., et al.: Overcoming catastrophic forgetting in neural networks. Proc. Natl. Acad. Sci. **114**(13), 3521–3526 (2017)
15. Lin, T.-Y., et al.: Microsoft COCO: common objects in context. In: Fleet, D., Pajdla, T., Schiele, B., Tuytelaars, T. (eds.) ECCV 2014. LNCS, vol. 8693, pp. 740–755. Springer, Cham (2014). https://doi.org/10.1007/978-3-319-10602-1_48
16. Liu, Z., et al.: Swin transformer: hierarchical vision transformer using shifted windows. In: Proceedings of the IEEE/CVF International Conference on Computer Vision, pp. 10012–10022 (2021)
17. Long, J., Shelhamer, E., Darrell, T.: Fully convolutional networks for semantic segmentation. In: Proceedings of the IEEE Conference on Computer Vision and Pattern Recognition, pp. 3431–3440 (2015)
18. Loshchilov, I., Hutter, F.: Decoupled weight decay regularization. arXiv preprint arXiv:1711.05101 (2017)
19. Peng, D., Liu, C., Liu, Y., Jin, L.: ViTEraser: harnessing the power of vision transformers for scene text removal with segMIM pretraining. In: Proceedings of the AAAI Conference on Artificial Intelligence, vol. 38, pp. 4468–4477 (2024)
20. Pratikakis, I., Zagoris, K., Barlas, G., Gatos, B.: ICFHR2016 handwritten document image binarization contest (H-DIBCO 2016). In: 2016 15th International Conference on Frontiers in Handwriting Recognition (ICFHR), pp. 619–623. IEEE (2016)
21. Ronneberger, O., Fischer, P., Brox, T.: U-net: convolutional networks for biomedical image segmentation. In: Navab, N., Hornegger, J., Wells, W.M., Frangi, A.F. (eds.) MICCAI 2015. LNCS, vol. 9351, pp. 234–241. Springer, Cham (2015). https://doi.org/10.1007/978-3-319-24574-4_28
22. Simonyan, K., Zisserman, A.: Very deep convolutional networks for large-scale image recognition. In: In 3rd International Conference on Learning Representations (ICLR 2015), pp. 1–14. Computational and Biological Learning Society (2015)
23. Sun, Y., Liu, J., Liu, W., Han, J., Ding, E., Liu, J.: Chinese street view text: large-scale Chinese text reading with partially supervised learning. In: Proceedings of the IEEE/CVF International Conference on Computer Vision, pp. 9086–9095 (2019)
24. Telea, A.: An image inpainting technique based on the fast marching method. J. Graphics Tools **9**(1), 23–34 (2004)

25. Vaswani, A., et al.: Attention is all you need. Adv. Neural Inf. Process. Syst. **30** (2017)
26. Wang, H., Cao, P., Wang, J., Zaiane, O.R.: UCTransNet: rethinking the skip connections in u-net from a channel-wise perspective with transformer. In: Proceedings of the AAAI Conference on Artificial Intelligence, vol. 36, pp. 2441–2449 (2022)
27. Wu, Y., He, K.: Group normalization. In: Proceedings of the European Conference on Computer Vision (ECCV), pp. 3–19 (2018)
28. Xu, X., Zhang, Z., Wang, Z., Price, B., Wang, Z., Shi, H.: Rethinking text segmentation: a novel dataset and a text-specific refinement approach. In: Proceedings of the IEEE/CVF Conference on Computer Vision and Pattern Recognition, pp. 12045–12055 (2021)
29. Yu, H., Wang, X., Niu, K., Li, B., Xue, X.: Scene text segmentation with text-focused transformers. In: Proceedings of the 31st ACM International Conference on Multimedia, pp. 2898–2907 (2023)

Modeling "Musical Words" for Efficient Sequence Search in Handwritten Music Manuscripts

Aitana Menárguez-Box(✉), Enrique Vidal, and Alejandro H. Toselli

PRHLT Research Center, Universitat Politècnica de València, Camí de Vera s/n, 46022 Valencia, Spain
{amenbox,evidal,ahector}@prhlt.upv.es

Abstract. Large-scale digitization has made vast collections of historical music manuscripts accessible, yet efficient search within them remains a major challenge. Retrieving melodic patterns from handwritten sources is central to musicological analysis, but current approaches typically rely on deterministic Optical Music Recognition (OMR) followed by symbol-level sequence matching. This pipeline is both error-prone and computationally expensive, limiting its scalability. In this work, we aim to improve retrieval efficiency by modeling recurring note patterns as "musical words", learned automatically using Byte Pair Encoding (BPE), and integrating them into a probabilistic indexing (PrIx) framework. By operating on these higher-level units instead of individual symbols, retrieval can be performed over shorter queries and with reduced computational cost. We evaluate the approach on three collections of handwritten chant manuscripts, comparing word-based indexing with conventional note-level search. The results show that BPE-derived "musical words" preserve retrieval effectiveness while significantly reducing query time, with gains becoming more pronounced under limited training data. These findings suggest that lexical modeling is a promising direction for scalable and robust retrieval in large collections of handwritten music, while remaining compatible with both probabilistic and conventional retrieval pipelines.

Keywords: Music score recognition and search · Probabilistic indexing · Melodic patterns · Musical words

1 Introduction

In recent years, the large-scale digitization of historical musical archives has significantly advanced the preservation of cultural heritage. Libraries and research institutions worldwide have made extensive collections of high-resolution manuscript images publicly available. However, the rapid growth of these digital repositories has outpaced advances in their organization and accessibility; collections remain largely unsearchable, creating a substantial gap between data availability and effective information access. Music Information Retrieval

F. Shafait et al. (Eds.): DAS 2026, LNCS 16851, pp. 282–298, 2027.
https://doi.org/10.1007/978-3-032-36207-0_17

(MIR) techniques, and in particular Optical Music Recognition (OMR), provide a technological bridge to address this limitation. Nevertheless, historical music manuscripts pose singular challenges, that need to be taken into account when addressing problems within this ambit, as described in [2,5,17].

In this context, the retrieval of *musical sequences* –such as motifs, themes, or recurrent liturgical formulas– plays a central role in musicological research. Traditionally, the search process relies on a two-step pipeline: first, a *deterministic* transcription of the manuscript is automatically generated (the 1-best hypothesis), and then a search is performed over this text-based representation. This approach suffers from two major drawbacks. First, deterministic recognition is highly susceptible to the propagation of errors; a single misclassified symbol can render a sequence unsearchable [4]. Second, performing sequence matching on a symbol-by-symbol basis across massive datasets is computationally demanding and scales poorly with the size of the collection.

To overcome these limitations, we propose a search framework based on *Probabilistic Indexing* (PrIx) [26, Cap. 3]. This avoids the pitfalls of deterministic transcription by considering the full probability distribution produced by the OMR model, as previously explored in Handwritten Text Recognition (HTR), where probabilistic lattices enable indexing beyond the 1-best sequence [27]. However, since the concept of a "word" is not defined in music as it is in text, we build on a previously studied method to group musical symbols into "musical words" using *Byte Pair Encoding* (BPE), allowing to operate over high-level units rather than individual notes.

While both PrIx and BPE segmentation have been studied independently, their combination raises a question: whether lexical representations can serve as efficient and effective indexing units in the absence of a well-defined notion of "word" in music. Here, we provide an evaluation of this question in the context of handwritten music retrieval.

Therefore, the main contributions of this work are:

1. We extend PrIx from HTR to OMR by indexing "musical words" (BPE-derived units), enabling lexical retrieval directly on manuscript images.
2. We provide a comprehensive comparison of search efficiency, showing that word-based indexing yields markedly lower query-time cost than traditional symbol-level sequence search.
3. We study data scarcity, analyzing how the amount of training data (FULL and SMALL) affects both optical recognition and retrieval when using BPE vocabularies, and show that word-based training leads to *similar or better* transcription accuracy than symbol-level training.
4. We assess *generalization* of the learned words beyond the source manuscript: vocabularies trained on a given dataset (VOR) transfer to distinct collections (EIN, SAL) without degrading recognition. This supports the viability of a shared lexical space for historical chant.

2 Related Work

2.1 Automatic Music Recognition

In MIR, the task of retrieving melodic fragments has traditionally relied on either symbolic transcriptions or manually curated databases [28]. A prime example of this is the search tool available within the *Cantus Database* [15]. This kind of tools, while robust for exploring hand-transcribed chant manuscripts, do not address the intricacies involved in working with *untranscribed manuscript images*, which constitute the vast majority of the millions of historical musical manuscripts in archives and libraries. Other approaches focus on audio-based or cross-modal retrieval [1,9], while interesting, are orthogonal to our *sheet-image-based* setting without audio sources.

More recent advancements, such as the OMMR4ALL[1] and the *Cantus Ultimus*[2] projects employ modern deep learning techniques to enhance Handwritten Music Recognition (HMR) capabilities. While these neural systems represent significant progress, they focus only on *one* (the most probable) recognition hypothesis, reducing the output to a single, irrevocable interpretation. While effective on clean data, these pipelines remain brittle in the presence of transcription errors and suffer from scalability issues: matching long sequences of single symbols is computationally expensive, especially as collection sizes grow.

In contrast, PrIx for HTR has shown strong gains in keyword spotting and document retrieval by exploiting the full posterior distribution produced by neural sequence models, rather than committing to a single hypothesis [25]. The success of PrIx in HTR suggests that similar benefits should transfer to HMR, where recognition ambiguity is equally pervasive. However, directly adopting PrIx for music raises an important challenge: unlike text, music does not possess an inherent lexical structure so words are not defined. This motivates the introduction of higher-level musical units, here called "musical words".[3]

2.2 Lexical Modeling of Music

The rationale for a musical lexicon is supported by decades of quantitative evidence connecting music and language via Zipf-like rank–frequency regularities. Zipf [33] already discussed musical distributions (e.g., melodic intervals, repetition distances) across composers, suggesting Zipfian behavior. With computational corpora, Manaris *et al.* showed in [3] that Zipf–Mandelbrot models over MIDI-derived metrics are informative for classification and perceived quality. Later, Zanette related musical structure to Simon's generative model of language, emphasizing the importance of context even when "words" are not predefined [32]. More recently, Perotti and collaborators provided evidence that Zipf's

[1] Available through https://ommr4all.informatik.uni-wuerzburg.de/en/.

[2] Available through https://cantus.simssa.ca/.

[3] We note that the term "musical word" is used as a convenient metaphor for recurrent groupings of symbolic units. This should not be interpreted as a strict musicological concept, as the definition of meaningful musical units typically involves perceptual, syntactic, and contextual factors.

law emerges in scores when "appropriate units" –such as chords or aggregated pitch events– are chosen, underscoring that unit selection is crucial for Zipfian regularities in music [18]. Complementary studies at the signal and symbolic levels (e.g., pitch/loudness fluctuations or interval statistics) further reinforce the presence of Zipf-like scaling in musical data [12,31].

Building on these observations, in a previous work [16] we explored the construction of musical words as higher-level units derived from note sequences, including unsupervised approaches based on Byte Pair Encoding (BPE). We showed that such representations yield structured musical encodings that exhibit Zipf-like behavior, supporting their suitability for musical modeling. Taken together, these findings motivate a lexical view of music. In our work, this perspective enables the transfer of PrIx from HTR to HMR: once a musical vocabulary is learned, PrIx can index and retrieve *words*, reducing query length and improving efficiency.

3 Methodology

3.1 Probabilistic Indexing

PrIx provides a mature and well-established framework to deal with the intrinsic word-level uncertainty generally exhibited by handwritten text in images (of historical manuscripts, in particular) [21,26,27,30].[4] It draws from Keyword Spotting (KWS) concepts but, rather than caring for "key" words, any element in an image which is likely enough to be interpreted as a word is detected and stored, along with its Relevance Probability (RP) and its location in the image.

PrIx in Music. In our setting, a music-symbol recognizer (details in Sect. 4.2) provides the posterior over all possible symbol sequences for a staff image. From this distribution, PrIx computes, for any staff image x and musical *note* v, the probability that a note appears at an ordinal position k in some transcription hypothesis. Each such event becomes a "*spot*" of the form:

$$\begin{aligned} s &= [\,x, v, k, \boldsymbol{b}, P\,], \\ P &\stackrel{\text{def}}{=} P(R \mid x, v, k), \quad k \in \mathbb{N} \end{aligned} \tag{1}$$

where $\boldsymbol{b}$ is a bounding box estimated as a by-product for the word v at the ordinal position k [26], R is the event that v occurs at position k in image x, and P its associated RP. In the end each image region x becomes represented as a set of PrIx spots. These constitute a searchable representation of the music: queries can then be answered by combining the RPs of the relevant spots.

[4] The book [26] provides a comprehensive, updated and consolidated digest of all the previous works on PrIx. In the sequel we will only cite this book for any PrIx topic, except if a previous publication may help contextualizing some matter.

Sequence Retrieval with PrIx. Once single-symbol (single-note) PrIx spots have been generated for each staff image, a melodic pattern (i.e., a sequence of notes) can be retrieved. Let $q = \langle q_1, \ldots, q_n \rangle$ be a query sequence, and let K denote the maximum ordinal position among all note-spots in a staff x. The RP that q appears in x is obtained by evaluating all possible starting positions for q. Following [26], we write:

$$P(R \mid x, q) \approx \max_{1 \leq k \leq K-n+1} P(R_{q,k}), \tag{2}$$

were $P(R_{q,k}$ is the Relevance Probability of x for the query q starting at ordinal position k. It can be approximated as:

$$P(R_{q,k}) \approx \min_{1 \leq j \leq n} P(R \mid x, q_j, k + j - 1). \tag{3}$$

In other words, the RP of a sequence is approximated by combining, through Boolean AND/OR operations, the RPs of its constituent symbols across all compatible ordinal alignments. This formulation, also used in [24], provides an effective approximation to the probability that the entire note sequence occurs within the staff.

With Eq. (3), the cost of sequence retrieval grows linearly with the length of the query and with the number of note positions to be examined. This note-by-note combination becomes increasingly expensive and sensitive to local uncertainty.

From Symbol-Level to Word-Level Retrieval. Once a vocabulary is defined, each musical word is treated as a PrIx symbol. A delimiter ensures consistent word boundaries, enabling indexing at the word level. Figure 1 provides an illustrative example of how word-level indexing operates.

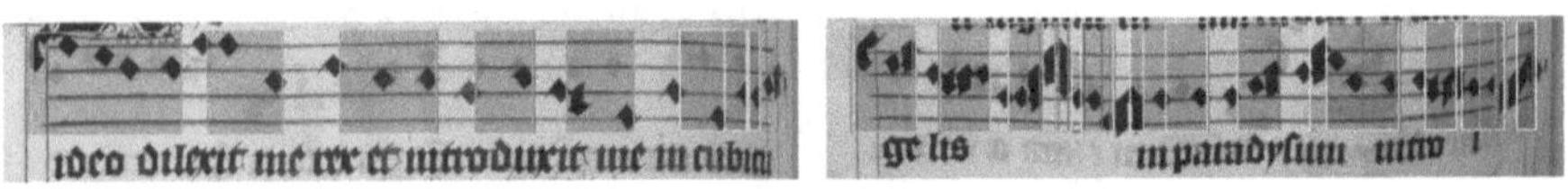

Fig. 1. Musical word PrIx spots for two staves from folio 150v of the VORAU-253 manuscript, illustrating how PrIx adapts to transcription and segmentation uncertainty. Simple or frequent musical patterns produce a few high-probability hypotheses (green), while visually degraded or uncommon patterns yield many alternative low-probability hypotheses (yellow to red). (Color figure online)

Under this representation, note sequence search operates over substantially shorter queries–now expressed as sequences of learned words. As a result, the number of AND/OR operations and the number of possible alignments to evaluate are greatly reduced. This leads to markedly more efficient retrieval, which is quantified in Sect. 4.3.

3.2 Modeling "Musical Words"

Following [16], to obtain higher-level musical units, we relay on Zipf concepts [32] and adopt Byte Pair Encoding (BPE) [10] as an unsupervised segmentation method over sequences of musical symbols. Here, BPE is applied to the VOR training transcripts by iteratively merging the most frequent adjacent pair of symbols, subject to two constraints: the pair must occur at least $o = 1$ times and the resulting unit must not exceed a maximum length of $l = 4$ notes. If no pair satisfies these conditions, training stops; the resulting list of applied merges is used to define a BPE vocabulary. So given the merge list obtained *only* from VOR, any sequence of notes (from *any* of the datasets used in this work) can be segmented deterministically by applying the merges in decreasing priority (earliest learned first) until no merge can be applied. This yields a consistent tokenization into BPE musical words without retraining BPE on new data.

Earlier experiments [16] show that such BPE-derived vocabularies tend to follow Zipf-like distributions, aligning with musically meaningful patterns. In our system, the resulting musical words serve both as recognition targets and as the indexing units for PrIx, ensuring coherence between optical modeling and lexical retrieval. Crucially, this process does not rely on any predefined musicological knowledge, making it appropriate for repertoires where such annotations are unavailable.

4 Experimental Setup

4.1 Data and Query Sets

We employ three collections of handwritten plainchant manuscript images: VORAU-253 (VOR) from the Austrian Centre for Digital Humanities at the Austrian Academy of Sciences, EINSIEDELN (EIN) from the Stiftsbibliothek Einsiedeln, and SALZINNES (SAL) from St. Mary's University–Patrick Power Library in Halifax. Musical transcripts for EIN and SAL were obtained from the *End-to-End Optical Music Recognition Resources* repository [8], provided in MEI format [22]. Corresponding chant records can also be consulted in the *Cantus Database* [15]. The VOR transcripts were manually produced in our laboratory, expanding upon a few fragments initially available in Volpiano notation [6,7].

Musical transcripts are available in geometrical format, so each note is represented by its *vertical* position on the staff. This means in what line (`l`) or space (`s`) (with a numerical bottom-to-top sorting) of the staff the given note is written. E.g., `s2` represents a note written on the second space of a staff.

Vorau (VOR). It constitutes our primary dataset and serves as the basis for exploring the transition from symbol-level to word-level indexing. All word vocabularies are learned *exclusively* from the *training* partition of VOR, using BPE applied to the symbolic ground-truth sequences. Example images of this manuscript can be found in Fig. 2.

To evaluate the effect of training-data availability, we consider two training partitions: VOR-FULL, the complete training partition, and VOR-small, a reduced subset of the training data designed to experiment with low-resource conditions. Importantly, the same partitions are annotated *with* and *without* BPE vocabularies, yielding symbol-based and word-based training targets for the optical model.

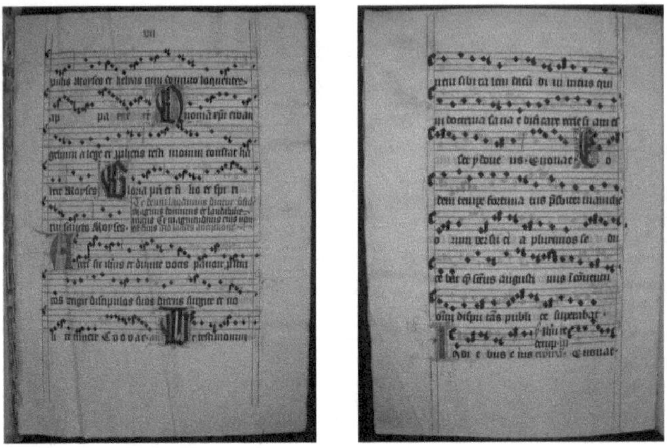

Fig. 2. Examples of page images from VORAU-253 music manuscript.

Einsiedeln (EIN) and Salzinnes (SAL). To evaluate generalization, we also test our word-based indexing approach on EIN and SAL. Both datasets share the same processing pipeline: their tokenization is made according to the resulting merges of the VORAU-253 BPE training. This allows us to assess whether learned musical words remain effective across different sources, scripts, and notational idiosyncrasies.

Table 1 summarizes the amount of data used in each experimental partition, including the number of staves, as well as statistics of the note-level sequences and their corresponding BPE segmentations (in terms of unique and running units). It is important to note that the BPE-word statistics reported here do not correspond to a change in the training targets. In all cases, training is performed at the symbol level, with sequences of musical symbols separated by an explicit delimiter (see Sect. 3.2). The reported BPE counts therefore reflect the distribution of these higher-level units, and are provided to characterize the resulting lexical structure.

Query Set. For searching in each dataset, we encode its test partition with the learned vocabulary according to the process described Sect. 3.2. After this, we collect the set of *unique* BPE words occurring in the test. Following the most commonly assumed setup in information retrieval and KWS experiments, the query set for each dataset is then defined as the set of these unique words. For fairness, queries for all symbol-level baselines are also defined by expanding each word query into its underlying sequence of notes.

Note that segmentation is not re-optimized at test time: it strictly follows the vocabulary learned on training data. While alternative segmentations are possible for a fixed vocabulary, the BPE merge order imposes a unique, reproducible segmentation consistent with training. A comparison of different segmentation strategies for musical sequences remains an open research question and is beyond the scope of this work.

Table 1. Datasets statistics for all experimental partitions. For each dataset and split, we report the number of staves, as well as the number of *unique* and *running* units for both note sequences and their corresponding BPE segmentations. Note that BPE-word counts refer to the tokenized sequences obtained after applying the learned BPE vocabulary, and are reported for analysis purposes; the optical models are trained at the symbol level with explicit separators between symbols (see Sect. 3.2).

Partition			Dataset			
			VOR-FULL	VOR-SMALL	EIN	SAL
Train	note seqs.	*unique*	27	25	24	19
		running	51 598	12 479	148 774	46 253
	BPE-words	*unique*	696	508	620	501
		running	19 283	4 669	52 937	17 671
		staves	1 933	455	4 751	1 671
Val.	note seqs.	*unique*	19	16	20	16
		running	6 548	3 991	19 426	5 912
	BPE-words	*unique*	324	298	474	297
		running	2 565	1 530	6 950	2 269
		staves	293	159	630	222
Test	note seqs.	*unique*	24	23	21	18
		running	38 239	37 822	21 372	7 361
	BPE-words	*unique*	568	568	491	332
		running	14 753	14 753	7 681	2 834
		staves	1 588	1 588	729	262

4.2 Optical Model for Music Recognition

The PrIx process departs from note sequence posteriors for each staff image. These were computed using a neural optical model trained on staff-line crops, combined with an n-gram language model following standard decoding practices for handwritten text recognition.

CRNN Optical Model. We employed a Convolutional-Recurrent Neural Network implemented with the *PyLaia* toolkit [20]. The convolutional frontend consists of four layers with 12, 24, 48, and 96 feature maps, each using 3×3 kernels and *LeakyReLU* activations; a 2×2 max-pooling layer is applied after the third convolution. The recurrent backend comprises three bidirectional LSTM layers with 256 units each. A final softmax layer predicts one of the 26 musical symbols in the alphabet (plus an explicit inter-symbol "space" and a non-character token). Compared to the earlier architecture in [7], this model uses a different CRNN configuration and explicitly models inter-symbol spacing.

Training. The model is trained with the Adam optimizer [13] using mini-batches of eight samples and a learning rate of $7 \cdot 10^{-4}$. The objective function is the *Connectionist Temporal Classification* (CTC) loss [11], and training stops when the symbol error rate on the development set fails to improve for 30 consecutive epochs.

Language Modeling and Decoding. To improve sequence consistency, we use a 4-gram language model with Kneser–Ney smoothing [14], estimated from the training transcripts using the *SRILM* toolkit [23]. Optical and language models are combined using the *Kaldi* decoder [19] to generate symbol lattices. From these lattices, ordinal-position PrIx spots are computed following Algorithm 5.16 of [26].

All data, code, and configuration files required to reproduce these models will be made available upon reasonable request.

4.3 Metrics

We evaluate our system along three complementary aspects (i) optical recognition accuracy, (ii) computational efficiency of the retrieval stage, and (iii) information-retrieval effectiveness.

Optical Recognition Accuracy. To measure the quality of the optical model, we use a metric analogous to the Character Error Rate (CER) in HTR, adapted to musical symbols. Since the recognizer outputs sequences of notes, we quantify accuracy through the Symbol Error Rate (SER). Given a reference sequence R and a hypothesis H, SER is defined as:

$$\mathrm{SER} = \frac{d(R, H)}{|R|} \tag{4}$$

where $d(R, H)$ is the edit distance (minimum number of insertions, deletions, and substitutions required to transform H into R), and $|R|$ is the length of the reference note sequence.

To quantify uncertainty, we report 95% *Confidence Intervals* (CI) assuming independent symbol-level errors, computed by means of the conventional Gaussian approximation to the binomial distribution.

Retrieval Effectiveness. To assess the quality of the retrieved staves, we compute standard information retrieval metrics based on relevance probability thresholds. For a fixed RP threshold, *precision* is the ratio between correctly retrieved staves (*hits*) and the total retrieved staves. *Recall* is the ratio between hits and the total number of relevant staves in the dataset. Varying the RP threshold yields different precision-recall pairs, forming a Recall-Precision Curve (RPC). For each query, we compute the Average Precision (AP) as the area under its RPC. These measurements characterize the retrieval quality of the system across note-level and word-level indexing conditions.

Retrieval Efficiency. If a musical word is seen as a sequence, the sequence must be checked at every possible starting position in the staff. At each position, we must combine the Relevance Probabilities of the n notes in order (see Eq. 3), meaning the cost of the search grows at least *proportionally* to the product of the query length and the length of the image of the staff. In contrast, under BPE-word indexing, looking for a word means to look for a *single* BPE-derived unit. Retrieval at each candidate position reduces to *one* probability lookup – effectively a constant-time hash-table access to the PrIx spot associated with that word. This reduces considerably the number of computational steps, yielding a substantial theoretical speed-up, specially for long melodic patterns.

We later report empirical runtimes obtained from our implementation of the PrIx search engine. For each query, we record the wall-clock time (in milliseconds) required to compute the full RP map over all test staves, and we report search times over the entire query set. The measurement has been performed on a machine with an Intel Core i9-12900K processor at 5.20 GHz. These real-time measurements reflect not only the min / max operations but also the practical cost of indexing structures and memory access patterns.

5 Results and Discussion

5.1 Optical Recognition Results

Table 2 reports SER values, together with 95% confidence intervals, for models trained with note-level targets and with BPE-derived word targets across the different training datasets. For the EIN and SAL experiments, the BPE vocabulary was *not* retrained: instead, we applied the same merge list learned from the VOR training partition to the ground-truth transcripts of these datasets. By applying the learned merges directly to their lines, individual musical symbols in EIN and SAL were grouped into the corresponding VOR-derived musical words, ensuring that all models operate under a shared lexical space. This process was the same as the one described in Sect. 3.2.

This evaluation setup allows us to examine on one hand, how word-based training affects recognition accuracy compared to conventional note-level training, and on the other, whether a BPE vocabulary learned on VOR generalizes across different manuscripts in terms of optical recognition performance.

Table 2. *Symbol Error Rate* (SER, % ± 95% CI) for symbol-level (*notes*) and BPE-based (*BPE-words*) training across VOR (FULL/SMALL) and cross-manuscript evaluations on SAL and EIN.

Training Mode	Training Data			
	VOR-FULL	VOR-SMALL	SAL	EIN
notes	**1.45** ± 0.12	3.72 ± 0.19	6.97 ± 0.63	3.11 ± 0.23
BPE-words	1.56 ± 0.12	**3.34** ± 0.18	**6.91** ± 0.58	**2.99** ± 0.23

For both training sets of VOR, the error remains essentially stable across both training types. This indicates that introducing BPE-words does not degrade recognition quality when ample training data are available. When applying the VOR-trained vocabulary to other manuscripts, BPE-based training does not hinder performance; on the contrary, it leads to improvements on SAL (6.91% vs. 6.97%) and EIN (2.99% vs. 3.11%). These results indicate that BPE-words retain enough structural regularity across manuscripts to contribute positively to recognition, even when trained on a different source.

Crucially for our objectives, training with BPE-words preserves optical quality while enabling lexical indexing (which in turn drives the retrieval speed-ups reported later). To solidify these observations, a more granular statistical analysis and larger evaluation sets would be desirable to determine whether the observed differences are systematic.

5.2 Lexical Properties of Learned "Musical Words"

Before evaluating retrieval performance, we examine the statistical properties of the BPE-words. Figure 3 shows the Zipf curve for the GT BPE segmentation of the VOR test partition, along with a curve obtained using expected values of word frequencies estimated from PrIx word RPs [29].

Both curves display clear Zipf-like rank-frequency behavior, consistent with previous findings in [16] and analogous to natural language Zipf curves. Notably, the PrIx curve from the VOR-FULL model adheres closely to the GT distribution. This indicates that the BPE vocabulary induces a stable lexical structure that the recognizer can recover, supporting the use of musical words as the basis for lexical retrieval. It also confirns the accuracy of word frequency estimates derived from PrIx RPs reported in [29] for natural language handwritten text images.

5.3 Retrieval Accuracy Results

Our experimental setup assumes a retrieval task where the basic unit of search and ground-truth relevance is a musical word learned via BPE (Sect. 3.2). This formulation is inspired by text retrieval, where queries are words and a hit requires matching word boundaries. E.g., if the query in a text were the word

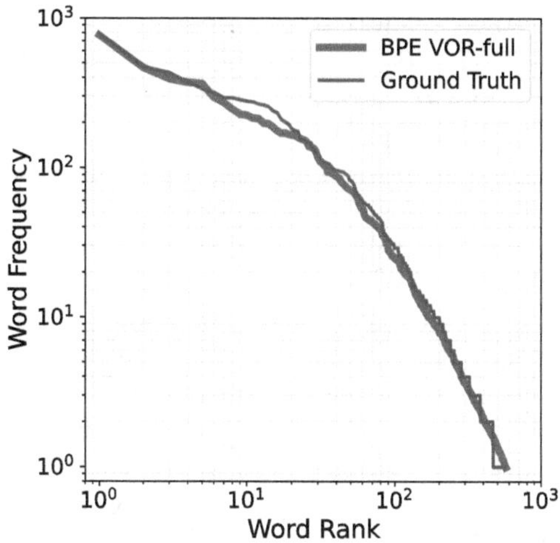

Fig. 3. Zipf curves (rank-frequency, log-log) for BPE musical words on VOR: GT BPE segmentation of the test set vs. expected word frequencies estimated from the PrIx created from the model trained with the VOR-FULL dataset.

"the", a retrieved spot inside the longer string "theorem" would *not* be considered a valid hit, because the boundaries of the retrieved substring do not match those of the intended word.

Under this assumption, we report retrieval effectiveness in terms of Average Precision for two indexing/search regimes: (i) note-level search (*notes*), and (ii) word-level search over BPE-derived musical units (*BPE words*). For each training configuration, the optical model is trained once and a corresponding PrIx is produced; the only difference between the two AP values per configuration is the evaluation protocol described below. To enable a fair comparison between symbol and word-level indexing, we evaluate under two complementary GT criteria:

Symbol-Exact. A retrieved staff is a *hit* if the predicted location contains the *exact sequence of individual symbols* specified by the query, aligned at the reported position, regardless of any word boundaries. Note that, under the lexical task definition, this protocol may credit matches that would not constitute valid words (analogous to accepting "theorem" for the word query "the").

Word-Exact. A retrieved staff is a *hit* if the predicted location contains the query *as a word*, i.e., respecting the BPE-defined word boundaries used to construct the test GT. This is the primary evaluation because the task is defined lexically: it is directly analogous to textual retrieval, where word queries must not match substrings of longer words.

The two evaluation criteria therefore reflect two distinct notions of relevance, corresponding to two retrieval paradigms: symbolic pattern matching and lexical retrieval. Neither criterion should be considered universally correct; each reflect a different view of what constitutes a meaningful match. Word-Exact aligns with a lexical perspective, while Symbol-Exact corresponds to sequence matching at the level of individual notes. This distinction highlights a broader modeling question: in the absence of a well-defined notion of word in music, the definition of relevance is itself non-trivial. Our evaluation is intended to make this choice explicit and to explore its practical implications. Table 3 presents the obtained AP figures for both BPE-word and note sequence queries. Besides, the corresponding average query times (in millisecond units) are reported (see Sec. 5.4 for details).

Word-level indexing delivers the intended gains: on VOR-FULL, BPE-word queries are about 2.6 times faster than note-sequence queries (9.20 ms vs. 24.08 ms), while achieving much higher AP under the lexical task (Word-Exact criteria: 0.80 vs. 0.20). Under Symbol-Exact criteria, note-level queries remain competitive or stronger (as expected, since they are scored at the symbol level), but when retrieval is framed lexically, BPE words align with the task and outperform. Importantly, these efficiency gains hinge on assuming musical words can be modeled; how to define them best is an active research question, yet our results show that pursuing lexical units is worthwhile because it substantially improves search efficiency without sacrificing recognition quality. Importantly, word-based indexing remains competitive even under Symbol-Exact evaluation, indicating that its effectiveness is not solely an artifact of the lexical criterion.

Table 3. Average Precision (AP) for note-level search (*note sequences*) and word-level search (*BPE words*) for the two different GT criteria mentioned above. Average query times expressed in milliseconds (ms) are also included. These are actual times measured on a real web interface for VORAU-253 melodic pattern search.

Training Set	Query Type	AP	GT criteria	Query Time (ms)
VOR-FULL	note sequences	0.96	symbol-exact	24.08
		0.20	word-exact	
	BPE words	0.89	symbol-exact	9.20
		0.80	word-exact	
VOR-SMALL	note sequences	0.97	symbol-exact	–
		0.20	word-exact	
	BPE words	0.69	symbol-exact	–
		0.70	word-exact	

5.4 Query-Time Performance

To experimentally confirm the theoretical search costs outlined in Sec. 4.3, we proceeded to measure the query times for both query types –BPE musical words and note-level sequences– using the PrIx search engine described in [26, Cap. 10]. Although this search engine follows a client-server architecture made of four main components ("Web Client", "Web Server", "Data Server", and "PrIx Server"), we have performed the search directly on the machine hosting the "PrIx Server" to avoid delay times strictly due to data transfer over the network.

Both the 452 musical word queries and the corresponding note sequence versions were searched using their respective PrIxs, which had previously been trimmed to have a similar density of 3.72[5] for a fair comparison. All queries were evaluated with a relevance-threshold of zero to measure the worst-case scenario, where all spots must be examined.

Figure 4 shows the scatter plot of query response time as a function of the number of PrIx spots retrieved. As observed in that figure, response time for musical word queries (red circles) is typically below 10 milliseconds and it exhibits a smooth trend, independent of the specific query or query length. Moreover, the number of retrieved spots is never very large and, above 50 retrieved spots, response time grows almost linearly with the number of retrieved spots, which is the optimal behavior expected. Conversely, in the case of queries with note sequences (colored squares), response times are clearly higher and do not follow a clear trend compared to the previous case. All of this is consistent with the global average query time figures reported in Table 3. To help understanding these complex query time trends, sequence query points in Fig. 4 are in different colors according to the number of notes involved in the query. Thus, the blue squares correspond to the four-note queries, which, of course, are the

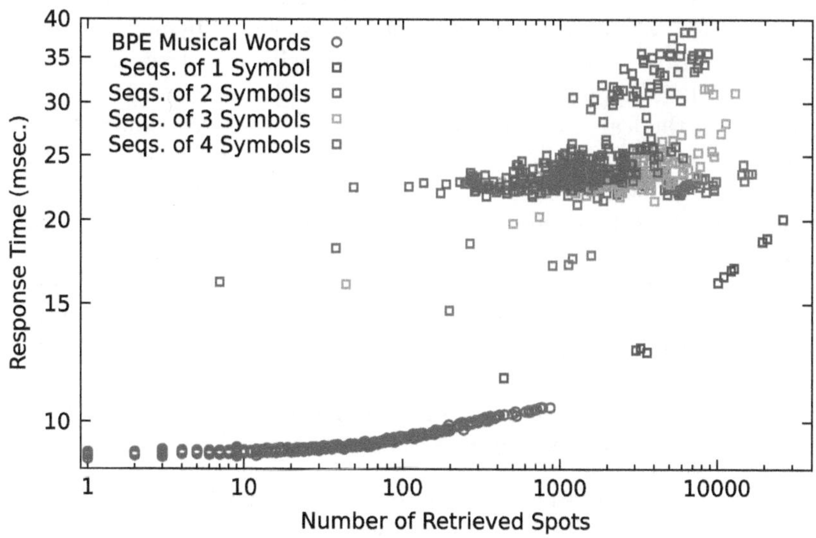

Fig. 4. Scatter plot of the total number of retrieved spots (using a search threshold of zero) versus the response time (in milliseconds) for both query types: the BPE Musical words (red circles) and the sequences of musical symbols (colored squares) grouped by the number of elements involved. Better seen in color.

[5] The PrIx density [26, Cap. 6] is calculated as the ratio between the total number of PrIx spots and the estimated running words, estimated in turn as the sum of the relevance probabilities of all the PrIx spots.

ones with the longest response times on average. These include, for example, the query <s2 s2 s2 s2>, which in this case has the longest response time because the note itself s2 has also the largest number of notes in the corresponding PrIx (26,000 spots), and searching for them for each starting and subsequent sequential positions is very time-consuming.

Overall, the time measurements follow closely the theoretical expectations derived from the PrIx combination rules: word-level retrieval scales as $O(K)$, while symbol-level retrieval grows at least as $O(nK)$, where n is the query length and K the number of candidate positions. We want to emphasize that the observed efficiency gains are not specific to PrIx retrieval. They stem from the reduction in query length and the use of higher-level indexing units. Therefore, similar improvements can be expected in previous retrieval pipelines based on deterministic recognitions, where sequence matching costs also scale with query length. In this sense, the benefits of operating on learned musical words are also useful to conventional 1-best-transcription-based search systems.

6 Conclusions and Future Work

This work introduced a lexical view of handwritten music retrieval and extends Optical Music Recognition (OMR) and *Probabilistic Indexing* (PrIx) from single notes to musical words learned with BPE. Across our datasets, word-level indexing delivered a clear *query-time efficiency gain* over note-by-note search, while *maintaining optical recognition accuracy* on par with note-level training (overlapping 95% CI's). Zipf curves for GT and PrIx-extracted words further support the stability of the learned lexical units, and applying a vocabulary trained on VOR directly to EIN and SAL showed that this shared lexical space transfers without degrading accuracy. In sum, we successfully demonstrated that treating music lexically is both coherent and practical: it preserves recognition quality and unlocks substantial retrieval speed-ups in large manuscript collections.

While here our definition of musical words is derived from monophonic melodic sequences, we believe the approach is not restricted to this setting. The same idea could be extended to richer musical representations, including duration tokens or polyphonic structures (e.g., chordal or harmonic units). Exploring such extensions introduces additional modeling challenges and remains an open direction for future work. We outline several directions to build on these results:

- Tokenization/segmentation. Systematically compare alternative musical word definitions (BPE settings, unigram/WordPiece, contour or duration-aware tokens) and study their effect on accuracy and efficiency.
- Segmentation uncertainty in PrIx. Model word-boundary probabilities jointly with PrIx so retrieval can marginalize over alternative segmentations.
- Statistical testing at scale. Paired analyses by staff/page and hierarchical models by manuscript to confirm small SER/AP differences and quantify cross-collection effects.
- Richer queries. Sentence-level queries and approximate matching over words (edits/gaps) for robust retrieval in noisy conditions.

Acknowledgments. This work is part of the I+D+i projects LIGHTVED (CIPROM/ 2023/017) funded by PROMETEO Program of the Generalitat Valenciana, and EUDEEP (PID2024161104OB-C21), funded by MICIU/AEI/10.13039/ 501100011033 and by FEDER, UE. The first author is supported by a predoctoral fellowship (CIACIF/2023/105) from the Generalitat Valenciana.

Disclosure of Interests. The authors have no competing interests to declare that are relevant to the content of this article.

References

1. Arzt, A., Böck, S., Widmer, G.: Fast identification of piece and score position via symbolic fingerprinting. In: ISMIR, pp. 433–438 (2012)
2. Bainbridge, D., Bell, T.: The challenge of optical music recognition. Comput. Humanit. **35**(2), 95–121 (2001)
3. Manaris, B., et al.: Zipf's law, music classification, and aesthetics. Comput. Music J. **29**(1), 55–69 (2005). http://www.jstor.org/stable/3681877
4. Burchardt, J.: Are searches in OCR-generated archives trustworthy? Jahrbuch für Wirtschaftsgeschichte/Econ. Hist. Yearb. **64**(1), 31–54 (2023)
5. Calvo-Zaragoza, J., Hajič Jr., J., Pacha, A.: Understanding optical music recognition. ACM Comput. Surv. **53**(4) (2020)
6. Calvo-Zaragoza, J., Toselli, A.H., Vidal, E.: Probabilistic music-symbol spotting in handwritten scores. In: 2018 16th International Conference on Frontiers in Handwriting Recognition (ICFHR), pp. 558–563. IEEE (2018)
7. Calvo-Zaragoza, J., Toselli, A.H., Vidal, E., Sánchez, J.A.: Music symbol sequence indexing in medieval plainchant manuscripts. In: 2019 International Conference on Document Analysis and Recognition (ICDAR), pp. 882–887 (2019)
8. DDMAL: End-to-end optical music recognition resources (2021). https://github.com/DDMAL/e2e-omr-resources. Accessed November 2025
9. Fremerey, C., Clausen, M., Ewert, S., Müller, M.: Sheet music-audio identification. In: ISMIR, pp. 645–650 (2009)
10. Gage, P.: A new algorithm for data compression. C Users J. **12**(2), 23–38 (1994)
11. Graves, A., Fernández, S., Gomez, F., Schmidhuber, J.: Connectionist temporal classification: labelling unsegmented sequence data with recurrent neural networks. In: Proceedings of the 23rd International Conference on Machine Learning (ICML), pp. 369–376. Association for Computing Machinery, New York (2006)
12. Hsü, K.J., Hsü, A.: Self-similarity of the "1/f noise" called music. Proc. Natl. Acad. Sci. **88**(8), 3507–3509 (1991)
13. Kingma, D.P., Ba, J.: Adam: a method for stochastic optimization. arXiv preprint arXiv:1412.6980 (2014)
14. Kneser, R., Ney, H.: Improved backing-off for m-gram language modeling. In: 1995 International Conference on Acoustics, Speech, and Signal Processing, vol. 1, pp. 181–184 (1995)
15. Lacoste, D., Bailey, T., Steiner, R., Koláček, J.: Cantus: a database for latin ecclesiastical chant (2011). Funded through the Digital Analysis of Chant Transmission project at Dalhousie University, Halifax, Nova Scotia, Canada (SSHRC 895-2023-1002)

16. Menárguez-Box, A., Vidal, E., Toselli, A.H.: Zipf's curves of plainchant melodies. In: Gonçalves, N., Oliveira, H.P., Sánchez, J.A. (eds.) IbPRIA 2025. LNCS, vol. 15938, pp. 383–395. Springer, Cham (2025). https://doi.org/10.1007/978-3-031-99568-2_31
17. Pacha, A., Hajič, J., Calvo-Zaragoza, J.: A baseline for general music object detection with deep learning. Appl. Sci. **8**(9) (2018)
18. Perotti, J.I., Billoni, O.V.: On the emergence of Zipf 's law in music. Phys. A: Stat. Mech. Appl. **549**, 124309 (2020)
19. Povey, D., et al.: The Kaldi speech recognition toolkit. In: IEEE 2011 Workshop on Automatic Speech Recognition and Understanding. No. CONF (2011)
20. Puigcerver, J.: Are multidimensional recurrent layers really necessary for handwritten text recognition? In: 2017 14th IAPR International Conference on Document Analysis and Recognition (ICDAR), vol. 01, pp. 67–72 (2017)
21. Puigcerver, J.: A probabilistic formulation of keyword spotting. Ph.D. thesis, Univertitat Politècnica de València (2018)
22. Roland, P.: The music encoding initiative (MEI). In: MAX2002. Proceedings of the First International Conference on Musical Application using XML, pp. 55–59 (2002)
23. Stolcke, A.: SRILM – an extensible language modeling toolkit. In: Proceedings of the 7th International Conference on Spoken Language Processing (ICSLP 2002) (2004)
24. Toselli, A.H., Vidal, E., Puigcerver, J., Noya-García, E.: Probabilistic multi-word spotting in handwritten text images. Pattern Anal. Appl. **22**, 23–32 (2019). https://doi.org/10.1007/s10044-018-0742-z
25. Toselli, A.H., Puigcerver, J., Vidal, E.: Two methods to improve confidence scores for lexicon-free word spotting in handwritten text. In: 2016 15th International Conference on Frontiers in Handwriting Recognition (ICFHR), pp. 349–354. IEEE (2016)
26. Toselli, A.H., Puigcerver, J., Vidal, E.: Probabilistic Indexing for Information Search and Retrieval in Large Collections of Handwritten Text Images. The Information Retrieval Series, vol. 49. Springer, Cham (2024)
27. Toselli, A.H., Vidal, E., Romero, V., Frinken, V.: HMM word graph based keyword spotting in handwritten document images. Inf. Sci. **370-371**, 497–518 (2016). https://www.sciencedirect.com/science/article/pii/S0020025516305461
28. Typke, R.: A survey of music information retrieval systems. In: Proceedings of the 6th International Conference on Music Information Retrieval (2021)
29. Vidal, E., Toselli, A.H.: Zipf curves and basic text analytics from untranscribed manuscript images. In: Barney Smith, E.H., Liwicki, M., Peng, L. (eds.) ICDAR 2024. LNCS, vol. 14806, pp. 271–288. Springer, Cham (2024). https://doi.org/10.1007/978-3-031-70543-4_16
30. Vidal, E., Toselli, A.H., Puigcerver, J.: Lexicon-based probabilistic indexing of handwritten text images. Neural Comput. Appl. **35**, 1–20 (2023)
31. Voss, R.P.: "I/f noise" in music and speech (1975)
32. Zanette, D.H.: Zipf's law and the creation of musical context. Music Sci. **10**(1), 3–18 (2006)
33. Zipf, G.K.: Human Behaviour and the Principle of Least Effort. Addison-Wesley, Boston (1949)

Reading Order Inference for Complex Document Layouts

Iddo Hakim(✉), Sharva Gogawale, Omer Ventura, Gal Grudka, Daria Vasyutinsky-Shapira, Berat Kurar-Barakat, and Nachum Dershowitz

School of Computer Science and AI, Tel Aviv University, Ramat Aviv, Israel
{iddoh,sharvag,omerventura,galgrudka}@mail.tau.ac.il,
{dariashap,berat,nachumd}@tauex.tau.ac.il

Abstract. Reading order inference remains a critical bottleneck in the digitization of complex historical manuscripts, where pages contain multiple spatially interleaved reading streams, the canonical example being the *Glossa Ordinaria* layout, in which a central text is surrounded by commentaries that wrap around it in non-rectangular, non-convex regions. We present a *training-free*, graph-based framework: each OCR text line becomes a node in a directed candidate-transition graph, edges are scored by a weighted additive ensemble of two lightweight language-model signals (causal language model conditional likelihood and BERT next-sentence prediction, NSP; a third sentence-embedding signal was evaluated but did not improve reading order), and the global reading order is recovered as a degree-constrained directed path cover. To avoid the cascading "edge-theft" failures of greedy edge selection, we propose a max-regret inference rule that prioritizes commitments with high opportunity cost. We evaluate on synthetic Glossa Ordinaria grid layouts, on *23 ALTO page geometries* (10 historical source pages plus mirrored and flipped variants), and on a *140-page multi-column English subset of OmniDocBench*, comparing our method against the canonical recursive XY-cut (PaddleOCR PP-StructureV3) and two LayoutReader variants (layout-only and text+layout) on identical inputs. On wrap-around Glossa layouts our method recovers 95% of ground-truth successor edges on average vs. XY-cut's 50%; on the OmniDocBench multi-column subset it reaches 88% macro edge accuracy versus XY-cut's 75% and LayoutReader's 25%. The LayoutReader baselines transfer poorly due to a word-level vs. line-level granularity mismatch. We additionally verify mirror-invariance under horizontal and vertical page reflections: Our method changes by less than 1 percentage point, classical XY-cut by 2 points, and LayoutReader-T by up to 8 points.

Keywords: Reading order inference · Document layout analysis · Non-Manhattan layouts · Training-free · Max-regret inference · Historical manuscripts · XY-cut · LayoutReader · Mirror invariance

F. Shafait et al. (Eds.): DAS 2026, LNCS 16851, pp. 299–316, 2027.
https://doi.org/10.1007/978-3-032-36207-0_18

1 Introduction

Reading order inference reconstructs the intended sequential flow of text elements detected by OCR. Accurate order is a prerequisite for searchable, continuous text streams and downstream document understanding; yet, it remains fragile whenever a page contains multiple spatially interleaved narratives (Fig. 1).

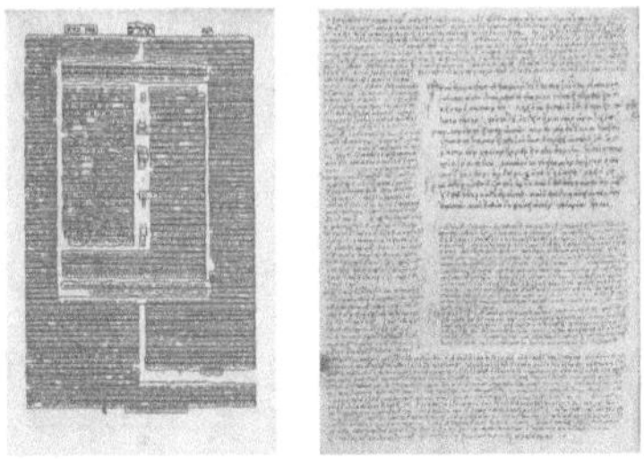

Fig. 1. Two examples of non-Manhattan layouts. *Left:* a printed Hebrew Bible page, main text flanked by an Aramaic translation, two commentaries wrapping around them, Masorah parva as abbreviated notes in an internal margin, and Masorah magna spanning the top and bottom margins, all automatically (and imperfectly) line-segmented. *Right:* a page of a manuscript (Codex Bodmer 25) of the Greek Bible with two regions of catena commentary.

For simple single-column pages, ordering is recovered by a top-to-bottom scan. Visually complex documents often contain *multiple partially independent reading streams*: parallel columns, marginalia, interlinear glosses, side notes, and figure captions, where *geometry alone is underdetermined*: several plausible successors may be nearby in space but belong to different narrative tracks.

Historical manuscripts amplify these difficulties. A canonical extreme is the *Glossa Ordinaria* page, where a central text is surrounded by multiple commentaries that wrap around it in non-rectangular, sometimes non-convex regions. Lines from different streams may be spatially interleaved, and human readers rely primarily on *semantic continuity* to remain within a stream. Classical geometric heuristics, including recursive XY-cut and proximity-based clustering, implicitly assume that spatial adjacency implies sequential adjacency, which often fails on such multi-text layouts [6–8]. We treat the canonical recursive XY-cut as our primary geometric baseline.

A natural alternative is to learn reading order from annotated documents, and modern multimodal models can be effective in-domain. But reading-order supervision is typically collected at the word/token level (e.g. ReadingBank for LayoutReader [14]), which differs from the OCR text-line atomic unit produced by historical OCR engines. As Sect. 6 shows, off-the-shelf LayoutReader variants, both layout-only and text-plus-layout (LayoutReader-T), transfer poorly to line-level historical inputs without retraining, even when fed the exact same boxes and text as our method. High-quality reading-order supervision for rare historical layouts is also expensive to obtain. This motivates *training-free* methods that exploit general linguistic priors already present in pretrained language models.

Goal. Recover reading order in complex, non-Manhattan, multi-stream layouts without annotated reading-order datasets or layout-specific fine-tuning.

Key Idea. Pretrained language models, even small ones, encode strong local coherence signals. We model a page as a directed graph of candidate "next-line" transitions; nodes are OCR lines and edge weights quantify semantic continuity. Given this scored graph, we search for a globally consistent set of successor relations under the constraint that each line has at most one predecessor and one successor. Throughout, we isolate layout inference from OCR noise by evaluating on known line boxes with controlled text; end-to-end robustness to OCR errors is out of scope (Sect. 9).

Contributions

1. **Graph formulation.** Reading order as a degree-constrained directed path cover over OCR lines, enabling multiple disjoint reading streams under one-predecessor/one-successor constraints.
2. **Training-free semantic scoring.** A weighted additive ensemble of causal LM conditional likelihood and BERT next-sentence prediction. We additionally evaluate sentence-embedding cosine similarity and find it does not improve reading order (Sect. 7).
3. **Regret-based inference.** A max-regret edge-selection algorithm that reduces greedy myopia by prioritizing decisions with high opportunity cost.
4. **Expanded evaluation across synthetic, historical, and public-benchmark layouts.** Synthetic "Glossa Ordinaria layout" pages [12] created by interlacing distinct Project Gutenberg books[1] into non-convex topologies; *23 ALTO page geometries* [5] from 10 historical source pages and their mirrored and flipped variants, a substantial expansion over prior small-scale studies; and a 140-page English multi-column subset of OmniDocBench [9] for direct comparison on a public benchmark.
5. **Multi-baseline comparison.** We benchmark against the canonical recursive XY-cut [7] (PaddleOCR PP-StructureV3), LayoutReader [14] (layout-only), and LayoutReader-T (text+layout), on identical OCR-line inputs, and verify mirror-invariance under horizontal and vertical page reflections.

2 Related Work

Reading order inference sits at the intersection of page layout analysis and document structure recovery. While many documents admit a near-linear scan, visually complex pages may contain multiple interleaved reading streams (marginalia, parallel commentaries) for which geometric adjacency is an unreliable proxy for sequential adjacency.

[1] https://www.gutenberg.org.

2.1 Geometric Heuristics

Early reading-order approaches rely on spatial regularities: projection-profile and recursive partitioning (XY-cut) [7], nearest-neighbor clustering (Docstrum) [8], and efficiency-tuned XY-cut variants [6]. These work well on Manhattan layouts but degrade when streams are spatially interleaved and region boundaries are non-convex, as in many historical manuscripts. We use the canonical recursive XY-cut as implemented in PaddleOCR's PP-StructureV3 pipeline as our geometric baseline (Sect. 6), so comparisons reflect a standard, citable implementation rather than a re-implementation.

2.2 Learning-Based and Semantic Approaches

Recent methods learn reading order as sequence generation or as pairwise relation prediction over layout entities, with multimodal inputs and supervised datasets [13–15]. LayoutReader [14] is trained on ReadingBank, a corpus of *word-level* layout and reading order from rendered DOCX files; we evaluate both its layout-only and text-plus-layout (LayoutReader-T) configurations on our OCR-line and OmniDocBench paragraph inputs (Sect. 6), where the word-to-line and word-to-paragraph granularity mismatch becomes a concrete limitation. End-to-end multimodal parsers such as Dots.OCR and PaddleOCR-VL operate at line granularity but bundle detection, recognition, and reading order into a single pipeline and do not expose their reading-order module for evaluation on externally supplied boxes, so a clean apples-to-apples comparison is not currently possible. Such systems are typically accurate in-distribution but depend on labeled reading order and may require fine-tuning for rare historical layouts. In NLP, sentence ordering and discourse coherence use local continuity cues and global optimization [1,4], but document reading order differs in three ways: the units are OCR lines (not sentences), the problem is constrained by 2D candidate adjacency, and multi-stream outputs (sets of disjoint paths) are often required rather than a single permutation. Our approach is training-free, uses pretrained language models as local semantic oracles, and combines them with explicit graph inference under degree constraints.

2.3 Reading-Order Representation and Evaluation

Reading order may be non-linear or partially specified, and evaluation becomes non-trivial when segmentation differs across systems [2]. Public benchmarks such as OmniDocBench [9] cover diverse modern documents (academic papers, financial reports, multi-column layouts) and evaluate ordering over document components or paragraph-like regions rather than OCR lines. For our wrap-around, non-Manhattan target setting, region decomposition is itself unreliable: regions may merge or split incorrectly even when individual line boxes remain usable (a Sayre's-paradox-like situation, where correct region ordering needs correct regions, but finding them depends on the reading streams). We therefore evaluate directly over OCR lines and successor relations on our 23-page ALTO corpus, and complement this with an evaluation on the multi-column English subset of OmniDocBench at its native paragraph granularity (Sect. 7.4). The graph formulation operates on (bbox, text) tuples regardless of unit size, so both granularities are handled with the same model. ALTO and similar archival formats can encode layout structure and optionally include ordering, but reading order is frequently missing or unreliable in complex manuscripts, motivating automatic inference.

3 Problem Formulation

We assume an OCR system produces a set of N text lines.

3.1 Candidate Transition Graph

We build a directed candidate graph $G = (V, E_{\text{cand}})$ with $V = \{1, \ldots, N\}$. An edge $(u, v) \in E_{\text{cand}}$ denotes that line v is a plausible successor of line u. Candidate generation controls both runtime and ambiguity; we use permissive "next-column" candidate sets to stress-test inter-column continuation decisions.

- *In artificial grid layouts*, each node belongs to a discrete column. For each node u, we connect it to *all* nodes in the next column to the right. This produces a dense candidate set in which geometry alone cannot determine which inter-column transition is the correct continuation.
- *In realistic layouts*, where nodes are derived from OCR text bounding boxes, we mimic the same stress test using bounding boxes: for each line u, we connect it to *all* lines below u (same-column successors) and to *all* lines whose left edge lies to the right of u's right boundary (next-column successors). This permissive set makes both same-column continuations and wrap-around "next-column" transitions candidates, so geometry alone cannot determine the correct successor in complex page geometries.

3.2 Degree Constraints and Path Cover

Reading order within each stream is a directed path: each line has at most one successor and at most one predecessor. We seek a subset $E \subseteq E_{\text{cand}}$ that maximizes the total score:

$$\max_{E \subseteq E_{\text{cand}}} \sum_{(u,v) \in E} S(u, v) \quad \text{s.t.} \quad \deg^+(u) \leq 1, \ \deg^-(v) \leq 1 \ \forall u, v. \tag{1}$$

where $S(u, v)$ is the semantic continuity score for the edge. Under the degree constraints alone, a maximizing edge set decomposes into a collection of disjoint directed paths and (in general) directed cycles. Because cycles do not correspond to valid reading orders, our inference procedures explicitly avoid cycle creation and return an acyclic path cover, possibly with multiple disjoint streams.

4 Training-Free Semantic Edge Scoring

For each candidate edge (u, v) we compute a semantic continuity score $S(u, v)$ using pretrained models *without fine-tuning.* Scores are computed independently per edge and cached, decoupling expensive model inference from the downstream search algorithm.

4.1 Causal Language Model Conditional Likelihood

Let a causal language model (CLM) define token log-likelihood $\log P_{\text{CLM}}(\cdot)$ under a fixed tokenizer. We instantiate the CLM with EleutherAI/pythia-410M and compute all CLM scores on a GPU in inference mode without fine-tuning. Given two line fragments t_u and t_v, we score the per-token continuation likelihood of t_v conditioned on t_u:

$$s_{\text{clm}}(u, v) = \frac{1}{|t_v|} \log P_{\text{CLM}}(t_v \mid t_u),$$

where $|t_v|$ denotes the number of tokens in t_v. In practice, we compute this by concatenating the token sequences $[t_u; t_v]$, running one forward pass, and summing log-probabilities only over the token positions belonging to t_v (i.e., masking out the prefix t_u) [10].

Context Truncation. Because causal models have a finite context window, we truncate t_u to the most recent L tokens (we use a fixed L across all experiments) so that $[t_u; t_v]$ fits within the model context.

Optional Frequency Normalization. Rare-token artifacts can inflate s_{clm} by over-rewarding unlikely but highly specific continuations. We therefore optionally normalize by the unconditional likelihood of t_v:

$$\widetilde{s}_{\text{clm}}(u, v) = s_{\text{clm}}(u, v) - \kappa \cdot \frac{1}{|t_v|} \log P_{\text{CLM}}(t_v), \tag{2}$$

with $\kappa \in [0, 1]$ controlling the correction strength. When $\kappa = 0$, no normalization is applied. We use a different symbol (κ) here than for the ensemble blending weights below (α, β, γ, see Sect. 7) to make the two roles distinct.

4.2 Next Sentence Prediction

BERT's next sentence prediction (NSP) head produces $p_{\text{nsp}}(u, v) \in (0, 1)$ indicating whether t_v is a plausible continuation of t_u [3]. NSP is the binary sentence-continuity objective on which BERT was pretrained, which makes it a standard, training-free probe for precisely the adjacency relation a reading-order edge encodes. We convert this to an additive score in log space:

$$s_{\text{nsp}}(u, v) = \log\big(\max(\varepsilon, p_{\text{nsp}}(u, v))\big),$$

where ε is a small constant to avoid $\log(0)$.

4.3 Sentence Embedding Similarity

We compute embeddings h_u, h_v using a sentence-transformer encoder [11] and define:

$$s_{\text{sim}}(u, v) = \cos(h_u, h_v).$$

Since cosine similarity can be negative, we use it directly as a bounded additive signal (no logarithm). We include this signal for completeness; the ablation in Sect. 7 shows it does not improve reading order, so all reported results set $w_{\text{sim}} = 0$.

4.4 Weighted Additive Ensemble

We combine all signals with a weighted additive ensemble:

$$S(u, v) = w_{\text{clm}} \cdot \widetilde{s}_{\text{clm}}(u, v) + w_{\text{nsp}} \cdot s_{\text{nsp}}(u, v) + w_{\text{sim}} \cdot s_{\text{sim}}(u, v) - \lambda \cdot d_{\text{Man}}(u, v),$$

where $w_{\text{clm}}, w_{\text{nsp}}, w_{\text{sim}} \geq 0$ and $\lambda \geq 0$. We use Manhattan distance between bounding-box centers, $d_{\text{Man}}(u, v) = |c_x(u) - c_x(v)| + |c_y(u) - c_y(v)|$, as an optional regularizer that discourages long-distance "teleportation" edges in very dense candidate graphs. Unless stated otherwise, we set $\lambda = 0$ and rely on candidate gating for locality.

Fixed-Parameter Transfer. The ensemble weights are selected once by a sweep on the synthetic grid task (Sect. 7) and then *transferred unchanged* to all ALTO and OmniDocBench experiments. The sweep selects $w_{\text{sim}} = 0$ (the embedding never helped and degraded results as its weight grew, Sect. 7), so the deployed scorer uses only $(w_{\text{clm}}, w_{\text{nsp}})$. We do not re-tune per dataset or per document: the Sect. 6 numbers are a fixed-parameter transfer test, not in-sample fits.

5 Inference Algorithms

Given candidate edges E_{cand} and scores $S(u, v)$, we infer a degree-constrained directed path cover: each node has at most one successor and at most one predecessor, yielding multiple disjoint reading streams. The maximum-weight degree-constrained acyclic path cover of Eq. (1) is a combinatorial problem we do not solve to optimality; the algorithms below are fast heuristics, and max-regret in particular carries no optimality guarantee, only reordering which decisions are committed first.

5.1 Greedy Selection Baselines

We consider two greedy constructions that repeatedly select high-scoring edges subject to feasibility: (i) *local greedy*, which processes nodes in a fixed scan order and assigns each node its best feasible successor, and (ii) *global greedy*, which repeatedly commits the highest-scoring feasible edge overall. While fast, greedy methods often fail by *edge theft*: an early incorrect edge consumes a node's in-degree, making the correct predecessor later unreachable and causing cascading errors.

Algorithm 1 Max-regret inference (cycle-avoiding path cover)

Input: candidate edges E_{cand}, scores $S(u, v)$
Output: edge set E (acyclic path cover)
$E \leftarrow \emptyset$; initialize all nodes as unassigned-in and unassigned-out
while there exists an unresolved node with at least one feasible outgoing candidate **do**
  for all unresolved nodes u **do**
    Let $\mathcal{C}(u) = \{v : (u, v) \in E_{\mathrm{cand}}$ and (u, v) is feasible$\}$
    Sort $\mathcal{C}(u)$ in descending order of $S(u, v)$ to get $v_1, v_2, \ldots$
    Regret$(u) \leftarrow S(u, v_1) - S(u, v_2)$ {or 0 if $|\mathcal{C}(u)| < 2$}
  end for
  $u^\star \leftarrow \arg\max_u \text{Regret}(u)$ {tie-break by $S(u, v_1)$}
  Commit the best feasible edge $(u^\star, v)$ by scanning $v_1, v_2, \ldots$ until one does not create a cycle
  Add the committed edge to E and mark $u^\star$ assigned-out and v assigned-in
end while
return E

5.2 Max-Regret Edge Selection

To reduce greedy myopia, we prioritize decisions with high opportunity cost. For each unresolved node u, let v_1 and v_2 denote the top two *feasible* outgoing candidates by score, where feasibility means: (a) u has no assigned successor, (b) v has no assigned predecessor, and (c) adding (u, v) would not create a directed cycle in the current partial solution. We define regret as:

$$\text{Regret}(u) = S(u, v_1) - S(u, v_2). \tag{3}$$

If a node has fewer than two feasible candidates, we set $\text{Regret}(u) = 0$ so that low-ambiguity nodes (e.g., boundary nodes with a single remaining option) do not dominate the schedule.

At each iteration, we select the node with maximum regret and commit its best feasible edge. If the best edge would form a cycle, we fall back to the next-best feasible candidate for that node.

Complexity. With N nodes and M candidate edges, each iteration selects one edge and updates feasibility. A straightforward implementation runs in $O(M \log d)$ time if per-node candidate lists are pre-sorted (d is average out-degree), with near-constant-time cycle checks using union-find-style successor tracing.

6 Experimental Setup

To isolate layout inference from OCR recognition noise, we evaluate on pages where the set of line bounding boxes is known and text content is controlled. We use three complementary settings: (i) synthetic grid layouts for controlled ablations over topology and candidate density, (ii) authentic historical page geometries extracted from ALTO XML [5] that preserve realistic, irregular line placements and non-rectangular region boundaries, and (iii) a multi-column English subset of the OmniDocBench public benchmark [9] for an apples-to-apples comparison on a community-recognized dataset.

6.1 Synthetic Grid Complex Layouts

We instantiate layouts on an $R \times C$ grid partitioned into disjoint regions (text streams) with non-convex wrap-around topologies (L-, H-, Y-, and O-shapes). Nodes in each region are populated with consecutive lines from a distinct Project Gutenberg book, producing spatially interleaved but semantically independent streams. Ground-truth reading order inside each region is a fixed deterministic traversal (row-major or column-major, held constant per experiment).

6.2 Historical Geometries from ALTO XML

We parse `TextLine` elements (and polygons when available) from ALTO XML. Our corpus comprises *23 page geometries*: 10 historical source pages, 2 horizontal mirrors, 10 vertical (top-bottom) flips, and one two-column ALTO as a Manhattan sanity check. The mirror and flip variants are produced by reflecting bounding-box coordinates and re-deriving each group's reading order under a column-major edge-aligned clustering rule, preserving the *semantic* reading order of the source page while breaking directional bias; they are used in Sect. 7.3 as a mirror-invariance test.

Each node is populated with consecutive Project Gutenberg words up to the node's box width; the text then continues into the next node in the region. RTL source layouts (e.g., Hebrew) are canonicalized to LTR by mirroring x-coordinates ($x' = W - x$). The synthetic grid generator and the line-level reading-order annotations for all 23 ALTO pages will be released publicly.

6.3 Metrics

We report *edge accuracy*: the fraction of non-terminal lines whose predicted successor matches the ground-truth successor. The last line of each stream is excluded since it has no successor. Errors are decomposed into *same-stream skips* (predicted successor stays in the correct stream but skips the immediate GT successor) and *cross-stream links* (predicted successor belongs to a different stream). Across the 23-page ALTO corpus, edge accuracy is computed over thousands of GT successor edges and tens of thousands of candidate transitions (e.g., Fig. 2 alone contributes 5,783 candidate edges). For OmniDocBench we additionally report the Normalized Edit Distance metric defined in the OmniDocBench paper [9] for direct comparability against its published leaderboard.

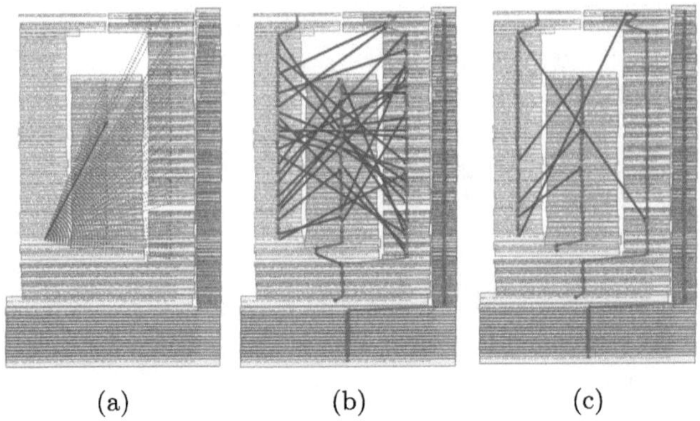

Fig. 2. Next-column stress test on an ALTO-based graph (5,783 candidate edges). (a) Candidate generator connects each line to all lines to the right of its right boundary. (b) Basic regret produces many spurious links. (c) Additive ensemble regret recovers the critical inter-stream continuation.

7 Results

We compare our full method, the *Weighted Additive Ensemble with Max-Regret inference*, against baselines spanning both the geometric and learning-based traditions. On synthetic grids we use a *Column Concatenation* baseline (nodes grouped by column, ordered top-to-bottom within columns, concatenated left-to-right). On ALTO pages we benchmark against three literature baselines: *XY-Cut* [7] as implemented in PaddleOCR PP-StructureV3 (`recursive_xy_cut`); *LayoutReader* (LR) [14], the layout-only LayoutLMv3 model (`hantian/layoutreader`); and *LayoutReader-T* (LR-T), the text+layout seq2seq variant from `nielsr/layoutreader-readingbank`, fed the same OCR-line boxes and the same line text that our pipeline uses. All four methods are evaluated on identical line-level inputs per page.

7.1 Synthetic Grid Results

On the 8×8 instance, performance rises from 50 to 59.2 ± 0.4 correctly recovered successor edges out of 61 (geometry-only $\rightarrow$ ours). On 16×16, it rises from 230 to 235.4 ± 8.2 out of 253 (Table 1). The smaller margin on the larger grid is expected: as the synthetic layout grows, a larger fraction of the page becomes trivially recoverable by column-wise ordering, leaving less headroom for semantic re-ranking.

7.2 ALTO Results

Table 2 reports edge-accuracy averages over the 11 pages in our main ALTO test corpus (10 historical source pages plus Sanity2Col), grouped by reading-order regime. The 10 vertical-flip and 2 horizontal-mirror variants are reserved for the robustness analysis in Sect. 7.3.

A clear three-regime picture emerges. On *Manhattan-ceiling* pages (Asher, Pesach_v2, O_IE104261952, Sanity2Col), XY-Cut recovers 100% of GT edges; our method averages 96.0%, within 11 percentage points on each individual page. These pages confirm that the XY-Cut implementation is sound and expose the only regime in which our method is not the top performer: the semantic ensemble occasionally re-orders adjacent same-column lines when short line-level text fragments make local continuation probabilities ambiguous.

On *wrap-around/Glossa Ordinaria* pages (Lieberman, Y_IE20854727, O_IE102667218, Y_IE30700205), where the reading order does not respect simple whitespace partitions, the gap between our method and XY-Cut is large: 94.8% vs. 49.7% on average. On Lieberman we recover *100.0%* of GT edges vs. XY-Cut's *39.6%*; on Y_IE20854727, *98.9%* vs. *40.5%*. The failure mode of XY-Cut on these pages is the bridging-gloss mechanism analyzed in Sect. 8.

On *mixed/near-Manhattan* pages (Berakhot2B, L_IE104330339, Vatican Palatina), XY-Cut recovers 91–99% of GT edges and our method matches it within ±4 percentage points. Across all 11 pages, the LayoutReader variants are uncompetitive: LR averages 22.5% (range 0 – 55), LR-T 19.8% (range 6–34). The text channel does not close the gap; the granularity mismatch with ReadingBank's word-level supervision is structural (Sect. 8).

Table 1. Synthetic grid: geometry-only baseline vs. ours, mean ± std over 5 seeds.

Dataset	GT edges	Column Concatenation		Ours	
		Correct	Acc. (%)	Correct	Acc. (%)
8×8	61	50	82.0	59.2 ± 0.4	97.0 ± 0.7
16×16	253	230	90.9	235.4 ± 8.2	93.0 ± 3.2

Table 2. ALTO 11-page corpus: per-regime average edge accuracy (%). Best per row in **bold**.

Regime	n	Ours	XY-Cut	LR	LR-T
Manhattan-ceiling	4	96.0	**100.0**	27.9	26.6
Mixed/near-Manhattan	3	95.3	**96.1**	15.7	18.3
Wrap-around/Glossa	4	**94.8**	49.7	22.3	14.1
All 11	11	**95.4**	80.5	22.5	19.8

7.3 Mirror-Invariance Robustness

Reading-order methods should be *geometry-equivariant*: mirroring or flipping a page should not change which line follows which under the (mirrored) reading order. Methods trained on left-to-right English may carry a directional prior that violates this. We evaluate this on two horizontal mirrors (Y_IE20854727, Y_IE30700205) and 10 vertical flips of the source pages; ground-truth reading order is re-derived under a column-major rule with edge-aligned clustering so the *semantic* order is preserved. The worst-case $|\Delta\text{Acc.}|$ across mirrors is 0.6 percentage points for our method, 2.1 percentage points for XY-Cut, 8.7 percentage points for LR, and 7.5 percentage points for LR-T. LR-T *is not* mirror-invariant, consistent with its LTR-English training distribution and a concrete instance of the directional-prior limitation we discuss in Sect. 8.

7.4 Public-Benchmark Evaluation: OmniDocBench

We further evaluate on a 140-page English multi-column subset of OmniDocBench [9] (92 double-column, 48 three-column; sources: 38 academic papers, 36 exam papers, 32 newspapers, 17 magazines, 15 books, 2 textbooks; 2,085 ground-truth successor edges). OmniDocBench annotates reading order at the paragraph-block level rather than the OCR-line level that is the primary target of this paper, so we evaluate at its native granularity (each annotated block is one node in our graph). The graph formulation operates on (bbox, text) tuples regardless of unit size, so no methodological change is required; the same ensemble weights from the synthetic-grid sweep are transferred unchanged.

Table 3 reports per-source edge accuracy alongside the Normalized Edit Distance metric used by the OmniDocBench paper itself [9]. Two observations consistent with the ALTO picture: (i) our method wins by a wide margin on the subsets where geometric partition struggles, exam papers (+54.7 percentage points over XY-Cut) and academic literature (+13 percentage points), which are the multi-column academic-style pages identified as challenging in prior reviews. (ii) XY-Cut wins on strongly rectilinear layouts (newspapers, three-column books) where its projection profile finds clean gutters. LayoutReader underperforms uniformly, even on subsets close to its ReadingBank training distribution (academic literature: 37.2%), confirming the granularity-mismatch argument of Sect. 8: word-level supervision does not transfer to coarser units, whether OCR lines or paragraph blocks.

The OmniDocBench leaderboard is populated entirely by end-to-end image-input systems[2]; ours is the first reported result for the training-free (bbox + text) regime, with NED 0.24 on the multi-column subset and 0.10 on exam papers. We do not claim parity with the image-input leaderboard (e.g., MinerU at 0.08): those systems solve a different, full-image task, whereas our method uses only boxes and text.

[2] MinerU, Mathpix, Marker; GOT-OCR, Nougat; GPT-4o, Qwen2-VL-72B, InternVL2-76B [9]. Best reported English NED 0.08 (MinerU), 0.12 (Qwen2-VL-72B), 0.13 (GPT-4o).

Table 3. OmniDocBench English multi-column subset (140 pages, 2,085 GT successor edges). Edge accuracy (%, higher is better) and Normalized Edit Distance (NED, lower is better) per data source. Edge accuracy is over the 140 pages all methods processed; NED is over the full 142-page subset. Best per row in **bold**.

Source	n	Edge accuracy (%)			NED	
		Ours	XY-Cut	LR	Ours	XY-Cut
academic_lit.	38	**93.4**	80.4	37.2	**0.14**	0.27
exam_paper	36	**96.9**	42.2	10.9	**0.10**	0.62
newspaper	32	74.2	**96.2**	8.9	0.51	**0.07**
magazine	17	**89.1**	82.6	26.0	0.33	**0.30**
book	15	82.0	**85.8**	53.7	0.19	**0.18**
textbook	2	85.7	**100.0**	36.9	0.13	**0.00**
Macro	140	**88.0**	75.3	24.6	**0.24**	0.30

7.5 Ablations

We sweep (α, β, γ) over 120 configurations on the 16×16 synthetic grid with 5 seeds and adopt $\alpha = 1.0, \beta = 0.2$, transferred unchanged to all ALTO and OmniDocBench experiments; the additive ensemble with max-regret recovers 235.4 ± 8.2 of 253 GT edges (93.0%). Performance is much more sensitive to α and β than to γ in the tested range. The swept optimum in fact sets $\gamma = 0$: the sentence-embedding term gives no improvement, and when we force it active with a real similarity matrix its effect is at best neutral and degrades accuracy monotonically as γ grows (across the ALTO layouts edge accuracy never rises and falls steadily at larger γ). The two language-model signals therefore carry the method, the causal-LM term (α) dominant and NSP (β) a consistent further gain, so the deployed scorer omits the embedding. With the additive scores fixed at these weights, greedy edge selection recovers only 141.6 ± 7.8 of 253 GT edges (56.0%), below the geometry-only baseline (90.9%) and far below max-regret (93.0%). Replacing greedy with max-regret gains +37.0 percentage points on average. The error decomposition in Table 4 shows that greedy produces 27.8 cross-stream and 85.6 same-stream errors per run; max-regret reduces both to 4.4 and 9.2 respectively.

Table 4. Greedy vs. max-regret inference on the 16×16 grid using additive ensemble scores ($\alpha = 1.0$, $\beta = 0.2$), averaged over 5 seeds.

Method	Correct	Acc. (%)	Cross-stream	Same-stream
Column Concatenation baseline	230	90.9	–	–
Greedy inference	141.6 ± 7.8	56.0	27.8	85.6
Max-Regret inference	235.4 ± 8.2	93.0	4.4	9.2

Runtime. On a single NVIDIA A40, per-page inference averages 93.5 s (median 88 s; 10.5 – 290 s) and peaks at 6.8 GB. Cost is dominated by per-edge semantic scoring, not the path search, and scales with candidate-edge density, so dense pages with many short, adjacent lines are the most expensive.

8 Analysis

This section summarizes dominant failure mechanisms in dense multi-stream layouts and explains why max-regret inference improves robustness and why the baselines we compare against fail in structurally specific ways.

8.1 Greedy vs Max-Regret

Constrained greedy enforces the same degree constraints as our main solver but commits edges by absolute score (the globally best feasible edge at each step). In dense candidate graphs this is vulnerable to *premature commitment*: an early high-scoring but incorrect edge (u, v) claims target v, renders v's true predecessor infeasible, and displaces other nodes onto weaker alternatives, triggering cascades of cross-stream links rather than isolated local mistakes. Max-regret retains the same feasibility constraints but reorders which decisions to commit first, prioritizing high-opportunity-cost sources (nodes whose best and second-best outgoing candidates differ greatly in score, Eq. (3)) because delaying these "must-take" commitments would let a competing node claim their best target.

8.2 Baseline Failure Modes

XY-Cut Needs a Whitespace Gutter that Does Not Exist. Recursive XY-cut [7] alternates horizontal and vertical projection-profile splits, choosing a cut wherever there is a contiguous run of background larger than a configured gap threshold. This requires a clean whitespace gutter fully separating the cells to be cut. On Glossa Ordinaria and similar bridging-gloss pages, one or more boxes (a wide gloss line, a marginal commentary that wraps across the gutter, or a full-width banner) intersect every candidate vertical cut, so the projection profile never reaches background between would-be columns. The recursion then either fails to split or splits on the wrong axis, after which every box in the misclassified region receives an incorrect successor in the terminal scan order. Our method bypasses both failure modes because it never decomposes the page into rectangles; a bridging row or shared whitespace band is just one more candidate node.

LayoutReader is Trained at the Wrong Granularity. LayoutReader [14] and the LR-T text+layout variant are trained on ReadingBank, which provides reading-order supervision over *words*. Our inputs are OCR text *lines* or paragraph blocks, each spanning many words and several pixels of vertical extent, so box geometry, input sequence length, and box-to-token alignment all differ from training. We observe a consistent 6–34% edge-accuracy band across all ALTO pages with LR-T no better than LR despite being fed text. The mismatch is structural, not semantic; the text channel cannot make a word-level seq2seq emit correct line or paragraph successors. The same mechanism explains the mirror-invariance failure in Sect. 7.3: a model whose training distribution is LTR-English at word granularity has no opportunity to learn reading order as a function of *relative* layout.

8.3 Semantic Failure Modes

Three failure modes recur. *Context-insensitive continuation*: fragments score high simply because they are broadly probable, not because they truly follow t_u; the marginal subtraction in Eq. (2) discounts globally likely fragments and emphasizes context-specific lift. *Cross-stream hallucination*: generic cues (quotation marks, dialogue verbs) appear across multiple streams; combining NSP with CLM provides a complementary signal that reduces these merges at high candidate density. *Semantic teleportation*: permissive candidate graphs allow semantically similar but spatially distant lines to link; candidate gating bounds this in practice.

9 Limitations

Computational Cost and Score Overlap. Our approach is training-free but not cost-free: scoring candidate edges requires running pretrained models over many line pairs, with cost scaling in the number of candidates per node. Candidate gating and score caching make this tractable in practice, but very dense candidate graphs (weak geometric constraints or all-to-all candidates) remain expensive. Performance is also bounded by *score overlap*: short or semantically generic line fragments may produce near-identical scores for correct and incorrect successors, making some errors irreducible without additional signals such as typography, indentation, or learned layout priors. This is most acute for very short segments (marginalia, page numbers, running heads) and degraded OCR, where every textual score weakens at once; fusing the semantic scores with layout and visual cues, so the ensemble can fall back on geometry where language is uninformative, is a natural extension.

Scope of Evaluation. We deliberately isolate layout inference from OCR noise by using clean line boxes and controlled English text. End-to-end behaviour under realistic OCR error rates therefore remains uncharacterized; a CER/WER sweep on our populated ALTO pages is the most direct next-step robustness study. The semantic stack (CLM + NSP) is also validated only on English: right-to-left scripts, top-to-bottom and vertical writing systems, and historical language distributions with poor tokenizer coverage are out of scope, though the graph formulation and inference rule are unchanged under such substitutions. Our public-benchmark evaluation covers the multi-column English subset of OmniDocBench (Sect. 7.4) at its native paragraph granularity; extending this to the full OmniDocBench corpus, to non-English subsets, and to line-level evaluations on other public benchmarks is straightforward future work. Broader collections of annotated manuscripts and complex business documents, such as the Tobacco-800 collection, would further stress-test robustness on diverse real-world layouts. Our ensemble weights ($w_{\text{clm}}, w_{\text{nsp}}$) are likewise selected once on the synthetic grid and transferred unchanged; per-layout-family tuning could improve individual regimes at the cost of the cleanest transfer story.

Baseline-Comparison Gap. End-to-end document parsers such as Dots.OCR and PaddleOCR-VL operate at line granularity but bundle detection, recognition, and reading order in a single pipeline, and do not expose their reading-order module for evaluation on externally supplied OCR boxes. A clean apples-to-apples comparison on identical inputs is therefore not currently possible; doing so would require either reimplementing their reading-order head or running both systems on their own full-pipeline outputs, with the attendant segmentation differences and partial-credit issues.

10 Conclusion

We presented a training-free framework for reading order inference in complex, non-Manhattan layouts. OCR lines become nodes in a candidate transition graph; edges are scored by a weighted additive ensemble of CLM likelihood and NSP.

On top of these local scores, a max-regret inference strategy outperforms greedy selection by prioritizing high-opportunity-cost decisions and reducing catastrophic "edge theft" failures under one-predecessor/one-successor constraints.

We evaluate across synthetic grids, 23 ALTO historical geometries, and a 140-page OmniDocBench multi-column subset [9], against the canonical recursive XY-cut (PaddleOCR PP-StructureV3) and two LayoutReader variants. Our method dominates on layouts where geometric partition is unreliable (Glossa, exam papers, academic), reaching 88.0% accuracy (NED 0.24) on OmniDocBench versus XY-cut's 75.3% (0.30) and LayoutReader's 24.6%, while structurally rectilinear layouts (newspapers, clean Manhattan) remain in XY-cut's strength regime. Our pipeline is also mirror-invariant to within ± 0.6 percentage points under horizontal and vertical reflections, whereas LayoutReader-T is not.

Future work will focus on integrating lightweight visual cues, such as indentation and font styles, improving robustness against OCR noise, expanding support for multilingual historical texts, and extending OmniDocBench to the full corpus and other public benchmarks.

Acknowledgments. We thank the reviewers for their suggestions. This research was funded in part by the European Union as part of "MiDRASH" [https://www.midrash.eu] (ERC project no. 101071829, with principal investigators: Daniel Stökl Ben Ezra, EPHE-PSL; Nachum Dershowitz, Tel Aviv University; Judith Olszowy-Schlanger, EPHE-PSL; and Avi Shmidman, Bar-Ilan University). Views and opinions expressed are, however, those of the authors only and do not necessarily reflect those of the European Union or the European Research Council Executive Agency. Neither the European Union nor the granting authority can be held responsible for them.

References

1. Barzilay, R., Lapata, M.: Modeling local coherence: an entity-based approach. Comput. Linguist. **34**(1), 1–34 (2008). https://doi.org/10.1162/coli.2008.34.1.1
2. Clausner, C., Pletschacher, S., Antonacopoulos, A.: The significance of reading order in document recognition and its evaluation. In: Proceedings of the 12th International Conference on Document Analysis and Recognition (ICDAR), pp. 688–692 (2013). https://doi.org/10.1109/ICDAR.2013.141
3. Devlin, J., Chang, M.W., Lee, K., Toutanova, K.: BERT: pre-training of deep bidirectional transformers for language understanding. In: Proceedings of the NAACL-HLT, pp. 4171–4186. ACL (2019). https://doi.org/10.18653/v1/N19-1423
4. Li, J., Hovy, E.: A model of coherence based on distributed sentence representation. In: Proceedings of the EMNLP, pp. 2039–2048. ACL (2014). https://doi.org/10.3115/v1/D14-1218
5. Library of Congress: ALTO: Technical metadata for layout and text objects (2022). https://www.loc.gov/standards/alto/
6. Meunier, J.L.: Optimized XY-cut for determining a page reading order. In: Proceedings of the 8th International Conference on Document Analysis and Recognition (ICDAR), pp. 347–351 (2005). https://doi.org/10.1109/ICDAR.2005.182
7. Nagy, G., Seth, S.C.: Hierarchical representation of optically scanned documents. In: Proceedings of the 7th International Conference on Pattern Recognition (ICPR), vol. 1, pp. 347–349 (1984)
8. O'Gorman, L.: The document spectrum for page layout analysis. IEEE TPAMI **15**(11), 1162–1173 (1993). https://doi.org/10.1109/34.244677
9. Ouyang, L., Qu, Y., Zhou, H., Zhu, J., et al.: OmniDocBench: benchmarking diverse pdf document parsing with comprehensive annotations. In: Proceedings of the CVPR (2025)
10. Radford, A., Wu, J., Child, R., Luan, D., Amodei, D., Sutskever, I.: Language models are unsupervised multitask learners. Technical report, OpenAI (2019)
11. Reimers, N., Gurevych, I.: Sentence-BERT: sentence embeddings using Siamese BERT-networks. In: Proceedings of the EMNLP-IJCNLP, pp. 3982–3992. ACL (2019). https://doi.org/10.18653/v1/D19-1410
12. Rozenberg, M., Munk, M., Kainan, A.: A Talmud page as a metaphor of a scientific text. Int. J. Qualitative Methods **5**(4), 30–44 (2006). https://doi.org/10.1177/160940690600500403
13. Wang, R., Fujii, Y., Bissacco, A.: Text reading order in uncontrolled conditions by sparse graph segmentation. In: Fink, G.A., Jain, R., Kise, K., Zanibbi, R. (eds.) ICDAR 2023. LNCS, vol 14192, pp. 3–21. Springer, Cham (2023). https://doi.org/10.1007/978-3-031-41731-3_1

14. Wang, Z., Xu, Y., Cui, L., Shang, J., Wei, F.: LayoutReader: pre-training of text and layout for reading order detection. In: Proceedings of the EMNLP, pp. 4735–4744. ACL (2021). https://doi.org/10.18653/v1/2021.emnlp-main.389
15. Xu, Y., Li, M., Cui, L., Huang, S., Wei, F., Zhou, M.: LayoutLM: pre-training of text and layout for document image understanding. In: Proceedings of the 26th ACM SIGKDD, pp. 1192–1200 (2020). https://doi.org/10.1145/3394486.3403172

IndicPage-OCR: Robust Low-Resource Adaptation for Multi-Script Indic Handwritten Page Recognition

Shaon Bhattacharyya(✉), Ajoy Mondal, and C. V. Jawahar

CVIT, International Institute of Information Technology, Hyderabad, Hyderabad, Telangana, India
shaon.b@research.iiit.ac.in, {ajoy.mondal,jawahar}@iiit.ac.in

Abstract. Handwritten OCR for Indic scripts remains challenging due to complex character structures, large inter-writer variation, and real-world degradations such as blur, uneven illumination, and scanning artifacts. Freely deployable OCR systems often lag behind commercial engines, especially in low-resource multi-script settings. In this work, we introduce **IndicPage-OCR**, a parameter-efficient adaptation framework built on GOT-OCR 2.0 for handwritten page recognition in *Hindi, Bengali, Tamil,* and *Telugu.* Instead of full end-to-end fine-tuning, we freeze the visual encoder and adapt only the language decoder using Low-Rank Adaptation (LoRA), enabling efficient script-specific specialization under limited data and compute. To handle long handwritten document sequences, we propose a *progressive learning strategy* that scales supervision from words to lines and finally full pages, combined with realistic augmentations for robust page-level recognition. Experiments show substantial reductions in Word Error Rate (WER) and Character Error Rate (CER), narrowing the performance gap between commercial and freely deployable OCR systems by approximately **80% in WER** and over **90% in CER**, while consistently outperforming general-purpose foundation-style baselines. These results demonstrate that decoder-focused adaptation with progressive training is an effective and scalable solution for low-resource Indic handwritten-page OCR. Code and resource are available at: https://indicpageocr.github.io/.

Keywords: OCR · Indic scripts · Vision-Language Models · parameter-efficient fine-tuning · progressive learning · handwritten page digitization

1 Introduction

Optical Character Recognition (OCR) has undergone a major transformation over the past decade, evolving from rule-based and handcrafted feature pipelines to deep neural architectures that jointly model visual and linguistic information [4,13]. Sequence-based recognizers, such as CNN-RNN models with CTC decoding, significantly improved robustness across scripts and writing styles, while

F. Shafait et al. (Eds.): DAS 2026, LNCS 16851, pp. 317–333, 2027.
https://doi.org/10.1007/978-3-032-36207-0_19

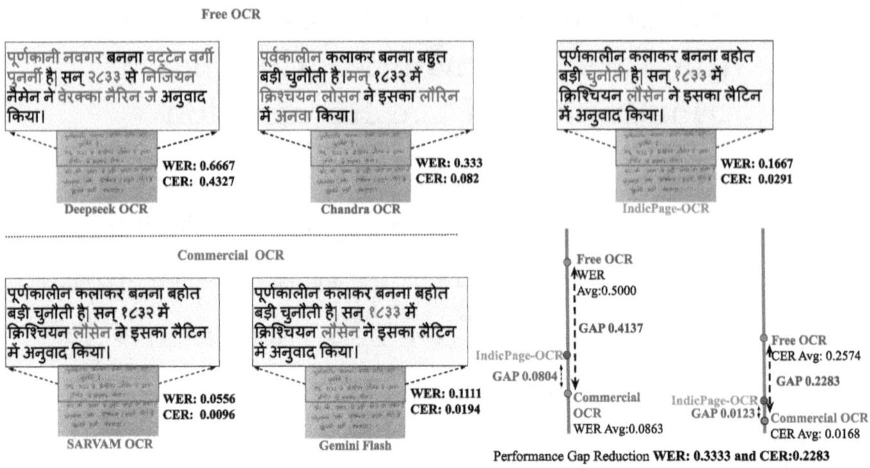

Fig. 1. Comparison of free, commercial, and adapted OCR systems on handwritten Bengali pages. Commercial engines demonstrate substantially lower WER and CER than freely available models, while our adapted approach significantly reduces this performance gap.

transformer-based architectures further enhanced long-range dependency modeling and parallelization [10]. Recently, vision-language and image-to-text generation models have enabled end-to-end page-level transcription without explicit segmentation [7].

Although OCR technology has progressed significantly, a pronounced accuracy gap remains between commercial engines and freely available or open-source systems, particularly for handwritten page-level recognition of Indic scripts. Proprietary solutions typically use large-scale private datasets, extensive engineering refinement, and continuous optimization pipelines, enabling them to achieve substantially lower WER and CER. In contrast, publicly deployable OCR models often struggle under the same conditions, exhibiting markedly higher error rates, which is especially evident for handwritten Bengali pages, as shown in Fig. 1. This discrepancy extends beyond benchmark comparisons and has tangible implications for real-world digitization efforts. In academic, governmental, and archival contexts, reliance on commercial APIs may be constrained by costs, data privacy concerns, licensing restrictions, or the need for offline processing. These limitations underscore the urgent need for adaptable, open OCR frameworks capable of achieving commercial-level performance without dependence on proprietary ecosystems.

Handwritten OCR for Indic languages presents unique challenges. Scripts such as Hindi, Bengali, Tamil, and Telugu contain large character inventories, compound glyphs, visually similar characters, and complex diacritic placement. Writer variability further amplifies intra-class variation, while page-level recognition introduces additional difficulties, including long text sequences, irregular spacing, skewed lines, and acquisition noise from scanning or mobile capture. Furthermore, domain-specific datasets such as notebooks, forms, or institutional documents are often limited in size, making full model retraining computationally expensive and prone to overfitting.

Foundation-style OCR and vision-language models provide promising solutions to these challenges. Pre-trained on large-scale corpora, such models incorporate strong visual representations and implicit language priors that can help resolve ambiguous or degraded handwritten strokes. GOT-OCR 2.0 [17], for instance, is designed for unified end-to-end OCR generation and can directly process long-form documents. However, direct application of such models to handwritten Indic pages does not automatically eliminate the observed performance gap. Therefore, effective and efficient domain adaptation is necessary.

In this work, we explicitly aimed to reduce the performance gap between commercial and freely deployable OCR systems for handwritten Indic page digitization. We propose **IndicPage-OCR**, a parameter-efficient adaptation of a foundation OCR model tailored for multi-script handwritten page transcriptions. Instead of full fine-tuning, we adopt Low-Rank Adaptation (LoRA) [5], freezing the vision encoder and fine-tuning only the text decoder, enabling effective script-specific adaptation while retaining strong pretrained visual features.

To address instability in long-sequence page training, we introduce a progressive learning strategy that gradually exposes the model to increasing levels of reading complexity (word → line → page). Combined with realistic augmentation tailored to document acquisition noise, IndicPage-OCR significantly improves the recognition accuracy and robustness of handwritten Indic scripts.

Extensive experiments across Hindi, Bengali, Tamil, and Telugu demonstrated that IndicPage-OCR markedly reduces both WER and CER compared to standard open OCR baselines, significantly closing the gap with commercial systems. These findings indicate that lightweight, targeted adaptation of foundation OCR models offers a scalable and cost-effective alternative for high-quality handwritten page digitization that does not depend on proprietary platforms.

Our main contributions are as follows:

- We propose **IndicPage-OCR**, a parameter-efficient adaptation of vision-language model for multi-script handwritten page digitization in Indic languages.
- We demonstrate that **decoder-only LoRA adaptation with a frozen vision encoder** is an efficient and effective strategy for adapting foundation OCR models to complex handwritten scripts.
- We introduce a **progressive learning strategy** for long-sequence handwritten page recognition, significantly improving training stability and convergence.
- As illustrated in Fig. 1, a substantial performance gap exists between freely deployable OCR systems and commercial solutions at the page level. IndicPage-OCR significantly narrows this gap, reducing the performance difference by up to **80% in WER** and over **90% in CER**, thereby bridging the openâĂŞcommercial divide while remaining fully open and cost-effective.

2 Related Work

2.1 Traditional and Indic OCR

Traditional OCR systems typically follow a multi-stage pipeline consisting of text localization, segmentation, and recognition. Deep learningâĂŞbased recognizers initially relied on CNN-RNN architectures trained with CTC loss, enabling sequence modeling without explicit character alignment [4]. Transformer-based recognizers later improved long-range dependency modeling and benefited from large-scale pretraining; TrOCR [10] is a representative example the demonstrates strong transfer to downstream OCR tasks.

Open-source OCR engines such as Tesseract [14], EasyOCR [6], and MMOCR [8] provide practical baselines for printed and scene text recognition. However, handwritten page-level recognition remains considerably challenging because of writer variability, long sequences, irregular spacing, and document degradation.

OCR for Indic scripts introduces additional complexity. Indic languages exhibit large character inventories, ligatures, diacritics, and visually similar glyphs, which increase ambiguity during recognition. Public datasets for Indic handwriting have historically been limited compared with Latin-script benchmarks. The IIIT-INDIC-HW-WORDS dataset is a notable resource supporting multi-script handwritten word recognition [3]. Despite these efforts, many Indic OCR approaches focus on word- or line-level recognition and often depend on segmentation heuristics when scaling full pages.

Data scarcity is a recurring bottleneck in handwriting recognition research. Recent work highlights the limited availability of large annotated resources for offline handwriting and proposes datasets and baselines to improve accessibility for HTR research [12]. In the context of Indic OCR, prior studies have explored adapting vision-language models for Hindi handwriting using parameter-efficient fine-tuning and progressive learning strategies, demonstrating that such approaches can be effective with limited supervision [2]. These findings motivate our focus on multi-script handwritten *page-level* digitization using an efficient adaptation strategy.

2.2 Vision-Language Models for OCR

Recent research has increasingly formulated OCR as an image-to-text generation problem using vision-language models (VLMs). Donut [7] demonstrated OCR-free document understanding by directly generating structured outputs from document images. Pix2Struct [9] and OFA [16] further show that unified multimodal sequence-to-sequence pretraining enables strong transfer across visually grounded language tasks. The transfer of large pretrained multimodal models to OCR has proven effective even without extensive OCR-specific supervision [11].

Recently, foundation-style OCR systems have aimed to process full pages holistically. GOT-OCR 2.0 proposes a unified end-to-end OCR model designed for long-form generation [17]. DeepSeek-OCR explores long-context OCR

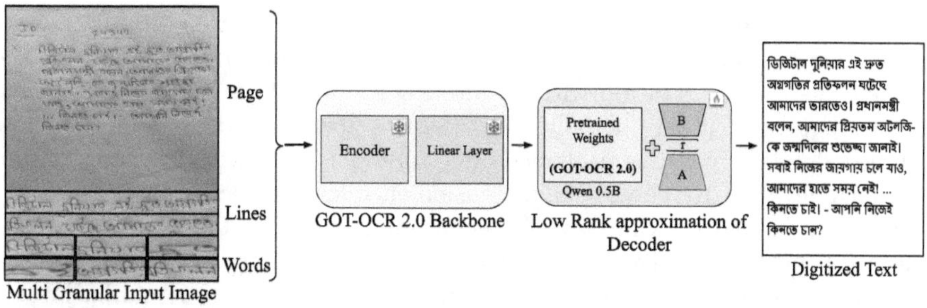

Fig. 2. Overview of the proposed handwritten page OCR pipeline. Multi-granular inputs (word, line, and page images) are processed by a frozen vision encoder and linear connector initialized from GOT-OCR 2.0. The Qwen-0.5B decoder was adapted using LoRA for language-specific fine-tuning. Progressive training from word to line to page enables stable long-context handwritten page transcription.

through optical token compression [18], and DeepSeek-OCR 2 extends this approach by reorganizing visual tokens based on structural semantics [19]. Large vision-language models such as Qwen2-VL and Qwen2.5-VL further highlight the trend toward holistic document parsing via large-scale multimodal pretraining [1,15].

Although VLM-driven OCR benefits from strong language priors that help resolve ambiguous handwritten strokes, adapting large pretrained models to specific scripts and domains remains challenging. Full fine-tuning is computationally expensive and unstable under limited supervision. In contrast, our work adopts parameter-efficient fine-tuning using LoRA [5] to specialize a foundation OCR model for multi-script handwritten page digitization, explicitly targeting the reduction of the performance gap with commercial systems.

3 Pipeline for Indic Handwritten OCR

3.1 Model Architecture and Initialization

Our overall pipeline is shown in Fig. 2. We adopt an encoder–decoder OCR architecture consisting of a **high-compression vision encoder**, **linear connector**, and **long-context text decoder**. Given an input image (resized to 1024×1024), the encoder produces a compact sequence of visual tokens, which are projected into the decoder embedding space through a connector. The decoder then autoregressively generates the transcription.

To obtain a strong starting point for page-level OCR, we initialized all components from the publicly available **GOT-OCR 2.0** checkpoint [17]. In our experiments, the decoder backbone corresponds to **Qwen-0.5B** [15]. This initialization provides robust visual representations and long-context decoding capability, forming the basis for adapting the model to Indic handwritten pages.

Although initialized from GOT-OCR 2.0, we show that decoder-only parameter-efficient fine-tuning suffices for handwritten Indic page-level OCR.

The pretrained encoder captures robust visual representations, while script-specific composition and long-context modeling are primarily handled by the decoder, enabling adaptation without altering the visual backbone.

Although built upon GOT-OCR 2.0 with a Qwen-based decoder, we show that decoder-only parameter-efficient fine-tuning is sufficient for handwritten Indic page-level OCR. The pretrained vision encoder provides robust visual and layout representations, while script-specific composition and long-context modeling are primarily handled by the decoder. Thus, effective adaptation can be achieved by modifying only the decoder, without updating the visual backbone.

3.2 Decoder Adaptation with LoRA

We perform language-wise adaptation using Low-Rank Adaptation (LoRA) [5]. In all experiments, we kept the **vision encoder frozen** and fine-tuned only the **text decoder** using LoRA, while the linear connector remained unchanged. This design reduces training cost, stabilizes optimization under limited data, and allows script-specific adaptation to be learned primarily in the decoder.

LoRA applies a low-rank update to the weight matrix $W \in \mathbb{R}^{d \times k}$:

$$W + \Delta W = W + BA, \tag{1}$$

where $B \in \mathbb{R}^{d \times r}$ and $A \in \mathbb{R}^{r \times k}$ are trainable matrices and r is the rank. During training, the update is scaled by α/r:

$$h = Wx + \frac{\alpha}{r}(BA)x. \tag{2}$$

We applied LoRA to the decoder's linear layers (excluding the LM head) with rank $r = 8$ and scaling $\alpha = 32$. A separate adapted model was trained for each language to ensure script-specific specialization while maintaining a consistent configuration across languages.

3.3 Progressive Learning (Word → Line → Page)

As shown in Fig. 2, we adopted a multi-granular training strategy that progressed from *word-level* inputs to *line-level* inputs and finally to *page-level* images.

Page-level handwritten OCR is inherently a long-sequence generation problem. If the model is trained directly on full pages from the beginning, it must simultaneously learn script-specific character composition, spacing patterns, and long-context decoding over hundreds to thousands tokens. Under limited supervision, this often leads to unstable early optimization. To mitigate this difficulty, we gradually increased reading complexity during training.

The word-level phase enables the decoder to rapidly adapt to the visual-to-token mapping of the script. At this stage, the model primarily learns character shapes, diacritics, conjunct forms, and short-range token statistics without the burden of long-range dependency.

Next, line-level training introduces consistent spacing patterns and local reading orders, encouraging the decoder to utilize a wider context window while maintaining manageable sequence lengths. This intermediate step bridges the gap between clean word crops and the irregular structure of complete handwritten pages.

Finally, page-level fine-tuning exposes the model to long-range dependencies such as line transitions and paragraph continuity. At this stage, the decoder has already internalized script-specific tokens and local context, allowing it to focus on stable long-form transcription. Page images also introduce realistic layout variations and acquisition noise, further improving robustness.

Overall, the word→line→page progression reduces early training difficulty, improves convergence stability, and enables direct full-page transcription without requiring external word or line detection during the inference phase.

3.4 In-House Dataset and Augmentation for Handwritten Robustness

We used an in-house handwritten dataset collected from practical digitization workflows for four Indic languages. The raw data are available at the page level (Hindi: 23,912 pages; Bengali: 7,189 pages; Telugu: 6,121 pages; Tamil: 5,454 pages). In addition to pages, we also have **existing word-level annotations**, which we use to construct word crops and derive line samples by grouping words along the same-text line. Using these word and line samples, we formed a large training pool and applied augmentation to expand diversity. Overall, the word- and line-level data together form approximately **12 million** samples after augmentation, whereas page images are retained for the final stage of progressive learning and for page-level evaluation.

To improve robustness under real capture conditions, we applied a handwriting-focused augmentation pipeline to the training samples. This includes blur (e.g., Gaussian/defocus), noise injection, contrast/brightness changes, compression artifacts, and mild geometric distortions (rotation/skewing). These transformations simulate common degradations in scanned or camera-captured handwritten pages and reduce overfitting to clean samples. For training and validation, we used a combined total of approximately **1.5 million** samples (train+val) drawn from the augmented pool. For inference and final reporting, we evaluated a fixed page-level test set of **500 pages per language** (Hindi, Bengali, Telugu, and Tamil).

3.5 Training Setup and Evaluation Metrics

All the language models were fine-tuned using the same configuration to ensure comparability across scripts. The high-compression vision encoder was kept frozen, and only the LoRA parameters in the Qwen-0.5B decoder were optimized. Training is performed on a single **NVIDIA RTX 6000 Ada** GPU with a batch size of **4**. We used the AdamW optimizer with cosine learning rate scheduling and enabled gradient checkpointing to support long page sequences efficiently. Across all languages, training typically converged by approximately **200K** steps.

Table 1. Page-level OCR performance on **500** handwritten test pages for Hindi and Bengali. The WER and CER were reported (lower is better). Best results are shown in bold, second-best in italic, and our model is underlined.

Hindi			Bengali		
Model	WER	CER	Model	WER	CER
Sarvam OCR	**0.0682**	**0.0419**	Sarvam OCR	**0.0542**	**0.0194**
Gemini Flash	*0.0708*	*0.0447*	Gemini Flash	*0.0584*	*0.0211*
Chandra OCR	0.2310	0.1600	Chandra OCR	0.3614	0.1243
DeepSeek-OCR	4.1910	3.7953	DeepSeek-OCR	6.9971	8.0158
Qwen2-VL-2B	2.3004	2.0898	Qwen2-VL-2B	1.7860	1.5792
Qwen2.5-VL-3B	4.9442	4.2353	Qwen2.5-VL-3B	2.7089	3.8575
olmOCR	4.9754	4.6725	olmOCR	2.0924	1.6342
PaddleOCR	0.9612	0.8413	PaddleOCR	-	-
Tesseract	0.8333	0.4911	Tesseract	0.9632	0.8122
EasyOCR	0.6154	0.3322	EasyOCR	0.6671	0.2941
IndicPage-OCR (Ours)	0.0823	0.0466	IndicPage-OCR (Ours)	0.0719	0.0221

For evaluation, we report standard OCR error metrics at both the character and word levels: **Character Error Rate (CER)** and **Word Error Rate (WER)**. Both metrics are computed using normalized edit distance:

$$CER = \frac{S_c + D_c + I_c}{N_c}, \quad WER = \frac{S_w + D_w + I_w}{N_w}, \tag{3}$$

where S, D, and I denote the number of substitutions, deletions, and insertions, respectively, and N represents the total number of ground-truth characters (N_c) or words (N_w). Lower CER and WER values indicate better transcription quality.

Inference Efficiency. Inference requires **12.42 s/page** (0.08 pages/sec) with **5 GB** peak memory on an RTX 6000 Ada GPU, confirming the practical deployability of decoder-only adaptation for full-page OCR.

4 Results

4.1 Comparison with Existing OCR Systems

Tables 1 and 2 present page-level OCR results across four Indic languages. Commercial systems such as Sarvam OCR consistently achieve the lowest WER and CER across scripts, reflecting the advantages of large proprietary training resources and highly optimized production pipelines. Gemini Flash also delivers strong performance, closely tracking Sarvam in most cases.

The performance of freely deployable and foundation-style OCR systems, the performance varies substantially. Prompting-based approaches using Qwen-VL models, DeepSeek-OCR, and olmOCR exhibited significantly higher error rates, particularly on long handwritten pages. This highlights the limitations of

Table 2. Page-level OCR performance on **500** handwritten test pages for Tamil and Telugu. WER and CER were reported (lower is better). Best results are shown in bold, second-best in italic, and our model is underlined.

Tamil			Telugu		
Model	WER	CER	Model	WER	CER
Sarvam OCR	**0.0382**	**0.0189**	Sarvam OCR	**0.0331**	**0.0221**
Gemini Flash	*0.0428*	*0.0207*	Gemini Flash	*0.0377*	*0.0246*
Chandra OCR	0.5567	0.3389	Chandra OCR	0.8858	0.5502
DeepSeek-OCR	1.4124	1.9058	DeepSeek-OCR	25.6800	17.1386
Qwen2-VL-2B	4.6471	4.1617	Qwen2-VL-2B	5.6157	4.0936
Qwen2.5-VL-3B	6.7536	6.2870	Qwen2.5-VL-3B	1.1500	5.7156
olmOCR	4.7085	3.9433	olmOCR	2.3767	1.5628
PaddleOCR	0.9542	0.6723	PaddleOCR	0.9324	0.7533
Tesseract	0.9553	0.7065	Tesseract	0.9876	0.6411
EasyOCR	-	-	EasyOCR	0.8161	0.5391
IndicPage-OCR (Ours)	<u>0.0855</u>	<u>0.0266</u>	IndicPage-OCR (Ours)	<u>0.0426</u>	<u>0.0292</u>

zero-shot or prompt-based adaptation for complex multiline handwritten transcription in Indic scripts.

IndicPage-OCR substantially improved all open and foundation-style baselines across Hindi, Bengali, Tamil, and Telugu. While commercial systems remain ahead in absolute error rates, the gap between IndicPage-OCR and commercial OCR is markedly smaller than that margins observed between commercial systems and generic VLM-based approaches.

A closer look at the language-wise results reveals interesting variations. Telugu achieved the lowest WER among the four languages (0.0416), indicating relatively stable word-level recognition. This may be attributed to clearer character separation and comparatively consistent stroke patterns in the collected pages. In contrast, Tamil showed the highest WER among the four models (0.0844), suggesting greater difficulty at the word level, likely because of dense character formations and tightly connected strokes that increase substitution and insertion errors in long sequences.

Hindi and Bengali fall between these extremes, with Hindi at 0.0813 WER and Bengali at 0.0721 WER. Although both scripts contain complex conjunct formations and diacritic structures, Bengali achieves a slightly better CER (0.0226) than Hindi (0.0455), indicating stronger character-level stability but slightly higher word-level segmentation sensitivity in Hindi. This suggests that small character-level ambiguities in Hindi propagate more strongly at the word-boundary level.

The differences between WER and CER trends across languages also reflect writer variability in the dataset. Variations in stroke thickness, spacing irregularity, and baseline alignment contribute differently to the word-level aggregation of character errors across scripts. Importantly, despite these script-specific

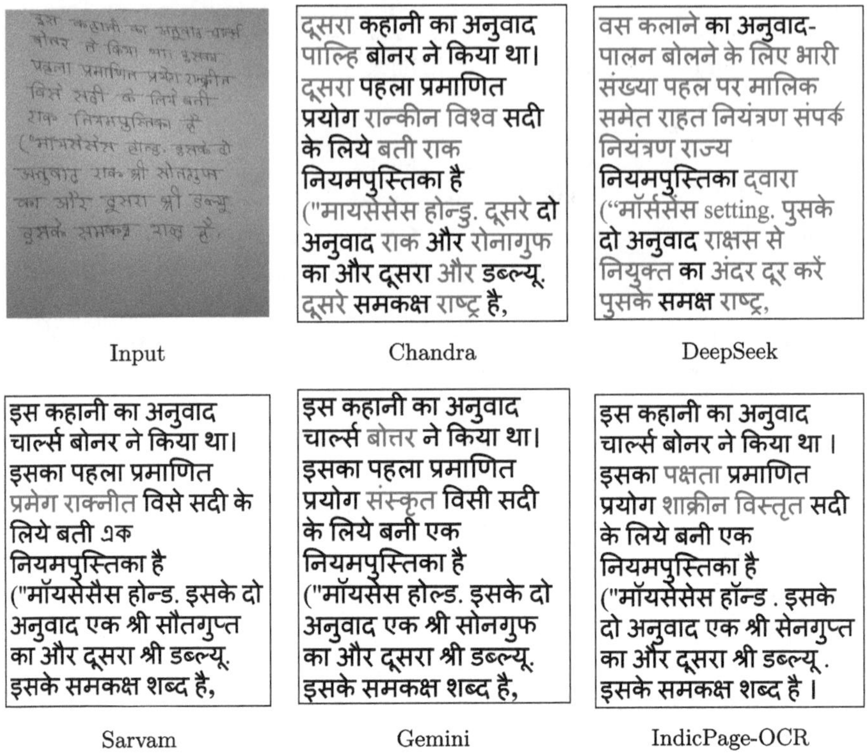

Fig. 3. Qualitative comparison on a handwritten Hindi page. Top row (left to right): Input, Chandra OCR, DeepSeek-OCR. Bottom row: Sarvam OCR, Gemini Flash, IndicPage-OCR (ours).

and writer-dependent challenges, the same decoder-only adaptation strategy maintained competitive performance across all four languages without language-specific architectural tuning.

Overall, these results demonstrate that IndicPage-OCR generalizes effectively across structurally diverse scripts, and that parameter-efficient adaptation significantly narrows the performance gap with commercial systems while preserving consistent multi-script behavior.

4.2 Qualitative Analysis

The qualitative comparisons in Figs. 3, 4, 5 and 6 reveal clear differences in transcription behavior across systems. Chandra OCR and DeepSeek-OCR frequently produce word substitutions, inconsistent handling of conjunct characters, and diacritic placement errors, particularly in longer handwritten passages where errors accumulate across multiple lines. Commercial systems such as Sarvam and Gemini maintain strong structural consistency and lexical stability throughout a page.

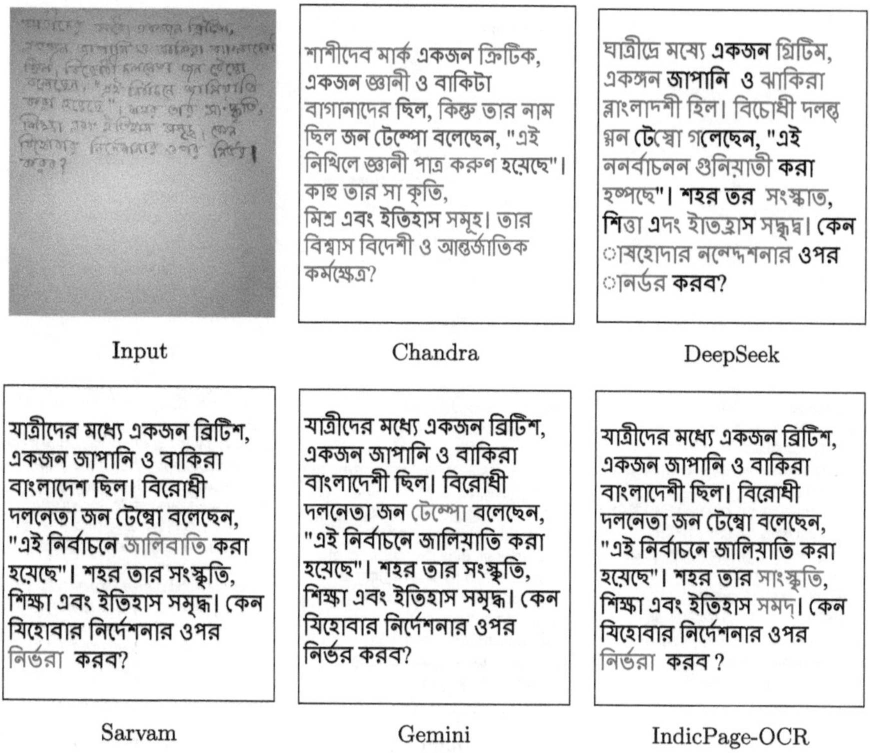

Fig. 4. Qualitative comparison on a handwritten Bengali page. Top row (left to right): Input, Chandra OCR, DeepSeek-OCR. Bottom row: Sarvam OCR, Gemini Flash, IndicPage-OCR (ours).

IndicPage-OCR demonstrated a stable character composition, improved long-sequence coherence, and reliable preservation of script-specific glyph structures across Hindi, Bengali, Tamil, and Telugu. Although minor errors remain in challenging regions, the model maintains readable and structurally faithful transcriptions, even under dense handwriting and line variability. These qualitative observations reinforce the quantitative findings, indicating that decoder-level adaptation combined with progressive training enhances the robustness of multi-line handwritten page transcription.

5 Ablation Study

5.1 Effect of Progressive Learning

To evaluate the contribution of progressive learning, we compared two training strategies: (1) direct page-level fine-tuning using LoRA for 200K steps, and (2) the proposed progressive word→line→page training strategy.

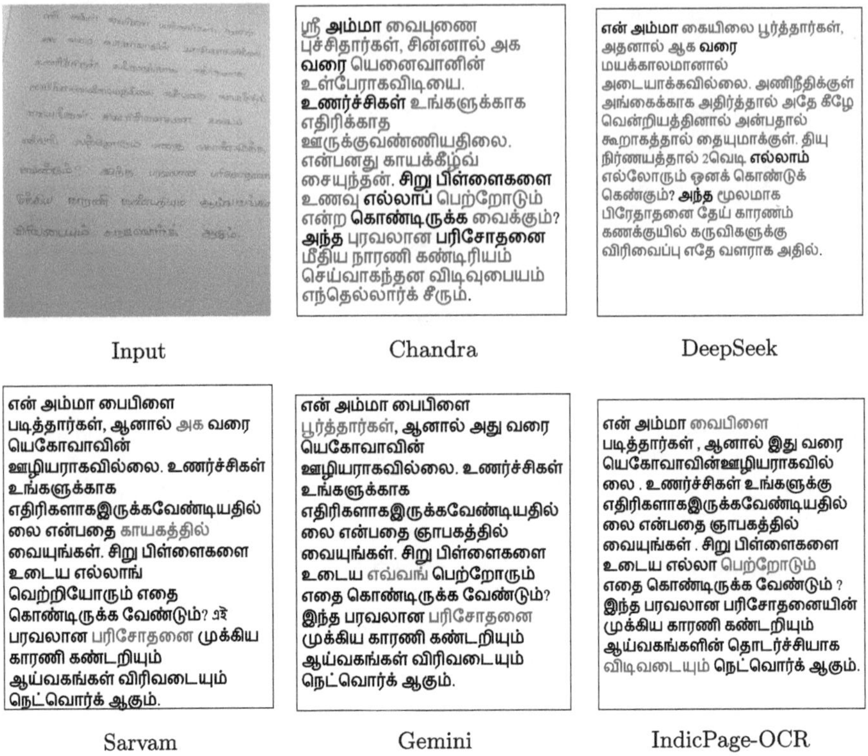

Fig. 5. Qualitative comparison on a handwritten Tamil page. Top row (left to right): Input, Chandra OCR, DeepSeek-OCR. Bottom row: Sarvam OCR, Gemini Flash, IndicPage-OCR (ours).

In both settings, the vision encoder was frozen and only the decoder was adapted using identical LoRA configurations.

Table 3. Ablation study on the effect of progressive learning for page-level handwritten OCR. We compare direct page-only LoRA fine-tuning (200K steps) against the proposed word→line→page progressive training strategy, keeping all other settings identical. WER and CER are reported per language (lower is better).

Language	Page-only Training		Progressive Training	
	WER	CER	WER	CER
Hindi	0.521	0.312	**0.0813**	**0.0455**
Bengali	0.488	0.284	**0.0721**	**0.0226**
Tamil	0.612	0.352	**0.0844**	**0.0261**
Telugu	0.437	0.261	**0.0416**	**0.0283**

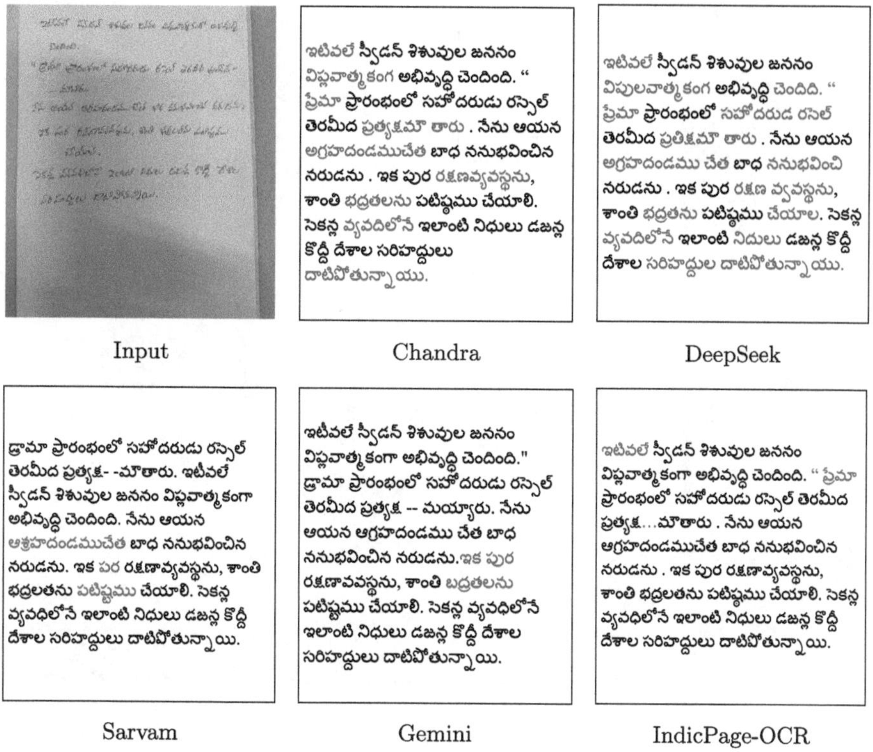

Fig. 6. Qualitative comparison on a handwritten Telugu page. Top row (left to right): Input, Chandra OCR, DeepSeek-OCR. Bottom row: Sarvam OCR, Gemini Flash, IndicPage-OCR (ours).

Direct page-level training leads to unstable optimizations and substantially higher error rates across all languages. The model struggles to simultaneously learn script-specific character composition and long-context generation when exposed only to full-page sequences at the beginning.

In contrast, progressive learning significantly reduced both WER and CER for every scripts as shown in Table 3. The word-level phase enables reliable visual-to-token alignment, the line-level phase improves local context modeling, and the final page-level phase stabilizes long-sequence decoding. Consistent improvements across structurally different scripts indicate that progressive learning is critical for effective adaptation of foundation OCR models to handwritten page transcriptions.

5.2 Noise Sensitivity Analysis

To evaluate robustness under realistic corruptions, we apply controlled degradations at test time, including blur, additive noise, JPEG compression, illumination

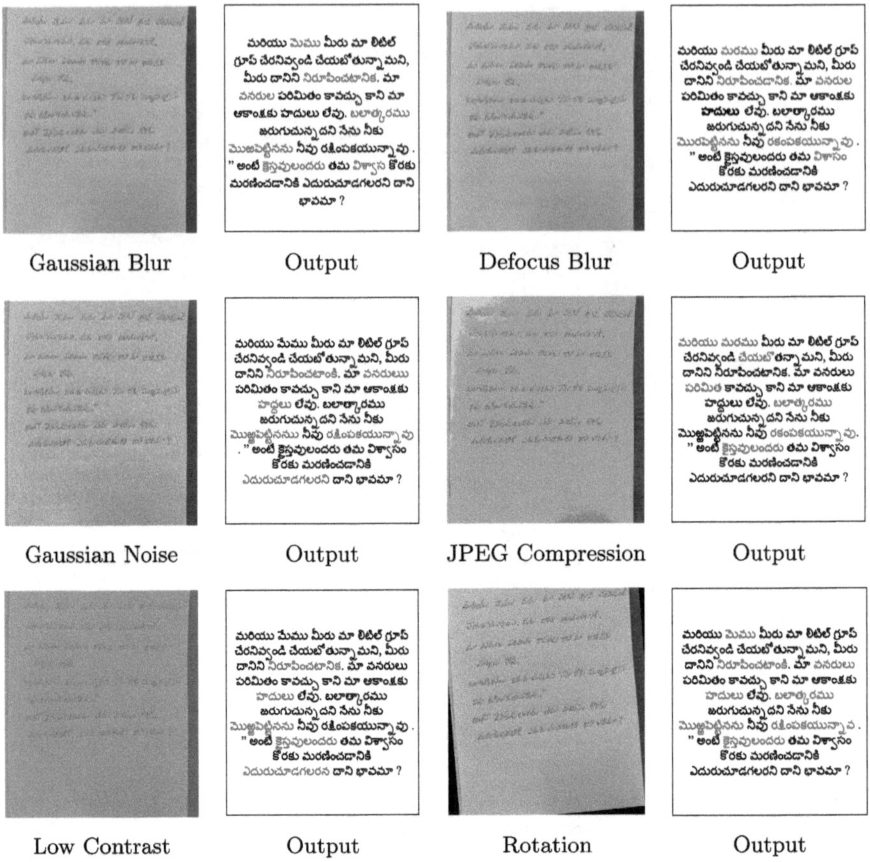

Fig. 7. Qualitative robustness evaluation of IndicPage-OCR on Telugu handwritten pages under controlled corruptions. Each corrupted input image is shown with its corresponding OCR output. The model remains stable under moderate degradations such as Gaussian noise and JPEG compression, while blur and illumination changes introduce more noticeable degradation, consistent with the quantitative trends in Table 4.

shifts, and rotation. Quantitative results are reported in Table 4, and qualitative examples are shown in Figs. 7 and 8.

Under clean conditions, IndicPage-OCR achieves stable performance across scripts (WER 0.042âĂŞ0.084). Blur introduces the largest degradation, particularly defocus blur (e.g., Hindi WER increases from 0.081 to 0.145 and Tamil from 0.084 to 0.158), as reduced stroke sharpness directly affects character boundary clarity. Gaussian noise causes moderate degradation (e.g., Bengali WER from 0.072 to 0.103), indicating reasonable tolerance to pixel-level perturbations. JPEG compression also increases errors, especially in dense handwritten regions, while illumination shifts result in comparatively smaller performance

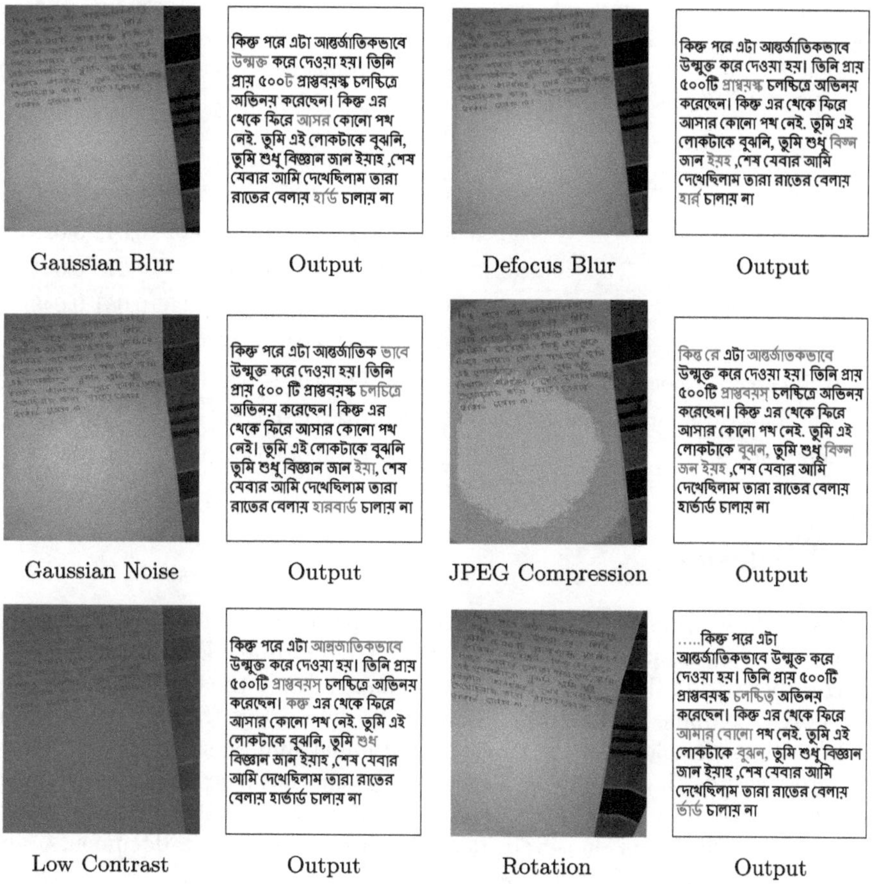

Fig. 8. Qualitative robustness evaluation of IndicPage-OCR on Bengali Handwritten pages under controlled corruptions. Each corrupted input image is shown alongside its corresponding OCR output. The model remains stable under moderate degradations such as Gaussian noise and JPEG compression, while blur and illumination changes introduce more noticeable degradation, consistent with the quantitative trends in Table 4.

drops. Rotation and skew lead to predictable but controlled degradation, with transcription remaining readable across languages.

Overall, while blur and compression produce the most significant impact, the model remains relatively stable under moderate noise and geometric distortions. These results suggest that decoder-only LoRA adaptation combined with progressive learning improves both clean accuracy and robustness under realistic document degradations.

Table 4. Noise sensitivity analysis under controlled test-time degradations for each language. We evaluated Gaussian blur, defocus blur, additive noise, JPEG compression, illumination shift, and rotation to assess the robustness under realistic document corruptions. WER and CER are reported per language (lower is better).

Condition	Hindi		Bengali		Tamil		Telugu	
	WER	CER	WER	CER	WER	CER	WER	CER
Clean	0.081	0.046	0.072	0.023	0.084	0.026	0.042	0.028
Gaussian blur	0.132	0.071	0.118	0.055	0.146	0.078	0.073	0.041
Defocus blur	0.145	0.079	0.126	0.061	0.158	0.085	0.081	0.048
Gaussian noise	0.114	0.063	0.103	0.049	0.127	0.067	0.066	0.037
JPEG compression	0.142	0.076	0.128	0.062	0.149	0.081	0.091	0.054
Low contrast/illumination shift	0.093	0.051	0.082	0.031	0.078	0.024	0.051	0.030
Rotation/skew	0.119	0.066	0.108	0.051	0.132	0.072	0.070	0.040

6 Conclusion

In this work, we presented IndicPage-OCR, a parameter-efficient adaptation of GOT-OCR 2.0 for handwritten page-level OCR in Hindi, Bengali, Tamil, and Telugu. By freezing the visual encoder and adapting only the Qwen-0.5B decoder with LoRA, combined with a progressive word→line→page training strategy, the model achieves stable full-page transcription under limited supervision.

Across scripts, IndicPage-OCR narrows the gap between open and commercial OCR systems while remaining computationally efficient and deployable. Future work will focus on severe degradations, mixed-script documents, and layout-aware modeling for improved structural understanding.

Acknowledgments. This work is supported by the MeitY Government of India, through the NLTM Bhashini (https://bhashini.gov.in/) project.

References

1. Bai, S., et al.: Qwen2. 5-VL technical report. arXiv preprint arXiv:2502.13923 (2025)
2. Bhattacharyya, S., Ghosh, S., Deb, P., Mondal, A., Jawahar, C.: Adapting vision-language models for Hindi OCR. In: International Conference on Document Analysis and Recognition, pp. 577–594. Springer (2025)
3. Gongidi, S., Jawahar, C.V.: IIIT-INDIC-HW-WORDS: a dataset for INDIC handwritten text recognition. In: IEEE International Conference on Document Analysis and Recognition (2021). https://api.semanticscholar.org/CorpusID:236425882
4. Graves, A., Fernández, S., Gomez, F., Schmidhuber, J.: Connectionist temporal classification: labelling unsegmented sequence data with recurrent neural networks. In: Proceedings of the 23rd International Conference on Machine Learning, pp. 369–376 (2006)

5. Hu, E.J., et al.: LoRA: low-rank adaptation of large language models. In: ICLR (2022)
6. JaidedAI: Easyocr. https://github.com/JaidedAI/EasyOCR
7. Kim, G., et al.: OCR-free document understanding transformer. In: European Conference on Computer Vision, pp. 498–517. Springer (2022)
8. Kuang, Z., et al.: MMOCR: a comprehensive toolbox for text detection, recognition and understanding. In: Proceedings of the 29th ACM International Conference on Multimedia, pp. 3791–3794 (2021)
9. Lee, K., et al.: Pix2Struct: screenshot parsing as pretraining for visual language understanding. In: International Conference on Machine Learning, pp. 18893–18912. PMLR (2023)
10. Li, M., et al.: TROCR: transformer-based optical character recognition with pre-trained models. In: Proceedings of the AAAI Conference on Artificial Intelligence, pp. 13094–13102 (2023)
11. Lin, J., et al.: Transferring general multimodal pretrained models to text recognition. In: Findings of the Association for Computational Linguistics: ACL 2023, pp. 588–597 (2023)
12. Mondal, A., Tulsyan, K., Jawahar, C.: Bridging the gap in resource for offline English handwritten text recognition. In: International Conference on Document Analysis and Recognition, pp. 413–428. Springer (2024)
13. Shi, B., Bai, X., Yao, C.: An end-to-end trainable neural network for image-based sequence recognition and its application to scene text recognition. IEEE Trans. Pattern Anal. Mach. Intell. 2298–2304 (2016)
14. Smith, R.: An overview of the tesseract OCR engine. In: ICDAR, pp. 629–633 (2007)
15. Wang, P., et al.: Qwen2-VL: enhancing vision-language model's perception of the world at any resolution. arXiv preprint arXiv:2409.12191 (2024)
16. Wang, P., et al.: OFA: unifying architectures, tasks, and modalities through a simple sequence-to-sequence learning framework. In: International Conference on Machine Learning, pp. 23318–23340. PMLR (2022)
17. Wei, H., et al.: General OCR theory: towards OCR-2.0 via a unified end-to-end model. arXiv preprint arXiv:2409.01704 (2024)
18. Wei, H., Sun, Y., Li, Y.: DeepSeek-OCR: contexts optical compression. arXiv preprint arXiv:2510.18234 (2025)
19. Wei, H., Sun, Y., Li, Y.: DeepSeek-OCR 2: visual causal flow. arXiv preprint arXiv:2601.20552 (2026)

Enterprise and Industrial Document AI

Rethinking Text Color Estimation in Screen Documents: A Reliability-Aware Approach

Liang Zhang(✉), Zhenfeng Lan, Huajian Ni, Pei Guo, Qinjie Hu, Jie Liu, and Zhongxin Wang

Kingsoft Office Software, Inc., Wuhan, Hubei, People's Republic of China
{zhangliang13,lanzhenfeng,nihuajian,guopei,huqinjie,liujie22, wangzhongxin}@wps.cn

Abstract. Accurate text color estimation is essential for high-fidelity document restoration and editable reconstruction, but it becomes ill-defined for screen-rendered, screen-captured, and AI-generated documents because anti-aliasing and subpixel rendering make boundary pixels deviate from the intrinsic text color. We propose a reliability-aware structural approach, where reliability denotes how faithfully an observed pixel reflects the underlying intrinsic color rather than rendering-induced mixtures. The key insight is that interior stroke regions provide stable color evidence, while boundary regions are corrupted by foreground-background blending and color fringing. Based on this observation, we introduce a skeleton-guided and edge-aware framework that emphasizes reliable interior cues, suppresses ambiguous boundary supervision, and uses curriculum-based training for robust optimization. Experiments on synthetic, real screenshot, and AI-generated document benchmarks show that the proposed method improves color estimation accuracy and robustness across diverse rendering conditions, highlighting the importance of modeling pixel reliability for intrinsic text color estimation.

Keywords: Document Image Analysis · Text Color Estimation · Screen-Rendered Documents · Reliability-Aware Learning · Structure-Aware Modeling · Anti-Aliasing Artifacts

1 Introduction

With the rapid proliferation of AI-generated documents, screen-rendered content, and screen-captured images, restoring document images into editable and high-fidelity representations has become increasingly important in document analysis systems. Beyond textual content recognition, accurately recovering visual attributes such as layout, font, and color is essential for preserving visual fidelity and enabling downstream editing. Among these attributes, text color plays a critical role in maintaining visual consistency and semantic meaning in reconstructed documents.

F. Shafait et al. (Eds.): DAS 2026, LNCS 16851, pp. 337–353, 2027.
https://doi.org/10.1007/978-3-032-36207-0_20

Despite its importance, text color estimation has not been explicitly formulated or systematically studied as an independent problem. Existing approaches typically treat it as a byproduct of segmentation or rendering pipelines, without dedicated modeling or evaluation. This limitation becomes increasingly pronounced in modern document scenarios, where rendering artifacts and style diversity introduce significant challenges to recovering intrinsic text appearance.

In contrast to traditional scanned or camera-captured documents, text in contemporary digital pipelines is generated through anti-aliasing and subpixel rendering mechanisms [1,7,18]. As a result, many pixels—particularly those near text boundaries—do not correspond to the intrinsic text color, but instead represent mixtures of foreground and background colors induced by rendering processes. This effect leads to substantial intra-instance color variation even within a single text region.

As visually illustrated in Fig. 1, these rendering artifacts make boundary pixels unreliable for intrinsic color estimation, while interior stroke pixels remain comparatively stable.

These observations indicate that text color estimation is inherently ambiguous at the pixel level. Therefore, simply increasing model capacity is insufficient, since the uncertainty mainly originates from unreliable supervision rather than feature representation itself.

Existing methods mainly rely on pixel-wise regression, clustering, or segmentation-based aggregation [6,14], implicitly assuming uniform pixel reliability. Such assumptions often fail under strong anti-aliasing and rendering artifacts.

In this work, we revisit text color estimation from a **reliability-aware structural perspective**. We argue that color observations within text regions are inherently heterogeneous in reliability, and that robust estimation should explicitly distinguish between reliable and unreliable regions. Specifically, interior stroke regions—well approximated by text skeletons—tend to preserve stable and faithful color information, whereas boundary regions are significantly corrupted by rendering artifacts.

Based on this insight, we reformulate text color estimation as a reliability-aware inference problem and propose a **reliability-aware structural modeling framework**. Our approach explicitly leverages text geometry to guide color estimation. A skeleton-guided representation is used to capture reliable interior regions, while an edge-aware suppression mechanism mitigates the influence of unreliable boundary pixels. Furthermore, we introduce a curriculum-based training strategy that progressively down-weights ambiguous supervision, enabling more stable and effective learning.

Extensive experiments on synthetic, screen-captured, and AI-generated datasets demonstrate that the proposed method consistently improves accuracy and robustness under diverse rendering conditions, particularly for screen-rendered and screen-captured documents. More importantly, our results highlight that explicitly modeling supervision reliability is essential for intrinsic text color estimation and cannot be compensated by conventional learning-based approaches alone.

The main contributions of this work are summarized as follows:

- We show that text color estimation in modern document images is inherently ill-defined at the pixel level due to rendering-induced color mixing, and formulate it as a reliability-aware inference problem.
- We propose a reliability-aware structural modeling framework that leverages skeleton-guided interior regions as reliable supervision for intrinsic text color estimation.
- We design an edge-aware learning strategy with curriculum-based supervision weighting, which progressively suppresses ambiguous boundary signals during training.
- We conduct extensive experiments on synthetic, screen-captured, and AI-generated datasets, demonstrating consistent improvements over both learning-based and non-learning baselines.

2 Related Work

Existing studies related to this work fall into four broad directions: document restoration and editable reconstruction, text appearance modeling, text segmentation and matting, and rendering-aware text analysis.

Document restoration and editable reconstruction. Document analysis and restoration methods have made strong progress on scanned, photographed, and degraded documents, while born-digital and screen-captured documents have introduced new challenges beyond traditional degradation models. Representative efforts include born-digital text extraction [3], general restoration pipelines [8,24,25], screen-document analysis [2,19–21], and layered editable representations [15,23]. However, these works mainly focus on localization, restoration, or editing, and typically treat color as an auxiliary cue rather than a target attribute to be estimated precisely.

Text color and appearance modeling. Text color and appearance modeling has been explored through heuristic color estimation [11,14], generic color modeling and segmentation [6,12,16], and recent text-attribute prediction or document rendering systems [5,13,17]. These studies show the importance of color as part of text appearance, but most of them either rely on simplified global color assumptions, focus on natural-image settings, or exclude intrinsic text color estimation from rendered document images.

Distinction from text segmentation and matting. Text segmentation [22] and matting [10] estimate foreground support or alpha maps, but do not directly recover intrinsic text color. In rendered screen documents, boundary pixels may be valid foreground pixels while their observed colors are foreground-background mixtures. We therefore use text structure as a reliability cue to emphasize stable interior strokes, rather than treating masks as binary color selectors.

Rendering-aware text analysis. Rendering-aware studies further show that anti-aliasing and subpixel rendering substantially alter observed boundary colors and introduce direction-dependent color distortions [1,4,7,18]. This line of work is important because it explains why rendered text boundaries do not behave like clean foreground pixels, especially in screenshots and low-resolution settings. However, prior methods usually suppress such artifacts for recognition rather than explicitly model their impact on color estimation. This gap motivates our reliability-aware formulation, which distinguishes stable interior evidence from artifact-contaminated boundaries.

3 Problem Definition and Challenges

Given an input document image I, the goal of text color estimation is to recover the intrinsic color representation C for each text instance. In modern screen-rendered and AI-generated documents, however, this task is inherently ill-posed because observed text pixels are not equally reliable.

As illustrated in Fig. 1, boundary pixels are heavily corrupted by anti-aliasing, compression, and foreground-background blending, while subpixel rendering further introduces color fringes that do not belong to the intrinsic text color [2,18]. These effects become more severe for thin strokes, small fonts, and low-resolution screenshots, where unreliable boundary regions occupy a larger fraction of the text area. The challenge is further amplified by the limited availability of exact intrinsic-color annotations for non-synthetic data.

Therefore, text color estimation should not be modeled as uniform pixel-wise regression. Instead, it is better viewed as a **reliability-aware inference problem**, where stable interior regions provide stronger evidence than artifact-contaminated boundaries.

4 Methodology

4.1 Overview

We reformulate text color estimation as a **reliability-aware structural inference problem**: interior stroke regions provide reliable color cues, while boundary pixels are treated as uncertain observations affected by rendering artifacts.

Based on this formulation, we propose a skeleton-guided, edge-aware learning framework, termed **Multi-task Attention Fusion Network (MAFN)**, for robust text color estimation. The key idea of MAFN is to leverage structural priors to guide color inference under unreliable pixel supervision.

As illustrated in Fig. 2, the framework uses a shared feature extractor with two cooperative branches: a structural prediction branch for text masks and a color regression branch guided by the predicted structural attention.

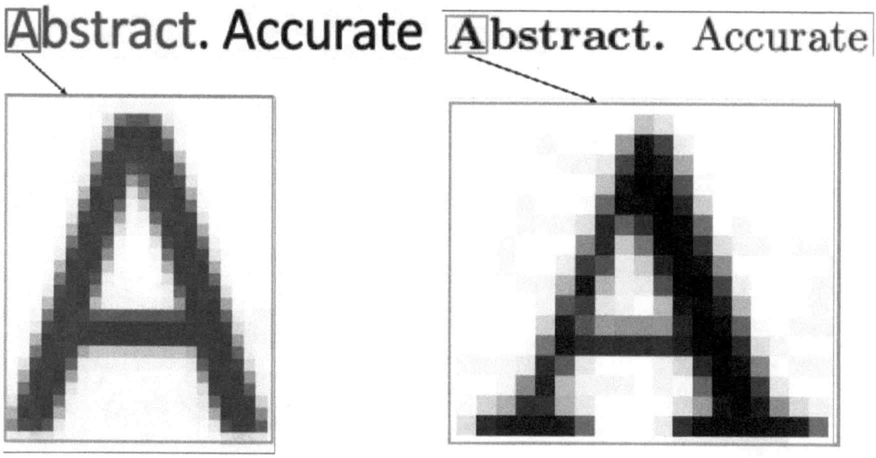

Fig. 1. Visualization of rendering-induced artifacts in text images. (a) Anti-aliasing produces smooth boundaries by blending foreground and background colors, leading to ambiguous boundary pixels. **(b)** Subpixel rendering introduces high-frequency color fringes (e.g., reddish/yellowish patterns) around character strokes, particularly in screen-captured documents. These effects result in unreliable pixel observations for intrinsic text color estimation.

4.2 Input Representation and Adaptive Feature Extraction

Given a text strip, the input is represented as

$$I \in \mathbb{R}^{3\times H\times W}, \tag{1}$$

where the height is fixed to $H = 48$ pixels and the width W is adaptive to preserve the text aspect ratio.

A shared convolutional backbone extracts features with asymmetric strides $(2, 1)$, reducing height from 48 to 6 while preserving horizontal resolution. The reduced feature map is then aggregated along the height dimension to obtain a column-aligned contextual feature

$$F_{\text{ctx}} \in \mathbb{R}^{C\times W}, \tag{2}$$

which remains horizontally aligned with the text layout. We use a lightweight ResNet-style backbone to focus the study on reliability-aware color estimation rather than backbone design.

4.3 Dual-Branch Network Architecture

The shared representation is fed into two heads for text localization and color estimation.

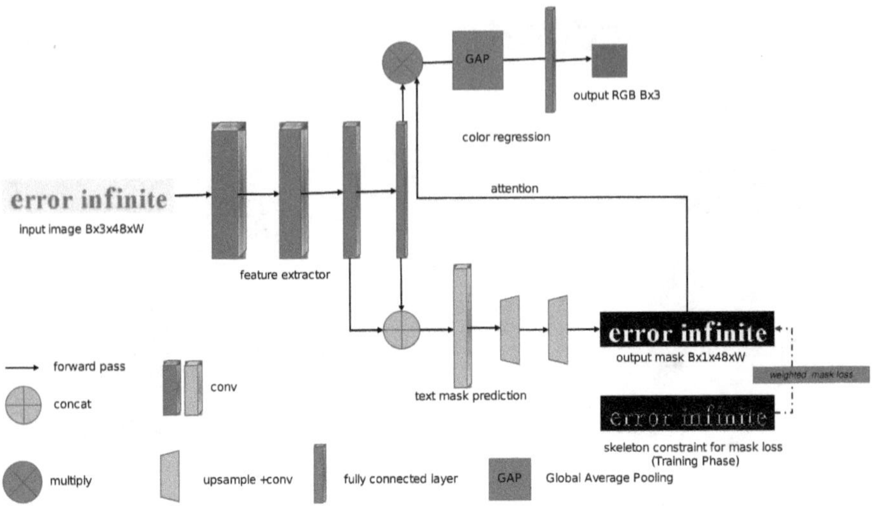

Fig. 2. Pipeline of the proposed MAFN method. The model takes a text image as input and simultaneously predicts a text mask and the text color. Key components include a Multi-scale Feature Extractor, a Text Mask Prediction module for structural awareness, and an Attention-guided Color Regression module for final estimation.

Text Mask Prediction Branch The mask branch predicts a pixel-level foreground probability map. Shallow and high-level features are concatenated and decoded with convolution and upsampling layers to restore the resolution to $48 \times W$.

A Sigmoid activation function is applied to produce the predicted text mask

$$M \in \mathbb{R}^{1 \times 48 \times W}, \tag{3}$$

where each value denotes foreground probability.

Color Regression Branch The color branch regresses the intrinsic text color. Instead of global pooling over all locations, it uses the mask-guided attention described below.

4.4 Explicit Mask-Guided Attention Mechanism

The predicted mask M is converted into a column-wise attention vector

$$A \in \mathbb{R}^{1 \times W} \tag{4}$$

is computed by averaging the mask probabilities along the height dimension:

$$A_i = \frac{1}{H} \sum_{j=1}^{H} M_{j,i}. \tag{5}$$

The contextual features are then aggregated as

$$V_{\text{global}} = \frac{\sum_{i=1}^{W} A_i \cdot F_{\text{ctx},i}}{\sum_{i=1}^{W} A_i + \epsilon}, \tag{6}$$

where $F_{\text{ctx},i}$ denotes the feature column at position i, and ϵ is used for numerical stability. The resulting vector is fed into an MLP to regress intrinsic text color. Since the mask is used as soft attention rather than a hard selector, moderate mask errors only adjust feature weights instead of discarding image evidence.

4.5 Loss Functions and Training Strategy

The overall optimization objective is defined as

$$L_{\text{total}} = L_{\text{mask}} + \lambda L_{\text{color}}, \tag{7}$$

where λ balances the segmentation and color regression tasks.

Spatially Weighted Mask Loss To reduce ambiguous supervision near anti-aliased boundaries, a spatial weight map

$$W \in \mathbb{R}^{H \times W} \tag{8}$$

is constructed from the ground-truth mask, assigning higher weights to skeleton/background pixels and lower weights to boundary pixels.

The mask loss is composed of a weighted binary cross-entropy loss and a Dice loss:

$$L_{\text{wbce}} = -\frac{1}{N} \sum_{i,j} W_{i,j} \left[y_{i,j} \log \hat{y}_{i,j} + (1 - y_{i,j}) \log(1 - \hat{y}_{i,j}) \right], \tag{9}$$

$$L_{\text{dice}} = 1 - \frac{2 \sum_{i,j} y_{i,j} \hat{y}_{i,j} + \epsilon}{\sum_{i,j} y_{i,j} + \sum_{i,j} \hat{y}_{i,j} + \epsilon}, \tag{10}$$

$$L_{\text{mask}} = L_{\text{wbce}} + L_{\text{dice}}, \tag{11}$$

where $y_{i,j}$ and $\hat{y}_{i,j}$ denote ground-truth and predicted mask values, and N is the number of pixels. The skeleton acts as a soft structural prior rather than a hard geometric constraint.

Color Regression Loss The color regression branch is supervised using an L_1 loss in Lab space:

$$L_{\text{color}} = \| C_{\text{pred}} - C_{\text{gt}} \|_1, \tag{12}$$

which encourages robust estimation of the intrinsic text color. Predicted colors are converted to RGB only for visualization and RGB-threshold accuracy.

Two-Stage Curriculum Learning Training is performed in two stages. Stage 1 uses uniform weights ($w_{\text{low}} = w_{\text{high}} = 1.0$) for stable morphology learning. Stage 2 reduces boundary weights ($w_{\text{low}} = 0.2$, $w_{\text{high}} = 1.0$), using an eroded center-aware mask to emphasize stroke interiors.

5 Dataset Construction

We construct a hybrid dataset of 152,842 text-strip images, with 140,703 samples for training and 12,139 for validation. It combines rule-based synthetic rendering, controlled real screenshots, and high-quality AI-generated text. For synthetic samples, intrinsic colors are recorded directly from rendering parameters. For real screenshots, text is created in presentation slides with pre-specified RGB colors before screen rendering and capture, so the ground truth is independent of any skeleton-based extraction.

For AI-generated text, exact rendering parameters are unavailable. We therefore obtain candidate reference colors using an independent offline pipeline: text erasure, difference-based foreground masking, morphological cleanup, and high-confidence interior-stroke averaging. Candidate labels are manually verified, and ambiguous samples such as gradients, multi-color text, transparency, or severe artifacts are removed. Non-synthetic samples with proxy labels are used only as lower-weight weak supervision to reduce the domain gap. We plan to release the generated dataset and annotation pipeline to support reproducibility.

6 Experiments

6.1 Experimental Setup

Datasets To comprehensively evaluate the proposed reliability-aware text color estimation method, experiments are conducted on a hybrid dataset consisting of synthetic and non-synthetic samples.

Training Set. The model is trained on the hybrid training split described above, including synthetic text strips and collected non-synthetic samples. The synthetic source, denoted as *StripeDataset*, includes diverse font styles, background textures, and text color combinations. The non-synthetic training subsets are incorporated to reduce the domain gap between synthetic and real rendering distributions. All training images are resized to a fixed height of 48 pixels, while the width is scaled proportionally and padded to preserve the aspect ratio.

Test Sets. Four benchmark datasets are used for evaluation:

- **Synthesize Clean:** clean synthetic samples without degradation, representing ideal rendering conditions.
- **Synthesize Degraded:** synthetic samples with blur, noise, compression, and resolution degradation.
- **Real Screenshots:** text regions cropped from screen captures rendered with sub-pixel anti-aliasing. The text is created in presentation slides with pre-specified RGB colors, and these assigned colors are used as ground truth. This subset represents the most challenging benchmark due to strong color fringing and foreground-background mixing.

- **AI-Generated Text:** a held-out benchmark containing approximately 8000 text images produced by high-quality generative models. Since exact rendering parameters are unavailable for generated images, candidate reference colors are extracted from high-confidence interior stroke pixels obtained by an offline text-erasure and difference-based masking pipeline, followed by manual verification. This subset represents a different rendering regime from **Real Screenshots**, with artifacts mainly arising from generation and anti-aliased boundary formation rather than screen-capture-induced subpixel color fringing. It is used to evaluate generalization across different document generation pipelines.

Label independence and verification. The four test benchmarks are kept separate from the training data at the source-image level. For **Synthesize Clean**, **Synthesize Degraded**, and **Real Screenshots**, the ground-truth colors are known from the RGB values specified before rendering and are therefore independent of any skeleton-based extraction procedure. For **AI-Generated Text**, the offline text-erasure and difference-based masking pipeline is used only to generate candidate reference labels, which are manually verified before evaluation. No test image or test-derived label is used for training, hyperparameter selection, or curriculum design. All compared methods are evaluated against the same reference labels, reducing the possibility that the evaluation is optimized toward the proposed method.

Implementation Details All experiments are implemented using the PyTorch framework.

A customized ResNet-style backbone [9] is adopted, consisting of four stages with 32, 64, 128, and 256 channels, respectively. After feature extraction, a global context aggregation module composed of average pooling and one-dimensional convolution (hidden dimension = 256) is applied to model long-range color dependencies along the text line.

The network contains two task-specific prediction heads:

- a **Mask Head** for pixel-level text region prediction, and
- a **Color Head** for mask-guided global color regression.

The overall training objective is defined as:

$$L = L_{\text{mask}} + \lambda L_{\text{color}}, \tag{13}$$

where L_{mask} is a weighted combination of binary cross-entropy loss and Dice loss, and L_{color} is an L_1 regression loss computed in the Lab color space. The weighting factor is set to $\lambda = 15.0$, and the mask loss weight is set to 1.0.

The network is optimized using Adam with an initial learning rate of 1×10^{-3} for 100 epochs. Training is performed on four NVIDIA GPUs with a batch size of 256. The checkpoint with the lowest validation CIEDE2000 score is selected for final evaluation. All images are resized to a fixed height of 48 pixels while preserving the aspect ratio, and variable-width samples are padded within each mini-batch. Training uses Lab-space color regression. Additional online synthetic generation is disabled for the benchmark experiments, while screen-rendering perturbation is enabled with probability 0.3.

We apply standard data augmentation during training, including mild rotation/affine perturbations, blur, and Gaussian noise, with probabilities between 0.2 and 0.3.

Training and validation splits. The benchmark training run uses four training sources: **Synthesize Clean**, **Synthesize Degraded**, **Real Screenshots**, and AI-generated text samples. The corresponding held-out validation sources are used only for model selection and reporting. No sample from the held-out validation or test splits is used during training.

It should be noted that this work focuses on intrinsic solid text color estimation, while gradient-colored or artistic text effects are beyond the scope of this study.

6.2 Evaluation Metrics

We report CIEDE2000 (ΔE_{00}), which measures perceptual color difference in the CIELAB space, together with MAE, RMSE, and thresholded accuracy. The MAE and RMSE in Lab space are defined as:

$$\mathrm{MAE} = \frac{1}{N}\sum_{i=1}^{N} |y_i^{\mathrm{pred}} - y_i^{\mathrm{gt}}|, \quad \mathrm{RMSE} = \sqrt{\frac{1}{N}\sum_{i=1}^{N}(y_i^{\mathrm{pred}} - y_i^{\mathrm{gt}})^2}, \tag{14}$$

where lower ΔE_{00}, MAE, and RMSE indicate better estimation. We further report Acc A/B/C, defined as the percentage of samples whose maximum per-channel RGB error is below 5, 10, and 15, respectively.

6.3 Quantitative Results

We evaluate text color estimation across four representative benchmarks, covering synthetic rendering, synthetic degradation, screen-rendered content, and AI-generated documents. Across all experiments, we aim to validate the central hypothesis that text color estimation benefits from explicitly modeling reliability heterogeneity, where structurally stable regions contribute more reliable supervision than boundary pixels.

Table 1 reports performance in terms of CIEDE2000 (ΔE_{00}), MAE and RMSE in Lab space, and three accuracy metrics based on the maximum per-channel RGB difference (Acc A/B/C). All results are averaged over the corresponding test splits.

Synthetic benchmarks. On clean synthetic data, the proposed method achieves low error (ΔE_{00}=1.61), indicating accurate recovery of intrinsic text color under controlled rendering conditions. When evaluated under synthetic

Table 1. Quantitative results of text color estimation on different datasets. Lower ΔE_{00}, MAE, and RMSE are better; higher Acc A/B/C are better. All numeric entries are rounded to two decimal places.

Dataset	ΔE_{00} ↓	Lab Error ↓		Accuracy (%) ↑		
		MAE	RMSE	Acc A (<5)	Acc B (<10)	Acc C (<15)
Synthesize Clean	1.61	1.49	2.23	90.90	98.50	99.70
Synthesize Degraded	2.24	2.22	3.54	76.10	93.90	98.10
Real Screenshots	3.08	3.40	10.96	69.23	85.78	91.38
AI-Generated Text	1.34	0.95	1.70	97.87	99.69	99.95

degradation, performance decreases moderately (ΔE_{00}=2.24), yet remains stable, demonstrating robustness to noise, compression, and perturbations that simulate real-world distortions.

Screen-rendered documents. The most challenging setting is the Real Screenshots subset, where subpixel rendering introduces strong color fringing and foreground-background mixing effects. Despite these severe artifacts, the proposed method maintains competitive performance (ΔE_{00}=3.08), highlighting its ability to handle rendering-induced ambiguity in pixel-level observations.

AI-generated documents. On the held-out **AI-Generated Text** subset, the method achieves the lowest error among the evaluated benchmarks (ΔE_{00}=1.34). Unlike **Real Screenshots**, which are mainly affected by ClearType-style subpixel color fringing [18], **AI-Generated Text** represents a different artifact regime characterized by generation-related anti-aliasing, boundary smoothing, and synthesis inconsistencies. These results indicate that the proposed method remains stable across different document rendering pipelines.

Overall, the results across all benchmarks support the central claim of this work: explicitly modeling structural reliability improves robustness under rendering-induced ambiguity. In particular, the largest performance degradation occurs on the **Real Screenshots** subset, which exhibits the strongest subpixel fringing and boundary blending. This behavior is consistent with our problem formulation: when boundary ambiguity becomes dominant, naive global aggregation is more easily biased, whereas reliability-aware weighting remains comparatively stable.

Perceptual interpretation of ΔE_{00}. Fig. 3 complements Table 1 by showing representative predictions together with their ground-truth colors and corresponding ΔE_{00} scores. Smaller perceptual errors consistently align with closer visual agreement between the predicted and target text colors.

Qualitative results across rendering regimes. Fig. 4 shows representative examples from three practical settings: clean rasterized text, AI-generated content, and screen-captured text with subpixel artifacts. Despite large variations in rendering quality and boundary contamination, the proposed method remains visually consistent by focusing on structurally reliable interior regions.

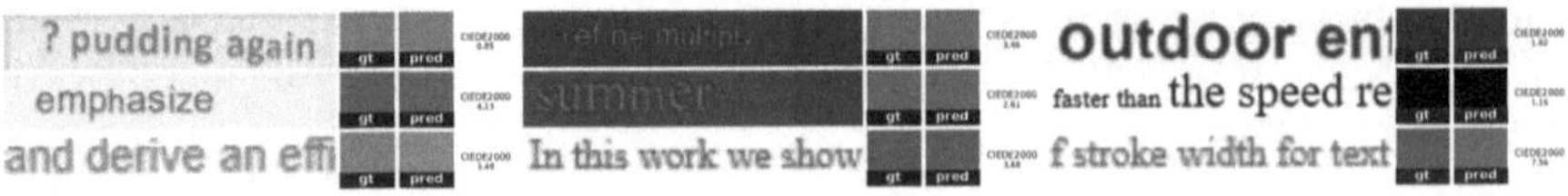

Fig. 3. Representative prediction examples with their corresponding ΔE_{00} scores. Lower perceptual error is associated with closer agreement between predicted and ground-truth text colors.

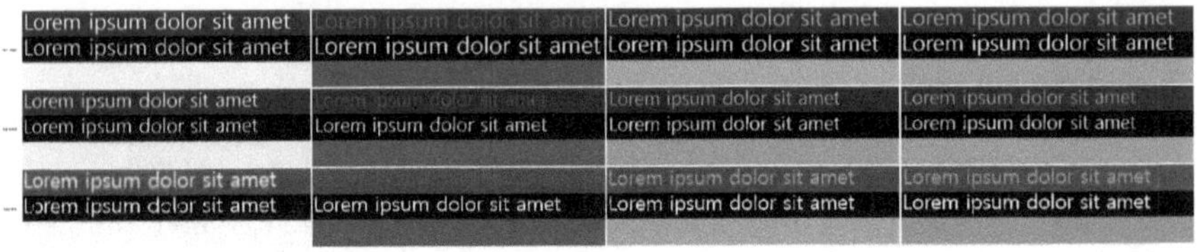

Fig. 4. Qualitative results on clean rasterized text, AI-generated content, and screen-captured text. The model remains stable across different rendering regimes and boundary ambiguities.

6.4 Comparison with Baselines

To test robustness beyond the degraded synthetic setting, we compare against heuristic and learned baselines on all four benchmarks: **Synthesize Clean**, **Synthesize Degraded**, **Real Screenshots**, and **AI-Generated Text**. The baselines cover statistical aggregation, clustering, morphology-based and skeleton-based heuristics, as well as two learned models: **ResNet18-Vanilla** [9] and a stronger **ResNet18-SegAware** baseline with joint foreground-mask prediction and mask-guided feature pooling. **ResNet18-SegAware** is our custom segmentation-aware ResNet18 baseline: it modifies ResNet18 to preserve denser features, predicts a foreground mask with a lightweight mask head, and uses the predicted mask for feature pooling before color regression. It is designed as a strong learned baseline, but does not use skeleton/center supervision, edge reliability weighting, or curriculum training.

Table 2 reports the complete comparison. On synthetic data, our method substantially outperforms the best heuristic baseline, especially under degradation (ΔE_{00}=2.24 vs. 10.28; Acc$<$ 5: 76.10% vs. 9.80%), showing that fixed aggregation rules remain sensitive to boundary corruption.

The same trend holds in real-world settings. On **Real Screenshots**, our method achieves the best perceptual error and threshold accuracy (ΔE_{00}=3.08, Acc$<$ 5 = 69.23%), where subpixel fringing strongly affects heuristic estimators. On **AI-Generated Text**, mask-derived heuristics can obtain low average error on clean high-contrast interiors, but our method gives the highest Acc$<$ 5 (97.87%), indicating more stable sample-level predictions.

Table 2. Complete baseline comparison across four datasets.

Type	Method	Syn. Clean		Syn. Deg.		Real Scr.		AI Text	
		ΔE_{00}	Acc	ΔE_{00}	Acc	ΔE_{00}	Acc	ΔE_{00}	Acc
Stat.	mean	11.32	1.70%	16.09	0.80%	23.86	0.00%	5.23	2.00%
	median	10.07	6.50%	15.27	1.50%	22.80	0.00%	2.40	45.09%
	trim. mean	10.76	2.90%	15.73	0.90%	23.40	0.00%	3.85	6.56%
Cluster	kmeans	8.34	9.40%	13.54	2.80%	20.88	0.00%	1.96	44.99%
	geo. median	9.98	8.10%	15.25	1.90%	22.84	0.00%	2.42	44.81%
	median cut	11.97	11.70%	16.90	4.60%	27.64	6.06%	6.78	36.47%
Morph.	core mean	6.37	32.10%	10.28	9.80%	17.05	0.00%	**1.25**	96.24%
	center-w.	10.46	3.80%	15.17	1.30%	21.62	0.00%	3.80	5.08%
Skel.	skel. mean	10.36	6.00%	15.40	2.30%	19.00	0.00%	2.41	17.03%
	skel. median	9.10	18.60%	14.46	6.60%	15.54	2.56%	1.30	84.40%
CNN	R18-Vanilla	8.29	4.10%	7.75	4.50%	3.71	45.69%	3.44	23.60%
	R18-SegAware	2.88	52.50%	3.29	45.50%	3.32	65.76%	2.16	87.70%
Ours	**Ours**	**1.61**	**90.90%**	**2.24**	**76.10%**	**3.08**	**69.23%**	1.34	**97.87%**

The learned baselines further clarify the contribution. **ResNet18-Vanilla** performs poorly in this setting because global regression does not explicitly identify reliable color evidence. **ResNet18-SegAware** improves with mask-aware pooling, but still trails our method on the main synthetic, degraded, and real screenshot settings, indicating that mask guidance alone is insufficient without reliability-aware treatment of edges, stroke interiors, and training progression (Fig. 5).

6.5 Ablation Study

To validate the effectiveness of each component in our reliability-aware design, we conduct controlled ablation studies on the **Synthesize Degraded** validation set (1,000 synthetic text strips with random degradation). The goal is to analyze how different design choices contribute to modeling structural reliability under boundary corruption and rendering artifacts.

We evaluate mean CIEDE2000 (ΔE_{00}) and Acc A (percentage of samples with maximum per-channel RGB error < 5). Lower ΔE_{00} and higher Acc A indicate better performance. The results are summarized in Table 3.

Effect of reliability modeling via mask supervision. We first study the role of explicit structural supervision. Removing the mask branch reduces the model to pure color regression, resulting in the worst performance (ΔE_{00}=2.78, Acc A=64.50%). This confirms that without explicit foreground localization, the model is highly affected by boundary contamination under degradation. Adding

Fig. 5. Qualitative comparison on a subset of the degraded validation set. Each row: original text strip, binary mask, and the predicted color from each baseline (statistical, clustering, morphology, skeleton) and our method (rightmost). Our method matches the perceived text color; baselines often show background or edge leakage (e.g., muddy or shifted colors).

joint mask supervision improves performance (ΔE_{00}=2.78 → 2.66), demonstrating that structural cues provide reliable guidance for isolating intrinsic text regions.

Effect of reliability-aware feature aggregation. Next, we analyze the impact of explicit mask-guided attention. Although mask prediction alone provides structural information, without explicit feature re-weighting the model still aggregates background-corrupted features, leading to suboptimal performance (ΔE_{00}=2.66). Incorporating explicit mask-guided attention reduces background interference and improves robustness, indicating that structural supervision must be coupled with feature-level reliability weighting.

Effect of progressive two-stage optimization. Finally, we examine the training strategy. Using only Stage 1 or Stage 2 leads to inferior performance (ΔE_{00}=2.46 and 2.37, respectively), suggesting that neither coarse alignment nor fine regression alone is sufficient under degradation. The proposed two-stage strategy achieves the best performance (ΔE_{00}=2.24), indicating that progressive refinement stabilizes optimization and improves robustness under noisy supervision.

Overall, the ablation results confirm that all three components—structural supervision, reliability-aware attention, and progressive training—are complementary and jointly necessary for robust text color estimation under degraded conditions.

Table 3. Controlled comparison on the **Synthesize Degraded** validation set (1,000 samples). ΔE_{00} rounded to two decimal places.

Configuration	ΔE_{00} ↓	Acc A (%) ↑	Remark
No mask branch	2.78	64.50	Color regression only.
+ Mask branch, no explicit attention	2.66	68.80	No explicit mask attention.
+ Explicit mask attention, Stage 1 only	2.46	70.70	After first-stage training.
+ Explicit mask attention, Stage 2 only	2.37	75.80	Fine color without Stage 1 init.
Ours (full)	**2.24**	**76.10**	Explicit attention + Stage 1→Stage 2.

7 Conclusion

In this work, we revisit the problem of text color estimation in modern document images and identify a fundamental limitation in conventional pixel-level modeling: intrinsic text color becomes ambiguous under contemporary rendering mechanisms such as anti-aliasing and subpixel rendering. This leads to unreliable supervision at stroke boundaries, making the task inherently ill-posed under standard assumptions.

To address this issue, we propose a reliability-aware formulation that explicitly distinguishes between structurally stable interior regions and ambiguous boundary pixels. Based on this insight, we develop a skeleton-guided, edge-aware framework that jointly performs text segmentation and color regression. The model emphasizes reliable stroke interiors while suppressing boundary-induced color contamination, and is further enhanced by a curriculum-based training strategy that progressively reduces the influence of unreliable supervision.

Extensive experiments on synthetic, screen-captured, and AI-generated benchmarks demonstrate that the proposed method consistently improves both accuracy and robustness across diverse rendering conditions, particularly in challenging screen-rendered scenarios with strong anti-aliasing artifacts.

More broadly, this work suggests that text appearance modeling in modern document images should move beyond uniform pixel assumptions toward reliability-aware representations that account for rendering-induced ambiguity.

Limitations and scope. This study focuses on dominant intrinsic *solid* text color estimation as a fundamental and verifiable setting. Gradient-filled, outline-styled, multi-colored, and heavily stylized text require richer appearance representations than a single global RGB value, such as spatially varying color fields, separate fill/outline colors, or compositing-aware parameters. For AI-generated text, exact rendering parameters are unavailable; we therefore use high-confidence interior-stroke reference labels with manual verification, while the synthetic and controlled real screenshot benchmarks use independently specified RGB colors as ground truth. These limitations define the intended scope of the present work and motivate future extensions toward richer appearance modeling and more strictly annotated non-synthetic benchmarks.

Failure cases. The method may fail when the single solid-color assumption is violated, such as gradient-filled, outline-styled, transparent, shadowed, or heavily stylized text. It may also be less reliable for extremely thin strokes or very low-resolution text, where interior stroke evidence is too small. Although the mask is used as soft attention, severe mask errors or AI-generation artifacts may still bias color estimation.

Future work will extend the reliability-aware formulation from single solid-color estimation to richer text appearance models, including gradients, fill-stroke decomposition, transparency, and multi-color effects, as well as integrating it into end-to-end editable document reconstruction frameworks.

Disclosure of Interests. The authors have no competing interests to declare that are relevant to the content of this article.

References

1. Bias, R.G., Larson, K., Huang, S.C., Aumer-Ryan, P.R., Montesclaros, C.: An exploratory study of visual and psychological correlates of preference for onscreen subpixel-rendered text. J. Am. Soc. Inform. Sci. Technol. **61**(4), 745–757 (2010)
2. Chen, C., Lin, L., Chen, Y., Li, B., Zeng, J., Huang, J.: Cma: a chromaticity map adapter for robust detection of screen-recapture document images. In: Proceedings of the IEEE/CVF Conference on Computer Vision and Pattern Recognition, pp. 15577–15586 (2024)
3. Chen, K., Yin, F., Liu, C.L.: Effective candidate component extraction for text localization in born-digital images by combining text contours and stroke interior regions. In: 12th IAPR Workshop on Document Analysis Systems (DAS), pp. 352–357. IEEE (2016)
4. Einsele-Aazami, F.: Recognition of Ultra Low Resolution, Anti-Aliased Text with Small Font Sizes. Ph.D. thesis, Université de Fribourg, Fribourg, Switzerland (Jul 2008). https://folia.unifr.ch/unifr/documents/300924, doctoral thesis. Degree supervisor: Rolf Ingold. Date: 03 Jul 2008. Pagination: X, 150 p
5. Gao, Y., Lin, Z., Liu, C., Zhou, M., Ge, T., Zheng, B., Xie, H.: Postermaker: towards high-quality product poster generation with accurate text rendering. In: Proceedings of the Computer Vision and Pattern Recognition Conference, pp. 8083–8093 (2025)
6. Garcia-Lamont, F., Cervantes, J., López, A., Rodriguez, L.: Segmentation of images by color features: a survey. Neurocomputing **292**, 1–27 (2018)
7. Gleichman, S., Ophir, B., Geva, A., Marder, M., Barkan, E., Packer, E.: Detection and segmentation of antialiased text in screen images. In: 2011 International Conference on Document Analysis and Recognition, pp. 424–428. IEEE (2011)
8. Guan, S., et al.: Prep-ocr: a complete pipeline for document image restoration and enhanced ocr accuracy. In: Proceedings of the 63rd Annual Meeting of the Association for Computational Linguistics (Volume 1: Long Papers), pp. 15413–15425 (2025)
9. He, K., Zhang, X., Ren, S., Sun, J.: Deep residual learning for image recognition. IEEE (2016)
10. Kim, B., et al.: Zim: zero-shot image matting for anything. In: Proceedings of the IEEE/CVF International Conference on Computer Vision, pp. 23828–23838 (2025)

11. Koizumi, S., Horiuchi, T.: Tinting component extraction from gray-scale image for color estimation. In: Proceedings of the IAPR Conference on Machine Vision Applications (IAPR MVA 2002), December 11-13, 2002, Nara-ken New Public Hall, Nara, Japan. pp. 180–183 (2002). http://b2.cvl.iis.u-tokyo.ac.jp/mva/proceedings/CommemorativeDVD/2002/papers/2002180.pdf
12. Kubota, Y., Yoshida, S., Inami, M.: Apparent color picker: color prediction model to extract apparent color in photos. Front. Signal Process. **3**, 1133210 (2023)
13. Kumar, R., Jinka, J.S., Sarvadevabhatla, R.K.: Textar: textual attribute recognition in multi-domain and multi-lingual document images. In: International Conference on Document Analysis and Recognition, pp. 280–296. Springer (2025)
14. Lienhart, R.W., Wernicke, A.: Estimating text color and segmentation of images (Oct 29 2002), uS Patent 6,473,522
15. Lin, Z., et al.: Mildedit: reasoning-based multi-layer design document editing. arXiv preprint arXiv:2601.04589 (2026)
16. Muratbekova, M., et al.: Color models in image processing: A review and experimental comparison (2025). arXiv:2510.00584 arXiv preprint
17. Peng, Y., et al.: Bizgen: advancing article-level visual text rendering for infographics generation. In: Proceedings of the Computer Vision and Pattern Recognition Conference, pp. 23615–23624 (2025)
18. Sheedy, J., Tai, Y.C., Subbaram, M., Gowrisankaran, S., Hayes, J.: Cleartype subpixel text rendering: preference, legibility and reading performance. Displays **29**(2), 138–151 (2008)
19. Wachenfeld, S., Fleischer, S., Klein, H.U., Jiang, X.: Segmentation of very low resolution screen-rendered text. In: Ninth International Conference on Document Analysis and Recognition (ICDAR 2007), vol. 2, pp. 1153–1157. IEEE (2007)
20. Wachenfeld, S., Klein, H.U., Jiang, X.: Annotated databases for the recognition of screen-rendered text. In: Ninth International Conference on Document Analysis and Recognition (ICDAR 2007), vol. 1, pp. 272–276. IEEE (2007)
21. Xu, X., Zhou, J., Zhang, H.: Screen-rendered text images recognition using a deep residual network based segmentation-free method. In: 24th International Conference on Pattern Recognition (ICPR), pp. 2741–2746. IEEE (2018)
22. Ye, M., et al.: Hi-sam: marrying segment anything model for hierarchical text segmentation. IEEE Trans. Pattern Anal. Mach. Intell. **47**(3), 1431–1447 (2025)
23. Yin, S., et al.: Qwen-image-layered: Towards inherent editability via layer decomposition. arXiv preprint arXiv:2512.15603 (2025)
24. Zhang, J., Peng, D., Liu, C., Zhang, P., Jin, L.: Docres: A generalist model toward unifying document image restoration tasks. In: Proceedings of the IEEE/CVF Conference on Computer Vision and Pattern Recognition, pp. 15654–15664 (2024)
25. Zhang, L., He, Y., Zhang, Q., Liu, Z., Zhang, X., Xiao, C.: Document image shadow removal guided by color-aware background. In: Proceedings of the IEEE/CVF Conference on Computer Vision and Pattern Recognition, pp. 1818–1827 (2023)

Decoupled Prototype Memory Mechanism for Complex Table Recognition

Di Wu[1,2], Liangrui Peng[1,2(✉)], and Tianqi Zhao[1,2]

[1] Department of Electronic Engineering, Tsinghua University, Beijing, China
{d-wu24,ztq24}@mails.tsinghua.edu.cn, penglr@tsinghua.edu.cn
[2] Beijing National Research Center for Information Science and Technology, Tsinghua University, Beijing, China

Abstract. In the mobile Internet era, table recognition is crucial for information extraction in data-intensive applications. Complex table images captured with mobile devices under varying viewpoints, lighting, and noise make reliable structure and content reconstruction highly challenging. Existing methods often separate detection, structure recognition, and content recognition into independent stages, which suffer from inconsistent supervision and error accumulation. Recently, multi-modal models have provided promising end-to-end solutions, yet they still struggle with long-range layout dependency modeling and contextual alignment in complex image scenarios. A decoupled prototype memory mechanism is proposed for multi-modal table recognition. Placed between the visual encoder and the language decoder, the module enhances encoder features with reusable table prototypes before markup generation. It decouples prototype assignment from feature retrieval: key memory is biased toward stable prototypes for matching and routing, while value memory retrieves more adaptive feature representations for enhancing the current input. A table structure-aware positional encoding further decomposes table positions into learnable row and column embeddings to model grid layouts. The memory module adopts a long–short memory design, consolidating reusable prototypes during training and retrieving input-specific prototypes during inference. We also introduce a synthesis method for complex-layout table data with rich content, position, LaTeX, and HTML annotations. Experiments on public and self-built datasets demonstrate the effectiveness of the proposed method.

Keywords: Table Recognition · Image-to-markup · Prototype Memory

1 Introduction

Tables are structured carriers that organize data in a two-dimensional grid. By aligning semantics across rows, columns, and headers, tables present and support efficient retrieval of information. Tables appear widely in scientific papers, financial reports, business documents, and web pages [16]. Converting unstructured

F. Shafait et al. (Eds.): DAS 2026, LNCS 16851, pp. 354–370, 2027.
https://doi.org/10.1007/978-3-032-36207-0_21

table images into machine-readable structured representations is a core task in document intelligence. Such conversion is important for database queries, spreadsheet manipulation, web-table question answering, and information extraction from table images. Current table-recognition work often focuses on subtasks such as table detection, table structure recognition, and table content recognition. In practice, many applications require directly converting a table image into a renderable markup language. This need motivates the end-to-end image-to-markup task [5], which restores structured markup from a table image and thus subsumes the above subtasks. Image-to-Markup avoids intermediate subtask pipelines and explicit OCR modules, simplifying the recognition flow [13]. Models such as the EDD model [32] proposed in PubTabNet have shown progress on this task.

However, current research still has limitations. Current end-to-end models also do not fully exploit the strong structural priors of tables. Tables have regular grid layouts, clear row/column alignment, and rowspan/colspan properties. Mainstream encoder–decoder models treat table images as generic natural images and rely on self-attention to learn spatial relations implicitly. These models lack explicit modeling of row–column positions and cell boundaries. Some methods add hard-coded structural priors; this can boost accuracy but limits generalization, since new table types require new priors.

This work introduces a memory mechanism into table recognition. A read–write external memory stores reusable prototype features. Memory networks have been applied to visual recognition before, but existing designs are not tailored to tables. Tables exhibit distinctive traits: a strict two-dimensional row–column topology, clear hierarchical scales, and highly reusable semantic prototypes.

A major issue is the lack of high-quality datasets suited for end-to-end training. Public datasets often provide structure or content labels, but these labels are merely aggregates of traditional subtask annotations and do not include complete, directly renderable markup. Moreover, many dataset images are clean renderings cropped from PDF or Word files. They lack simulations of common degradations observed in photographed documents, such as lighting variations, perspective distortions, and blurring. As a result, models trained on these datasets generalize poorly to real-world captures. To address the limitations above, this paper presents a large-scale end-to-end table-recognition dataset, Table2Markup, and a decoupled prototype memory module.

The main contributions are summarized as follows:

(1) A structure- and content-decoupled prototype memory mechanism is proposed for complex table recognition in multimodal models. Key memory stores stable structural prototypes for matching and routing, while value memory returns content-aware prototype features for representation enhancement. Together with table structure-aware positional encoding, this design explicitly models row–column layouts and improves recognition of complex tables.
(2) A Slot-Attention-based prototype consolidation strategy is introduced for memory updating. During training, Slot Attention periodically aggregates

reusable table prototypes and writes them into the memory bank as corpus-level long-range knowledge. During inference, visual tokens retrieve input-specific memory features from the learned memory bank, improving convergence and robustness across diverse table types.

(3) To support the training process of the above table recognition methods, a method to synthesize a highly realistic dataset of complex-layout tables, Table2Markup, is proposed. The annotations of Table2Markup include table-line coordinates and detailed cell information. The detailed cell information covers cell text, start row and column indices, rowspan/colspan information, and absolute cell coordinates. Table2Markup supports both LaTeX and HTML markup.

On the PubTabNet benchmark, the proposed method achieves a TEDS score of 91.1%, outperforming existing methods such as EDD. On the proposed Table2Markup dataset, the model attains TEDS scores of 93.3% for HTML generation and 97.6% for LaTeX generation, demonstrating the method's effectiveness and generalization.

2 Related Work

2.1 Table Recognition

Table-recognition tasks are usually divided into three types: (i) Table Detection (TD), which locates table bounding boxes in documents or images; (ii) Table Structure Recognition (TSR), which recovers table layout; and (iii) Table Content Recognition (TCR), which extracts table structure and cell contents. This staged paradigm is easy to implement in practice but causes error accumulation between stages and prevents joint end-to-end optimization. Early methods [23,24] follow this staged design. The EDD model [32] proposed by PubTabNet is an early example of an end-to-end table recognition approach, using two decoders to generate table structure and content tokens separately. CascadeTabNet [19] developed an integrated detection-and-segmentation pipeline based on a cascaded mask-RCNN and a high-resolution network. TableFormer [17] extended EDD by adding a cell-object detection decoder to avoid training a custom OCR decoder. More general encoder–decoder models [12,27], such as OmniParser, have been proposed to unify multiple table recognition tasks into a single decoder, simplifying architecture and promoting knowledge sharing across tasks. However, these end-to-end models do not fully exploit the strong structural priors of tables. We therefore propose a novel table recognition method with a decoupled prototype memory mechanism, which leverages structural priors of tables without being constrained by fixed layouts, while enabling fully end-to-end table recognition.

2.2 Deep Learning with Memory Mechanism

Memory mechanisms in deep learning were originally proposed to enhance neural networks with long-term storage and retrieval capabilities. By introducing a differentiable external memory, they allow models to retain key information across samples or over time, addressing the limitations of conventional networks in modeling long dependencies [9,22]. Early representative works include Memory-Augmented Neural Networks [22] and Neural Turing Machines [9]. The former introduced an explicit read–write mechanism, while the latter adopted differentiable addressing for end-to-end optimization. In computer vision, memory mechanisms have been applied to video understanding for maintaining temporal consistency [10]. Matching Networks [26] and Prototypical Networks [25] leverage memory prototypes to improve few-shot learning. Learnable external memory offers advantages in capturing long-term dependencies, cross-sample relationships, and high-level semantic patterns. However, most existing memory mechanisms are designed for temporal modeling or instance-level reasoning and lack structures tailored to highly organized visual objects such as tables. To address this, we propose a prototype memory mechanism specifically designed for tables. It can be integrated into a single-decoder end-to-end recognition model, explicitly modeling the row–column structure and semantic alignment of tables. This design maintains end-to-end trainability while enhancing structural perception and generalization for complex table layouts.

2.3 Public Table Datasets

Existing public table datasets are generally constructed separately for TD, TSR, and TCR tasks. Early datasets such as Marmot [7] and ICDAR 2013 [8] provide high-quality annotations but are small and thus insufficient for training modern data-driven models. TableBank [15] adopts a weakly supervised pipeline to generate table annotations from Word and LaTeX sources, producing a large dataset for TD and TSR. General layout datasets, such as PubLayNet [33], supply table-region labels but typically lack cell-level text annotations. PubTabNet [32], SciTSR [4], and FinTabNet [31] annotate both structure and content and extend to more complex table formats; however, they mainly focus on clean rendered tables. The recent TabRecSet [30] dataset provides end-to-end annotated in-the-wild document images, but its annotations remain merely a combination of TD, TSR, and TCR tasks. The conventional staged decomposition of tasks used by current datasets is not well-suited for training end-to-end image-to-markup models. The proposed dataset offers three advantages: (i) In addition to standard TR labels, it provides directly renderable structured annotations that support end-to-end training; (ii) It includes both LaTeX and HTML markup to cover academic and web use cases and to support markup conversion; (iii) It spans both clean rendered and realistic synthetic table images, increasing its applicability.

3 Methodology

3.1 Network Architecture

This work proposes a prototype memory mechanism tailored for tables. Figure 1 shows the architecture of TabDec. The input table image is first encoded by a visual encoder into feature maps at three scales: 16×16, 32×32, and 64×64. Each scale gets its own table-aware positional encoding module. Row and column embeddings are learned and summed to produce an explicit row–column index encoding for every spatial location. These position-aware multi-scale features feed into the corresponding prototype memory modules. Each memory module holds a prototype memory bank initialized by Slot Attention. Slot Attention clusters input features into a fixed number of prototype vectors. During training, the memory bank is dynamically maintained by a momentum-based update rule, and a sparse retrieval mechanism reads the prototypes most relevant to the input features. Retrieved memory features are added to the original features. A linear projection layer maps the fused features to the embedding space of the language model. Finally, the projected visual embeddings are inserted as visual tokens into the input sequence of Qwen-0.5B [2].

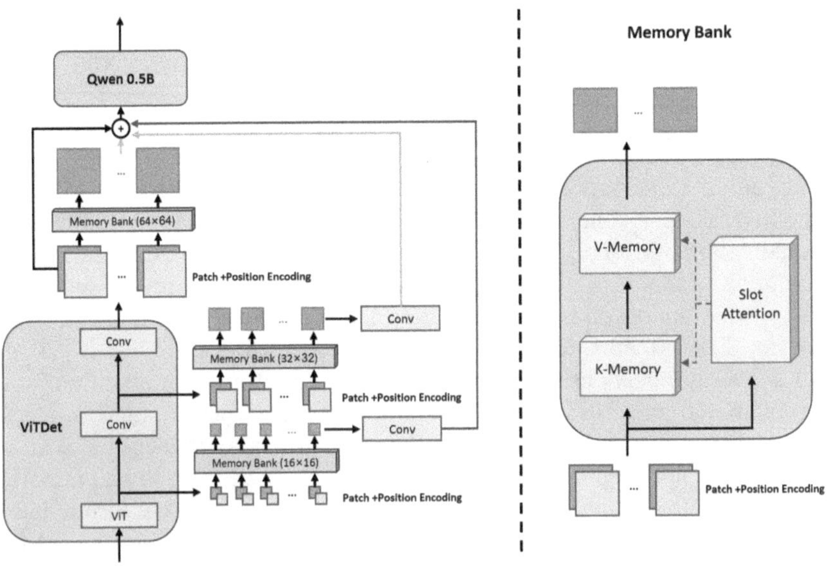

Fig. 1. TabDec: multi-scale table features encoded with row–column positional embeddings are sent to Slot-Attention–initialized prototype memory banks, dynamically maintained; memories sparsely retrieve relevant prototypes and fuse with original features.

3.2 Decoupled Prototype Memory

Modeling Prototype Memory for Tables. The prototype memory mechanism for document recognition treats the generation of local visual tokens as a generative process governed by a set of latent prototypes. Let the visual feature of a patch be $q \in \mathbf{R}^D$. Assume a discrete latent variable $z \in 1, \ldots, M$ indexes the prototypes, with prior $p(z = m) = \pi_m$. Conditioned on a prototype, q follows a Gaussian likelihood:

$$p(q \mid z = m) = \mathcal{N}(q; \mu_m, \sigma_q^2 I) \tag{1}$$

where μ_m denotes the centroid vector of prototype m. Given an observation q, the posterior over z is

$$p(z = m \mid q) = \frac{\pi_m \exp\left(-\frac{1}{2\sigma_q^2}|q - \mu_m|^2\right)}{\sum_j \pi_j \exp\left(-\frac{1}{2\sigma_q^2}|q - \mu_j|^2\right)} \tag{2}$$

Dropping constants independent of m and rewriting in a dot-product form gives a softmax-dot posterior:

$$p(z = m \mid q) = \text{softmax}_m\left(q^\top k_m + b_m\right) \tag{3}$$

Thus, common cross-attention weights admit a probabilistic interpretation as a learnable approximation of the posterior probability that the observation belongs to each prototype.

The attention value-weighted sum:

$$\mathbf{E}[y \mid q] = \sum_m p(z = m \mid q) v_m \tag{4}$$

corresponds to the posterior predictive. Each prototype's instantiation vector is denoted by v_m.

Compared with general documents, tables have stronger row–column structural priors, while the instance content within cells is more variable. To better store the structure template in memory, the original single prototype vector μ_m is explicitly disentangled into two parts for representation and update: a set of keys $K = k_m$ used for posterior assignment, and a set of values $V = v_m$ used for posterior prediction. Concretely, the read stage keeps the probabilistic semantics but replaces the original expression with

$$\alpha_m(q) = \text{softmax}_m\left(q^\top k_m + b_m\right) \tag{5}$$

$$\mathrm{r}(q) = \sum_m \alpha(q) v_m \tag{6}$$

where k_m and v_m are no longer equivalent.

Asymmetric Memory Updates. For the decoupled key and value memories, we design separate initialization and update strategies. Key memory is initialized with low variance to encourage memory slots to converge to stable common patterns. Value memory is initialized with high variance to increase slot diversity and better capture the varied fine-grained details in tables. For updates, key memory uses a slow exponential moving average (EMA) to keep its features relatively stable. Value memory uses a fast EMA to adapt more quickly to changing features.

3.3 Slot-Based Long-Range Memory Update

Although table layouts and topics vary, their basic structural components, such as captions, headers, data cells, and separators, are limited and reusable. We assume any table image can be composed from a finite set of structural components:

$$\mathbf{I} = \mathcal{F}\left(\bigcup_{i=1}^{N} \phi(\mathbf{p}_i, c_{\pi(i)})\right) \tag{7}$$

Here $\mathbf{p}_i$ denotes the i-th local region, $c_{\pi(i)}$ its component type, $\phi(\cdot)$ describes the instantiated form of that component at position $\mathbf{p}_i$, and $\mathcal{F}(\cdot)$ is the composition function. Style changes mainly affect $\phi(\cdot)$, while the component set $\mathcal{C} = c_1, \ldots, c_K$ remains relatively stable. Considering this model, table recognition can be seen as learning a component dictionary $\mathbf{C}$ and an assignment matrix $\mathbf{A}$ that explain all local features:

$$\min_{\mathbf{C},\mathbf{A}} |\mathbf{X} - \mathbf{AC}| F^2 + \lambda R(\mathbf{A}) \tag{8}$$

Here $\mathbf{X} \in \mathbf{R}^{N \times D}$ is the feature matrix, $\mathbf{C} \in \mathbf{R}^{K \times D}$ is the component dictionary, and $R(\cdot)$ is a sparsity regularizer. This optimization can be viewed as a differentiable approximation to an expectation–maximization (EM) algorithm: the E-step computes responsibilities of local features for components, and the M-step updates component centers.

Slot Attention provides an end-to-end trainable, differentiable decomposition for this optimization. Its iterative process can be interpreted as an approximate solver for the following objective:

$$\mathbf{C}^{(t+1)}, \mathbf{A}^{(t+1)} = \arg\min_{\mathbf{C},\mathbf{A}} |\mathbf{X} - \mathbf{AC}|_F^2 + \tau \cdot \mathrm{KL}(\mathbf{A}|\mathbf{A}^{(t)}) \tag{9}$$

The KL-divergence term stabilizes iterations, and the temperature parameter τ controls assignment sharpness. After several iterations, each slot converges to a stable structural prototype that represents a reusable component class in tables. During early training, the clustering centers learned by Slot Attention are used to initialize memory slots. Slot Attention periodically updates the memory module as a long-range memory unit.

3.4 Table Structure-Aware Positional Encoding

To precisely capture the inherent row–column grid of tables, we use a table-aware positional encoding (TPE). This method decomposes positional information into separate learnable row and column encodings. Given the feature map $\mathbf{F} \in \mathbf{R}^{H \times W \times D}$, the encoding is added directly to the original feature to obtain the enhanced feature:

$$\mathbf{F}'_{i,j} = \mathbf{F}_{i,j} + \mathbf{P}_{i,j} \tag{10}$$

The positional encoding is defined as:

$$\mathbf{P} = \mathbf{E}_{\mathrm{row}} \oplus \mathbf{E}_{\mathrm{col}} \tag{11}$$

Where the row encoding $\mathbf{E}_{\text{row}}$ and column encoding $\mathbf{E}_{\text{col}}$ are generated from parameter matrices $\mathbf{W}_{\text{row}} \in \mathbf{R}^{R_{\max} \times D}$, $\mathbf{W}_{\text{col}} \in \mathbf{R}^{C_{\max} \times D}$ by table lookup and broadcast addition: $\mathbf{E}_{\text{row}}(i,j) = \mathbf{W}_{\text{row}}(i)$, $\mathbf{E}_{\text{col}}(i,j) = \mathbf{W}_{\text{col}}(j)$.

4 The Table2Markup Dataset

This section outlines our self-built dataset, Table2Markup, and the data collection and annotation pipeline. Existing table datasets still follow the traditional TD, TSR, and TCR task divisions and lack annotations that map tables to structured markup languages. Such table-to-markup annotations are essential for training image-to-markup models. Therefore, a method is proposed to build a table dataset from LaTeX sources, and the dataset Table2Markup in our work is constructed based on that method.

4.1 Dataset Overview

The Table2Markup dataset contains two parts: clean rendered table images and simulated photographic table images. Clean images are high-quality renderings produced directly from the LaTeX source. Simulated in-the-wild images are obtained by applying various degradations to the clean images. Table2Markup comprises approximately 31,000 table images, of which 4,479 are clean, and each clean image yields six augmented images. Sources cover multiple arXiv categories, including computer science (cs.), physics (physics.), and mathematics (math.). At the structural level, tables average 7.2 rows $\times$ 4.1 columns, and 14.0% of cells involve rowspan or colspan. Figure 2 shows the joint distribution of row and column counts.

The synthesis process can be used to construct larger, cross-domain table datasets.

4.2 Construction and Annotation Pipeline

LaTeX sources were downloaded in bulk from arXiv.org; table-related parts were extracted and rendered into table images. We developed a multi-level, noise-robust extraction strategy to handle the complexity and diversity of LaTeX tables in academic papers. A breadth-first traversal visits the document tree to find environments such as 'tabular', 'table', 'longtable', and 'subtable', and extracts complete source fragments. We recursively parse nested structures, handle conditional compilation, and extract tables from all possible branches. We also preserve table context and support multi-file tables. The raw LaTeX code is normalized and rendered using the native LaTeX toolchain [20] to ensure the rendered images match the visual presentation in the original papers.

As part of the dataset preparation, we built an automated annotation pipeline that combines LaTeX-source inference with rendered-image analysis. The pipeline produces high-precision annotations for table structure, cell content, and end-to-end LaTeX and HTML labels. End-to-end labels, cell contents, and relative cell positions are obtained directly from the parsed LaTeX

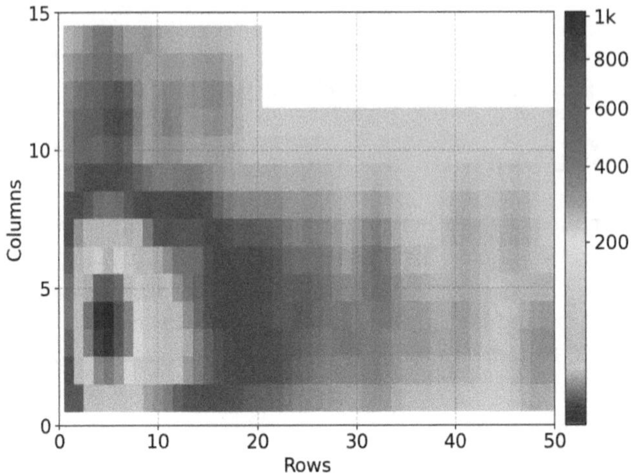

Fig. 2. Distribution of table counts across different row and column sizes in Table 2Markup.

source. To get absolute cell coordinates, we assign distinct background colors to cells, re-render the table, and detect the colored regions. Unlike standard image-segmentation or vision-model-based methods, these absolute coordinates are deterministic, derived from rendering rules. For Web use, we also provide HTML annotations for each table. We convert LaTeX table code to semantic HTML5 [11] table markup using Pandoc [6].

Figure 3(d) shows an example annotation from the Table 2Markup. Each image annotation contains: geometric coordinates of table lines and cell boxes; row and column indices with rowspan/colspan information; header vs. body classification; cell text content; and the corresponding LaTeX source used for rendering, as well as HTML structural annotations.

To improve model generalization in real scenes, we augmented rendered images to create a simulated in-the-wild photo dataset for end-to-end recognition. Random perspective transforms and small rotations simulate viewpoint tilt and camera shake. Brightness and contrast are adjusted, and random shadows and vignetting are simulated to emulate non-uniform lighting in real photos. Gaussian blur and directional motion blur model out-of-focus and camera motion. Subtle texture overlays mimic real paper surfaces. Random-quality JPEG compression introduces compression artifacts that occur in practical image storage. Figure 4 shows examples of simulated in-the-wild images obtained from these augmentations. This pipeline yields ample real-like table images while saving substantial manual effort and time.

File	Size	Ratio Favor
KE-MLM	1603	0.45
Donald Trump	840	0.41
Joe Biden	763	0.50
PStance	20914	0.48
Bernie Sanders	6161	0.56
Donald Trump	7709	0.46
Joe Biden	7044	0.44
SCD	4901	0.59
Abortion	1915	0.56
Barack Obama	985	0.53
Gay rights	1375	0.64
Marijuana	626	0.71

(a) Original image

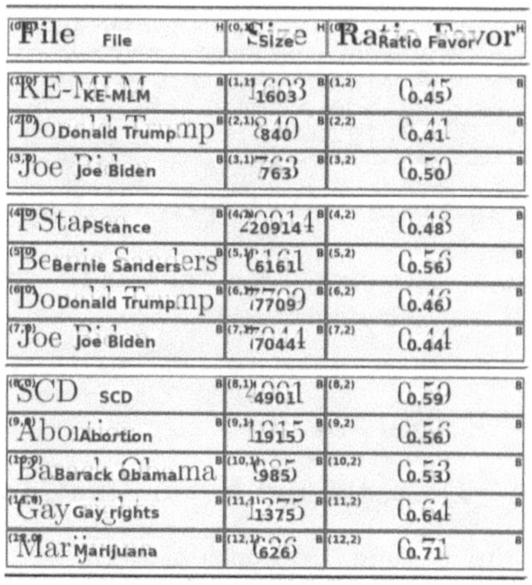

(b) Annotation image

File	Size	Ratio Favor
KE-MLM	1603	0.45
Donald Trump	840	0.41
Joe Biden	763	0.50
PStance	20914	0.48
Bernie Sanders	6161	0.56
Donald Trump	7709	0.46
Joe Biden	7044	0.44
SCD	4901	0.59
Abortion	1915	0.56
Barack Obama	985	0.53
Gay rights	1375	0.64
Marijuana	626	0.71

(c) LaTeX-rendered image

File	Size	Ratio Favor
KE-MLM	1603	0.45
Donald Trump	840	0.41
Joe Biden	763	0.50
PStance	20914	0.48
Bernie Sanders	6161	0.56
Donald Trump	7709	0.46
Joe Biden	7044	0.44
SCD	4901	0.59
Abortion	1915	0.56
Barack Obama	985	0.53
Gay rights	1375	0.64
Marijuana	626	0.71

(d) HTML-rendered image

Fig. 3. Example table annotations: (a) original table image; (b) schematic of table-structure and content annotations—red lines indicate table borders; red and blue boxes denote header and body cell locations, respectively; the top-left corner of each cell shows its relative row and column indices; the green text at the center denotes cell content; (c) table image rendered from LaTeX annotations; (d) table image rendered from HTML annotations.

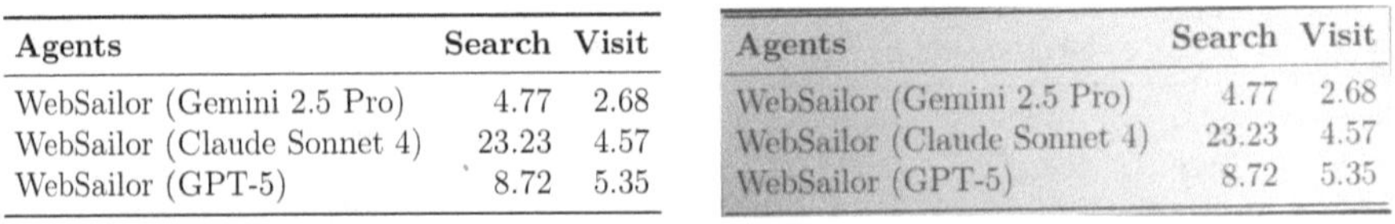

Agents	Search	Visit
WebSailor (Gemini 2.5 Pro)	4.77	2.68
WebSailor (Claude Sonnet 4)	23.23	4.57
WebSailor (GPT-5)	8.72	5.35

(a) Original image

Agents	Search	Visit
WebSailor (Gemini 2.5 Pro)	4.77	2.68
WebSailor (Claude Sonnet 4)	23.23	4.57
WebSailor (GPT-5)	8.72	5.35

(b) Augmented image

Fig. 4. Example of a simulated in-the-wild image derived from a rendered image via data augmentation.

5 Experiments

5.1 Datasets and Evaluation Metrics

Training and evaluation were carried out on our self-built dataset, Table2Markup, and the public dataset PubTabNet. Table 2Markup contains 31,353 table images and their LaTeX and HTML representations. The set includes 4,479 clean rendered renderings and 26,874 realistic synthetic images. PubTabNet is currently the largest public table dataset, containing about 568k images with corresponding structured HTML presentation. PubTabNet is automatically generated by matching XML and PDF representations from the PubMed Central [21] Open Access subset (PMCOA).

Tree Edit Distance-based Similarity (TEDS) [32] is used as the primary evaluation metric. TEDS parses a table sequence into a tree, computes the tree-edit distance between the prediction and the ground truth, and normalizes the result to produce a similarity score. In addition to TEDS, we report BLEU [18], METEOR [3], edit distance [14], F1-score, precision, and recall as auxiliary metrics.

5.2 Implementation Details

To save training resources, we initialized the model from the pretraining checkpoint of GOT [28]. During fine-tuning, the encoder is frozen, while the projection layer, the memory module, and the decoder are trained. Training ran on eight Huawei Ascend 910B accelerators with a batch size of 2. The total number of model parameters is 0.6B, and the memory module contains 15M parameters when the number of memory slots is set to 512. After loading the pretrained weights, the model was trained for 20000 steps with a learning rate of 1e-5, taking about 5 h.

5.3 Ablation Study

Ablation Study of Model Components. To validate the contribution of each component, we conducted an ablation study on the PubTabNet dataset; results appear in Table 1. Specifically, the following settings were evaluated:

Memory: add the memory module;

KV-mem: add the decoupled key–value memory;

Table 1. Ablation study on PubTabNet dataset

Model	TEDS		
	Simple	Complex	All
Baseline	89.4	81.0	85.2
Memory	92.2	81.0	86.6
KV-mem	91.9	83.9	87.9
Memory + TPE	87.0	87.0	87.0
KV-mem + TPE	91.6	**88.9**	90.3
KV-mem + TPE + Slot	**94.5**	87.7	**91.1**

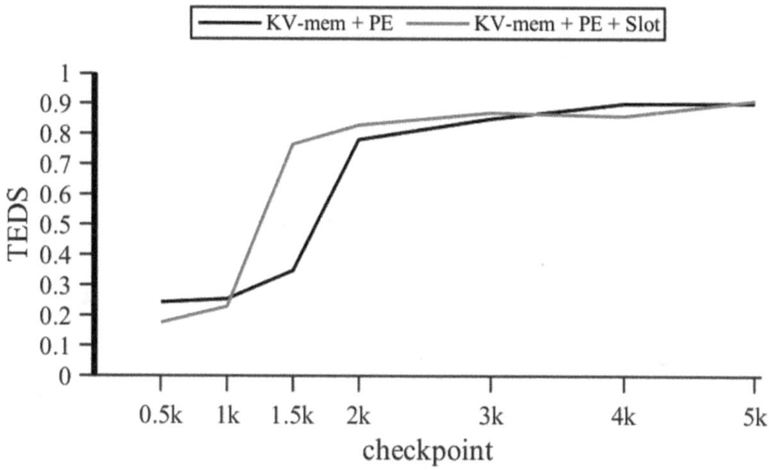

Fig. 5. Ablation study of slot attention on the PubTabNet dataset. Compare model performance with and without slot attention at multiple training checkpoints.

Memory+TPE: add the memory module together with table positional encoding (TPE);

Memory+TPE+Slot: add Slot Attention on top of Memory+TPE for staged initialization.

Introducing the memory module noticeably improves handling of simple tables. Adding the decoupled memory and the table positional encoding further improves performance on complex table types. When positional encoding is added alone, performance on simple tables can drop. This may be because simple tables resemble ordinary text layouts, where the row/column constraints bring less benefit, and the imposed spatial restrictions can hurt. Combining memory and TPE recovers this gap: TEDS for complex tables rises to 88.9, while TEDS for simple tables remains high at 91.6. The model achieves a TEDS of 94.5 on simple samples and 87.7 on complex samples, with an overall TEDS of 91.1. This represents the best overall performance.

Table 2. Ablation study of different memory sizes on PubTabNet dataset

Memory Size	TEDS		
	Simple	Complex	All
128	92.8	72.5	82.6
256	94.1	85.0	89.5
512	**94.5**	**87.7**	**91.1**
1024	94.4	84.0	89.2
2048	91.9	77.2	84.6

Table 3. Experiment results on PubTabNet dataset

Model	TEDS		
	Simple	Complex	All
Tabula	78.0	57.8	67.9
Traprange	60.8	49.9	55.4
Camelot	80.0	66.0	73.0
Acrobat Pro	68.9	61.8	65.3
EDD	91.2	85.4	88.3
TabDec(Ours)	94.5	87.7	91.1

Figure 5 shows that adding Slot Attention speeds up convergence and improves the model's ability to adapt to small-sample table datasets.

Ablation Study of Memory Size. The prototype memory module contains multiple memory slots to store memory information. To find the optimal memory capacity for table data, an ablation study is conducted by varying the number of slots. Results appear in Table 2. The table memory module achieves its best performance with a memory size of 512. When the slot count is too small, the limited capacity cannot cover the diverse structural components and content patterns in tables. When the slot count is too large, performance also drops, possibly because excessive slots increase optimization difficulty and some slots become ineffective or unstable prototypes. On document datasets, the original memory module achieves its best results with a memory size of 2048, which is higher than the 512 optimal for table data. This reflects the more complex layouts in documents and shows that choosing a memory capacity suited to the task is key to improving performance.

5.4 Comparison with Other Methods

Comparison with Other Methods on PubTabNet Dataset. On the PubTabNet dataset, the proposed method is compared with several existing approaches, including traditional table tools (Tabula, TrapRange, Camelot,

Table 4. Performance comparison on Table2Markup (clean rendered)

Model	Edit Distance$\downarrow$		F1-score$\uparrow$		Precision$\uparrow$		Recall$\uparrow$		BLEU$\uparrow$		METEOR$\uparrow$		**TEDS**$\uparrow$	
	HTML	LaTeX	HTML	LaTeX	HTML	LaTeX	HTML	LaTeX	CHTML	LaTeX	HTML	LaTeX	HTML	LaTeX
Qwen3-VL [1]	85.3	57.9	3.0	48.3	2.4	45.2	4.8	57.8	0.1	36.2	2.3	47.0	3.9	20.6
DeepSeek-OCR [29]	75.2	-	7.7	-	5.9	-	13.1	-	0.4	-	8.1	-	61.1	-
GOT [28]	-	27.3	-	61.1	-	60.7	-	61.8	-	51.2	-	66.8	-	88.9
TabDec(Ours)	10.5	6.2	73.1	88.9	76.1	89.1	70.8	88.7	65.8	86.1	72.1	91.1	93.3	97.6

Table 5. Performance comparison on Table2Markup (realistic synthetic)

Model	Edit Distance$\downarrow$		F1-score$\uparrow$		Precision$\uparrow$		Recall$\uparrow$		BLEU$\uparrow$		METEOR$\uparrow$		**TEDS**$\uparrow$	
	HTML	LaTeX	HTML	LaTeX	HTML	LaTeX	HTML	LaTeX	CHTML	LaTeX	HTML	LaTeX	HTML	LaTeX
Qwen3-VL [1]	85.3	64.3	2.6	43.5	2.2	42.8	4.0	51.2	0.1	30.9	0.2	40.6	3.9	19.6
DeepSeek-OCR [29]	78.1	-	7.0	-	6.2	-	8.8	-	0.4	-	5.0	-	23.3	-
GOT [28]	-	32.2	-	58.6	-	57.9	-	59.5	-	47.3	-	63.5	-	84.5
TabDec(Ours)	10.7	9.2	74.6	84.5	77.1	84.7	72.7	84.3	65.7	80.1	73.2	87.7	93.1	95.5

Acrobat Pro) and the deep-learning end-to-end model EDD. Results are shown in Table 3. On simple samples, TEDS reaches 94.5%; on complex samples, TEDS is 87.7%; the overall TEDS is 91.1%, a 2.8%-point improvement over EDD.

Comparison with Other Methods on Image2Markup. The Table2-Markup dataset is used to evaluate the general vision–language model Qwen3-VL [1], OCR-focused models DeepSeek-OCR [29] and GOT [28], and TabDec. Results on clean rendered table images are shown in Table 4, and results on realistic synthetic table images are shown in Table 5. Because Table2Markup provides exact markup annotations, some existing models lack sufficient precision on table structure and fonts. On clean rendered images, TabDec achieves a TEDS of 93.3% for HTML and 97.6% for LaTeX, showing a clear advantage. The model also adapts well to realistic synthetic tables, achieving TEDS of 93.1% for HTML and 95.5% for LaTeX.

5.5 Visualization

Figure 6 shows rendered examples of TabDec recognition results on different table images. The markup generated by TabDec can be directly rendered back into clear table images, showing that the model effectively learns the mapping from table images to structured markup. For tables with rowspan/colspan, sparse layouts, varied fonts, and real-world photographic degradations, TabDec preserves the overall row–column structure, cell organization, and main textual content. These qualitative results further show that the proposed decoupled prototype memory improves structural perception for complex table layouts and enhances the interpretability and renderability of end-to-end table recognition.

	HaluSum	RepliQA	SciQAG
Llama-3.2-3B-Instruct			
Llama-3.2-3B-Instruct	22.98	34.71	39.52
NTP (no instruct)			39.41
MTP (instruct)	22.43	35.47	39.66
MPP (instruct)	**23.02**	35.58	40.11
NLKG (instruct)	23.11	**36.83**	39.88
Qwen3-1.7B			
Qwen3-1.7B	18.85	28.73	37.54
NTP (no instruct)	19.02	28.8	37.10
MTP (instruct)	18.60	**31.71**	**38.55**
MPP (instruct)	18.76	30.42	37.94
NLKG (instruct)	**19.55**	29.4	38.30

(a) Sample (i)

	HaluSum	RepliQA	SciQAG
Llama-3.2-3B-Instruct			
Llama-3.2-3B-Instruct	22.98	34.71	39.52
NTP (no instruct)			39.41
MTP (instruct)	22.43	35.47	39.66
MPP (instruct)	**23.02**	35.58	40.11
NLKG (instruct)	23.11	**36.83**	39.88
Qwen3-1.7B			
Qwen3-1.7B	18.85	28.73	37.54
NTP (no instruct)	19.02	28.80	37.10
MTP (instruct)	18.60	**31.71**	**38.55**
MPP (instruct)	18.76	30.42	37.94
NLKG (instruct)	**19.55**	29.40	38.30

(b) Recognition Result (i)

	Japanese	**English**	**Chinese**
# Dialogues	299	300	298
Total dialogue time	108:13:34	119:56:12	108:05:12
Total used dialogue time	49:13:39	27:20:31	25:04:53
# Backchannel utterances	58800 (34.4%)	24612 (28.4%)	21182 (27.5%)
# Non-backchannel utterances	112177	62158	55955
Total BC time [s]	29253.73 (16.5%)	11006.73 (11.2%)	7695.41 (8.5%)
Total non-BC time [s]	147965.49	87424.67	82598.55

(c) Sample (iii)

	Japanese	**English**	**Chinese**
# Dialogues	299	300	298
Total dialogue time	108:13:34	119:56:12	108:05:12
Total used dialogue time	49:13:39	27:20:31	25:04:53
# Backchannel utterances	58800 (34.4%)	24612 (28.4%)	21182 (27.5%)
# Non-backchannel utterances	112177	62158	55955
Total BC time [s]	29253.73 (16.5%)	11006.73 (11.2%)	7965.41 (8.5%)
Total non-BC time [s]	147965.49	87424.67	82598.55

(d) Recognition Result (iii)

Fig. 6. Comparison of example table images and TabDec recognition results.

6 Conclusion

Tables are a key medium for organizing and presenting structured data in documents. Accurately recognizing table content from document images is critical for information retrieval, data analysis, and document intelligence. Current data annotations and model designs for table recognition have limitations.

A prototype memory module tailored for tables is also proposed. The module decouples structural and content features and is integrated into end-to-end recognition models. This design substantially improves layout understanding and recognition accuracy. The work explores a learnable way to inject table structural priors into neural networks, advancing models' ability to understand structured documents.

An automated annotation pipeline is developed that combines LaTeX-source inference with rendered-image analysis to produce high-precision annotations of table structure, cell content, and end-to-end LaTeX and HTML markup. Based on this pipeline, the dataset Table2Markup is built. Table2Markup includes high-quality rendered table images and a large set of simulated photographic table images generated by applying realistic degradations to the renderings. It fills a gap in existing resources by providing end-to-end markup annotations and supports training of image-to-markup models.

Future work will enforce the syntactic correctness of the generated markup through constraints and related strategies. In addition, the dataset will be further enlarged to cover more domains and a wider variety of table types, and its effectiveness will be validated on external real-world table datasets.

Acknowledgements. This work was supported by Huawei Noah's Ark Lab. The authors thank Yuyang Li for useful discussions. The authors also thank the anonymous reviewers for their valuable comments that improved the quality of the paper.

References

1. Bai, J., et al.: Qwen3-vl technical report (2025). arXiv preprint arXiv:2511.21631
2. Bai, J., et al.: Qwen technical report (2023). arXiv preprint arXiv:2309.16609
3. Banerjee, S., et Lavie, A.: METEOR: an automatic metric for MT evaluation with improved correlation with human judgments. In: Proceedings of the ACL Workshop on Intrinsic and Extrinsic Evaluation Measures for Machine Translation and/or Summarization (2005)
4. Chi Z., Huang H., Xu H., et al.: Complicated table structure recognition (2019). arXiv preprint arXiv:1908.04729
5. Deng, Y., Kanervisto, A., Ling, J., Rush, A.M.: Image-to-markup generation with coarse-to-fine attention. In: International Conference on Machine Learning, pp. 980–989. PMLR (2017)
6. Dominici, M.: An overview of Pandoc. In: TUGboat **35**(1), 44–50 (2014)
7. Fang J., Tao, X., Tang, Z., Qiu, R., et Liu, Y.: Dataset, ground-truth and performance metrics for table detection evaluation. In: 2012 10th IAPR International Workshop on Document Analysis Systems, pp. 445–449. IEEE (2012)
8. Göbel, M., Hassan, T., Oro, E., et Orsi, G.: ICDAR 2013 table competition. In: 2013 12th International Conference on Document Analysis and Recognition. IEEE (2013)
9. Graves, A., Wayne, G., et Danihelka, I: Neural turing machines. arXiv:1410.5401 (2014)
10. Han, T., Xie, W., Zisserman, A.: Memory-augmented dense predictive coding for video representation learning. In: European Conference on Computer Vision, pp. 312–329. Springer, Cham (2020)
11. Hickson, I., David, H.: Html5. In: W3C working draft WD-Html5-20110525, p. 53 (2011)
12. Hou, Q., Jun W.: TABLET: table structure recognition using encoder-only transformers. In: International Conference on Document Analysis and Recognition. Springer, Cham (2025)
13. Kim, G., Hong, T., Yim, M., Park, J., et al.: Donut: document understanding transformer without ocr. arXiv:2111.15664 (2021)
14. Lcvenshtcin, V. I.: Binary coors capable or "correcting deletions, insertions, and reversals." In: Soviet physics-doklady (1966)
15. Li, M.: Tablebank: Table benchmark for image-based table detection and recognition. In: Proceedings of the Twelfth Language Resources and Evaluation Conference (2020)
16. Lu, W., Zhang, J., Fan, J., Fu, Z., Chen, Y., Du, X.: Large language model for table processing: A survey. Front. Comp. Sci. **19**(2), 192350 (2025)
17. Nassar, A., Livathinos, N., Lysak, M., et Staar, P.: Tableformer: table structure understanding with transformers. In: Proceedings of the IEEE/CVF Conference on Computer Vision and Pattern Recognition (2022)
18. Papineni, K., Roukos, S., Ward, T., et Zhu, W. J.: Bleu: a method for automatic evaluation of machine translation. In: Proceedings of the 40th annual meeting of the Association for Computational Linguistics (2002)

19. Prasad, D., Gadpal, A., Kapadni, K., Visave, M., Sultanpure, K.: CascadeTabNet: an approach for end to end table detection and structure recognition from image-based documents. In: Proceedings of the IEEE/CVF Conference on Computer Vision and Pattern Recognition Workshops (2020)
20. Rahtz, S.: The pdfTEX user manual (2018)
21. Roberts, R. J.: PubMed Central: The GenBank of the published literature. In: Proceedings of the National Academy of Sciences, 98(2), 381–382 (2001)
22. Santoro, A., Bartunov, S., Botvinick, M., Wierstra, D., et Lillicrap, T: Meta-learning with memory-augmented neural networks. In: International Conference on Machine Learning. PMLR (2016)
23. Schreiber, S., Agne, S., Wolf, I., Dengel, A., Ahmed, S.: Deepdesrt: deep learning for detection and structure recognition of tables in document images. In: 14th IAPR International Conference on Document Analysis and Recognition (ICDAR), vol. 1, IEEE (2017)
24. Siddiqui, S., Fateh, I., Rizvi, S., Dengel, A., Ahmed, S.: Deeptabstr: deep learning based table structure recognition. In: 2019 International Conference on Document Analysis and Recognition (ICDAR), IEEE (2019)
25. Snell, J., Swersky, K., et Zemel, R: Prototypical networks for few-shot learning. In: Advances in Neural Information Processing Systems, 30 (2017)
26. Vinyals, O., Blundell, C., Lillicrap, T., et Wierstra, D.: Matching networks for one shot learning. In: Advances in Neural Information Processing Systems, 29 (2016)
27. Wan, J., Song, S., Yu, W., Liu, Y., Cheng, W., Wan, J., et al.: Omniparser: a unified framework for text spotting key information extraction and table recognition. In: Proceedings of the IEEE/CVF Conference on Computer Vision and Pattern Recognition (2024)
28. Wei, H., Liu, C., Chen, J., Wang, J., Kong, L., et al.: General ocr theory: Towards ocr-2.0 via a unified end-to-end model. arXiv preprint arXiv:2409.01704 (2024)
29. Wei, H., Sun, Y., et Li, Y.: Deepseek-ocr: Contexts optical compression. arXiv preprint arXiv:2510.18234 (2025)
30. Yang, F., Hu, L., Liu, X., Huang, S., et Gu, Z.: A large-scale dataset for end-to-end table recognition in the wild. In: Scientific Data 10.1:110 (2023)
31. Zheng, X., Burdick, D., Popa, L., Zhong, X., Wang, N.: Global table extractor (gte): a framework for joint table identification and cell structure recognition using visual context. In: Proceedings of the IEEE/CVF Winter Conference on Applications of Computer Vision, pp. 697–706. (2021)
32. Zhong F., ShafieiBavani E., Jimeno Y.: Image-based table recognition: data, model, and evaluation. In: European Conference on Computer Vision, pp. 564-580. Springer (2020)
33. Zhong, Xu., Jianbin T., and Antonio J.: Publaynet: largest dataset ever for document layout analysis. In: 2019 International conference on document analysis and recognition (ICDAR). IEEE (2019). https://doi.org/10.1109/ICDAR.2019.00166

A Recognition-Aware Multimodal Framework for Actual and Intended Text Transcription of Child Handwriting

M. S. Sumi Suresh[1(✉)], Sahana Rangasrinivasan[1,2], Abbie Olszewski[3], Srirangaraj Setlur[1], and Venu Govindaraju[1]

[1] Department of Computer Science and Engineering, University at Buffalo, New York, USA
{sumisure,srangasr,setlur,govind}@buffalo.edu
[2] Department of Computer Science and Engineering, Amrita School of Computing, Amrita Vishwa Vidyapeetham, Amritapuri, India
[3] College of Education and Human Development, University of Nevada-Reno, Reno 89557, NV, USA
aolszewski@unr.edu

Abstract. Children's handwriting exhibits substantial visual and linguistic variability, including malformed or mirrored letters, spelling errors, and irregular spacing that reflect developmental learning processes. In educational assessment and learning disability analysis, it is important to distinguish between the exact text a child wrote and the text they intended to write. However, existing handwritten text recognition (HTR) systems typically produce a single normalized transcription, often removing developmentally meaningful evidence through autocorrection or language-driven normalization. In this work, we introduce a dual-level transcription framework for child handwriting that generates two complementary outputs from the same sample: (i) an actual-text transcription preserving the child's exact written form, and (ii) an intended-text transcription providing a minimally corrected, context-aware interpretation of meaning. The framework uses a recognition-aware two-stage architecture consisting of an actual-text predictor followed by a multimodal sequence-to-sequence intended-text predictor. Experiments on child handwriting data show that the proposed method accurately preserves the child's written form while producing a readable, intended interpretation without excessive grammatical rewriting. Compared with conventional HTR systems and large language or vision-language models, our approach achieves stronger actual-text transcription performance and competitive intended-text prediction while requiring substantially fewer computational resources. This work establishes dual-level child handwriting transcription as a behavior-aware recognition framework with applications in educational assessment and learning disability analysis.

Keywords: Child Handwriting · Dual-level transcription · Handwritten Text Recognition · Multimodal Large Language Models

F. Shafait et al. (Eds.): DAS 2026, LNCS 16851, pp. 371–388, 2027.
https://doi.org/10.1007/978-3-032-36207-0_22

1 Introduction

Handwriting produced by children differs substantially from adult writing as it reflects an ongoing developmental process involving motor control, visual perception, and emerging orthographic knowledge [7]. As a result, child handwriting often contains irregular letter shapes, inconsistent spacing, phonetic spelling errors, omissions, mirrored letter forms, and other non-standard constructions [5]. These patterns are not merely recognition noise; they may provide meaningful evidence for educational assessment and for the analysis of learning disabilities such as dyslexia and dysgraphia [5].

In practice, this creates a distinction between two related but different aspects of written production: the *exact* text the child wrote and the text the child *intended* to write. In educational and clinical settings, both are important. The former preserves developmentally meaningful spelling and writing patterns, while the latter provides a readable interpretation of the child's likely linguistic intent. Existing handwritten text recognition (HTR) systems, however, are designed to produce only a single transcription, and inadvertently fail to preserve this distinction. Additionally, models trained on relatively regular adult handwriting often struggle to generalize to the visual variability of child handwriting, while language-driven recognizers tend to normalize non-standard forms into fluent text. As illustrated in Fig. 1, existing recognition systems typically resolve this variability by favoring standardized linguistic forms, thereby removing behavioral evidence embedded in the original writing.

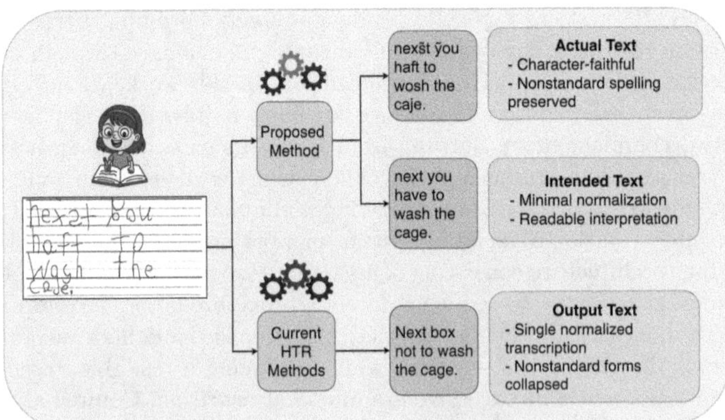

Fig. 1. Comparison of existing HTR and the proposed method for dual-level child handwriting transcription.

To address this limitation, this work formulates child handwriting recognition as a *dual-level transcription* problem. Given a handwritten input, the system produces both (i) an *actual-text* transcription that preserves the observable written form and (ii) an *intended-text* transcription that provides a minimally corrected interpretation of the likely meaning. This formulation separates visually grounded recognition from controlled linguistic interpretation, avoiding unconstrained normalization or full grammatical rewriting.

To realize this formulation, a two-stage recognition-aware multimodal framework is proposed. The first stage predicts actual text while extracting visual sequence features and recognition confidence. The second stage uses these signals in a multimodal sequence-to-sequence decoder to predict the intended text. This design keeps intended-text prediction grounded in the original handwriting signal rather than relying solely on post-hoc language correction. Experimental results on child handwriting data show that the proposed framework preserves child-produced writing patterns while also providing readable interpretations of intended text. In addition, the proposed method achieves strong performance in actual-text transcription and promising performance in intended-text prediction, compared to both conventional HTR baselines and much larger foundation-model baselines. The main contributions of this paper are as follows:

- **Introduction of a dual-level transcription setting** for child handwriting, consisting of actual-text transcription and intended-text prediction.
- **Proposal of a two-stage recognition-aware multimodal framework** that uses actual text, visual sequence features, and recognition confidence to generate intended text.
- **Development of a multimodal intended-text decoder** that combines text memory, visual memory, and confidence embedding for recognition-aware correction.
- **Empirical validation** showing strong actual-text performance and promising intended-text prediction relative to conventional HTR baselines and much larger foundation-model baselines.

2 Literature Review

Handwritten Text Recognition (HTR) is typically formulated as a sequence recognition task that maps visual input to a single textual transcription [9]. Classical approaches combine convolutional feature extraction with recurrent modeling and Connectionist Temporal Classification (CTC) decoding [11,15]. More recent transformer-based architectures leverage attention mechanisms and large-scale pretraining to capture long-range dependencies and contextual structure [12,14]. These methods have achieved strong performance on standardized handwriting benchmarks and document transcription tasks. Language modeling components are often integrated into recognition pipelines to improve fluency and reduce ambiguity. While effective for adult handwriting and well-formed orthography, such integration assumes a single transcription objective and does not explicitly distinguish between visually faithful reproduction and linguistic interpretation. Recent multimodal foundation models jointly process visual and textual inputs and have been applied to document understanding and handwriting interpretation [1,6]. These systems can generate fluent and semantically coherent descriptions of handwritten content. However, their training objectives emphasize semantic plausibility and linguistic fluency, and generated outputs are not constrained to remain character-faithful to the written input.

Child Handwriting Recognition: Research on child handwriting recognition has largely focused on improving robustness to developmental variability [17]. Approaches address irregular character formation, spacing instability, and nonstandard spelling through modified recognition pipelines, including joint recognition–segmentation methods and preprocessing strategies for atypical letter shapes [8,16]. While these advances improve transcription accuracy and structural precision, they remain centered on producing a single output sequence and do not explicitly distinguish between written production and intended meaning.

Text Normalization and Correction: Neural sequence-to-sequence models and large language models have demonstrated strong performance in grammatical error correction and orthographic normalization [2,3]. These systems typically operate on transcribed text and aim to produce standardized, well-formed output. Correction may involve substitution, insertion, or restructuring of tokens, with optimization objectives centered on linguistic well-formedness rather than visual fidelity.

3 Problem Discussion and Motivation

Child handwriting differs fundamentally from adult writing in that spelling errors, letter distortions, mirrored characters, irregular spacing, incorrect capitalization, and letter transpositions often reflect developmental learning stages rather than random noise. In educational and clinical settings, such deviations are not merely transcription errors; they may serve as behavioral indicators of phonological awareness, orthographic knowledge, motor planning, or learning difficulties. Existing OCR and HTR systems, however, are typically optimized to recover linguistically well-formed text and therefore tend to normalize or auto-correct atypical child handwriting patterns. As a result, developmentally meaningful evidence may be removed during transcription.

Why Exact Actual-Text Transcription Matters: In child handwriting analysis, exact written form is often more informative than a normalized interpretation. Preserving handwriting features such as spelling errors, mirrored letters, and incorrect capitalization is essential for applications such as learning disability screening, progress monitoring, and educational assessment [17]. In these settings, the error itself may be diagnostically meaningful. Existing HTR models are not well suited to this requirement: models trained on relatively uniform adult handwriting often fail to generalize to children's irregular letter forms, while models with strong language priors tend to auto-correct visually grounded deviations into fluent standard text [16]. For educational assessment, effective recognition must therefore prioritize visual faithfulness over linguistic normalization, preserving the deviations that reflect a child's learning profile.

Why Intended-Text Interpretation Is Also Needed: Although preserving the original written form is critical, the intended meaning is also important in practice. In educational assessment and clinical evaluation, practitioners such as speech-language pathologists (SLPs) and occupational therapists (OTs) routinely distinguish between what a child *wrote* and what the child *intended* to write. This distinction helps separate written execution from linguistic intent and supports more accurate interpretation of spelling, phonological, and handwriting-related difficulties. In many cases, the intended meaning cannot be reliably inferred from the written output alone, particularly when the handwriting contains severe spelling deviations, poorly formed letters, or mirrored characters. A separate intended-text representation, therefore, provides a readable interpretation without discarding the original behavioral evidence.

Why Full Grammatical Correction Is Inadequate: A fully normalized or grammatically rewritten sentence is often inappropriate for child handwriting analysis. Expectations for grammar, vocabulary, sentence structure, and writing conventions vary substantially across grade levels and developmental stages. An irregularity in the writing may therefore reflect a genuine difficulty in writing, a developmentally typical error, or incomplete language knowledge appropriate for the child's age. Full grammatical correction can introduce words, punctuation, or restructured phrasing that are not visually grounded in the handwritten input, thereby removing behavioral evidence present in the original writing. Large language models further amplify this problem, as they often improve fluency by paraphrasing or restructuring sentences rather than applying minimal corrections. Such rewriting weakens the boundary between observed handwriting and inferred meaning and can obscure developmental patterns embedded in the original writing, as illustrated in Fig. 2.

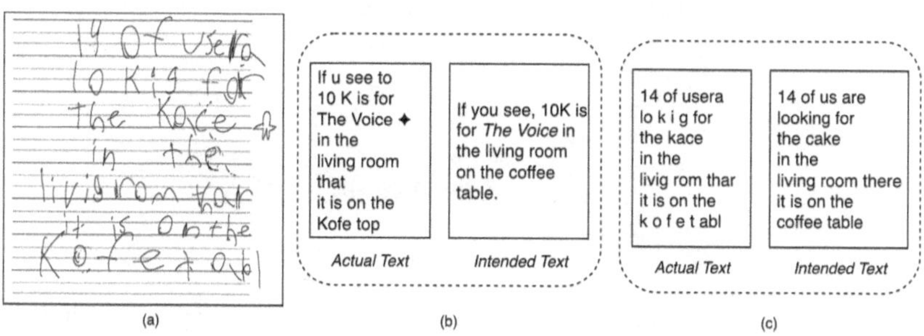

Fig. 2. Actual and intended text recognition results using GPT-5.2. (a) The child handwriting sample, (b) Results obtained from GPT-5.2, (c) The ground truth.

Why Dual-Level Transcription Is Necessary: These limitations motivate a dual-level formulation of child handwriting transcription. Rather than forcing a single output to satisfy both fidelity and readability, the task is decomposed into two complementary representations:

- *Actual Text:* a character-faithful transcription that preserves spelling, spacing, capitalization, and structural irregularities present in the child's writing;
- *Intended Text:* a minimally corrected, context-aware interpretation that clarifies the likely intended meaning without full grammatical rewriting.

This separation preserves the boundary between recognition and inference. The actual text remains visually grounded and behaviorally informative, while the intended text provides a readable interpretation for educational assessment, learning-disability screening, and downstream analysis. The distinction between actual and intended transcription is, therefore, not merely a modeling choice but a necessary design principle for behavior-aware document understanding.

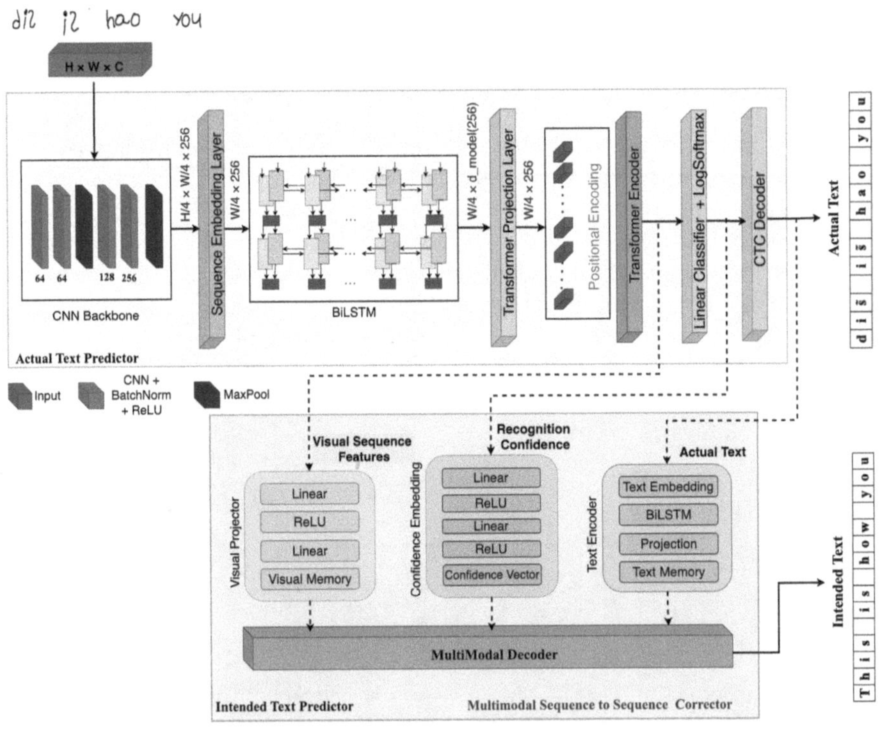

Fig. 3. Architecture of the proposed two-stage framework for actual and intended text prediction.

4 Methodology

We propose a line-level, two-stage recognition-aware framework that predicts both the *actual text* written by the child and the *intended text* reflecting the child's likely target expression while preserving behavioral cues. Page-level samples are first segmented into individual handwritten lines, and intended-text prediction remains grounded in the original handwriting signal rather than relying solely on post-hoc language correction.

Figure 3 illustrates the proposed architecture. The upper branch corresponds to the *actual-text predictor*, where given a handwritten line image I, the framework first performs actual text transcription using a contextual sequence recognizer. This stage produces three outputs: i) the predicted actual text ($\hat{\mathbf{y}}^a$), (ii) intermediate visual sequence features ($\mathbf{V}$), and (iii) a recognition confidence score ($\mathbf{C}$). The lower branch in Fig. 3 corresponds to the *recognition-aware multimodal sequence-to-sequence corrector*, which predicts the intended text. It receives three inputs from the first stage: visual memory derived from image features, text memory derived from predicted actual text, and confidence embedding derived from the recognition score. These three representations are fused in the multimodal decoder to generate the intended-text sequence. In Fig. 3, solid arrows denote the main forward prediction path, while dashed arrows indicate intermediate outputs from the actual-text predictor that are reused by the intended-text predictor.

Formally, given an input handwritten line image $\mathbf{I}$, the first stage predicts an actual text sequence $\hat{\mathbf{y}}^a$, while the second stage predicts an intended text sequence $\hat{\mathbf{y}}^i$:

$$\hat{\mathbf{y}}^a = f_{\text{act}}(\mathbf{I}), \qquad \hat{\mathbf{y}}^i = f_{\text{int}}(\hat{\mathbf{y}}^a, \mathbf{V}, \mathbf{C}),$$

where $\mathbf{V}$ denotes the visual sequence features extracted from the first stage and $\mathbf{C}$ denotes the recognition confidence.

4.1 Actual Text Predictor

The actual-text transcription module is designed to faithfully preserve the child's written form, including spelling irregularities, non-canonical letter formations, and mirrored character variants, without applying language-level normalization. It consists of four components: (i) a convolutional feature extractor, (ii) a sequence embedding layer, (iii) a bidirectional recurrent encoder, and (iv) a Transformer-based contextual encoder with CTC transcription as shown in Fig. 3.

The input image $\mathbf{I} \in \mathbb{R}^{H \times W \times C}$ is first processed by a CNN backbone with four convolutional blocks. Each block consists of convolution, batch normalization, ReLU activation, and max-pooling, producing feature channels of 64, 64, 128, and 256, respectively. This backbone extracts local stroke-level and character-level visual features from the handwritten line.

The resulting feature map is converted into a width-wise sequential representation by collapsing the vertical feature dimension at each horizontal position. This transformation preserves the left-to-right ordering of handwritten content and produces a feature sequence across the line. Thereafter, a linear projection maps each time-step feature to a 256-dimensional sequence embedding.

The sequence is then passed through a two-layer bidirectional LSTM (BiLSTM). The BiLSTM produces a 256-dimensional output representation, with 128 hidden units in each direction. This recurrent encoder captures local bidirectional dependencies along the handwriting line, which is useful for modeling irregular child handwriting where neighboring characters and strokes may be ambiguous.

The BiLSTM output is then projected to the Transformer dimension and refined using a Transformer encoder. The Transformer encoder contains four layers, four attention heads, a feed-forward dimension of 512, and a dropout rate of 0.1. Positional encoding is added before the Transformer encoder to preserve sequence order. The output of the Transformer encoder is used as the visual sequence representation $\mathbf{V}$. A linear classifier followed by log-softmax produces frame-wise character probabilities, and the actual-text sequence is obtained using CTC decoding:

$$\hat{\mathbf{y}}^a = \text{CTCDecode}(\mathbf{L}),$$

where $\mathbf{L}$ denotes the frame-wise log-probability sequence. To preserve child-specific writing behaviors, the recognition vocabulary extends the standard character set with dedicated symbols for mirrored character variants (e.g., left-right mirrored, top-bottom mirrored, and combined variants). These variant tokens are mapped back to the corresponding character representation rather than being collapsed into the canonical character form. This allows the actual-text transcription to retain visually meaningful writing patterns that may be important for downstream behavioral analysis.

In addition to the predicted actual text, the actual-text predictor provides two outputs to the intended-text predictor: the visual sequence representation $\mathbf{V}$ and a recognition confidence score $\mathbf{C}$. The confidence score is computed by averaging the maximum posterior character probability across valid time steps:

$$C = \frac{1}{T} \sum_{t=1}^{T} \max_{k} \exp(\mathbf{L}_{t,k}).$$

Thus, the actual text predictor outputs three elements for each input line image: the actual text $\hat{\mathbf{y}}^a$, the visual sequence features $\mathbf{V}$, and the confidence score C.

4.2 Intended Text Predictor

The second stage predicts the intended text by correcting the actual text using recognition-aware multimodal evidence. As illustrated in Fig. 3, the intended-text predictor receives three outputs from the actual-text predictor: (i) visual sequence features $\mathbf{V}$, (ii) recognition confidence score C, and (iii) predicted actual text $\hat{\mathbf{y}}^a$. These inputs are processed by three branches: a visual projector, a confidence embedding module, and a text encoder.

Visual Projector: The visual sequence features $\mathbf{V}$ are mapped to the intended-text decoder space using a lightweight projection network composed of two linear

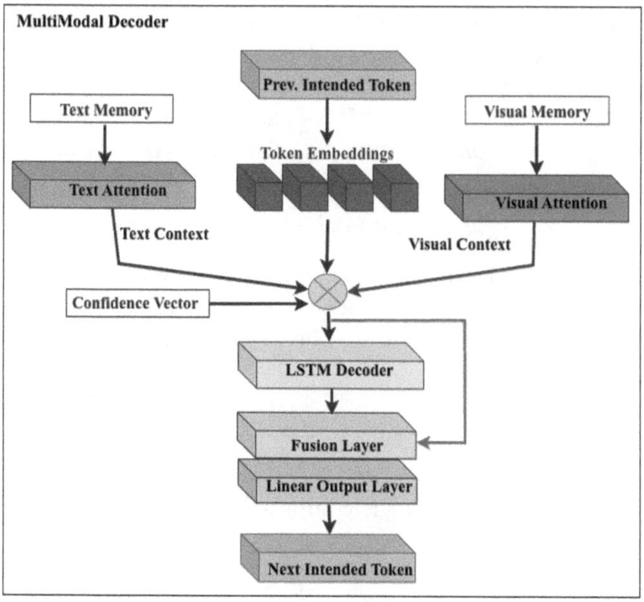

Fig. 4. Detailed architecture of the multimodal decoder used in the intended text predictor.

layers with a ReLU activation, producing the visual memory $\mathbf{M}^v$. This branch allows the intended-text predictor to access image-derived sequence evidence rather than relying only on the predicted actual text.

Confidence Embedding: The scalar recognition confidence $\mathbf{C}$ is transformed into a learned confidence vector $\mathbf{e}^c$ using a small multilayer perceptron. This branch allows the decoder to adapt its behavior according to the reliability of the actual-text prediction. For example, when the confidence is low, the decoder can rely more on contextual and visual evidence.

Text Encoder The predicted actual text $\hat{\mathbf{y}}^a$ is converted into character-level token embeddings and passed through a BiLSTM to encode contextual dependencies. A projection layer maps the BiLSTM outputs to a common text-memory space $\mathbf{M}^t$. This memory provides a contextual representation of the predicted actual text and is used by the multimodal decoder through attention.

In our implementation, the text embedding, hidden-state, text-memory, and visual-memory dimensions are set to 256, while the confidence embedding dimension is 32.

Multimodal Decoder. A detailed view of the multimodal decoder is shown in Fig. 4. The decoder generates the intended text autoregressively, one token at a time. At decoding step j, the previous intended token y^i_{j-1} is embedded and used as the decoder input e_{j-1}.

The decoder attends separately to text memory and visual memory using additive attention. Let $\mathbf{s}_{j-1}$ denote the decoder hidden state from the previous decoding step, and let $\mathbf{M}^t = \{\mathbf{M}_1^t, \mathbf{M}_2^t, \ldots, \mathbf{M}_N^t\}$ denote the text memory produced by the text encoder, where $\mathbf{M}_n^t$ is the memory vector at the n-th text position. The text attention computes an alignment score between the decoder state and each text-memory element using additive attention:

$$e_{j,n}^t = \mathbf{v}_t^\top \tanh(\mathbf{W}_m^t \mathbf{M}_n^t + \mathbf{W}_s^t \mathbf{s}_{j-1}),$$

where $\mathbf{W}_m^t$ and $\mathbf{W}_s^t$ are learnable projection matrices for the text memory and decoder state, respectively, and $\mathbf{v}_t$ is a learnable attention vector that maps the joint representation to a scalar score. The attention weights are then obtained by normalizing these scores:

$$\alpha_{j,n}^t = \frac{\exp(e_{j,n}^t)}{\sum_{k=1}^N \exp(e_{j,k}^t)}.$$

Finally, the text context vector $\mathbf{c}_j^t$ is computed as the weighted sum of the text memory:

$$\mathbf{c}_j^t = \sum_{n=1}^N \alpha_{j,n}^t \mathbf{M}_n^t.$$

Similarly, visual attention produces weights $\alpha_{j,m}^v$ over visual memory and computes the visual context vector $\mathbf{c}_j^v$ as a weighted sum. The token embedding, text context, visual context, and confidence vector are concatenated and provided as input to the LSTM decoder:

$$\mathbf{u}_j = [\mathbf{e}_{j-1}; \mathbf{c}_j^t; \mathbf{c}_j^v; \mathbf{e}^c],$$
$$\mathbf{s}_j = \mathrm{LSTM}(\mathbf{u}_j, \mathbf{s}_{j-1}).$$

Fusion Layer: The model performs an additional fusion step after the LSTM decoder. Specifically, the decoder output is concatenated again with the attention contexts, token embedding, and confidence vector:

$$\mathbf{f}_j = [\mathbf{s}_j; \mathbf{c}_j^t; \mathbf{c}_j^v; \mathbf{e}_{j-1}; \mathbf{e}^c].$$

This fused representation is passed through a nonlinear projection layer:

$$\tilde{\mathbf{h}}_j = \phi(\mathbf{W}_f \mathbf{f}_j + \mathbf{b}_f),$$

where $\phi(\cdot)$ denotes a ReLU activation function.

This fusion stage re-integrates the decoder state with the multimodal contextual cues before prediction, allowing the model to make the next-token decision using both sequential decoder dynamics and external conditioning signals. Finally, the next-token distribution is obtained through a linear output layer:

$$p(y_j^i \mid y_{<j}^i, \hat{\mathbf{y}}^a, \mathbf{V}, \mathbf{C}) = \mathrm{Softmax}(\mathbf{W}_o \tilde{\mathbf{h}}_j + \mathbf{b}_o).$$

The intended text prediction is generated autoregressively until the end-of-sequence token is produced:

$$\hat{\mathbf{y}}^i = \{\hat{y}_1^i, \hat{y}_2^i, \ldots, \hat{y}_J^i\}.$$

4.3 Training Objective

The actual text predictor is trained using the CTC loss:

$$\mathcal{L}_{act} = \mathcal{L}_{CTC}(L, \mathbf{y}^a),$$

where $\mathbf{y}^a$ denotes the ground-truth actual text. The intended text predictor is trained using teacher forcing with a token-level cross-entropy loss:

$$\mathcal{L}_{int} = -\sum_{j=1}^{J} \log p(y_j^i \mid y_{<j}^i, \hat{\mathbf{y}}^a, \mathbf{V}, \mathbf{C}),$$

where y_j^i is the target intended token at decoding step j. Since intended-text labels are unavailable for training samples, pseudo-labels generated by a LLaMA-based teacher are used as supervision for intended-text prediction. The teacher is used only during training and is not part of the inference-time architecture.

During training, the decoder input sequence is prefixed with a start-of-sequence token and the prediction target is suffixed with an end-of-sequence token. This formulation enables variable-length intended-text prediction, allowing the intended output to differ in length from the actual text sequence. The decoder is initialized using multimodal context derived from the text encoder states, pooled visual memory, and confidence embedding. During intended-text training, the actual-text recognizer is kept fixed, and its predicted text, visual features, and confidence scores are used only as conditioning inputs for the intended-text predictor.

5 Experimental Setup

Experiments were conducted to evaluate actual-text transcription and intended-text prediction on child handwriting data. All models were implemented in PyTorch and trained on a high-performance computing server equipped with one NVIDIA A100 GPU (80 GB) running CUDA 12.2.

Datasets: The actual-text predictor was first pretrained on the IAM Handwriting Database using its standard train/validation/test split. The proposed framework was then finetuned on a private child handwriting corpus[1] that consists of 500 handwriting samples collected from elementary school students in Grades K–5 using an e-ink tablet [17]. These samples were segmented into 2,437 line-level instances, which were split in a writer-independent manner into 70% training, 10% validation, and 20% testing sets. In contrast to IAM, this dataset contains substantial visual and linguistic variability, including incomplete strokes, mirrored letters, irregular spacing, phonetic spellings, non-standard capitalization, and developmental writing errors. All child-handwriting samples were manually

[1] The data in this dataset was collected under a protocol approved by the Institutional Review Board (IRB) at the University of Nevada, Reno. We are exploring options to release a de-identified version of the corpus publicly.

annotated with actual text by preserving the visible written form without correcting spelling, capitalization, spacing, or word choice. Visually non-standard or mirrored character forms were retained in the corresponding actual-text annotation. The test set was additionally annotated with intended text, defined as a minimally corrected interpretation of the child's likely target expression. Since intended-text labels were unavailable for training and validation, the intended-text predictor was trained using teacher-generated pseudo-labels. The LLaMA-based teacher was used only for training supervision and is not part of the inference-time model.

Baseline Models: For actual-text transcription, the proposed model was compared with representative handwriting and OCR baselines that span recurrent, sequence-based, and transformer-based recognition architectures. Classical recurrent baselines include RNN+CTC and CRNN, which combine convolutional feature extraction with bidirectional recurrent encoders and CTC-based decoding [10,18]. Sequence-based recognizers include ASTER, MASTER, and SRN, which incorporate rectification, attention, or semantic reasoning modules originally developed for irregular scene text recognition [13,19,20]. Transformer-based OCR baselines include ViTSTR, TrOCR, and E-TrOCR, representing fully transformer-based and large-scale pretrained recognition systems [4,12,16]. Except for TrOCR, all baselines are pretrained on IAM and then fine-tuned on the child handwriting dataset. TrOCR is initialized from the publicly released `trocr-base` checkpoint, which is already pretrained on large-scale printed and handwritten text corpora; therefore, we fine-tune it directly on the child handwriting dataset and do not report IAM results for TrOCR in our pipeline.

To evaluate both actual- and intended-text prediction, we compare with LLM/VLM baselines, including Llama3, GPT-4o, Qwen, Phi-3 Vision, InternVL2-8B, and LLaVA-OneVision. These models are included because standard HTR systems do not jointly preserve *actual* child-written text and predict *intended* text. For all LLM/VLM baselines, a single prompt requested two outputs: *actual text*, defined as a literal transcription of the handwriting, and *intended text*, defined as the minimally corrected text with only spelling, spacing, or case edits. The prompt instructed the model not to rewrite heavily, not to add new content, to preserve punctuation and word order, and to keep the best literal guess in *actual text* when uncertain.

Evaluation Metrics: The actual-text predictor is trained with CTC loss, while the intended-text predictor is trained with token-level cross-entropy under teacher-guided decoding. Performance is evaluated using Character Error Rate (CER) and Word Error Rate (WER) for both actual-text transcription and intended-text prediction. These two evaluations are reported separately, since actual transcription measures fidelity to the child's written form, whereas intended transcription measures recovery of the underlying intended meaning.

Table 1. Comparison of actual-text transcription performance on the IAM and child handwriting datasets.

Model	IAM Dataset				Child Handwriting Dataset			
	Validation		Test		Validation		Test	
	CER	WER	CER	WER	CER	WER	CER	WER
RNN+CTC	20.35	50.54	12.27	39.52	13.56	41.24	13.51	39.31
CRNN	12.74	40.51	9.27	33.09	13.10	41.31	13.21	40.95
ASTER	73.59	97.80	61.84	81.40	62.18	85.25	56.50	80.23
MASTER	57.98	83.74	44.84	67.25	45.86	71.00	40.68	68.39
SRN	57.44	81.38	45.25	66.21	46.68	71.54	41.54	66.03
ViTSTR	29.16	66.14	21.96	57.45	19.30	51.35	18.37	50.33
TrOCR	–	–	–	–	17.34	41.20	14.88	36.95
E-TrOCR	18.27	46.41	10.97	35.62	14.06	40.97	13.90	40.12
Proposed	**10.17**	**34.22**	**7.60**	**28.57**	**10.92**	**35.89**	**9.89**	**28.94**

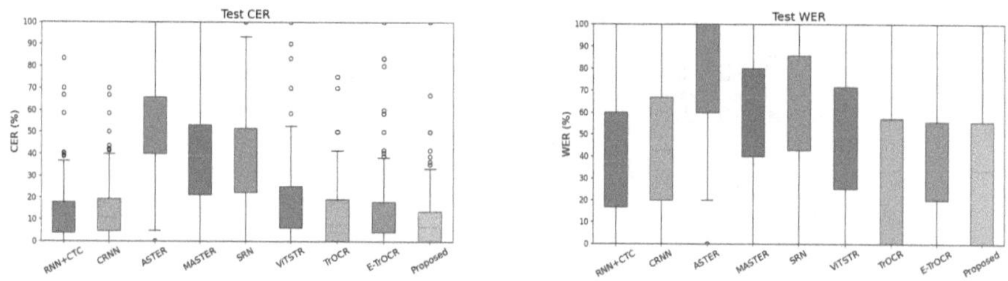

(a) (b)

Fig. 5. Distribution of actual-text recognition errors on the child handwriting test set across all baseline models. (a) Character Error Rate (CER) distribution. (b) Word Error Rate (WER) distribution.

6 Results and Discussions

6.1 Actual Text Predictor Performance

Table 1 summarizes the performance of the proposed actual text predictor and baseline models on both the IAM dataset and the child handwriting dataset. Overall, the proposed model achieves the lowest CER and WER across all evaluations, indicating strong transcription accuracy in both the adult-handwriting pretraining domain and the more challenging child-handwriting target domain.

On IAM, classical recurrent baselines such as RNN+CTC and CRNN perform reasonably well, whereas scene-text models such as ASTER, MASTER, and SRN perform poorly, suggesting limited transfer from scene-text recognition to line-level handwriting. Transformer-based OCR models improve performance by modeling longer-range dependencies more effectively, with TrOCR and E-TrOCR emerging as the strongest baselines. However, the proposed model consistently achieves the best validation and test performance on IAM, demonstrating that its actual-text prediction pipeline learns a strong and transferable handwriting representation.

A similar pattern is observed on the child handwriting dataset. Despite the substantial variability in children's writing, including irregular spacing, incomplete strokes, and developmental spelling variations, the proposed model achieves the best test performance in both CER and WER. Figure 5 supports this observation by showing lower median error and a tighter spread for the proposed method, indicating improved robustness and more consistent preservation of the exact written form produced by children.

Table 2. Comparison of foundation-model baselines and the proposed model on the child handwriting test set for actual-text and intended-text prediction.

Model	Actual Text		Intended Text	
	CER	WER	CER	WER
Qwen	23.31	38.56	13.25	24.72
Phi-3 Vision	24.24	36.67	56.40	64.87
InternVL2-8B	54.10	76.06	25.90	46.89
LLaVA-OneVision	22.66	41.44	18.47	32.41
Llama 3	8.81	24.22	13.04	23.04
GPT-4o	8.32	23.49	10.28	21.06
Proposed	9.89	28.94	19.22	30.22

6.2 Intended Text Predictor Performance

Since the proposed method achieves the best actual-text recognition performance among conventional HTR baselines in Table 1, it is used as the first-stage recognizer for intended-text experiments. Table 2 compares the complete dual-output framework with large language and vision-language baselines on the child handwriting test set for both actual-text transcription and intended-text prediction. Among all compared models, GPT-4o achieves the best overall performance, followed closely by Llama 3. These results indicate the strong generalization ability of large foundation models when applied to child handwriting understanding. However, these models are substantially heavier and require considerably greater computational and memory resources than the proposed lightweight architecture.

The proposed model achieves competitive actual-text transcription performance, with test CER and WER that remain reasonably close to those of Llama 3 and GPT-4o, while outperforming several other multimodal baselines. This suggests that the proposed architecture is effective at preserving the child's exact written form even without relying on a very large foundation model.

For intended-text prediction, GPT-4o and Llama 3 again provide the strongest results, reflecting their stronger language modeling and correction capabilities. At the same time, some baselines that perform relatively well on actual-text transcription do not necessarily yield equally strong intended-text predictions. In particular, Phi-3 Vision shows substantially degraded intended-text performance, suggesting a tendency to over-correct or generate unstable normalized outputs. Conversely, Qwen performs relatively weakly on actual-text transcription but produces noticeably better intended-text results, indicating that stronger language-level correction can partially compensate for weaker literal transcription.

Although the proposed model does not surpass the strongest commercial and large open-weight baselines on intended-text prediction, it remains competitive relative to several vision-language alternatives while being much lighter and more task-specific. This trade-off is important in practice, as it shows that a compact recognition-aware model can provide strong actual-text fidelity together with reasonable intended-text prediction, without the heavy computational cost of large foundation models.

Model	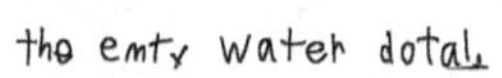			
	Actual Text	**Intended Text**	**Actual Text**	**Intended Text**
Qwen	the empty water dotal.	the empty water dotal	especially because he is alone.	especially because he is alone.
Phi-3 Vision	the berty warfed	the berty warfed	C5P6S 5a11b 6t 6t	C5P6S 5a11b 6t 6t
InternVL2-8B	tho emt x water dotal,	the emt x water dotal,	CSRSSallu6 b.Cause hC 15 aalone.	CSRSSallu6 b.Cause hC 15 alone.
LLaVA-OneVision	the entity water dotal.	the entity water dotal.	Crispally because he is alone.	Crispally because he is alone.
Llama 3	the emty water dotal,	the empty water total,	espressally because he 15 alone.	especially because he is alone.
GPT-4o	the emty water botel.	the empty water bottle.	especally because he is alone.	especially because he is alone.
Proposed	the emty water b̃otal.	the empty water total,	esPessally because he is alone.	espesally because he is alone.

Fig. 6. Qualitative comparison on child handwriting samples.

Qualitative Analysis. Figure 6 shows qualitative comparisons on challenging child handwriting samples. The examples contain poorly formed letters, missing or distorted strokes, mirrored characters, and non-standard spellings, making both actual-text and intended-text prediction difficult.

GPT-4o and Llama 3 are generally the strongest models for intended-text prediction, as their language modeling ability helps recover the likely intended meaning even when the written form is noisy. However, both models can still struggle at the actual-text level when the child's handwriting is visually ambiguous. In contrast, several vision-language baselines, especially LLaVA-OneVision, tend to autocorrect or normalize child-produced spellings already in the actual-text transcription. This behavior reduces fidelity to the original writing, which is problematic when the goal is to preserve the exact form written by the child.

The examples also indicate that mirrored or severely malformed letters remain difficult for all models, including GPT-4o and Llama 3, suggesting that these cases are driven by visual ambiguity rather than language context alone. The proposed model provides a more balanced behavior: it better preserves the child's original spelling and writing pattern in actual-text transcription, while also producing promising intended-text predictions that are reasonably close to stronger large-model baselines.

7 Conclusion

This paper introduced a dual-level transcription framework for child handwriting that distinguishes between the exact text written by the child and the text the child likely intended to write. By separating actual-text transcription from intended-text prediction, the proposed approach preserves developmentally meaningful writing patterns while still providing a readable interpretation of the handwritten content. A recognition-aware two-stage architecture was proposed to support this formulation, consisting of an actual-text predictor and a multimodal sequence-to-sequence intended-text predictor conditioned on the recognized text, visual sequence features, and recognition confidence. This design enables intended-text generation to remain grounded in visual evidence while avoiding excessive normalization. Experiments showed that the proposed method achieves strong actual-text transcription performance and promising intended-text prediction on challenging child handwriting data. The results demonstrate the value of treating child handwriting recognition as a dual-level problem rather than a single normalized transcription task. This work provides a behavior-aware foundation for future handwriting analysis systems in educational and clinical settings. Future work will explore stronger visually grounded intended-text prediction for severely ambiguous child-written forms, as well as extension of the framework toward downstream educational applications such as behavioral indicator extraction and progress monitoring.

Acknowledgments. This material is based upon work supported by the National AI Research Institutes program of the U.S. National Science Foundation and the Institute of Education Sciences, U.S. Department of Education, under Award #2229873 (National AI Institute for Exceptional Education). This work was also supported in part by a Google Foundation Research Grant. The views expressed herein are those of the authors and do not necessarily reflect the views of the National Science Foundation, the Institute of Education Sciences, the U.S. Department of Education, or Google.

References

1. Achiam, J., et al.: Gpt-4 technical report (2023). arXiv:2303.08774 arXiv preprint
2. de Araújo, S.S., Bezerra, B.L.D., de Sousa Neto, A.F.: A proposal of post-ocr spelling correction using monolingual byte-level language models. In: Proceedings of the 2025 ACM Symposium on Document Engineering, pp. 1–4 (2025)
3. de Araújo, S.S., Bezerra, B.L.D., de Sousa Neto, A.F., Zanchettin, C.: A proposal for post-ocr spelling correction using language models. In: Latinx in AI@ NeurIPS 2024 (2024)
4. Atienza, R.: Vision transformer for fast and efficient scene text recognition. In: International Conference on Document Analysis and Recognition, pp. 319–334. Springer (2021)
5. Baggett, M., Diamond, L.L., Olszewski, A.: Dysgraphia and dyslexia indicators: analyzing children's writing. Interv. Sch. Clin. **59**, 319–330 (2024)
6. Bai, S., et al.: Qwen2. 5-vl technical report. arXiv preprint arXiv:2502.13923 (2025)
7. Chung, P.J., Patel, D.R., Nizami, I.: Disorder of written expression and dysgraphia: definition, diagnosis, and management. Translational Pediatrics **9**(1), S46 (2020)
8. Corbillé, S., Anquetil, É., Fromont, É.: Precise segmentation for children handwriting analysis by combining multiple deep models with online knowledge. In: International Conference on Document Analysis and Recognition, pp. 236–252. Springer (2023)
9. Garrido-Munoz, C., Rios-Vila, A., Calvo-Zaragoza, J.: Handwritten text recognition: a survey (2025). arXiv:2502.08417 arXiv preprint
10. Graves, A., Liwicki, M., Fernández, S., Bertolami, R., Bunke, H., Schmidhuber, J.: A novel connectionist system for unconstrained handwriting recognition. IEEE Trans. Pattern Anal. Mach. Intell. **31**(5), 855–868 (2008)
11. Kang, L., Toledo, J.I., Riba, P., Villegas, M., Fornés, A., Rusinol, M.: Convolve, attend and spell: an attention-based sequence-to-sequence model for handwritten word recognition. In: German Conference on Pattern Recognition, pp. 459–472. Springer (2018)
12. Li, M., et al.: Trocr: transformer-based optical character recognition with pre-trained models. In: Proceedings of the AAAI Conference on Artificial Intelligence, vol. 37, pp. 13094–13102 (2023)
13. Lu, N., Yu, W., Qi, X., Chen, Y., Gong, P., Xiao, R., Bai, X.: Master: multi-aspect non-local network for scene text recognition. Pattern Recogn. **117**, 107980 (2021)
14. Parres, D., Anitei, D., Paredes, R.: Handwritten document recognition using pre-trained vision transformers. In: International Conference on Document Analysis and Recognition, pp. 173–190. Springer (2024)
15. Puigcerver, J.: Are multidimensional recurrent layers really necessary for handwritten text recognition? In: 2017 14th IAPR international conference on document analysis and recognition (ICDAR), vol. 1, pp. 67–72. IEEE (2017)
16. Rangasrinivasan, S., MS, S.S., Setlur, S., Jayaraman, B., Govindaraju, V.: From scribbles to text: a novel transformer-based recognition model for child handwriting. In: International Conference on Document Analysis and Recognition, pp. 115–131. Springer (2025)
17. Rangasrinivasan, S., Suresh, S., Olszewski, A., Setlur, S., Jayaraman, B., Govindaraju, V.: Ai-enhanced child handwriting analysis: a framework for the early screening of dyslexia and dysgraphia. SN Comput. Sci. **6**(5), 1–26 (2025)
18. Shi, B., Bai, X., Yao, C.: An end-to-end trainable neural network for image-based sequence recognition and its application to scene text recognition. IEEE Trans. Pattern Anal. Mach. Intell. **39**(11), 2298–2304 (2016)

19. Shi, B., Yang, M., Wang, X., Lyu, P., Yao, C., Bai, X.: Aster: An attentional scene text recognizer with flexible rectification. IEEE Trans. Pattern Anal. Mach. Intell. **41**(9), 2035–2048 (2018)
20. Yu, D., Li, X., Zhang, C., Han, J., Liu, J., Ding, E.: Towards accurate scene text recognition with semantic reasoning networks. In: 2020 IEEE/CVF Conference on Computer Vision and Pattern Recognition (CVPR), pp. 12110–12119 (2020)

Handwritten Devanagari Character Recognition: Conjunct Consonants Dataset Creation and Benchmarking with Transfer Learning

Sudip Tiwari, Satyasa Khadka, Sandhya Baral, and Sanjeeb Prasad Panday(✉)

Pulchowk Campus, Institute of Engineering, Lalitpur, Nepal
{077bct085.sudip,077bct078.satyasa,077bct076.sandhya}@pcampus.edu.np, sanjeeb@ioe.edu.np

Abstract. Handwritten optical character recognition (OCR) for the low-resource Devanagari script has advanced rapidly, yet conjunct consonants remain a critical bottleneck due to complex ligatures and diverse writing styles. We introduce HDCOCO, a curated dataset of 50,020 handwritten samples spanning 41 conjunct classes, collected across multiple institutions and public venues to capture wide demographic and stylistic variation. We validate HDCOCO through embedding-space analyses, PCA/t-SNE visualization, centroid-distance difficulty scoring, and stratified splits, confirming class separability and fair evaluation. Using consistent transfer-learning protocols, we benchmark VGG-11/16 and ResNet-34/50, identifying ResNet-34 as optimal (SOTA results of 99.15% on HDCOCO, 99.57% on DHCD). Our results demonstrate that modern CNNs with careful augmentation provide reliable recognition for challenging conjuncts, and HDCOCO meaningfully extends prior resources focused on simple characters. We release preprocessing and validation procedures to facilitate reproducible research on diacritics, additional conjuncts, and deployment models, providing a foundation for state-of-the-art Devanagari OCR on real-world handwriting.

Keywords: Devanagari · Optical character recognition · conjunct consonants · handwriting recognition · transfer learning · CNN · Benchmarks

1 Introduction

Optical Character Recognition (OCR) remains a cornerstone technology for enabling the digitization and searchability of handwritten and printed materials. Handwritten character recognition for the Latin script, driven in part by

---

Dataset available at https://doi.org/10.5281/zenodo.20742112.
Code and preprocessing pipeline available at our GitHub repository.
S. Tiwari, S. Khadka and S. Baral—These authors contributed equally to this work.

F. Shafait et al. (Eds.): DAS 2026, LNCS 16851, pp. 389–405, 2027.
https://doi.org/10.1007/978-3-032-36207-0_23

क	का	कि	की	कु	कू	कृ	कॄ	कॢ	कॣ
ɐ	aː	i	iː	u	uː	r̩	r̩ː	l̩	l̩ː
के	कै	को	कौ	कं	कः	कॅ	कॉ		
eː	aj	oː	aw	õ	h	æ	ɒ		

Fig. 1. Devanagari Consonants with Diacritics.

the global prevalence of English with approximately 372 million native speakers [25], has been extensively studied and has achieved near-human performance in many benchmarks. In contrast, the Devanagari script—used to write major South Asian languages such as Hindi, Marathi, Nepali, and Sanskrit—serves over 532 million speakers collectively [25]. Despite this widespread use, recognition of handwritten Devanagari text continues to lag behind due to the script's rich structural complexity, as shown in Fig. 1.

In languages written with complex Indic scripts such as Devanagari, OCR is critical for cultural preservation, digital archiving, and downstream applications in natural language processing. A defining challenge is the treatment of **conjunct consonants** (ligatures formed by combining successive consonants without intervening vowels). These conjuncts are ubiquitous in real-world text, encode essential phonetic distinctions, and often exhibit substantial stylistic variation across writers. Existing datasets and benchmarks, such as the *Devanagari Handwritten Character Dataset (DHCD)* [1] and related works [2], primarily emphasize isolated vowels, numerals, or simple consonants. As a result, current OCR systems fail to generalize when confronted with conjunct consonants, leading to significant drops in accuracy in practical handwritten text.

Despite steady progress in CNN-based approaches [3,25], three persistent gaps remain:

1. **Underrepresentation of conjunct consonants:** Publicly available datasets contain no samples of such frequent ligatures, leaving OCR systems undertrained for these forms.
2. **Lack of demographic and stylistic diversity:** Most datasets are collected from limited age or educational groups, restricting robustness to natural handwriting variation.
3. **Incomplete benchmarks:** Few studies provide systematic baselines across multiple modern CNN architectures using a controlled experimental protocol.

To address these gaps, we make the following contributions:

- **HDCOCO Dataset:** We introduce *HDCOCO* [26] (Handwritten Devanagari COnjunct COnsonants), a balanced corpus of 50,020 samples covering 41 conjunct classes. Data were collected from schools and public exhibitions to capture the broad demographic diversity.

- **Rigorous Dataset Validation:** We perform embedding-based visualization (PCA/t-SNE), centroid-distance difficulty scoring, and stratified train–test splits to confirm class separability and evaluation fairness.
- **Transfer-Learning Benchmarks:** We benchmark four representative CNN architectures—VGG-11, VGG-16, ResNet-34, and ResNet-50—under a unified transfer-learning framework.
- **Reproducibility Protocol:** We release preprocessing pipelines and training procedures, providing a strong and transparent baseline for future research on Devanagari OCR.

By filling a critical dataset gap and establishing controlled benchmarks, our work advances the recognition of Devanagari conjunct consonants and strengthens the foundation for scalable OCR systems in low-resource languages.

Both the HDCOCO dataset and the accompanying code are publicly released to support reproducible research; we kindly ask that users cite this paper and the dataset DOI [26] when using HDCOCO in their work.

2 Literature Review

Handwritten character recognition in Indic scripts sits at the intersection of pattern recognition, document analysis, and deep learning. For *Devanagari*, the challenge is amplified by its composite character system: vowels, consonants, and compound forms such as conjunct consonants and diacritic combinations. Effective recognition demands both robust preprocessing pipelines and high-capacity classifiers.

2.1 Preprocessing and Feature Extraction

Early work emphasized handcrafted features combined with statistical classifiers. Techniques such as Histogram of Oriented Gradients (HOG) [5], Scale-Invariant Feature Transform (SIFT) [14], Local Binary Patterns (LBP) [16], and corner detectors like Features from Accelerated Segment Test (FAST) [19] and Speeded Up Robust Features (SURF) [4] were widely used to encode stroke structure and texture [17,21]. These features were typically paired with classifiers including quadratic discriminants, k-Nearest Neighbors, Support Vector Machines (SVMs), and multilayer perceptrons. Reported accuracies ranged from 80–95%, but performance degraded on noisy inputs and complex ligatures, reflecting the limits of feature engineering.

2.2 Deep Learning Architectures

Architectures ranging from early CNNs such as LeNet-5 [13], AlexNet [12], and VGG [22] to deeper and more sophisticated designs like Inception [24], ResNet [10], DenseNet [11], and the Vision Transformer (ViT) [7] have successively advanced image classification performance. These networks progressively replaced hand-engineered features with learned hierarchical representations, offering robustness to handwriting variability.

For Devanagari, Acharya et al. [2] applied deep CNNs to large-scale recognition, reaching 98.47% accuracy. Aneja et al. [3] systematically benchmarked multiple CNNs (AlexNet, DenseNet, VGG, Inception), using transfer learning to attain up to 99% accuracy on 46 classes. More recent studies [25] explored lightweight CNNs such as LeNet variants for resource-constrained settings, balancing accuracy (92–95%) with low computational cost. While these works demonstrated the power of CNNs, their focus remained on simple vowels and consonants, with conjunct consonants largely absent.

2.3 Datasets and Resource Gaps

Several public datasets have supported progress in handwritten Devanagari character recognition:

- **Devanagari Handwritten Character Dataset (DHCD)** (Acharya and Gyawali, 2015) [1]—92k images covering 46 classes (vowels, consonants, numerals).
- **Devanagari (Nepali) Handwritten Character Dataset (NHCD)** (Pant et al., 2012) [18]—Nepali handwritten characters collected from multiple individuals.
- **Handwritten Devanagari Characters-Vowels & Numerals Dataset (HDCVND)** (Duddela and Mehta, 2020) [20]—vowels and numerals.
- **Handwritten Marathi Devanagari Characters Dataset (HMDCD)** (Deore, 2018) [6]—handwritten Marathi characters.
- **Comprehensive Devanagari Compound Characters (CDCC)** (Meenakshi et al., 2024) [15] - 36k images with 600 samples per class; includes 10 numerals, 13 vowels, 17 similar-looking consonants and 20 compound characters.

While the early corpora (DHCD, NHCD, HDCVND, HMDCD) catalyzed progress, they share a common gap: the near-total absence of conjunct consonants (CC), consonants with diacritics (CD), and conjunct consonants with diacritics (CCD). Table 1 highlights this imbalance, showing that despite tens of thousands of samples for basic vowels, consonants, and numerals, coverage of conjunct forms has remained at zero for years.

Only very recently has this gap begun to be addressed. The **CDCC dataset** (Meenakshi et al., 2024) [15] introduces 36k images across 60 classes, including 20 compound characters. This is a welcome step, yet its conjunct coverage is extremely limited: just 600 samples per class and only 20 different characters, far below what is required to capture the real-world setting. In practice, OCR models trained on such small conjunct subsets remain brittle, failing to generalize in real-world application.

The implications are clear. Conjunct consonants are not rare edge cases; they are indispensable to the integrity of Devanagari scripts. Building robust OCR without them is akin to designing English OCR that recognizes letters but fails on common digraphs. CDCC's introduction shows the research community now

Table 1. Comparison of existing handwritten Devanagari datasets. Nearly all omit conjunct consonants (CC), consonants with vowel diacritics (CD), and conjunct consonants with vowel diacritics (CCD).

DS	V	C	N	CC	CD	CCD
DHCD total	0	72,000	20,000	0	0	0
DHCD per class	0	2,000	2,000	0	0	0
HDCVND total	16,250	0	22,500	0	0	0
HDCVND per class	1,250	0	2,250	0	0	0
NHCD total	2,652	7,380	2,880	0	0	0
NHCD per class	221	205	288	0	0	0
HMDCD total	1,200	3,600	1,000	0	0	0
HMDCD per class	100	100	100	0	0	0
CDCC total	7,800	10,200	6,000	12,000	0	0
CDCC per class	600	600	600	600	0	0

acknowledges the problem, but it also underscores how little ground has yet been covered. The field urgently needs large-scale, balanced conjunct consonant datasets that can support deep learning models at the same scale as DHCD and similar corpora did for simple characters. This gap has real consequences beyond dataset statistics: Gurung et al. [9] report that their CNN-based Devanagari HTR system struggled specifically on ligatures due to a lack of training samples, a failure mode HDCOCO directly targets.

2.4 Summary

In short, the literature establishes that: (1) handcrafted pipelines plateau well below usable accuracy; (2) modern CNNs with transfer learning achieve state-of-the-art results on isolated characters; and (3) despite CDCC, the availability of conjunct consonant data remains orders of magnitude too small for robust OCR. This paper takes the next step by introducing **HDCOCO**, a 50,020-sample dataset covering 41 conjunct consonants with rigorous validation and demographic diversity. Alongside, we benchmark leading CNN architectures under a controlled protocol, setting reproducible baselines to drive future research in this long-neglected area.

3 Methodology

3.1 Data Collection

A comprehensive dataset is critical for robust handwritten character recognition. Our collection process was designed to ensure both coverage of diverse conjunct consonants and inclusion of handwriting variations across demographics.

ख्य	ग्र	घ्र	ङ्ग	च्छ	ञ्ज
ट्ट	ठ्य	ड्ग	ढ्य	ण्ड	थ्य
द्ध	ध्य	न्त	प्र	ब्र	भ्य
म्य	य्य	र्म	ल्य	व्य	ष्ठ
ष्ट	स्त	ह्म	स्य	श्व	र्व
त्य	त्व	त्त	न्य	र्य	द्र
श्र	द्य	क्र	द्व	न्न	

Fig. 2. Selected 41 conjunct consonants for data collection.

क्ष	ज्ञ	त्र

Fig. 3. Three conjunct consonants already covered in DHCD [1].

Selection of Characters: The selection followed a two-step strategy:

1. Conjunct consonants with >5% frequency in Sanskrit texts, as identified by Stiehl [23].
2. The most frequently used conjunct for each base consonant, as suggested by Thatte [25] and Stiehl [23].

This yielded 44 candidate classes. Three were excluded as they are already part of DHCD [1], leaving 41 novel conjunct consonants.

Figure 2 shows the final 41 characters, and Fig. 3 highlights 3 CCs excluded.

Collection Process: Data were gathered from volunteers in three phases using standardized A4 sheets with 50 grid boxes. Together these phases provided broad demographic diversity. Representative samples are shown in Fig. 4.

(a) Data collected from schools.

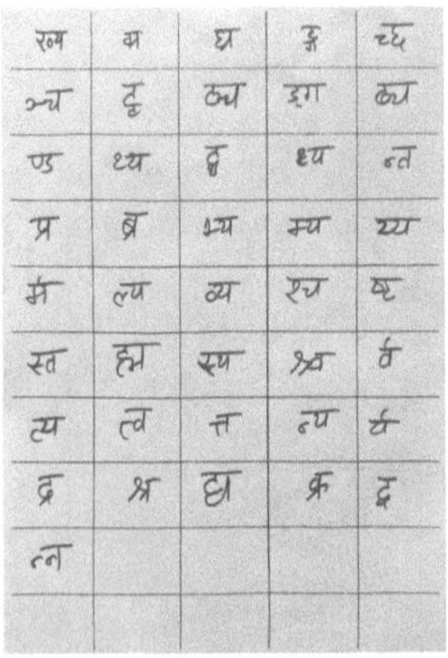

(b) Data collected from exhibition visitors.

Fig. 4. Representative handwriting samples.

3.2 Preprocessing

Sheets were scanned and cropped into individual images using grid segmentation with tunable offsets. Misaligned or unclear samples were discarded (Fig. 5). Images were converted to grayscale, binarized with Otsu's thresholding, and resized to 64×64 pixels using Lanczos interpolation (Fig. 6). To improve robustness, augmentations (random rotation $\pm 10°$, random cropping, normalization to ImageNet statistics) were applied (Figs. 7–8).

Fig. 5. Cropped characters extracted from a data sheet.

(a) Grayscale cropped samples

(b) Resized binarized images

Fig. 6. Intermediate results in the preprocessing pipeline.

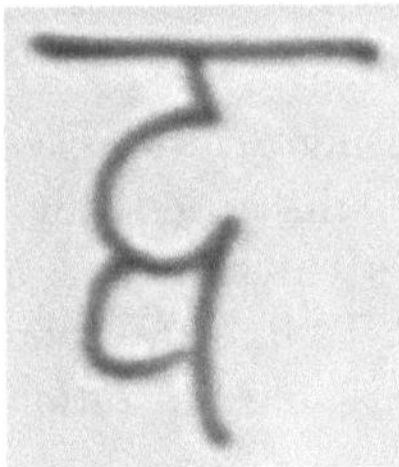

Fig. 7. Original sample before augmentation.

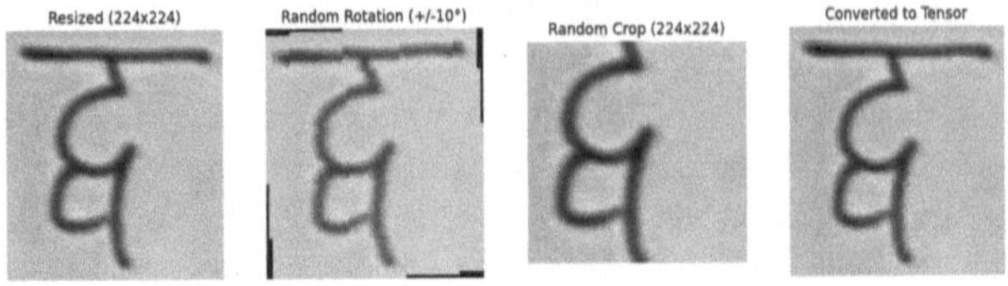

Fig. 8. Augmented samples (rotation and cropping).

3.3 Dataset Validation

We use "validation" here to mean confirming that HDCOCO is well-posed for benchmarking, i.e., that its class labels correspond to visually separable categories and that train–test splits yield a fair, unbiased evaluation. We conducted a multi-stage validation:

1. **Data Loading:** PyTorch's `ImageFolder` was used for organization. Images were resized to 224×224, converted to 3-channel grayscale, and normalized.
2. **Embeddings:** Features were extracted using ResNet18 (head removed).
3. **Visualization:** PCA+t-SNE projections (2D, 3D) showed clear inter-class separation and intra-class cohesion (Fig. 9).
4. **Difficulty Scoring:** Euclidean distance to class centroid was used as a complexity metric, yielding a long-tailed distribution (Fig. 10).
5. **Splitting:** StratifiedShuffleSplit ensured class balance. Histograms across multiple seeds confirmed matched difficulty distributions (Fig. 11).

The t-SNE/PCA clustering (Fig. 9) shows limited inter-class overlap, confirming that our 41 classes are visually distinguishable rather than ambiguous. The long-tailed difficulty-score distribution (Fig. 10) indicates HDCOCO retains genuine handwriting variation rather than being artificially clean. Finally, seed-stable train–test difficulty distributions (Fig. 11) confirm our splits are unbiased, so the reported accuracies are not an artifact of an easy test set.

3.4 Model Selection and Training

Following Aneja et al. [3], four CNNs were chosen: VGG11, VGG16, ResNet34, ResNet50. Pretrained ImageNet weights were fine-tuned with transfer learning. Training was performed on Kaggle with dual NVIDIA Tesla T4 GPUs. Architectures were first validated on DHCD before benchmarking on HDCOCO to establish reproducible baselines.

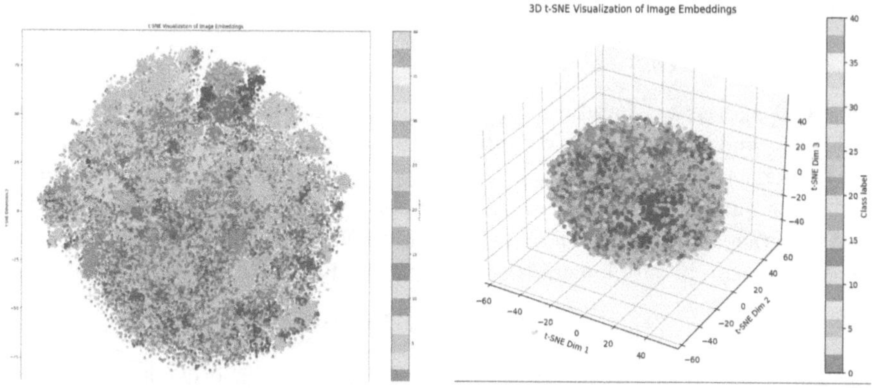

(a) 2D t-SNE

(b) 3D t-SNE

Fig. 9. t-SNE visualization of embeddings.

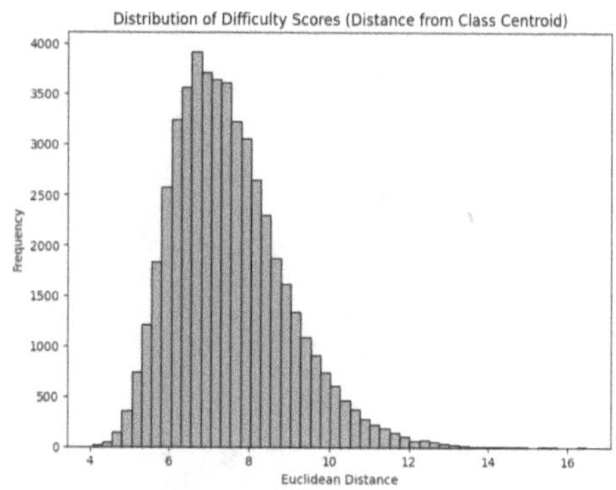

Fig. 10. Distribution of difficulty scores.

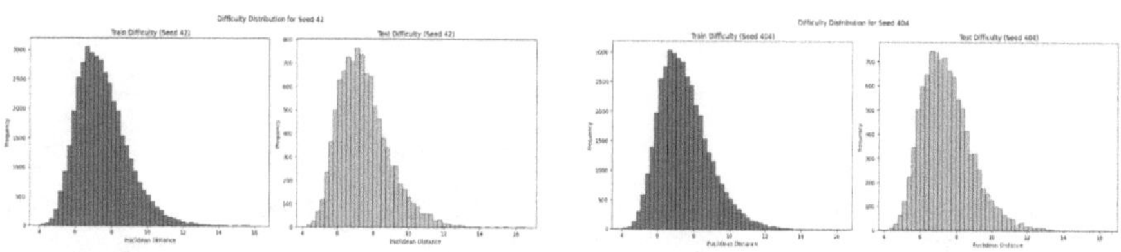

(a) Train and Test Difficulty Score Histogram for seed = 42.

(b) Train and Test Difficulty Score Histogram for seed = 404.

Fig. 11. Comparison of Difficulty Score Distributions for Train and Test Splits.

4 Results

4.1 HDCOCO Dataset Summary

HDCOCO aggregates handwriting from diverse sources (two schools, a public exhibition, and independent volunteers) to capture demographic and stylistic variation. Each of the 41 conjunct classes was initially targeted at ∼1,800 samples. After quality control (grid misalignment, stroke loss, duplicates), usable counts averaged ∼1,200/class. We then enforced class balance at **1,220 samples per class** via supplemental collection, yielding a final dataset of

$$41 \times 1,220 \;=\; \textbf{50,020 images.}$$

This balancing avoids skewed training dynamics and ensures that evaluation metrics are not dominated by high-frequency classes.

4.2 Benchmarking on DHCD (Protocol Validation)

Before evaluating conjuncts, we reproduced transfer-learning baselines on DHCD to verify the training protocol and hyperparameter ranges. All models achieved high accuracy; notably:

VGG-11: 97.03% test accuracy (Fig. 12).

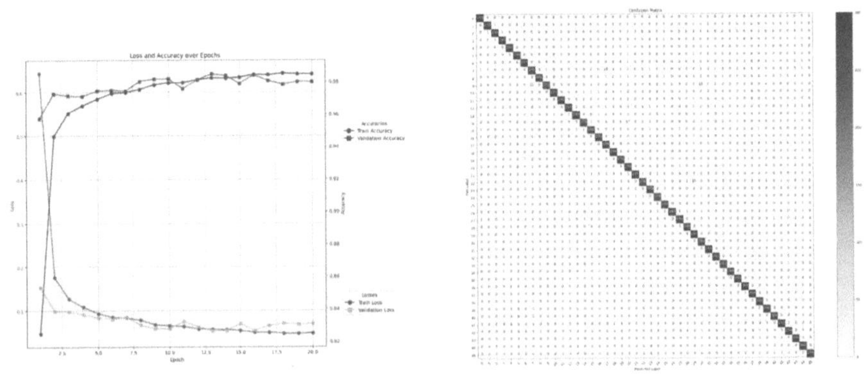

(a) Train/Test Accuracy and Loss for VGG-11 on DHCD.

(b) Confusion Matrix for VGG-11 on DHCD.

Fig. 12. Results of VGG-11 on DHCD.

VGG-16: 99.57% test accuracy (Fig. 13).

ResNet-34: 99.57% test accuracy (Fig. 14).

These results align with prior reports and confirm that our pipeline (preprocessing, augmentation, optimization) is competitive and stable.

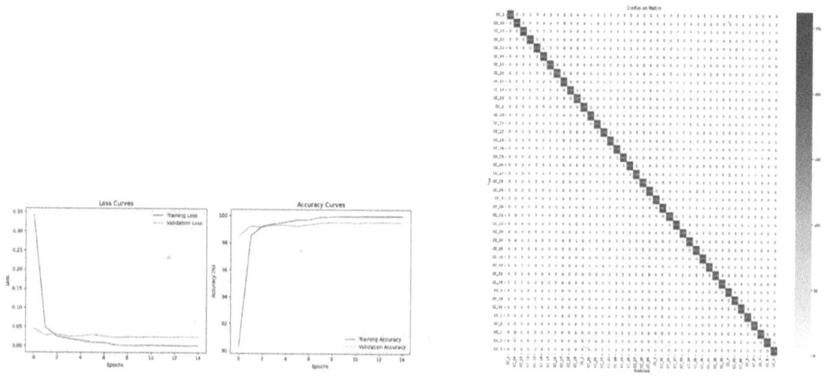

(a) Train/Test Accuracy and Loss for VGG-16 on DHCD.

(b) Confusion Matrix for VGG-16 on DHCD.

Fig. 13. Results of VGG-16 on DHCD.

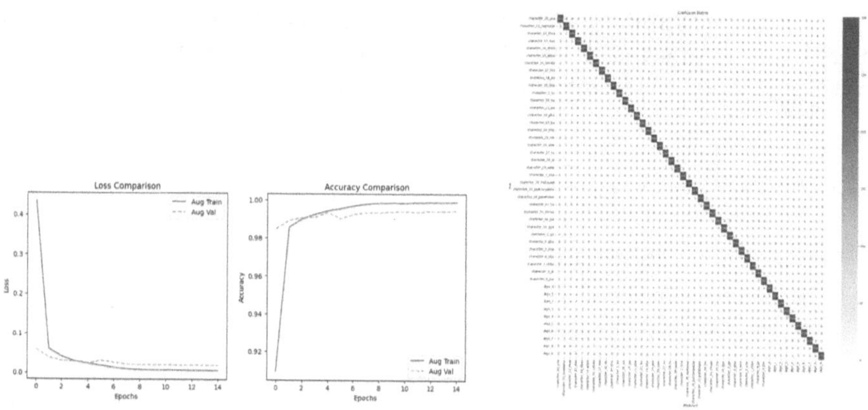

(a) Train/Test Accuracy and Loss for ResNet-34 on DHCD.

(b) Confusion Matrix for ResNet-34 on DHCD.

Fig. 14. Results of ResNet-34 on DHCD.

4.3 Benchmarking on HDCOCO (Conjunct Consonants)

We then fine-tuned the same models on HDCOCO using the validated protocol.

VGG-11: With standard augmentations and ImageNet initialization, VGG-11 converged smoothly (Fig. 15), reaching **98.75%** test accuracy. The confusion matrix (Fig. 15) indicates that residual errors are concentrated among visually similar ligatures (e.g., clusters with overlapping sub-strokes), which is expected given VGG's shallower receptive-field interactions.

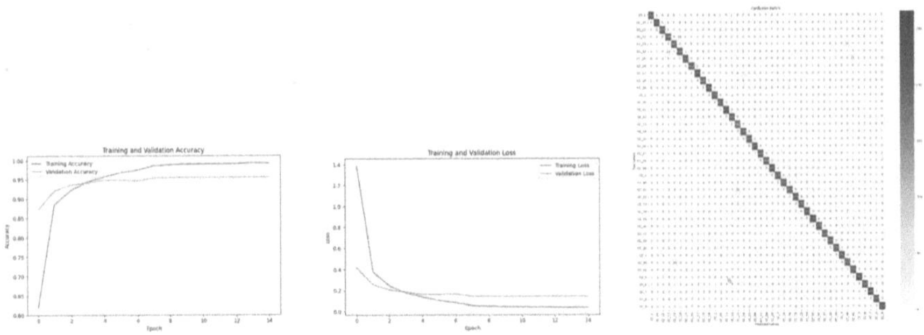

(a) Train/Test Accuracy plot (b) Train/Test Loss plot (c) Confusion Matrix

Fig. 15. Results of VGG-11 on HDCOCO.

ResNet-34: ResNet-34 provided the best accuracy–efficiency trade-off. Training completed in **33m56s** with strong generalization: **99.21%** validation and **99.15%** test accuracy (Fig. 16). Residual connections appear to help disentangle stroke junctions and broken ligatures without overfitting.

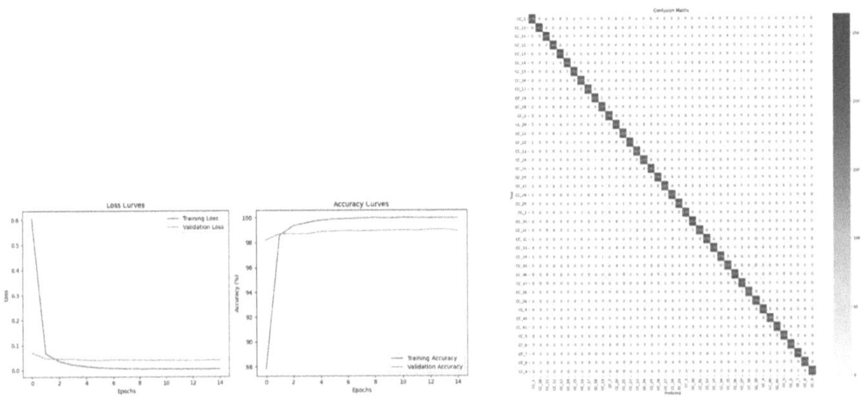

(a) Train/Test Accuracy and Loss plot (b) Confusion Matrix

Fig. 16. Results of ResNet-34 on HDCOCO.

VGG-16: VGG-16 attained **98.96%** test accuracy after **107m10s** (Fig. 17). Despite excellent training accuracy (99.93%), the gap to validation suggests sensitivity to subtle stroke perturbations that ResNet's skip connections handle better.

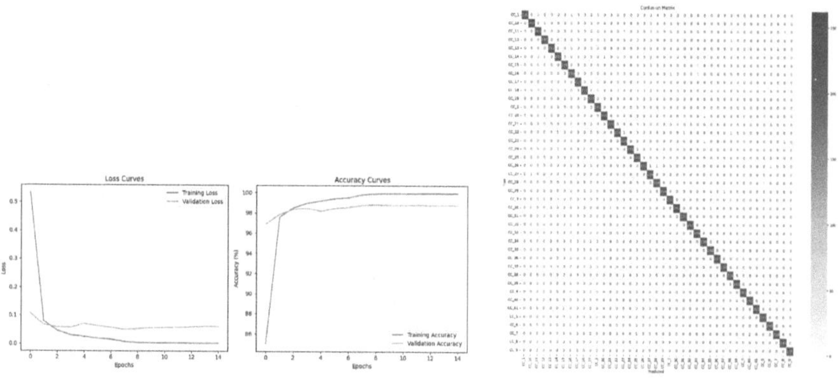

(a) Train/Test Accuracy and Loss plot (b) Confusion Matrix

Fig. 17. Results of VGG-16 on HDCOCO.

ResNet-50: ResNet-50 achieved **99.12%** test accuracy in **61m14s** (Fig. 18). Deeper capacity did not translate into a clear accuracy gain over ResNet-34—likely because class boundaries are already well captured at 34 layers once augmentation is adequate.

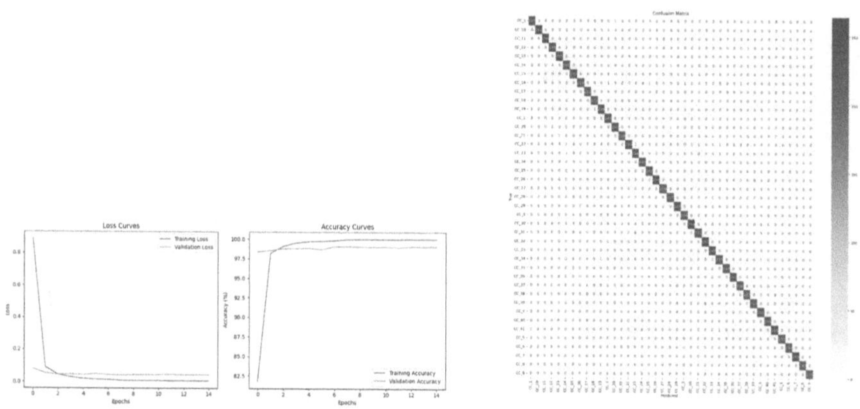

(a) Train/Test Accuracy and Loss plot (b) Confusion Matrix

Fig. 18. Results of ResNet-50 on HDCOCO.

4.4 Analysis

(1) Conjuncts are Learnable at Scale: All models exceed 98.7% on HDCOCO, indicating that balanced, diverse conjunct data unlocks performance comparable to simple letters on DHCD.

(2) Accuracy–Efficiency Frontier: *ResNet-34* is the preferred baseline: near-top accuracy with the lowest training time among the top performers. VGG-16 matches DHCD accuracy but trains substantially longer on HDCOCO with no corresponding test gain.
(3) Error Structure: Confusions cluster between classes sharing similar sub-ligatures or stroke orderings; reducing these errors will likely require (i) targeted augmentation along stroke topology, (ii) finer or higher input resolution, or (iii) hybrid CNN–attention blocks to model long-range stroke dependencies.
(4) Benchmark Value: Taken together, DHCD reproductions validate our pipeline, and HDCOCO results establish the first strong, reproducible baselines focused on conjunct consonants—precisely where public benchmarks have been weakest.

4.5 Summary Table

Table 2 reports the key metrics for all CNNs on HDCOCO. ResNet-34 achieved the best balance of accuracy and efficiency, while VGG-16 required substantially more training time with no corresponding gain in test accuracy. All models exceeded 98.7% test accuracy, confirming that conjunct consonants are learnable at scale.

Table 2. HDCOCO benchmarking summary (per-model end metrics).

Model	Train Acc (%)	Test Acc (%)	Training Time
VGG-11	99.20	98.75	n/a*
ResNet-34	99.92	**99.15**	**33m56s**
VGG-16	99.93	98.96	107m10s
ResNet-50	99.98	99.12	61m14s

*Training time not recorded under identical hardware constraints.

Our results are summarized in Table 3.

Table 3. Comparison of Test Accuracy on DHCD and HDCOCO dataset with previous benchmark results.

Model	Test Acc on DHCD		Test Acc on HDCOCO	
	Aneja '19	Ours	Aneja '19	Ours
VGG-11	99	97.03	–	98.75
VGG-16	98	**99.57**	–	98.96
ResNet-34	–	**99.57**	–	**99.15**
ResNet-50	–	–	–	99.12

5 Conclusion

This work addresses a critical gap in Devanagari OCR: the recognition of handwritten conjunct consonants. We introduced **HDCOCO**, a balanced dataset of 50,020 images spanning 41 conjunct classes, collected across diverse demographics to capture natural handwriting variability. Through rigorous validation (embedding separability, centroid-based difficulty scoring, stratified splits), we confirmed the dataset's suitability for benchmarking.

Using transfer learning, we systematically evaluated VGG-11, VGG-16, ResNet-34, and ResNet-50 on both DHCD and HDCOCO. Our results demonstrate state-of-the-art performance: **ResNet-34 achieved 99.57% accuracy on DHCD and 99.15% on HDCOCO**, outperforming prior baselines while maintaining computational efficiency. These benchmarks establish HDCOCO as the first large-scale, validated resource for conjunct consonant recognition and provide reproducible reference points for future OCR research.

In summary, HDCOCO and its baselines fill a longstanding gap in Devanagari OCR, extending robust recognition beyond simple vowels and consonants to the complex conjunct forms essential for real-world applications.

6 Future Scope

Despite these contributions, several challenges remain:

- **Dataset expansion:** Many conjunct consonants, consonants with diacritics (CD), and conjunct+diacritic combinations (CCD) are still absent. Extending HDCOCO to include these forms is essential for real-world utility (e.g., downstream improvement in full-text recognition) within a segmentation-and-recognition pipeline as proposed by Gurung et al. [9]. Notably, Gurung et al. [9] report that ligatures and conjuncts remain a primary failure mode in CNN-based Devanagari HTR due to insufficient training data, precisely the gap HDCOCO is designed to address.
- **Model generalization:** Future work can explore vision transformers, hybrid CNN–attention architectures, and self-/few-shot learning to improve robustness on visually similar ligatures and low-resource classes.
- **Synthetic data:** Stroke-based rendering or GAN-based augmentation [8] could address data scarcity for rare conjuncts and improve model resilience.

By tackling these directions, researchers can build on HDCOCO to achieve more accurate, scalable, and deployment-ready OCR systems for Devanagari and related Indic scripts.

References

1. Acharya, S., Gyawali, P.: Devanagari handwritten character dataset. UCI Machine Learning Repository (2015). https://doi.org/10.24432/C5XS53

2. Acharya, S., Pant, A.K., Gyawali, P.K.: Deep learning based large scale handwritten devanagari character recognition. In: 2015 9th International Conference on Software, Knowledge, Information Management and Applications (SKIMA), pp. 1–6 (2015). https://doi.org/10.1109/SKIMA.2015.7400041
3. Aneja, N., Aneja, S.: Transfer learning using CNN for handwritten Devanagari character recognition. In: 2019 1st International Conference on Advances in Information Technology (ICAIT), pp. 293–296 (2019). https://doi.org/10.1109/ICAIT47043.2019.8987286
4. Bay, H., Tuytelaars, T., Van Gool, L.: SURF: speeded up robust features. In: Leonardis, A., Bischof, H., Pinz, A. (eds.) ECCV 2006, Part I. LNCS, vol. 3951, pp. 404–417. Springer, Heidelberg (2006). https://doi.org/10.1007/11744023_32
5. Dalal, N., Triggs, B.: Histograms of oriented gradients for human detection. In: 2005 IEEE Computer Society Conference on Computer Vision and Pattern Recognition (CVPR'05), vol. 1, pp. 886–893 (2005). https://doi.org/10.1109/CVPR.2005.177
6. Deore, S.P., Pravin, A.: Devanagari handwritten character recognition using fine-tuned deep convolutional neural network on trivial dataset. Sādhanā **45**(1), 1–13 (2020). https://doi.org/10.1007/s12046-020-01484-1
7. Dosovitskiy, A., et al.: An image is worth 16x16 words: transformers for image recognition at scale (2021). https://arxiv.org/abs/2010.11929
8. Goodfellow, I., et al.: Generative adversarial networks. Commun. ACM **63**(11), 139–144 (2020). https://doi.org/10.1145/3422622
9. Gurung, B., Thapa, M., Shrestha, R., Sthapit, R.: Digitization of Devanagari handwritten text using CNN. Int. J. Eng. Technol. Infrast. Dev. **2**(1), 81–105 (2025). https://doi.org/10.3126/injet-indev.v2i1.82456, https://www.nepjol.info/index.php/injetindev/article/view/82456
10. He, K., Zhang, X., Ren, S., Sun, J.: Deep residual learning for image recognition. In: 2016 IEEE Conference on Computer Vision and Pattern Recognition (CVPR), pp. 770–778 (2016). https://doi.org/10.1109/CVPR.2016.90
11. Huang, G., Liu, Z., Van Der Maaten, L., Weinberger, K.Q.: Densely connected convolutional networks. In: 2017 IEEE Conference on Computer Vision and Pattern Recognition (CVPR), pp. 2261–2269 (2017). https://doi.org/10.1109/CVPR.2017.243
12. Krizhevsky, A., Sutskever, I., Hinton, G.E.: Imagenet classification with deep convolutional neural networks. In: Proceedings of the 26th International Conference on Neural Information Processing Systems - Volume 1. NIPS'12, Red Hook, NY, USA, pp. 1097–1105. Curran Associates Inc. (2012)
13. Lecun, Y., Bottou, L., Bengio, Y., Haffner, P.: Gradient-based learning applied to document recognition. Proc. IEEE **86**(11), 2278–2324 (1998). https://doi.org/10.1109/5.726791
14. Lowe, D.G.: Distinctive image features from scale-invariant keypoints. Int. J. Comput. Vision **60**(2), 91–110 (2004). https://doi.org/10.1023/B:VISI.0000029664.99615.94
15. Meenakshi, Premjith, B., Sowmya, V., Jyotish Lal, G.: a new comprehensive dataset and deep learning approach for devanagari handwritten character recognition with special attention to compound characters. In: Rivera, G., Pedrycz, W., Moreno-Garcia, J., Sánchez-Solís, J.P. (eds.) Innovative Applications of Artificial Neural Networks to Data Analytics and Signal Processing. Studies in Computational Intelligence, vol. 1221, pp. 417–437. Springer, Cham (2024). https://doi.org/10.1007/978-3-031-69769-2_17, https://doi.org/10.1007/978-3-031-69769-2_17

16. Ojala, T., Pietikäinen, M., Mäenpää, T.: Multiresolution gray-scale and rotation invariant texture classification with local binary patterns. IEEE Trans. Pattern Anal. Mach. Intell. **24**(7), 971–987 (2002). https://doi.org/10.1109/TPAMI.2002.1017623
17. Pal, U., Wakabayashi, T., Kimura, F.: Comparative study of devnagari handwritten character recognition using different feature and classifiers. In: 2009 10th International Conference on Document Analysis and Recognition, pp. 1111–1115 (2009). https://doi.org/10.1109/ICDAR.2009.244
18. Pant, A., Panday, S., Joshi, S.: Off-line Nepali handwritten character recognition using multilayer perceptron and radial basis function neural networks, pp. 1–5 (2012). https://doi.org/10.1109/AHICI.2012.6408440
19. Rosten, E., Drummond, T.: Machine learning for high-speed corner detection. In: Leonardis, A., Bischof, H., Pinz, A. (eds.) ECCV 2006, Part I. LNCS, vol. 3951, pp. 430–443. Springer, Heidelberg (2006). https://doi.org/10.1007/11744023_34
20. Sai Prashanth, D., Kumar Mehta, R.V.: Handwritten Devanagari characters - vowels and numerals (38,750 isolated images + 9,200 isolated images). Mendeley Data, V4 (2023). https://doi.org/10.17632/pxrnvp4yy8.4
21. Sharma, N., Pal, U., Kimura, F., Pal, S.: Recognition of off-line handwritten devnagari characters using quadratic classifier. In: Kalra, P.K., Peleg, S. (eds.) ICVGIP 2006. LNCS, vol. 4338, pp. 805–816. Springer, Heidelberg (2006). https://doi.org/10.1007/11949619_72
22. Simonyan, K., Zisserman, A.: Very deep convolutional networks for large-scale image recognition (2015). https://arxiv.org/abs/1409.1556
23. Stiehl, U.: Die devanagari-grundzeichen (the basic characters of Devanagari) (2007). https://www.sanskritweb.net/deutsch/devanagari.pdf
24. Szegedy, C., et al.: Going deeper with convolutions. In: 2015 IEEE Conference on Computer Vision and Pattern Recognition (CVPR), pp. 1–9 (2015). https://doi.org/10.1109/CVPR.2015.7298594
25. Thatte, R.: Handwritten Devanagari character recognition using deep learning (2022). https://doi.org/10.47611/harp.228
26. Tiwari, S., Khadka, S., Baral, S.: Handwritten Devanagari conjunct consonants (HDCOCO) dataset (2025). https://doi.org/10.5281/zenodo.20742112

Historical, Symbolic and Camera-Based Documents

Edge-Aware Relation Prediction for Symbol-Level Structure Refinement in Online Handwritten Mathematical Expressions

Thanh-Nghia Truong[1(✉)], Hung Tuan Nguyen[1], Cuong Tuan Nguyen[2], and Masaki Nakagawa[1]

[1] Tokyo University of Agriculture and Technology, Tokyo 184-8588, Japan
{fw7852,fx7297}@go.tuat.ac.jp, nakagawa@cc.tuat.ac.jp
[2] Vietnamese-German University, Ho Chi Minh City, Vietnam
cuong.nt2@vgu.edu.vn

Abstract. Handwritten mathematical expression (HME) recognition requires not only recognizing symbols but also determining their two-dimensional structural relations. Typically, a symbol relation recognizer followed by a 2D context-free grammar parser produces a primitive graph, where each node is a recognized symbol and each edge is a relation between two nodes. In this paper, we focus on symbol-level structure refinement building upon the primitive graph from an online HME. First, we formulate refinement as a multi-class relation prediction problem over candidate edges that are extracted from the primitive graph. Second, we combine primitive relation priors, confidence scores, and pairwise symbol geometry as edge features for an edge-aware graph neural network based on message passing and edge classification. Third, we introduce two auxiliary losses for edge existence and positive-edge relation prediction, to address relation class imbalance in the primitive graphs. Evaluation results on the test sets of the competition on recognition of online HME (CROHME) in 2014, 2016, and 2019 demonstrate that the proposed method substantially improves over the primitive graph baseline and over prior binary link-prediction refinement. In summary, high-recall primitive graphs provide a useful starting point for symbol-level structure refinement and explicit structure recovery in online HMEs.

Keywords: online handwritten mathematical expression recognition · graph neural network · graph refinement · edge prediction

1 Introduction

Recognizing handwritten mathematical expressions (HMEs) remains challenging because a correct interpretation depends not only on symbol identities, but also on the two-dimensional structural relations among symbols. Superscripts, subscripts, fractions, roots, and nested layouts are inherently spatial. As a result, a

F. Shafait et al. (Eds.): DAS 2026, LNCS 16851, pp. 409–424, 2027.
https://doi.org/10.1007/978-3-032-36207-0_24

system that outputs only a one-dimensional sequence may achieve strong transcription accuracy while still obscuring the explicit structure needed for editing, semantic post-processing, or graph-based downstream reasoning [1,2]. We focus on online HME recognition to improve its performance in educational scenarios such as inputting a formula into Moodle [3] and automatic scoring of HME answers [4].

Earlier work on online HME recognition addressed this problem through explicit structural modeling. Grammar-based and parser-based methods used stochastic grammars, structural analysis, and tree construction to recover mathematical layout from online handwriting [5–8]. These approaches established the importance of explicit symbolic structure, but their overall performance depends strongly on the quality of symbol segmentation and relation hypotheses.

Recent studies have revisited the explicit structure learning problem using deep neural networks. Nguyen et al. [9] proposed a temporal classification framework with global context for symbol segmentation, symbol recognition, and relation classification. Truong et al. [10] then proposed learning a Symbol Relation Tree (SRT) more directly from online trajectories. On the other hand, graph neural networks (GNNs) have gained attention in HME recognition, as graph representations provide a natural formulation of the relational structure of mathematical expressions [11,12].

The most relevant prior work for the present study is the symbol-level graph refinement framework of Nguyen et al. [13]. Their method consists of two main stages. First, a bidirectional long short-term memory network (BLSTM)-based primitive recognizer and a 2D context-free grammar (2D-CFG) parser construct a primitive symbol graph. Next, a GNN-based binary link prediction model determines whether primitive edges should be retained or removed. The method improves structure recognition, although the final relation label remains largely dependent on the primitive relation assignment. In practice, primitive candidate edges are often noisy in both edge existence and relation type.

In this paper, we attempt to overcome this limitation by reformulating symbol-level graph refinement as a multi-class relation prediction problem. Instead of predicting only keep/discard decisions, the proposed model directly predicts one of the structural relation classes or the `None` class for every candidate edge. This allows the model to correct both structural connectivity and relation labels within a unified framework.

To support this formulation, we propose an edge-aware graph neural network based on graph attention with edge features. Each candidate edge is represented by three complementary features: (1) the primitive relation hypothesis, (2) its confidence score, and (3) pairwise geometric features between the connected symbols. We further introduce auxiliary losses for edge existence and positive-edge relation prediction to regularize training on sparse and imbalanced primitive graphs.

We evaluate the proposed method on datasets from the competition on recognition of online handwritten mathematical expressions (CROHME), using the CROHME 2019 training set for training and the CROHME 2014, 2016, and

2019 test sets for evaluation. Our primary evaluation target is the expression-level exact-match rate (Exp.) for the labeled structure graph in the Stroke Label Graph (SLG) representation.

The contributions of this paper are as follows:

1. Reformulate symbol-level graph refinement for online HME recognition from binary link prediction to multi-class relation prediction.
2. Propose an edge-aware relation prediction network using primitive relation priors, confidence scores, and symbol geometry in graph message passing.
3. Introduce auxiliary objectives for edge existence and positive-edge relation prediction, and show that relation-aware graph refinement substantially improves SLG exact match over both the primitive graph and a binary link-prediction baseline.

2 Related Work

This section reviews prior work on online HME recognition, grouped into explicit structural methods, sequence-based neural approaches, neural structure learning, and graph-based methods.

2.1 Explicit Structural Methods for Online HMEs

Explicit structure modeling has a long history in mathematical expression recognition. Grammar-based systems have used stochastic grammars, hidden Markov models, and structural parsing rules to model two-dimensional notation [5,6]. Tapia and Rojas [14] used minimum spanning tree construction with symbol dominance, while Le and Nakagawa [8] improved structural analysis for online handwritten mathematical expressions using explicit symbolic parsing. These approaches demonstrate the importance of recovering two-dimensional structure, but they typically separate segmentation, recognition, and structural analysis, which can amplify upstream errors.

2.2 Sequence-Based Neural HME Recognition

The dominant recent trend in HME recognition has been end-to-end sequence generation. Online or offline handwritten expressions are encoded and decoded into markup [15–18]. These models have achieved strong expression-level accuracy on the CROHME benchmarks, but their outputs are primarily linear strings. As discussed in the recent survey of Truong et al. [1], encoder-decoder approaches are highly effective for transcription, yet they do not always expose an explicit symbol graph or tree that can be edited or constrained directly.

2.3 Neural Structure Learning for Online HMEs

Several works on online HME recognition have remained explicitly structure-aware. Nguyen et al. [9] introduced a temporal classification approach that jointly predicts symbol segmentation, symbol classes, and spatial relations by leveraging global context from a deep bidirectional recurrent model. Truong et al. [10] extended this direction by directly learning a Symbol Relation Tree (SRT) from online trajectories. These methods motivate our setting because they treat relation prediction as a first-class component rather than a by-product of sequence decoding.

2.4 Graph-Based HME Recognition

Graph neural networks are a natural fit for mathematical notation because they can model pairwise and higher-order structural dependencies. Wu et al. [11] introduced a graph-to-graph (G2G) formulation for online HME recognition at the stroke level, avoiding explicit symbol segmentation. Tang et al. [19] proposed graph reasoning for offline HME recognition, while Xie et al. [12] extended graph attention networks (GAT) to form edge-weighted graph attention models (EGAT) that jointly classify nodes and edges in graph-structured HME recognition.

Compared with stroke-level graph formulations, symbol-level graphs reduce graph size and may be easier to refine once a primitive graph with high edge recall has been constructed. Nguyen et al. [13] explicitly explored this setting by applying GNN-based binary link prediction to symbol-level primitive graphs. Their work is the closest antecedent to ours. Our method differs in one key respect: we treat graph refinement as multi-class relation prediction rather than binary pruning, which lets the network revise both edge existence and relation labels.

3 Method

This section presents the proposed symbol-level graph refinement method, including the problem formulation, overall framework, input graph features, network architecture, and training objectives.

3.1 Problem Formulation

Let an online handwritten mathematical expression be represented by a candidate symbol graph $G_c = (V, E_c)$, where each node $v \in V$ corresponds to a primitive symbol, and each candidate edge $e_{ij} \in E_c$ corresponds to a possible spatial relation between symbols i and j. The goal of graph refinement is to recover the labeled ground-truth graph G^* in the SLG format. A candidate edge that exists in the ground-truth graph is defined as a positive edge, otherwise it is a negative edge. Instead of binary keep/discard decisions, we cast refinement as a multi-class edge classification over:

$$\mathcal{R} = \{\texttt{Right}, \texttt{Above}, \texttt{Below}, \texttt{Sup}, \texttt{Sub}, \texttt{Inside}, \texttt{None}\}.$$

Edges predicted as `None` are removed from the final graph, while all other predictions define the SLG output.

3.2 Overview

The overall system consists of two stages as shown in Fig. 1. The upstream stage follows the primitive graph construction used in the binary link-prediction framework of Nguyen et al. [13]. A primitive graph consists of primitive symbols and primitive relations. In the refinement stage, we propose an edge-aware GAT relation network denoted as **EA-GAT-Rel**, which nodes are trace features from primitive symbols and edges are extracted from primitive relations. Specifically, the EA-GAT-Rel model is composed of an edge-aware GAT encoder and a relation decoder. The encoder uses three stacked edge-aware GAT layers with edge features derived from primitive relation labels, confidence scores, and geometric features between connected symbols. The relation decoder employs a fully connected layer combining with a softmax layer to predict a relation class for each directed candidate edge. This paper focuses on the refinement stage, while the primitive stage is reproduced from the prior study [9].

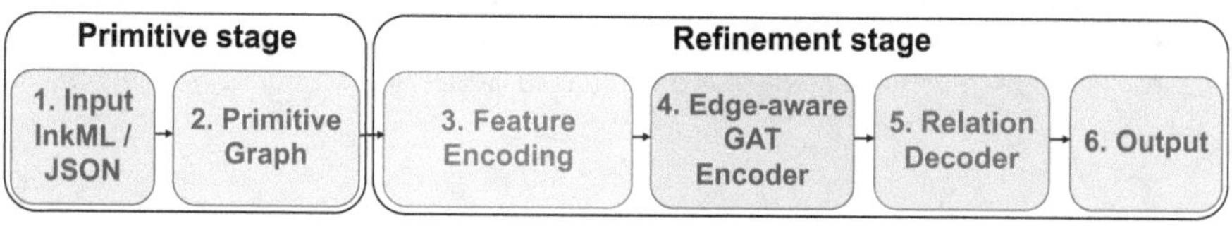

Fig. 1. Overview of the proposed symbol-level graph refinement pipeline.

3.3 Input Graph and Features

The primitive graph is constructed by the upstream symbol-level graph construction proposed by Nguyen et al. [13]. It is the input graph for our proposed EA-GAT-Rel model. For completeness, Fig. 2 summarizes the upstream primitive-graph construction stage. Online strokes are grouped into recognized symbol candidates. A BLSTM-based primitive recognizer first predicts symbol hypotheses and initial relation hypotheses between temporally consecutive symbols according to the writing order. A 2D-CFG parser then builds a more complete primitive graph. At the level-1 pairwise parsing stage, the parser considers all pairs of recognized symbols, not only temporally consecutive ones. For a non-consecutive pair, a temporary input sequence containing only the strokes of the two symbols is passed through the BLSTM-based primitive recognizer to obtain a primitive relation hypothesis and a confidence score. Consequently, each node in the primitive graph stores a primitive symbol predicted by the upstream model, while each candidate edge consists of (1) a primitive relation hypothesis, (2) a confidence score from the primitive predictor, and (3) geometric features derived from the corresponding symbol pair.

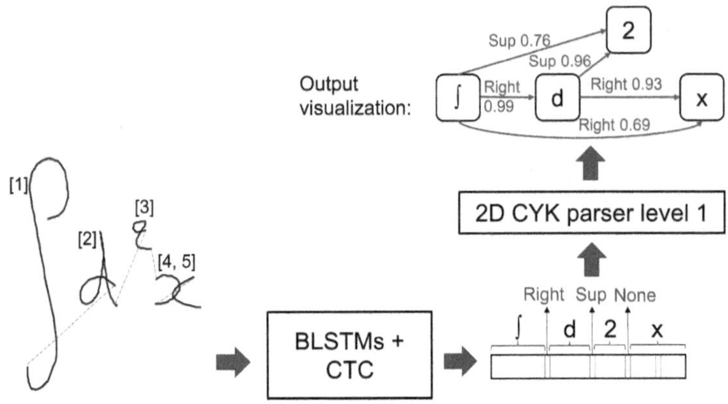

Fig. 2. Upstream primitive-graph construction stage. The primitive graph follows the parsing pipeline used by Nguyen et al. [13]; Online stroke sequences are first processed by a BLSTM-based primitive recognizer to obtain symbol candidates and initial relation hypotheses. A 2D-CFG parser then considers pairwise symbol combinations and converts the resulting hypotheses into a primitive symbol graph, whose nodes represent primitive symbols and whose edges store relation labels and confidence probabilities for downstream refinement.

Primitive Relation Prior. The primitive stage assigns a relation type to each emitted primitive edge, which is then used as a candidate edge by the refinement model.

Primitive Confidence. The upstream parser also provides a confidence score for the primitive relation. Rather than treating the primitive relation as certain, we expose its confidence to the graph network. In practice, the one-hot relation vector and the scalar confidence can be combined into a weighted relation representation.

Pairwise Symbol Geometry. For each candidate edge, we compute geometric descriptors for the corresponding symbol pair, including relative offsets, absolute offsets, Euclidean distance, bounding-box size statistics, and horizontal/vertical overlaps. These features capture layout regularities that are especially important for distinguishing horizontal, vertical, and containment-like relations in mathematical notation.

3.4 Edge-Aware Relation Prediction Network

Figure 3 illustrates the proposed symbol-level edge-aware graph attention network for relation prediction. We refine the primitive graph using stacked edge-aware graph attention (EA-GAT) layers. Let $h_i^{(\ell)}$ denote the node embedding of symbol i at layer ℓ, and let a_{ij} denote the edge feature vector for candidate edge (i, j). The edge feature vector a_{ij} is constructed once from the primitive relation, confidence, and geometric features. It is a static input of every edge-aware graph layer. During the message passing process, only node embeddings are updated. The update performed by a generic edge-aware graph layer is formulated as:

$$h_i^{(\ell+1)} = \mathrm{GNN}\big(h_i^{(\ell)}, \{h_j^{(\ell)}, a_{ij}\}_{j \in \mathcal{N}(i)}\big), \tag{1}$$

where $\mathcal{N}(i)$ denotes the set of neighboring nodes of node i in the primitive symbol-level graph. Since the edge features participate in both attention and message passing, the update of node i depends not only on neighboring node embeddings but also on the candidate relation information carried by the connecting edges. This allows the model to preserve reliable primitive relations while revising those that are inconsistent with the surrounding graph context and symbol geometry.

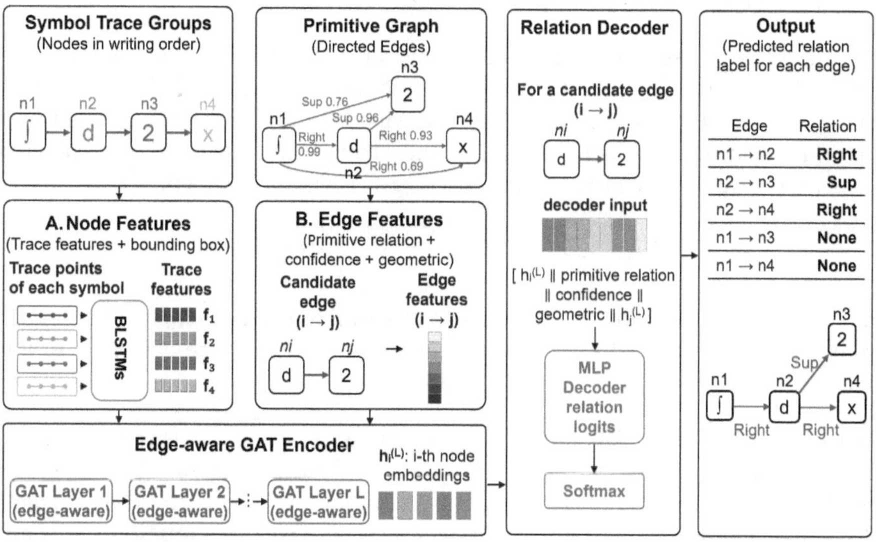

Fig. 3. Architecture of the proposed symbol-level edge-aware graph attention network for relation prediction.

After the last graph layer, a symbol relation feature vector z_{ij} of the candidate edge (i, j) is constructed by:

$$z_{ij} = [h_i^{(L)} \parallel a_{ij} \parallel h_j^{(L)}], \tag{2}$$

where $\|$ denotes concatenation. The relation decoder predicts a final relation label of every candidate edge.

$$p(y_{ij} \mid G_c) = \text{softmax}(\text{MLP}(z_{ij})) \tag{3}$$

Unlike binary link prediction, the relation decoder independently predicts the relation label of a candidate edge instead of inheriting it from the primitive graph.

In our experiment, the proposed refinement network is implemented in PyTorch Geometric using three stacked GATv2Conv layers with edge features, followed by an MLP-based decoder for relation prediction on candidate edges.

3.5 Training Objectives

The primary objective is the multi-class cross-entropy loss over the edge relation labels:

$$\mathcal{L}_{\text{rel}} = - \sum_{(i,j)\in E_c} \log p(y^*_{ij} \mid E_c), \tag{4}$$

where $y^*_{ij} \in \mathcal{R}$ is the ground-truth relation label of the candidate edge (i, j). Note that primitive graphs contain few positive edges compared with many negative candidate edges, and their relation labels are also imbalanced across classes. To improve training stability, we add two auxiliary losses.

Edge-Existence Loss. We define an auxiliary binary objective to predict whether a candidate edge belongs to the ground-truth SLG graph, regardless of its relation label. This term separates structural edge selection from fine-grained relation classification, and helps mitigate the strong imbalance between positive edges and abundant negative candidate edges. Formally, we define:

$$\mathcal{L}_{\text{exist}} = - \sum_{(i,j)\in E_c} \left[b^*_{ij} \log q_{ij} + (1 - b^*_{ij}) \log(1 - q_{ij}) \right], \tag{5}$$

where $b^*_{ij} \in \{0, 1\}$ indicates whether the candidate edge (i, j) exists in the ground-truth graph, and $q_{ij} = p(y_{ij} \neq \texttt{None} \mid G_c)$ is the edge-existence probability.

Positive-Edge Relation Loss. Because the primary relation loss is computed over all candidate edges, it is dominated by the abundant `None` class. To address this imbalance, we introduce an additional relation loss restricted to positive edges. This loss encourages better discrimination among structural relation types and complements the binary existence loss, which only predicts the presence of edges. The positive-edge relation loss is defined as follows:

$$\mathcal{L}_{\text{pos}} = - \sum_{(i,j)\in E_c : b^*_{ij}=1} \log p(y^*_{ij} \mid E_c). \tag{6}$$

The complete objective is summarized as follows:

$$\mathcal{L} = \mathcal{L}_{\text{rel}} + 0.5\, \mathcal{L}_{\text{exist}} + 0.5\, \mathcal{L}_{\text{pos}}. \tag{7}$$

4 Experiments

This section reports recognition experiments, including the dataset, evaluation protocol, baseline comparisons, and comparison with state-of-the-art systems.

4.1 Dataset and Evaluation Protocol

We train on the CROHME 2014 training set and evaluate on the CROHME 2014, 2016, and 2019 test sets. The input of the refinement model is not the raw online trajectory itself but a primitive graph with symbol and candidate-relation hypotheses produced by an upstream primitive recognizer. The output is a refined labeled graph in SLG format. The SLG format used in CROHME represents an expression as a symbol-level graph. Object entries define the symbols, and edge entries determine directed structural relations between pairs of symbols, to one of `Right`, `Above`, `Below`, `Sup`, `Sub`, and `Inside`.

We organize the reported metrics by evaluation level. First, **candidate coverage** measures whether ground-truth structural edges are present in the input primitive graph before refinement. Second, structure recognition rate (**Struct.**) counts an expression as correct only if the complete predicted graph topology matches the ground truth without considering the predicted symbol labels. Finally, our primary metric is the expression-level exact-match rate (**Exp.**), for the final labeled structure graph. **Exp.** requires the complete graph topology and all relation labels must match.

A practical issue in modular graph refinement is that not every raw sample reaches the refinement stage successfully. To make this explicit, we report two evaluation scenarios:

- **Valid-sample evaluation** focuses only on test samples with usable primitive graphs that include all symbols and edges of the ground-truth graph. This scenario measures the effectiveness of the graph refinement stage itself. Here, "valid" refers to the availability of a graph input for refinement, not to the complexity of the expression or the number of symbols it contains.
- **All-sample evaluation** measures the performance of the whole pipeline using all test samples. Test expressions skipped before refinement, for example because no usable primitive graph is produced, are counted as failures in expression-level recognition rates.

The valid-sample evaluation isolates the refinement module from upstream graph-construction failures, while the all-sample setting reflects system coverage more faithfully. Note that the reduced number of valid samples does not imply that these samples contain fewer symbols or are structurally simpler; it only indicates that the upstream stage failed to produce a usable graph representation for the remaining samples. This dual reporting clarifies the distinction between refinement quality and pipeline coverage.

4.2 Main Results and Comparison with Baselines

We compare the proposed method against the following baselines.

- **Primitive graph**: direct evaluation of the upstream primitive graph, which is also the primitive graph used by the refinement models.
- **Binary link prediction**: a symbol-level graph refinement model that predicts whether a candidate edge should be kept or removed. If an edge is kept, its relation label is inherited from the primitive graph. This is the formulation used by Nguyen et al. [13]. We use this method as the primary baseline because it is the closest prior work to our setting: both methods refine symbol-level primitive graphs produced by an upstream recognizer and parser, but Nguyen et al. perform binary link prediction, whereas our method performs multi-class relation prediction.
- **Proposed edge-aware relation prediction**: the full model described in Sect. 3.

Tables 1 and 2 summarize the input coverage and expression-level recognition rates on the CROHME test sets. Because all refinement models use the same primitive graph as input, Candidate Coverage is reported once in Table 1 rather than repeated for each model.

First, Table 1 presents valid sample and candidate coverage rates. The valid sample rate denotes the proportion of HMEs for which the upstream stage produces usable primitive graphs. Valid sample rates range from 88% to 90%, indicating that approximately 10% of the input samples are excluded when computing the final expression recognition rate. Although the input primitive graphs provide high coverage of the ground-truth structure, with candidate coverage about 98% on all three test sets, it does not guarantee accurate structure recognition.

Second, Table 2 demonstrates the primitive graph baseline performance at 10.82%, 7.31%, and 6.85% of expression-level rate, and 11.85%, 7.40%, and 7.22% structure recognition rate on CROHME 2014, 2016, and 2019, respectively. This indicates that the primitive graphs still contain many redundant or mislabeled edges and therefore require substantial refinement.

Third, the binary link-prediction method improves considerably over the primitive graph by removing redundant candidate edges. It reaches the expression-level rates of 39.70%, 36.33%, and 37.65%, and the structure recognition rates of 41.77%, 37.71%, and 39.04% on CROHME 2014, 2016, and 2019, respectively. Note that the performance of the primitive graph baseline and the binary link-prediction method is evaluated only on the valid samples.

Finally, the proposed edge-aware multi-class relation prediction model (EA-GAT-Rel) achieves the highest results among the compared baselines. In the valid-sample evaluation, it reaches Exp. of **70.66%**, **70.29%**, and **73.27%** on CROHME 2014, 2016, and 2019, respectively, with corresponding Struct. of 74.80%, 74.53%, and 77.52%. Under all-sample evaluation, the proposed model still achieves Exp. of **62.27%**, **62.07%**, and **66.06%**, showing that the gains are not limited to a narrowly filtered subset of valid samples.

Overall, these results show that edge coverage is already high in the upstream primitive stage, and that the main remaining challenge lies in correcting redundant and mislabeled primitive candidate edges. Compared with both the primitive graph and the binary link-prediction baseline, the EA-GAT-Rel model is substantially more effective because it directly predicts the final relation class for each candidate edge rather than relying on binary pruning alone.

Table 1. Evaluation set coverage and input candidate-graph coverage.

Metric	CROHME 2014	CROHME 2016	CROHME 2019
Raw samples	986	1147	1199
Valid samples	869	1013	1081
Valid sample rate (%)	88.13	88.32	90.16
Candidate coverage (%)	97.95	98.17	98.43

Note: Candidate coverage is computed on valid samples.

Table 2. Expression-level recognition rates (%) on CROHME test sets.

Model	Scenario	CROHME 2014		CROHME 2016		CROHME 2019	
		Exp.	Struct.	Exp.	Struct.	Exp.	Struct.
Primitive graph	Valid-sample	10.82	11.85	7.31	7.40	6.85	7.22
Binary link	Valid-sample	39.70	41.77	36.33	37.71	37.65	39.04
EA-GAT-Rel	Valid-sample	**70.66**	74.80	**70.29**	74.53	**73.27**	77.52
EA-GAT-Rel	All-sample	**62.27**	65.92	**62.07**	65.82	**66.06**	69.89

Note: Exp. requires exact graph topology and all relation labels to match, while Struct. evaluates exact graph topology only. Valid samples include only expressions with usable primitive graphs; all samples use the full raw test set as the denominator.

4.3 Comparison with State-of-the-Art Systems

Table 3 compares the proposed model with state-of-the-art and representative HME recognition systems on the CROHME benchmarks, where Gram, Graph, EncDec, and GNN represent Grammar-, Graph-, Encoder-decoder-, and GNN-based HME recognition systems, respectively. For all three CROHME test sets, i.e., CROHME 2014, 2016, and 2019, our method follows the same training protocol as the compared systems: a single model is trained on the CROHME 2014 training set and then evaluated on each test set without dataset-specific retraining.

For direct system-level comparison, we use the *all-sample* evaluation, in which expressions that do not produce a usable primitive graph are counted as failures

Table 3. HME recognition rates (%) compared with state-of-the-art and representative systems on the CROHME datasets.

System	Type	CROHME 2014		CROHME 2016		CROHME 2019	
		Exp.	Struct.	Exp.	Struct.	Exp.	Struct.
MyScript [20–22]	Gram	62.68	–	67.65	88.14	79.15	90.66
UPV/Wiris [20–22]	Gram	37.22	–	49.61	74.28	60.13	79.15
Tokyo [20–22]	Gram	25.66	–	43.94	61.55	39.95	58.22
Samsung [21,22]	Gram	–	–	65.76	–	79.82	89.32
MST [23]	Graph	26.88	–	–	–	–	–
Tree-BLSTM [24]	Graph	29.91	–	27.03	–	–	–
Tree-construction [10]	Graph	44.12	58.62	41.76	49.43	–	–
TAP [15]	EncDec	46.90	–	44.80	–	–	–
iFlyTek [22]	EncDec	–	–	57.02	–	80.73	91.49
G2G [11]	GNN	–	–	52.05	–	–	–
LGM-EGAT [12]	GNN	–	–	56.41	–	58.22	83.07
GGM-EGAT [12]	GNN	–	–	56.67	–	60.72	83.74
Nguyen et al. [13]	GNN	37.59	51.07	38.35	48.60	38.86	51.42
EA-GAT-Rel (valid-sample)	GNN	70.66	74.80	70.29	74.53	73.27	77.52
EA-GAT-Rel (all-sample)	GNN	62.27	65.92	62.07	65.82	66.06	69.89

Note: Our method is reported under two evaluation scenarios: *valid-sample*, computed on expressions for which a usable primitive graph is produced, and *all-sample*, where the full raw test set is used as the denominator and skipped expressions are counted as failures.

in expression-level recognition rates. This setting more faithfully reflects the overall behavior of the complete pipeline. We additionally report *valid-only* results in Table 3 to isolate the effectiveness of the graph refinement stage itself.

Compared with previously reported graph-based and GNN-based systems, the proposed method shows clear gains in both expression-level and structure-level recognition rates. In particular, relative to Nguyen et al. [13] using a binary link prediction, our system achieves substantially higher results on all three CROHME test sets. The structure recognition rates improve by at least 15% points, while the expression-level rates improve by at least 17 percentage points. These improvements indicate that multi-class relation prediction is markedly more effective than binary link pruning for refining primitive symbol graphs.

The proposed system also outperforms earlier explicit structure-recognition approaches such as MST-based parsing, Tree-BLSTM, and tree-construction methods, as well as previously reported GNN-based systems such as G2G, LGM-EGAT, and GGM-EGAT. However, the strongest competition systems and integrated recognition pipelines, such as MyScript and iFlyTek, still report higher expression-level results on some benchmarks. One possible reason is that these systems benefit from larger-scale training data.

5 Discussion

This section provides a deeper analysis of the proposed method beyond the quantitative results. We highlight the main insights gained from the experiments and discuss the limitations that remain in the current framework.

5.1 Main Insights

The experiment results support three main observations. First, the primitive graph alone is insufficient for accurate structure recognition. Despite high candidate coverage, the expression-level recognition rate remains suboptimal due to the presence of numerous redundant primitive edges or incorrectly assigned relation labels. This confirms that the main challenge is not only to emit candidate edges that cover the ground-truth structure, but also to refine the primitive graph into a structurally correct labeled representation.

Second, binary link prediction addresses only a part of the relation-refinement problem. Nguyen et al. [13] demonstrated that GNN-based edge pruning improved the primitive graph, but the relation label of each retained edge remained inherited from the primitive parser. In contrast, the proposed model predicts a relation label, including `None`, for each primitive candidate edge. This allows the model to correct both edge existence and relation labels within a unified framework, which explains the large gains over binary link prediction at the system level.

Third, the proposed model uses edge-level information during graph refinement by incorporating primitive relation priors, confidence scores, and pairwise symbol geometry. These cues provide additional context for the prediction of structural relations beyond node features alone. However, because no dedicated ablation study is conducted for these components, the observed gains should be attributed to the full refinement framework rather than to edge-aware modeling in isolation.

In summary, these findings support the use of GNN-based primitive graph refinement as a practical direction for online HME structure recognition. Once a high-recall primitive graph is available, the proposed edge-aware multi-class refinement framework can recover a substantial portion of the remaining structural errors within the available candidate-edge set. Future work should explore tighter integration between primitive parsing and graph refinement, more reliable primitive graph construction, and stronger global structural constraints during decoding.

5.2 Limitation

The proposed method still depends on the quality of the upstream primitive graph. Although the 2D-CFG parser can consider arbitrary recognized-symbol pairs, the refinement stage only receives the primitive edges that are emitted in the primitive graph. Therefore, any ground-truth relation that is not emitted as a primitive edge lies outside the refinement search space. The refinement stage can suppress redundant primitive edges and relabel incorrect primitive relations, but it cannot introduce a missing edge. Consequently, the final performance of the proposed refinement stage is inherently upper-bounded by the candidate-edge recall of the upstream primitive graph construction stage. In addition, although the proposed model substantially improves graph exact matching, it does not explicitly enforce global structural constraints during decoding. Consequently, some remaining errors may arise from graph configurations that are locally plausible but globally inconsistent.

Furthermore, the remaining gap to the strongest competition systems highlights an important system-level trade-off. Our pipeline explicitly separates primitive parsing from graph refinement, so errors from the upstream primitive stage would propagate to the final output. The pipeline, however, offers a clear advantage of an explicit labeled graph that is interpretable and structurally editable than a directly generated linear markup string. From this perspective, the current results suggest that graph refinement is no longer the main bottleneck once a high-coverage primitive graph is available. Consequently, next improvements will likely depend on the robustness and coverage of primitive graph construction.

6 Conclusion

This paper presented an edge-aware graph neural network for symbol-level graph refinement in online HME recognition. Given a primitive graph from an upstream primitive stage, the proposed method treats primitive relations as candidate edges and directly predicts relation classes for those edges, forming a symbol relation graph. By reformulating refinement from binary link prediction to multi-class relation prediction, the method can jointly correct edge existence and edge labels within the available primitive-edge set. Experiments on the CROHME 2014, 2016, and 2019 test sets show clear improvements over both the primitive graph baseline and prior binary link-prediction refinement, indicating that edge-aware relation modeling with primitive priors, confidence, and geometry is an effective strategy for primitive graph refinement in online HME recognition.

Acknowledgment. This work is partially supported by MEXT/JSPS KAKENHI Grant Number JP24H00738.

References

1. Truong, T.N., Nguyen, C.T., Zanibbi, R., Mouch'ere, H., Nakagawa, M.: A survey on handwritten mathematical expression recognition: the rise of encoder-decoder and GNN models. Pattern Recognition **153**, 110531 (2024). https://doi.org/10.1016/j.patcog.2024.110531
2. Xie, Y., et al.: ICDAR 2023 CROHME: competition on recognition of handwritten mathematical expressions. In: Fink, G.A., Jain, R., Kise, K., Zanibbi, R. (eds.) ICDAR 2023. LNCS, vol. 14188, pp. 553–565. Springer, Cham (2023). https://doi.org/10.1007/978-3-031-41679-8_33

3. Truong, T.N., Nguyen, H.T., Nakagawa, M., Yasuno, F.: Integrating handwriting input interface and recognition into Moodle for handwritten answers in Japanese, English, mathematical and chemical formulas, vol. 124, pp. 98–105 (2025). https://ken.ieice.org/ken/paper/20250301TcIi/eng/
4. Asakura, T., et al.: Automatic scoring system for digitized handwritten answers. IEEE Access **14**, 25949–25967 (2026). https://doi.org/10.1109/ACCESS.2026.3663589
5. Yamamoto, R., Sako, S., Nishimoto, T., Sagayama, S.: On-line recognition of handwritten mathematical expressions based on stroke-based stochastic context-free grammar. In: Proceedings of the Tenth International Workshop on Frontiers in Handwriting Recognition, pp. 249–254 (2006)
6. Álvaro, F., Sánchez, J.A., Benedı, J.M.: Recognition of on-line handwritten mathematical expressions using 2D stochastic context-free grammars and hidden markov models. Pattern Recogn. Lett. **35**, 58–67 (2014). https://doi.org/10.1016/j.patrec.2012.09.023
7. Awal, A.M., Mouch'ere, H., Viard-Gaudin, C.: A global learning approach for an online handwritten mathematical expression recognition system. Pattern Recogn. Lett. **35**, 68–77 (2014). https://doi.org/10.1016/j.patrec.2012.10.024
8. Le, A.D., Nakagawa, M.: A system for recognizing online handwritten mathematical expressions by using improved structural analysis. Int. J. Doc. Anal. Recogn. **19**(4), 305–319 (2016). https://doi.org/10.1007/s10032-016-0272-4
9. Nguyen, C.T., Truong, T.-N., Nguyen, H.T., Nakagawa, M.: Global context for improving recognition of online handwritten mathematical expressions. In: Lladós, J., Lopresti, D., Uchida, S. (eds.) ICDAR 2021. LNCS, vol. 12822, pp. 617–631. Springer, Cham (2021). https://doi.org/10.1007/978-3-030-86331-9_40
10. Truong, T.N., Nguyen, H.T., Nguyen, C.T., Nakagawa, M.: Learning symbol relation tree for online handwritten mathematical expression recognition. In: Wallraven, C., Liu, Q., Nagahara, H. (eds.) ACPR 2021. LNCS, vol. 13189, pp. 307–321. Springer, Cham (2022). https://doi.org/10.1007/978-3-031-02444-3_23
11. Wu, J.W., Yin, F., Zhang, Y.M., Zhang, X.Y., Liu, C.L.: Graph-to-graph: towards accurate and interpretable online handwritten mathematical expression recognition. In: Proceedings of the AAAI Conference on Artificial Intelligence, vol. 35, pp. 2925–2933 (2021). https://doi.org/10.1609/AAAI.V35I4.16399
12. Xie, Y., Zanibbi, R., Mouchère, H.: Local and global graph modeling with edge-weighted graph attention network for handwritten mathematical expression recognition. Pattern Recogn. **178**, 113410 (2026). https://doi.org/10.1016/j.patcog.2026.113410
13. Nguyen, C.T., Nguyen, N.T., Dao, T.H.M., Nguyen, H.M.N., Dinh, H.T.: Link prediction graph neural networks for structure recognition of handwritten mathematical expressions. In: Jin, L., Zanibbi, R., Eglin, V. (eds.) ICDAR 2025. LNCS, vol. 16225, pp. 279–291. Springer, Cham (2025). https://doi.org/10.1007/978-3-032-09368-4_17
14. Tapia, E., Rojas, R.: Recognition of on-line handwritten mathematical expressions using a minimum spanning tree construction and symbol dominance. In: Lladós, J., Kwon, Y.-B. (eds.) GREC 2003. LNCS, vol. 3088, pp. 329–340. Springer, Heidelberg (2004). https://doi.org/10.1007/978-3-540-25977-0_30
15. Zhang, J., Du, J., Dai, L.: Track, attend, and parse (TAP): an end-to-end framework for online handwritten mathematical expression recognition. IEEE Trans. Multimedia **21**(1), 221–233 (2019). https://doi.org/10.1109/TMM.2018.2844689

16. Zhao, W., Gao, L., Yan, Z., Peng, S., Du, L., Zhang, Z.: Handwritten mathematical expression recognition with bidirectionally trained transformer. In: Lladós, J., Lopresti, D., Uchida, S. (eds.) ICDAR 2021. LNCS, vol. 12822, pp. 570–584. Springer, Cham (2021). https://doi.org/10.1007/978-3-030-86331-9_37
17. Zhao, W., Gao, L.: CoMER: Modeling coverage for transformer-based handwritten mathematical expression recognition. In: Avidan, S., Brostow, G., Cissé, M., Farinella, G.M., Hassner, T. (eds.) ECCV 2022. LNCS, vol. 13688, pp. 392–408. Springer, Cham (2022). https://doi.org/10.1007/978-3-031-19815-1_23
18. Zhu, J., Gao, L., Zhao, W.: ICAL: Implicit character-aided learning for enhanced handwritten mathematical expression recognition. In: Barney Smith, E.H., Liwicki, M., Peng, L. (eds.) ICDAR 2024. LNCS, vol. 14898, pp. 21–37. Springer, Cham (2024). https://doi.org/10.1007/978-3-031-70549-6_2
19. Tang, J.M., Wu, J.W., Yin, F., Huang, L.L.: Offline handwritten mathematical expression recognition via graph reasoning network. In: Wallraven, C., Liu, Q., Nagahara, H. (eds.) ACPR 2021. LNCS, vol. 13188, pp. 17–31. Springer, Cham (2022). https://doi.org/10.1007/978-3-031-02375-0_2
20. Mouch'ere, H., Viard-Gaudin, C., Zanibbi, R., Garain, U.: ICFHR 2014 competition on recognition of on-line handwritten mathematical expressions (CROHME 2014). In: 2014 14th International Conference on Frontiers in Handwriting Recognition (ICFHR), pp. 791–796. IEEE (2014). https://doi.org/10.1109/ICFHR.2014.138
21. Mouch'ere, H., Viard-Gaudin, C., Zanibbi, R., Garain, U.: ICFHR 2016 CROHME: competition on recognition of online handwritten mathematical expressions. In: 2016 15th International Conference on Frontiers in Handwriting Recognition (ICFHR), pp. 607–612. IEEE (2016). https://doi.org/10.1109/ICFHR.2016.0116
22. Mahdavi, M., Zanibbi, R., Mouch'ere, H., Viard-Gaudin, C., Garain, U.: ICDAR 2019 CROHME + TFD: competition on recognition of handwritten mathematical expressions and typeset formula detection. In: 2019 International Conference on Document Analysis and Recognition (ICDAR), pp. 1533–1538. IEEE (2019). https://doi.org/10.1109/ICDAR.2019.00247
23. Hu, L., Zanibbi, R.: MST-based visual parsing of online handwritten mathematical expressions. In: 2016 15th International Conference on Frontiers in Handwriting Recognition (ICFHR), pp. 337–342 (2016). https://doi.org/10.1109/ICFHR.2016.0070
24. Zhang, T., Mouch'ere, H., Viard-Gaudin, C.: A tree-BLSTM-based recognition system for online handwritten mathematical expressions. Neural Comput. Appl. **32**(9), 4689–4708 (2020). https://doi.org/10.1007/s00521-018-3817-2

End-to-End Camera-Based Analysis of Egyptian Hieroglyphic Textual Artifacts

David Konieczny, Said Yasin, and Torsten Zesch(✉)

CATALPA – Center of Advanced Technology for Assisted Learning and Predictive Analytics, FernUniversität in Hagen, Hagen, Germany
said.yasin@studium.fernuni-hagen.de , torsten.zesch@fernuni-hagen.de

Abstract. We present an end-to-end system for analyzing Old Egyptian hieroglyphic textual artifacts. In contrast to previous approaches, we provide a fully open-source solution that spans all substeps including segmentation, reading order prediction, glyph classification, and translation. For this purpose, we collect and annotate datasets for hieroglyph segmentation, determination of the reading direction, and glyph classification. Our approach embeds the analysis pipeline into a camera-display setup with spatially aligned visual feedback for physical textual artifacts such as stelae or clay tablets.

A runnable demo application with source code and setup instructions is available at https://github.com/said702/Hieroglyph_Demo.

Keywords: Hieroglyph Recognition · Ancient Egyptian · Camera-Based Document Analysis

1 Introduction

Even for trained specialists, reading ancient Egyptian[1] texts is challenging, for several reasons related to the properties of Egyptian hieroglyphic writing. As a dead language with a limited surviving text corpus, the interpretation of Egyptian texts requires a comparatively high degree of contextual inference compared to modern languages. The known set of signs comprises several thousand distinct glyphs. If any new characters are found, they are only variants of already existing glyphs with the same or very similar meaning. In addition, Egyptian has relatively free reading order as the arrangement of hieroglyphs is quite variable and depends e.g. on the available space as well as aesthetic considerations, particularly with regard to how the inscription is arranged around pictorial representations (see Fig. 1).

Consequently, accurate translations by experts still demand considerable manual effort, taking into account subtle nuances in the script. Additionally, its use over a period of more than 3,000 years led to the development of

[1] Throughout the paper, we use the short form *Egyptian* to refer to the ancient Egyptian language, not the Arabic dialect spoken in modern Egypt.

F. Shafait et al. (Eds.): DAS 2026, LNCS 16851, pp. 425–440, 2027.
https://doi.org/10.1007/978-3-032-36207-0_25

Fig. 1. Color-annotated excerpt from the stela of Mery (Guillaume Blanchard, July 2004, wikimedia.org). Characters marked with rectangles are good indicators for detecting the reading direction. The numbering in section C describes exemplarily the exact sequence of the glyphs.

multiple language stages and dialects, making accurate translation even more difficult. We focus here on Middle Egyptian as the classical stage of ancient Egyptian literature. While many inscriptions have been documented in photographs or published only in non-digital formats in Egyptology, a comprehensive machine-readable digital dataset suitable for modern NLP methods is still lacking. The automated processing of ancient Egyptian documents could offer valuable insights into past lifestyles and the historical state of the world, including the evolution of technology, science, and religion as well as finally aiding in preservation for future generations [14].

To address this gap, we propose the first practically usable end-to-end pipeline that integrates glyph localization, reading-order inference, classification, and translation for spatially aligned visual feedback. The approach is designed to handle practical challenges such as diverse glyph arrangements, different writing surfaces and materials, and partially damaged inscriptions. Beyond supporting the scholarly work of Egyptologists, the system can enable broader future applications, for example in museums or for tourists.

In addition to the proposed end-to-end pipeline, we introduce newly collected and annotated datasets and release an open-source real-time demo application that integrates the full pipeline in a camera-display setup and renders predictions as a spatially aligned overlay.

2 The Egyptian Language and Hieroglyphic Script

As we do not expect all readers to be familiar with the topic, we provide a short intro into the Egyptian written language, which was used for about 3,000 years. It encompassed several linguistic stages and a total of four writing systems, the most well-known of which is the hieroglyphic script [2]. This well-preserved writing system was used for inscriptions on statues, historical records on walls, or religious texts such as hymns or ritual spells intended for eternity.

Glyphs. Hieroglyphic writing represents a combination of pictographic signs that convey what they depict visually, phonetic signs that stand for sounds and syllables, and determinatives that assign a specific meaning to a sequence of sounds. Therefore, in the majority of the available text corpora, around one thousand different (partially very similar) hieroglyphs with very different frequency of occurrence must be distinguished. Although about 200 signs are particularly common, the correct recognition of rare glyphs is crucial for the translation process since the same phonetic sequence can be assigned different meanings through various classification signs. During the separation of the glyphs from each other, it is important to note that some characters consist of multiple components such as the three parallel strokes in Fig. 2, and must therefore be treated as a group.

Additionally, a robust system must ensure that damage to the writing material such as chipped areas of a stone tablet or holes in papyrus is not mistakenly identified as potential characters.

Reading Order. Hieroglyphic texts can be written from right to left or vice versa, in rows or columns (in any case, the text is to be read from top to bottom). In Fig. 1, text section A and the C represent horizontal writing from right to left, while section B is to be read as a column. The left-to-right or right-to-left reading direction can usually be determined at a glance by observing hieroglyphs depicting humans, animals, or their limbs: such hieroglyphs always face the beginning of the text. Once the reading direction has been determined, the exact order must still be inferred from the arrangement of the signs, which is not trivial as shown in Example C by the black numbering.

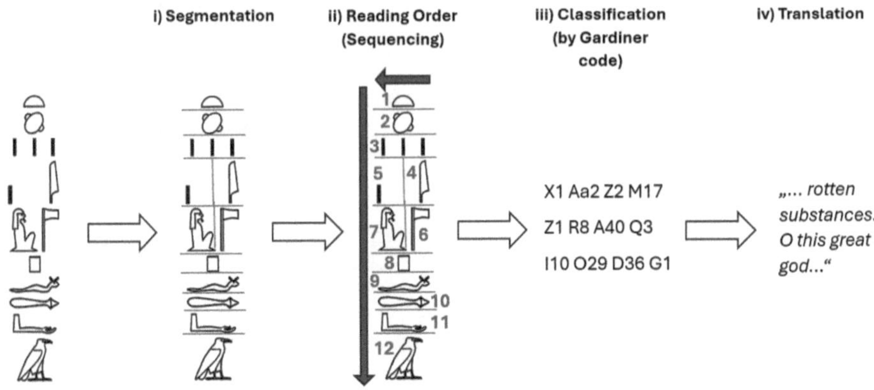

Fig. 2. Representation of the overall process from the photography of an inscription to its translation.

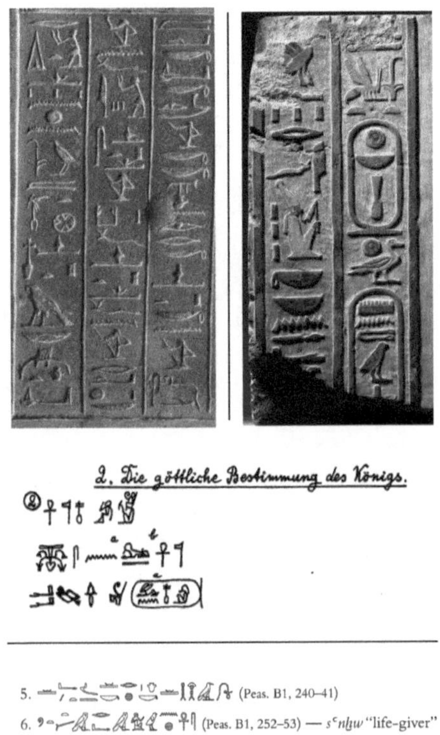

Fig. 3. Examples for each of the four script types. Crops shown from left to right, top to bottom: Stele of Pay from the Museo Egizio, relief of Nebhepetre Mentuhotep from the Metropolitan Museum, manuscript from Sethe's *Urkunden II* and hieroglyphic font from [2].

3 Steps for Translating Egyptian

The following steps are necessary to go from hieroglyph image to translation:

Segmentation First, the hieroglyphs to be recognized must be separated from one another (step i in Fig. 2).

Sequencing Next, the reading direction must be determined, and a sequence reflecting the correct reading order must be derived from the arrangement of the signs (step ii).

Classification The segmented and sequenced glyphs must be classified into a transcription system called *Gardiner code* - a historically developed and still widely used system of encoding hieroglyphs by a letter-number combination such as 'A15' or 'X1'. (step iii)

Translation Finally, a Gardiner sequence can be used for machine translation (step iv). Here, we apply a translation model by Cao et al. [5].

4 Related Work

To our knowledge, there is no practically usable system for translating directly from an arbitrary image of hieroglyphic text. While one might assume that recent vision-language models such as GPT-4V [16], Flamingo [1], and BLIP-2 [11] address this challenge, they mainly generate image-conditioned text and do not provide explicit, reliable symbol localization. However, spatially aligned augmented reality requires geometrically grounded glyph coordinates, necessitating an explicit localization stage. Therefore, we focus on an AR-oriented pipeline that couples explicit glyph localization with translation.

The methods and models developed to date mostly solve only partial problems or are tailored to specific cases, often lacking robust generalization capabilities (e.g. fixed reading direction). Close in spirit to our system is the open-source project Fabricius.[2] However, it is not fully end-to-end, as the application relies on manual image preprocessing steps like removing noise or inpainting incomplete hieroglyphs. Translation suggestions are provided based on the Gardiner code sequence by matching it against a dictionary, which results in low translation quality.

A few smartphone applications have been proposed, such as *Scriba* [13], *ARsinoë* [17], or *Aegyptos* [12]. However, their performance could not be further evaluated due to a lack of publicly available implementations. The system by Duque-Domingo et al. [6] only focuses on specific input types (e.g. cartouches a distinctive oval frame that exclusively contains royal names).

Translation-Only (Step iv). This branch of work treats Egyptian as a low-resource language and only focuses on the transcribed tokens to modern language translation part. To this end, [22] extracted around 100,000 Gardiner-code

[2] https://fabriciusworkbench.withgoogle.com/.

tokens, transcription, POS tags, and German[3] (in a few cases English) translations for sentences from the Thesaurus Linguae Aegyptiae (TLA).[4] [5] used the pre-trained M2M-100 multilingual model, which enables translations between 100 modern languages, to generate English equivalents for the numerous German translations. The research group published the model's source code, which was successfully tested and is reused later in the context of this work.

Transcription-Only (Steps i to iii). [8] addressed the segmentation of hieroglyphs using a retrieval-based approach with the help of saliency maps. [15] applied a similar approach to handwritten hieroglyphic texts by the German Egyptologist Kurt Sethe [19]. [10] used Canny edge detection for segmentation and introduced a feature to determine the reading direction (left-to-right or right-to-left) by comparing single hieroglyphs with directional patches. The classification task, based on HOG feature extraction, was limited to 197 distinct signs. [4] used CNN-based segmentation and classification, but with an even smaller vocabulary of 40 classes.

Closest to our approach are Sobhy et al. [20], who translate scanned hieroglyphic images into English, and Fuentes-Ferrer et al. [9], who focus on segmentation and recognition of hieroglyphic texts. In contrast, our work integrates segmentation, reading-order prediction, glyph classification, translation, and camera-based spatial feedback into one end-to-end system.

5 Data

To capture the visual diversity of Egyptian hieroglyphic material, our datasets combine carved inscriptions, handwritten manuscripts, and computer-generated font renderings.

As a basis for all datasets, we collected images of hieroglyphic texts from various publicly accessible online sources: the Museo Egizio in Turin[5], the Metropolitan Museum in New York[6], and the British Museum in London[7], as well as images from the literature, such as those by [8], and excerpts from works by [2,7,19,21], and [3]. These were categorized according to script type, distinguishing between carved hieroglyphs without details (32 images), carved hieroglyphs with details (6 images), handwritten manuscripts (46 images), and computer-generated fonts (19 images). Figure 3 shows one example for each of the four types.

[3] In the history of Egyptology, especially in the 19th and early 20th centuries, German was a dominant language, with key works by scholars like Adolf Erman, Kurt Sethe, and Hermann Grapow, which led to the abundance of translations in German.

[4] https://thesaurus-linguae-aegyptiae.de.

[5] https://www.museoegizio.it/en/.

[6] https://www.metmuseum[.org/de.

[7] https://www.britishmuseum.org/.

5.1 Segmentation Data (Step I)

As training data for distinguishing between hieroglyphs and background, a set of annotated images had to be created from the image collection described in the previous section. Following the example of [4], the VGG Image Annotator[8] was used for annotation. In contrast to the polygonal framing of the hieroglyphs by those authors, the hieroglyphs were marked with rectangular bounding boxes, assigned to the class 'Hieroglyph' and labeled with the appropriate Gardiner code (see Fig. 4).

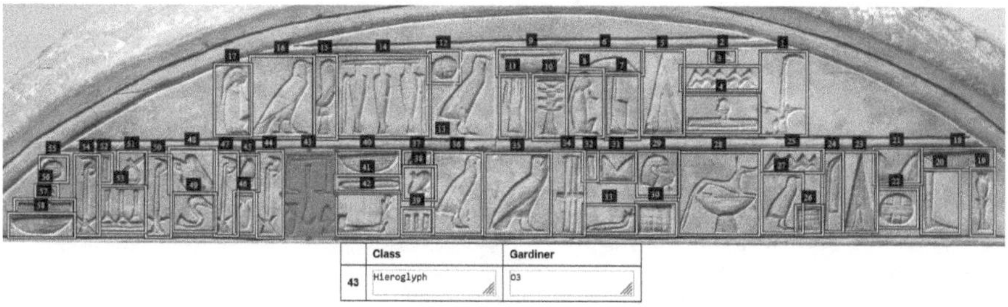

Fig. 4. Fully annotated partial image of the stela of Qemnen with attributes for a selected box. (https://collezioni.museoegizio.it/en-GB/material/Cat_1513)

That way, a total of 33 excerpts of hieroglyphic inscriptions were manually annotated. In addition, 10 already annotated images of the Pyramid Texts from the dataset by [8] were used. For the purpose of data augmentation, 30 additional variants altering the contrast and mirroring the images as well as the annotations were generated from these examples. In total, 56 annotated images with 14,100 labeled bounding boxes were available for training and 17 images with 5,812 annotations for testing.

5.2 Reading Direction Data (Step II)

In the implemented sequencing approach, the first step will be to determine whether the reading direction is left-to-right or right-to-left. For the human eye, identifying the orientation of the hieroglyphs is almost trivial by focusing on the direction in which human and animal signs are facing (cf. Fig. 1). To train a model to perform this task, 55 image excerpts each with left-to-right and right-to-left reading direction were collected and labeled accordingly with LR and RL. After data augmentation by creating 40 variants per image, the final dataset for the model training contained 4,400 images.

[8] https://www.robots.ox.ac.uk/~vgg/software/via/via.html.

5.3 Classification Data (Step III)

To provide a dataset with as much variation as possible for model training of a classifier, hieroglyph patches labeled with Gardiner codes from different sources were consolidated. The data includes (1) computer-generated depictions for each of the 1,072 characters in the Unicode block of standard hieroglyphs[9] using the Aegyptus font[10], as well as existing font images from additional sources in SVG[11], TIFF[12], and PNG[13] format; (2) approximately 9,700 patches from the dataset of [9]; (3) approximately 4,000 manually cropped, unprocessed patches from [8]; (4) approximately 2,700 patches from the self-annotated images for segmentation; (5) approximately 400 manually cropped, rare hieroglyphs from a manuscript of [7].

After further preprocessing steps like greyscaling, rescaling and harmonizing the data type as well as a slight reduction in the number of classes, the final dataset for the classifier contained 20,078 patches from 967 classes[14] with data augmentation by Albumentations[15]—focusing on balancing the very uneven class distribution found in real inscriptions—the dataset was expanded to approximately 316,000 examples with at least 120 variants per class (see selection in Fig. 5).

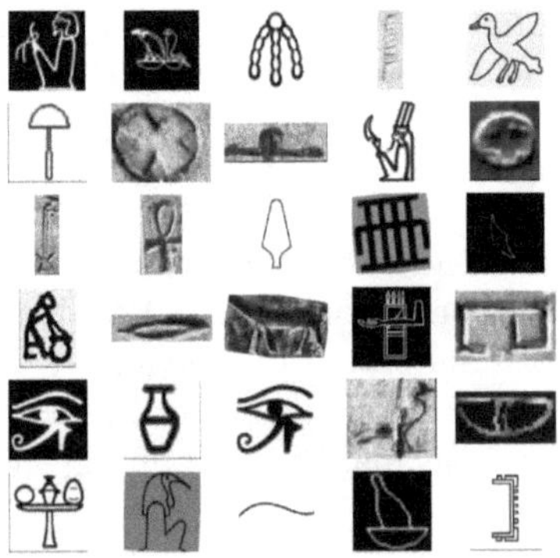

Fig. 5. 30 patches from the classifier dataset after data augmentation.

5.4 Translation Data (Step IV)

Note that we do not re-train this step, but rely on an existing translation model that was trained by [5].

[9] Section U+13000 to U+1342E.
[10] localfonts.eu.
[11] github.com/googleartsculture/workbench.
[12] github.com/fayrose/MiddleEgyptianDictionaryWebsite.
[13] github.com/fgimbert/Hieroglyphs.
[14] The list of selected classes is available at this CSV file on GitHub.
[15] https://github.com/albumentations-team/albumentations.

6 Implementation and Evaluation

The model choices were guided by the different requirements of the pipeline stages: segmentation requires accurate localization of complex and partially overlapping glyphs, whereas reading-direction prediction and glyph classification benefit from fast inference in the interactive camera-display setup.

6.1 Segmentation

We use the Detectron2[16] framework using a ResNet50 backbone with a Feature Pyramid Network. Following the example of [4][17] a version pre-trained on the COCO dataset was chosen. Initial tests showed that the model, without specific fine-tuning, recognized only a few hieroglyphs. Therefore, the model was fine-tuned using the above mentioned annotated dataset to distinguish between two classes: hieroglyph and background.

After numerous variations of the model parameters for training the object detection algorithm as well as adjustments to the predictor, the final model that was trained over 4,000 epochs with a learning rate of 0.01 achieved very solid average precision (AP) values across different object sizes (see Table 1) and correctly identified most glyphs even in more extended inscriptions. Particularly noteworthy was the model's ability to already identify compound signs such as the plural markers Z2 (three parallel strokes), Z3 (three vertically stacked strokes) or Z4 (pair of parallel strokes) as unified entities (cf. Fig. 6, selections of the bottom row). In addition, the model reliably recognizes hieroglyphs even when bounding boxes overlap (cf. the glyphs depicting a snake and hand to the left side of the second row of Fig. 6).

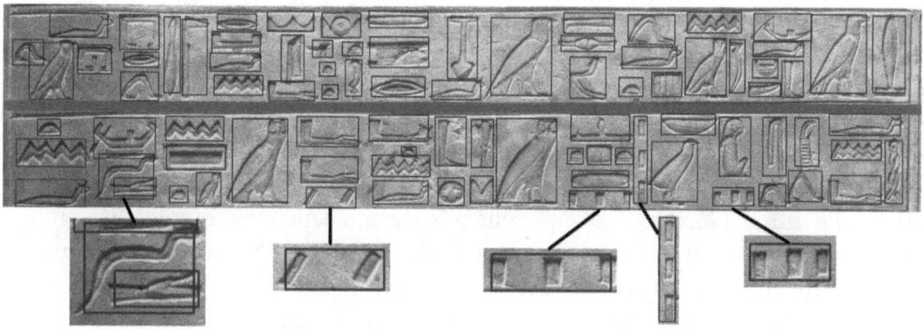

Fig. 6. Automatic segmentation and line separation in an excerpt from the Stele of Qemnen.

[16] https://github.com/facebookresearch/detectron2.

[17] According to the authors, their code was lost and could thus not be reused.

Table 1. Evaluation of the segmentation model

AP	AP50	AP75	APs	APm	APl
63.124	90.806	78.737	51.012	63.730	74.415

6.2 Reading Direction and Order

To classify the general reading direction, we train a lightweight YOLOv11 classifier (nano variant) on the full images labeled with either LR or RL. The model, only 3 MB in size, achieved an accuracy of 98.4% in determining the reading direction after just ten epochs of training (in a dataset with roughly equal class distribution). In the second step, the arrangement of the text, i.e. whether it is written in rows or columns, is determined using the bounding boxes identified during the segmentation step by the Detectron2 algorithm. Our method uses the Cohen–Sutherland algorithm to identify horizontal and vertical lines that do not intersect any bounding boxes, indicating gaps between text lines or columns (see the red bands in Fig. 6). Based on this layout, bounding boxes are grouped into sublists representing text columns or rows. In the third step, these sublists are sorted by the weighted center coordinates of the boxes according to their arrangement and detected reading direction. This approach is reliable when image quality is high and inscriptions are not severely distorted or damaged. However, complex layouts with nested subcolumns or multiple reading directions remain challenging and require further investigation.

6.3 Hieroglyph Classification

Using the sorted bounding boxes from the previous step, hieroglyph patches are cropped from the image and are then passed to the classifier. For classification, we train a YOLOv11 medium variant. The resulting model with a memory size of 22 MB achieved a Top-1 accuracy of 97.0% and a Top-5 accuracy of 99.9% after training for 15 epochs. Figure 7 shows the continued example from Fig. 6, now with bounding boxes labeled with the classification result.

A detailed error analysis revealed that misclassifications often occurred within the same Gardiner group, which suggests that the model does indeed semantically categorize similar signs correctly. For example, the confusion matrix showed that the frequently occurring hieroglyph G17 (owl) was classified correctly in 77.8% of cases. All other predictions were classified as G18 (double owl), which belongs to the same Gardiner group G. Additional examples affected by misclassifications are shown in Table 2.

6.4 Translation

We take reference translations from an excerpt of the TLA corpus [18]. Even small errors in the predicted Gardiner sequence can lead to significant changes in the translation, especially when key elements such as subjects or verbs are

```
Actual Gardiner Sequence:
U28 G1 P1 I9 D58 N42 N21 N35 T34 G17 D54 I9 Q3 X1 N1 O34 U1 G1 F36 I9 N17 N36 D21 N27
   X1 Z1 R14 X1 X1 N25 M17 D36 D21 N31 I9 N35 R8 O29a O1 D21 D54 I9 O4 G1 D54
I9 V28 N35 D36 T18 S29 D54 A40 Z2 V30 G43 Z3 R4 X1 Q3 X1 X4 Z2 G17 U23 D58 N26 O49
   D37 X1 N35 I9 D36 D36 Z4 G17 N35 N37 G17 X1 P3 I10 D46 X1 N35 I9

Predicted Gardiner Sequence:
U29 G1 P1 I9 D58 N42 N35↔Aa14 R8 G17 D54 I9 Q3 N1↔X1 O34 U1 G1 F36 I9 N17 N36 D21
   N27 X1 R14↔D2 N25↔X1↔X1 S29 D36 D21 N31 I9 N35 D58 O29a O1 D21 D54 I9 O4 G1 D54
I9 V28 N35 D36 T18 D54 S29 B1↔D55 V30↔Z3 G43 Z1 X1↔R4 Q3 O47↔X1 Z2 G17 A2 D58 N26
   O49 V30 X1 N35 I9 D36 D36 Z4 G17 N35 N37 G17 X1 P3 D46↔I10 X1 N35 I9
```

Fig. 7. Complete evaluation of a section from the Stele of Qemnen with predicted classes and comparison between actual and predicted Gardiner sequence.

Table 2. The 10 most frequent hieroglyphs in the training dataset that were classified with less than 100% precision. Strict precision means exact label match, while lenient precision also allows other (closely related) labels from the same Gardiner group.

Code	Rank	Samples	Precision (strict)	Precision (lenient)
G17	8	1356	.78	1.00
M23	16	1114	.33	1.00
D58	19	1058	.99	.99
W24	42	1024	.96	.98
Z11	62	872	.44	.44
N37	54	852	.94	1.00
O4	49	852	.87	.96
Z2	53	852	.94	.95
R4	26	798	.63	1.00
Aa1	51	752	.99	.99

altered by misclassified signs. However, in some cases, the semantics remained surprisingly intact despite multiple erroneous characters (See Fig. 8). The generated translation should therefore be treated as a suggestion rather than as an authoritative Egyptological translation, since it depends directly on the predicted Gardiner-code sequence.

Source: Stela of Nebipusenwosret	
AGS:	U1 D4 I9 F35 F35 F35 F13 N31 X1 Z2 E18 G17 O1 D21 X1 D54 I9 F35 D21 X1
ActTrans.:	"he may see the beauty of Wepwawet in his beautiful procession"
AutoTrans AGS:	"May he see the beauty of Wepwawet at his beautiful procession."
PGS:	U2 D4 I9 F35 F35 F35 F13 N31 I9 Z2 E19 G17 O1 D21 X1 D54 I9 D21<>F35 I9
AutoTrans PGS:	"May he see the beauty of his upuaut in his procession."
Comment:	Only the substitution of the word ending X1 with the possessive pronoun I9 slightly changes the meaning of the sentence.

Source: Stela of Qemnen	
AGS:	U28 G1 P1 I9 D58 N42 N21 N35 T34 G17 D54 I9 Q3 X1 N1
ActTrans.:	"May he traverse the firmament. May he journey through the sky."
AutoTrans AGS:	"May he cross the embalming place as he moves through the sky."
PGS:	U29 G1 P1 I9 D58 N42 N35 Aa14 R8 G17 D54 I9 Q3 N1<>X1
AutoTrans PGS:	"May he ascend, may the god traverse his sky."
Comment:	ActTrans and AutoTrans already differ in the object of their first sentence. The substitution of three codes in the middle of the sequence significantly alters the meaning of the translation.

Fig. 8. Application of the translation model created by [5] to excerpts from the stela of Nebipusenwosret (britishmuseum.org/collection/object/Y_EA101) and the stela of Qemnen. Abbreviations: AGS = Actual Gardiner Sequence, ActTrans = Actual Translation, AutoTrans = Automatic Translation, PGS = Predicted Gardiner Sequence.

7 Demo System

We combine all steps into a real-time Python application using a document camera (IPEVO V4K) to evaluate the practical applicability of automatic recognition of Egyptian hieroglyphs. Camera access is provided via the OpenCV library. The application is controlled via keyboard input, e.g. pressing the Enter key triggers the camera to take a snapshot of the current view. The captured image is processed through the proposed pipeline for recognizing and classifying individual hieroglyphs displaying the results on the screen in real time. To ensure smooth operation and prevent freezing during processing, the recognition pipeline is executed asynchronously in the background while the camera stream remains active. When run on a GPU, the pipeline processes and translates images within a few seconds, enabling efficient real-time interaction. This setup allows the output to be displayed sequentially. Figure 9 illustrates the sequential outputs of the demo system.

First, the system detects individual glyph regions and estimates the text arrangement (i.e., whether the hieroglyphs are written in rows or columns), which is visualized using guiding lines while on-screen status messages indicate the current processing stage. Next, the segmentation output highlights the detected glyph regions that are passed to the classifier. Finally, the recognized glyphs are displayed together with their assigned Gardiner codes, and the result can be shown as an augmented overlay that remains visible even under partial occlusion (e.g., by a finger). The translation is generated from the complete predicted

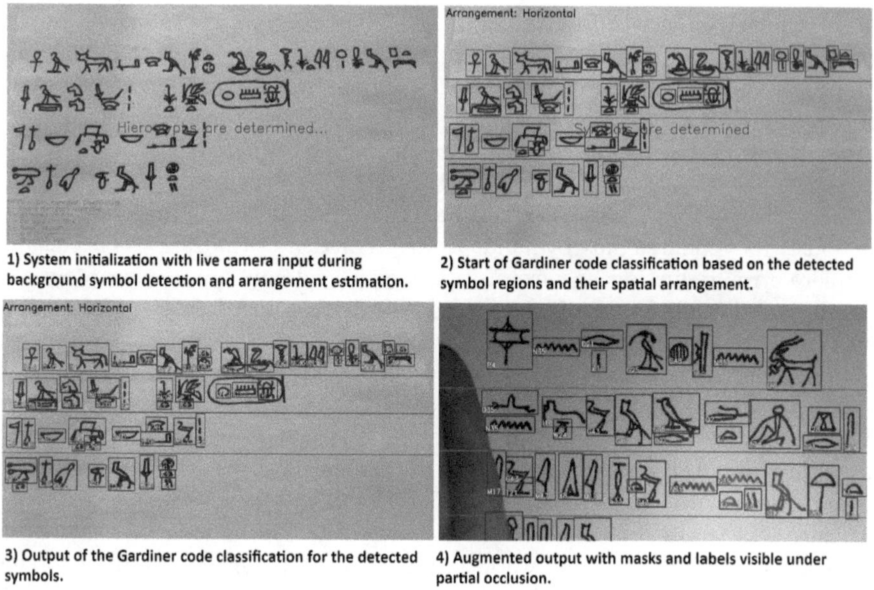

Fig. 9. Hieroglyph recognition pipeline using a document camera

Gardiner-code sequence, while the overlay mainly shows the localized glyphs and their Gardiner codes.

8 End-to-End Case Study

We conduct an end-to-end case study using the demo system. In the evaluation using 46 test cases—specially compiled pairs of images and their expected Gardiner sequences which were depicted from the Thesaurus Linguae Aegyptiae—the average correction requirement was around 40%, measured by the Levenshtein distance between the predicted and actual Gardiner code sequences, as shown in Fig. 7. There the Levenshtein distance is 24 compared to a sequence length of 86. A more detailed analysis revealed that approximately 70% of the errors were due to misclassifications, while the remaining 30% resulted from segmentation and sorting errors. As expected, the best results were achieved with clear and high-contrast images. More difficult were cases involving handwritten or carved inscriptions, particularly when they were written without sufficient detail, damaged, or poorly lit.

9 Conclusion

Our goal was to develop a camera-based prototype for analyzing hieroglyphs on real-world photographs. Due to the characteristics of the extinct Egyptian script, which differ significantly from modern languages and are compounded by the limited availability of data, the machine reading and translation of hieroglyphs still pose a major challenge for current algorithms. In contrast to previous works, the developed demo system addresses all four sub-tasks: segmentation for the automatic detection of hieroglyphs, sequencing for sorting them according to the reading direction, classification by their Gardiner code, and translation. The implementation shows promising results, especially for the digitization of texts previously documented only by hand, although further optimization is needed for complete error reduction. For example, errors in classification could potentially be reduced through further diversification of the training dataset or by using a large language model.

By making all relevant datasets and code available as open-source contributions, this work takes an important step towards the digitization and automated analysis of Egyptian script culture. The developed demo system not only facilitates the scholarly work of Egyptologists, but can also be used more broadly in the future in the form of apps or augmented reality devices—for example, for tourists or museums. The current system still has limitations that define important directions for future work. In particular, a larger user study involving Egyptologists and a more detailed treatment of complex layouts with multiple reading directions would further strengthen the evaluation.

References

1. Alayrac, J.B., et al.: Flamingo: a visual language model for few-shot learning. In: Proceedings of the 36th International Conference on Neural Information Processing Systems. NIPS '22, Curran Associates Inc., Red Hook, NY, USA (2022)
2. Allen, J.P.: Middle Egyptian: An Introduction to the Language and Culture of Hieroglyphs. 3 edn. Cambridge University Press (2014). https://doi.org/10.1017/CBO9781107283930
3. Allen, J.P.: Middle Egyptian Literature: Eight Literary Works of the Middle Kingdom. Cambridge University Press (2014)
4. Barucci, A., Cucci, C., Franci, M., Loschiavo, M., Argenti, F.: A deep learning approach to ancient Egyptian hieroglyphs classification. IEEE Access **9**, 123438–123447 (2021). https://doi.org/10.1109/ACCESS.2021.3110082
5. De Cao, M., De Cao, N., Colonna, A., Lenci, A.: Deep learning meets egyptology: a hieroglyphic transformer for translating Ancient Egyptian. In: Pavlopoulos, J., et al. (eds.) Proceedings of the 1st Workshop on Machine Learning for Ancient Languages (ML4AL 2024), pp. 71–86. Association for Computational Linguistics, Hybrid in Bangkok, Thailand and online (2024). https://doi.org/10.18653/v1/2024.ml4al-1.9, https://aclanthology.org/2024.ml4al-1.9/
6. Duque-Domingo, J., Herrera, P.J., Valero, E., Cerrada, C.: Deciphering Egyptian hieroglyphs: Towards a new strategy for navigation in museums. Sensors **17**(3) (2017). https://doi.org/10.3390/s17030589, https://www.mdpi.com/1424-8220/17/3/589
7. Erman, A., Grapow, H.: Wörterbuch der Aegyptischen Sprache – Band I. Akademie-Verlag, Berlin (1971)

8. Franken, M., van Gemert, J.C.: Automatic egyptian hieroglyph recognition by retrieving images as texts. In: Proceedings of the 21st ACM International Conference on Multimedia. p. 765–768. MM '13, Association for Computing Machinery, New York, NY, USA (2013). https://doi.org/10.1145/2502081.2502199
9. Fuentes-Ferrer, R., Duque-Domingo, J., Herrera, P.J.: Recognition of Egyptian hieroglyphic texts through focused generic segmentation and cross-validation voting. Appl. Soft Comput. **171**, 112793 (2025). https://doi.org/10.1016/j.asoc.2025.112793, https://www.sciencedirect.com/science/article/pii/S1568494625001048
10. Hossam, R., Salem, M.A.M.M., Elias, R.: Image based hieroglyphic character recognition. In: 2018 14th International Conference on Signal-Image Technology and Internet-Based Systems (SITIS), pp. 32–39. IEEE, Las Palmas de Gran Canaria, Spain (2018). https://doi.org/10.1109/SITIS.2018.00016
11. Li, J., Li, D., Savarese, S., Hoi, S.: Blip-2: Bootstrapping language-image pretraining with frozen image encoders and large language models (2023). https://arxiv.org/abs/2301.12597
12. Mohsen, S., Mansour, R., Bassem, A., Dessouky, B., Refaat, S., Ghanim, T.: Aegyptos: Mobile application for hieroglyphs detection, translation and pronunciation. In: 2023 International Mobile, Intelligent, and Ubiquitous Computing Conference (MIUCC), pp. 1–8 (2023). https://doi.org/10.1109/MIUCC58832.2023.10278338
13. Moustafa, R., Hesham, F., Hussein, S., Amr, B., Refaat, S., Shorim, N., Ghanim, T.: Hieroglyphs language translator using deep learning techniques (scriba). In: 2022 2nd International Mobile, Intelligent, and Ubiquitous Computing Conference (MIUCC), pp. 125–132. IEEE, Cairo, Egypt (2022). https://doi.org/10.1109/MIUCC55081.2022.9781784
14. Muñoz Sánchez, R.: When hieroglyphs meet technology: A linguistic journey through Ancient Egypt using natural language processing. In: Sprugnoli, R., Passarotti, M. (eds.) Proceedings of the Third Workshop on Language Technologies for Historical and Ancient Languages (LT4HALA) @ LREC-COLING-2024. pp. 156–169. ELRA and ICCL, Torino, Italia (2024). https://aclanthology.org/2024.lt4hala-1.18/
15. Nederhof, M.J.: Ocr of handwritten transcriptions of ancient egyptian hieroglyphic text. In: Altertumswissenschaften in a Digital Age: Egyptology, Papyrology and Beyond, Leipzig (2016)
16. OpenAI: GPT-4V(ision) System Card. Tech. rep., OpenAI (2023). https://cdn.openai.com/papers/GPTV_System_Card.pdf
17. Plecher, D.A., Eichhorn, C., Seyam, K.M., Klinker, G.: Arsinoë - learning egyptian hieroglyphs with augmented reality and machine learning. In: 2020 IEEE International Symposium on Mixed and Augmented Reality Adjunct (ISMAR-Adjunct), pp. 326–332 (2020). https://doi.org/10.1109/ISMAR-Adjunct51615.2020.00092
18. Richter, T.S., Hafemann, I., Fischer-Elfert, H.W., Dils, P.: Teilauszug der datenbank des vorhabens "Strukturen und Transformationen des Wortschatzes der ägyptischen Sprache" vom Januar 2018 (2018)
19. Sethe, K.: Urkunden der 18. Dynastie, Band I. Hinrichs, Leipzig (1909)
20. Sobhy, A., Helmy, M., Khalil, M., Elmasry, S., Boules, Y., Negied, N.: An ai based automatic translator for ancient hieroglyphic language–from scanned images to English text. IEEE Access **11**, 38796–38804 (2023). https://doi.org/10.1109/ACCESS.2023.3267981
21. Werning, D.A.: Das Höhlenbuch im Grab des Petamenophis (TT33): Szenen, Texte, Wandtafeln. Propylaeum (2020). https://doi.org/10.11588/propylaeum.615, https://books.ub.uni-heidelberg.de/propylaeum/catalog/book/615

22. Wiesenbach, P., Riezler, S.: Multi-task modeling of phonographic languages: Translating middle Egyptian hieroglyphs. In: Niehues, J., et al. (eds.) Proceedings of the 16th International Conference on Spoken Language Translation. Association for Computational Linguistics, Hong Kong (2019). https://aclanthology.org/2019.iwslt-1.33/

Text-Based Classification of Historical Notarial Documents from Noisy HTR Transcripts

Daniela Perdomo(✉), Veronica Romero, and Miriam Gil

Escola Tècnica Superior d'Enginyeria, Departament d'Informàtica, Universitat de València, Burjassot, Valencia, Spain
{daniela.perdomo,veronica.romero,miriam.gil}@uv.es

Abstract. Automatic classification of handwritten historical documents is an important task for improving access to archival collections, enabling efficient information retrieval, and supporting large-scale digital humanities research. This work investigates the feasibility of classifying historical manuscripts using transcriptions generated by Handwritten Text Recognition (HTR) systems and Transformer-based classification models. Although Transformer-based models have achieved strong performance in text classification, their robustness under realistic archival conditions characterized by transcription errors, orthographic variation, limited annotated data, and unknown classes remains insufficiently studied.

We present a text-based classification study that integrates a convolutional recurrent HTR system with Spanish Transformer-based classifiers trained on automatically generated transcriptions. Experiments are conducted on a corpus of eighteenth-century notarial deeds grouped into twelve document categories. The proposed evaluation considers the main difficulties of archival document analysis, including noisy HTR output, historical spelling variation, long document transcriptions, class imbalance, and open-set classification scenarios.

The study provides empirical evidence on the extent to which modern language models can exploit imperfect HTR transcripts for semantic document classification. The results show the potential of text-based approaches for organizing handwritten archival materials under realistic transcription conditions, contributing to scalable methods for document analysis, access, and retrieval in historical collections.

Keywords: Document Classification · Historical Manuscripts · Handwritten Text Recognition · Noisy Text · Spanish NLP · Transformers

1 Introduction

Recent advances in document digitization have increased the availability of historical manuscript collections in digital form. Libraries and archives have digitized vast amounts of paper-based material, creating repositories that preserve

F. Shafait et al. (Eds.): DAS 2026, LNCS 16851, pp. 441–458, 2027.
https://doi.org/10.1007/978-3-032-36207-0_26

valuable sources for historical, legal, socio-economic, and cultural research. However, digitization alone does not guarantee meaningful access to document contents. To make these collections searchable and usable, additional document analysis tasks are required, including automatic transcription, semantic classification, indexing, and information extraction.

This challenge is particularly relevant in notarial archives. Historical notarial deeds constitute a key primary source for studying legal practices, economic transactions, family relations, and social structures. Since most collections lack document-level semantic metadata, automatically identifying document types, such as powers of attorney, sales, leases, or testaments, is essential for improving access, retrieval, and large-scale analysis.

Traditionally, the semantic classification of such documents has been carried out manually by archival experts. However, manual classification is costly and does not scale to large digitized collections. If perfect transcriptions were available, document classification could be addressed using standard natural language processing (NLP) techniques applied to clean text [5,9,12]. In realistic archival settings, however, manually transcribing large manuscript collections is impractical. Handwritten Text Recognition (HTR) systems offer a scalable alternative by automatically converting handwritten images into text [26], but their output is often noisy, especially for historical manuscripts. For documents similar to those considered in this work, word recognition accuracies around 40–60% have been reported even when systems are trained on substantial annotated material [25].

The resulting HTR transcripts typically contain recognition errors, spelling variation, segmentation errors, and inconsistencies related to early modern orthography. These factors increase lexical sparsity and may affect downstream NLP models, which are commonly developed and evaluated on clean, grammatically regular text. Although Transformer-based language models have achieved strong results in many text classification tasks, their behavior when applied to noisy HTR transcripts of historical manuscripts remains insufficiently studied.

This work investigates whether semantic classification of historical notarial documents can be performed using only automatically generated HTR transcripts. To ensure comparability with previous research, we use the same dataset studied by Prieto et al. [16], which contains eighteenth-century notarial deeds from the *Archivo Histórico Provincial de Cádiz*. Their work addressed the classification of untranscribed handwritten document images using probabilistic indexing. In contrast, our study focuses on the complementary question of how far a text-based pipeline can go once imperfect HTR transcripts are available. This allows us to isolate the contribution of noisy textual content without relying on visual or layout features.

The paper makes the following contributions. First, it provides a systematic evaluation of Spanish Transformer-based language models for classifying historical notarial documents from noisy HTR transcripts. Second, it studies the role of lightweight linguistic normalization aimed at reducing variability caused by historical spelling and HTR artifacts. Third, it analyzes strategies for handling long notarial documents under the input-length constraints of Transformer models. Fourth, it evaluates two simple open-set classification strategies designed for realistic archival scenarios in which unknown document types may appear at inference time. Finally, it contextualizes the results with respect to previous work on the same dataset, highlighting the trade-off between image-based probabilistic indexing and text-based classification from HTR output.

2 Related Work

Document classification has been extensively studied in natural language processing, evolving from traditional bag-of-words and TF–IDF representations to neural approaches based on word embeddings and Transformer language models [4,12,14]. Under clean-text conditions, fine-tuned Transformer models have achieved strong performance in many classification scenarios [6,9,23]. However, these approaches assume that documents are available as reliable electronic text, an assumption that rarely holds for historical handwritten collections.

In historical document analysis, classification has often relied on visual, layout-based, or multimodal information. Early work explored handwritten document retrieval and classification using visual similarity between text lines [20]. More recent multimodal approaches combine visual features extracted from document images with textual representations obtained from OCR output [1]. Such methods can be effective, but they usually require aligned visual and textual information, more complex architectures, and additional computational resources. In large archival settings, these requirements may limit scalability and reproducibility, particularly when documents consist of multiple manuscript pages rather than single well-structured document images.

A particularly relevant line of work for historical manuscripts is content-based classification of untranscribed handwritten document images. Prieto et al. [16] addressed open-set classification of notarial image documents using probabilistic indexing. Instead of relying on explicit transcriptions, their method represents handwritten images through probabilistic word hypotheses and derives TF–IDF document representations from these estimates. This approach is well suited to scenarios where HTR transcripts are unavailable or too unreliable. Our work studies a complementary setting in which handwritten document images are first automatically transcribed using an HTR system, and then we evaluate how effectively modern language models can exploit these noisy textual representations for semantic document classification.

The increasing maturity of HTR systems has renewed interest in text-based processing of historical manuscripts. Modern HTR methods based on recurrent neural networks and, more recently, Transformer architectures have enabled large-scale transcription of handwritten collections [8,26]. Nevertheless, HTR output for historical manuscripts remains noisy and is affected by spelling variation, character substitutions, segmentation errors, and omissions. Several studies have shown that HTR errors can degrade downstream NLP tasks. For example, Hamdi et al. [7] showed that OCR noise has a strong negative impact on named entity recognition, while broader surveys on historical NLP emphasize the difficulty of processing imperfectly transcribed documents [5].

For historical archives, document-level classification can provide useful semantic metadata even when automatic transcriptions are not accurate enough for close reading. In this setting, the central question is whether imperfect HTR transcripts preserve enough information to identify the type of a document. The present work therefore focuses on text-only classification from automatically generated HTR transcripts, analyzing the behavior of Spanish Transformer models under realistic conditions of transcription noise, historical orthography, class imbalance, long documents, and unknown document types.

3 Content-Based Classification of Untranscribed Images

The content-based classification of handwritten documents consists of assigning a semantic category to a handwritten document according to its textual content. Formally speaking, let $\mathbf{x}$ be a handwritten document and C be a finite set of semantic classes. The goal is to find the most likely class $\hat{c} \in C$ that represents the content of the handwritten document. The optimal prediction is given by:

$$\hat{c} = \arg\max_{c \in C} P(c \mid \mathbf{x}), \tag{1}$$

where $P(c \mid \mathbf{x})$ denotes the posterior probability of class c given the image $\mathbf{x}$.

By further considering the word-sequence transcription of the document image, $\mathbf{w} = (w_1, w_2, ..., w_l)$, as a hidden variable in Eq. 1, the previous expression can be rewritten as:

$$\hat{c} = \arg\max_{c \in C} \sum_{\mathbf{w} \in W} P(c, \mathbf{w} \mid \mathbf{x}) = \arg\max_{c \in C} \sum_{\mathbf{w} \in W} P(c \mid \mathbf{w}, \mathbf{x})\, P(\mathbf{w} \mid \mathbf{x}) \tag{2}$$

where W is the set of all possible transcriptions, $P(\mathbf{w} \mid \mathbf{x})$ is the probability that image $\mathbf{x}$ is transcribed as the word sequence $\mathbf{w}$, and $P(c \mid \mathbf{w}, \mathbf{x})$ is the probability that the document belongs to class c given both $\mathbf{w}$ and $\mathbf{x}$.

Making the naive assumption that c depends only on $\mathbf{w}$, and approximating the sum by the dominating term, the expression can be rewritten as:

$$\hat{c} \approx \arg\max_{c \in C} \max_{\mathbf{w} \in W} P(c \mid \mathbf{w})\, P(\mathbf{w} \mid \mathbf{x}), \tag{3}$$

This equation can be approximated using a two-step approach. First, the most likely word sequence, $\hat{\mathbf{w}}$, for the given handwritten document is obtained. Then, the class probabilities are computed based on this transcription. Consequently, the final class prediction is given by:

$$\hat{c} \approx \arg\max_{c \in C} P(c \mid \hat{\mathbf{w}}), \quad \text{with} \quad \hat{\mathbf{w}} = \arg\max_{\mathbf{w} \in W} P(\mathbf{w} \mid \mathbf{x}). \tag{4}$$

In practice, the probability $P(\mathbf{w} \mid \mathbf{x})$ can be estimated using a HTR system, which maps the handwritten image $\mathbf{x}$ to its most likely word sequence $\hat{\mathbf{w}}$. Once $\hat{\mathbf{w}}$ has been obtained, the class posterior $P(c \mid \hat{\mathbf{w}})$ can be computed using a text classification model.

3.1 Automatic Transcription Technology

Modern HTR systems typically rely on deep learning architectures that integrate convolutional layers with recurrent components to perform visual sequence modeling. These models are usually trained using the Connectionist Temporal Classification (CTC) objective function. A common design consists of multiple convolutional layers followed by Bidirectional Long Short-Term Memory (BLSTM) units [2], which enable the network to model long-range dependencies along the text line. A final softmax layer produces posterior probability distributions over the set of characters defined in the training alphabet. This class of models is known as Convolutional Recurrent Neural Networks (CRNNs) [21].

Since the objective is to transcribe complete handwritten documents, the page images must first be segmented into individual text lines. In this work, page-to-line segmentation is performed using P2PaLA [18], a deep learning–based layout analysis system specifically designed for historical and handwritten documents. Once the page has been decomposed into line images, each line is processed by the HTR system, and the resulting line-level transcriptions are subsequently concatenated to obtain the full document transcription.

In this work, CRNN-based models are trained and decoded using the Laia Toolkit [17], a widely used framework for handwritten recognition. For each input line image, the trained CRNN produces a sequence of character posterior probability vectors. Although CRNNs are capable of implicitly learning certain lexical and contextual patterns, their performance can be improved by incorporating explicit language models such as character-level N-gram models. The learned N-gram constraints can be integrated with the CRNN outputs through different decoding strategies [2]. In this work, the N-gram models are represented as a stochastic finite-state transducer, whose transition probabilities are computed by combining the N-gram likelihoods with the character posterior probabilities estimated by the CRNN, appropriately normalized using character priors [2]. The resulting probabilistic transducer is then decoded using the Viterbi algorithm to obtain the most likely transcription of the input handwritten line.

3.2 Text Classification Using BERT

Recent advances in NLP have demonstrated that transformer based architectures achieve state-of-the-art performance across a wide range of text classification tasks. Among these models, BERT has become a widely adopted approach due to its capacity to learn deep contextualized text representations by jointly modeling left and right contexts through self-attention mechanisms [4].

In this work, BERT is used as the core of the text classification module. The transcribed text obtained from the handwritten document is first tokenized into subword units using the BERT tokenizer [27] and encoded as a sequence of input embeddings. These embeddings are then processed by a BERT model to generate contextualized representations. For classification, the embedding associated with the special `[CLS]` token is used as a global representation of the entire document, following the standard fine-tuning strategy proposed in [4]. This representation is passed through a task-specific linear layer followed by a softmax activation function, which produces a normalized probability distribution over the predefined set of semantic classes and enables the final class decision.

4 Experimental Framework

This section describes the dataset and the empirical methodology used in the experiments reported in Sect. 5.

4.1 Corpora

With the aim of conducting experiments comparable to previous work, the corpus used in this paper is the same dataset employed by Prieto et al. [16]. The dataset consists of two historical notarial protocol books dated 1723–1724, comprising a total of 555 deeds belonging to 41 different classes.

Historical notarial deeds constitute a particularly challenging benchmark because they combine long documents, strong lexical overlap between categories, formulaic expressions, severe class imbalance, and historical spelling variation. These characteristics differ from many historical collections where categories are visually or semantically more distinct.

Following the criteria established by Prieto et al. [16], classes with very few samples were excluded. Specifically, only those classes containing at least one deed in each book and a minimum of six deeds overall were considered. Under this criterion, 12 classes were retained, while the remaining 29 classes were grouped into an additional Reject class. Table 1 summarizes the class distribution and the number of deeds per class. It should be noted that the number of page images per deed varies considerably, ranging from two to several tens of pages.

The two books in this dataset are part of the collection compiled within the Carabela project [25], in which more than 150,000 images from the *Archivo General de Indias* (AGI) and the *Archivo Histórico Provincial de Cádiz* (AHPC) were processed to enable textual search over the digitized manuscripts. Specifically, the documents used in this paper belong to the AHPC collection.

It is important to note that the two books used in this work do not include manually corrected transcriptions or ground-truth text-line layout annotations. This limitation reflects a common archival scenario in which document-level labels may be available, while detailed transcription or layout ground truth is too costly to produce. The dataset was selected because it enables direct comparison with previous work on the same notarial collection. The only available ground-truth data for these books consists of the segmentation and classification of the deeds contained within them.

4.2 Methodology

This section describes the methodological framework adopted in this study. We present the end-to-end processing pipeline, with particular attention to the challenges posed by historical notarial documents.

Table 1. Class distribution of the 12 retained notarial categories.

Class	# Deeds
PA (Power of Attorney)	240
LP (Letter of Payment)	72
DB (Debenture)	44
LE (Lease)	32
TE (Testament)	29
SA (Sale)	21
RI (Risk)	17
CS (Census)	12
DP (Deposit)	10
ST (Statement)	9
CN (Cession)	6
TF (Treaty of Fact)	6
Total	498

Handwritten Text Recognition

The first step in our pipeline consists of training an HTR system. This training was carried out as described in the Carabela project [25], using 558 carefully transcribed page images from the collection. As detailed in Subsect. 3.1, the HTR system is based on a CRNN-based model trained and decoded using the Laia Toolkit [17]. The model used is composed of four convolutional layers and LeakyReLU as activation function, and four recurrent layers with 256 bidirectional long short-term memory units.

After training the model, the two historical notarial protocol books were processed. First, the text lines were automatically detected using P2PaLA [18] and then were automatically transcribed using the previously trained HTR model. Given that the information about the pages that belong to the different deeds is available, the transcriptions of all the lines belonging to the same deed were concatenated to obtain a document-level transcription.

Text Normalization

The resulting transcripts exhibit typical sources of variation and noise commonly found in historical manuscripts and amplified by HTR processes, including inconsistent spellings, archaic graphemes, missing diacritics, or character substitutions. Such heterogeneity increases lexical sparsity and obscures regular linguistic patterns, thereby complicating downstream NLP tasks. Previous research has highlighted the importance of structured normalization pipelines to mitigate these issues in historical corpora and to improve the stability and effectiveness of subsequent linguistic analysis [10].

Following this line of work, we apply a linguistically informed normalization pipeline tailored to early modern Spanish [15]. The pipeline consists of the following steps: Unicode normalization (NFKC) [24] and lowercasing; removal of isolated short tokens frequently produced by HTR segmentation errors; removal of foliation markers, page numbers, and non-linguistic structural artifacts; and lemmatization using a lightweight Spanish spaCy model. These transformations aim to reduce surface-level variability while preserving document classification.

Transformer-Based Document Classification

The normalized document transcriptions are used to fine-tune BERT-based classifiers under a supervised learning paradigm, minimizing a cross-entropy loss function. All model parameters are fine-tuned, allowing the classifiers to adapt the pre-trained linguistic knowledge to the specific characteristics of the document collection and the target classification task. It should be noted that the transcriptions used for fine-tuning may contain errors introduced during the HTR decoding process.

We evaluate five encoder-only Transformer configurations, corresponding to four Spanish or multilingual pre-trained model families, to assess their robustness to historical language variation and HTR noise:

- **BERTIN:** A Spanish language model derived from BERT. The current release highlights the top-performing RoBERTa-base architecture trained from scratch on the Spanish subset of the mC4 dataset using Flax [19].
- **BETO:** A Spanish BERT-based model comparable in size to BERT-Base and trained with Whole Word Masking (WWM). We evaluate both Hugging Face checkpoints: dccuchile/bert-base-spanish-wwm-cased and dccuchile/bert-base-spanish-wwm-uncased [3].
- **SpanBERTa:** A Spanish RoBERTa-based checkpoint distributed through the Hugging Face model hub. We include it as an additional Spanish encoder-only Transformer model. Since no peer-reviewed publication is available for this checkpoint, we cite the original RoBERTa architecture and identify the exact checkpoint through the model repository [13,22].
- **mRoBERTa:** A multilingual RoBERTa-based model pretrained from scratch on 35 European languages and programming code, distributed by the Barcelona Supercomputing Center through the Hugging Face model hub [11].

All models were fine-tuned under identical hyperparameter settings. Each model was trained for 50 epochs in each leave-one-out (LOO) iteration, using a maximum input length of 512 tokens, a per-device batch size of 8, gradient accumulation over 2 steps, and an effective batch size of 16. We used a learning rate of 3×10^{-5}, weight decay of 0.02, mixed-precision training, and a fixed random seed of 42. The models were implemented with the Hugging Face Transformers library and a PyTorch backend. Since the LOO protocol uses one document as a test sample in each iteration, no validation split was used inside each fold.

Open-Set Classification

In practical archival scenarios, document collections may include document types that are not represented in the training data. To avoid forcing predictions under such conditions, we complement the closed-set classification setup with two open-set classification (OSC) strategies.

Confidence-Based Rejection (OSC-C). In the first strategy, models are trained using the twelve known document classes under a standard closed-set formulation. At inference time, a confidence-based rejection mechanism is applied. Given an input document $\mathbf{x}$, the classifier outputs posterior probabilities $P(c \mid \mathbf{x})$ over the known classes. If the maximum posterior probability is below a threshold t, the prediction is rejected and the document is assigned to an unknown category; otherwise, the document is assigned to the most probable class.

This approach does not require unknown-class examples during training and implements a selective classification mechanism. The rejection threshold is evaluated through a systematic sweep in order to analyze the trade-off between coverage and classification performance on known classes.

Explicit Unknown Class (OSC-U). The second strategy introduces an explicit unknown (UNK) class, resulting in a 13-class classification problem. Documents not belonging to any of the twelve predefined categories are assigned to the UNK class during training, allowing the model to learn an explicit decision boundary for unknown document types. Unlike confidence-based rejection, this strategy treats open-set recognition as a supervised multi-class classification problem and relies on the availability of representative unknown documents during training.

Handling Long Documents

Transformer-based architectures impose a maximum input length of 512 tokens, which limits the direct processing of multi-page notarial deeds. In this paper we study two strategies to handle longer documents.

In the first strategy, each document is truncated to its first 512 tokens, and classification is performed using only this initial segment. This approach is motivated by the structure of notarial documents, in which key information about the document type is typically concentrated at the beginning of the text.

In the second strategy, rather than truncating documents to their initial tokens, we adopt a sliding-window segmentation strategy combined with probabilistic aggregation. Each normalized transcription is segmented into overlapping chunks of 512 tokens using a stride of 128 tokens. The overlap reduces boundary effects and preserves contextual continuity across fragments. During training, each chunk inherits the document-level label and is treated as an independent training instance. At inference time, the document is segmented identically. For document-level prediction, posterior probabilities $P(c \mid x_i)$ are computed for each chunk x_i. Instead of majority voting, we apply soft-voting aggregation by averaging softmax probabilities across all chunks. The final class corresponds to the maximum aggregated probability.

5 Experiments and Results

This section presents the experimental results. Accuracy and macro-averaged F1-score are used as the main evaluation measures. Macro-F1 is particularly relevant in this setting because the dataset is highly imbalanced. First, we compare the evaluated Transformer-based models in a closed-set classification (CSC) scenario using the 12 known classes. Next, we analyze the strategies proposed for handling long documents. Finally, we evaluate the two open-set classification (OSC) settings described in Sect. 4.2.

For comparability with previous work on the same dataset [16], all experiments are carried out following a LOO protocol. In each iteration, one document is held out for testing and the model is trained on the remaining documents. This protocol maximizes the use of the limited annotated data while providing one prediction for every document in the corpus.

5.1 Comparison of Transformer-Based Classification Models

Table 2 reports the global performance of the evaluated models. The experiments are conducted using only the samples belonging to the 12 known classes, and each document is represented by the first 512 tokens of its normalized HTR transcription. This setting evaluates whether the initial part of a noisy automatic transcript contains enough information to identify the document type.

Table 2. CSC performance under the LOO protocol using the 12 known classes.

Model	Accuracy	Macro-F1
BERTIN	0.831	0.677
BETO uncased	**0.876**	**0.780**
BETO cased	0.867	0.733
SpanBERTa	0.876	0.774
mRoBERTa	0.837	0.673

BETO uncased achieves the best macro-F1 score, followed closely by SpanBERTa. Although SpanBERTa obtains the same accuracy, BETO uncased provides a slightly better balance across classes, which is more relevant under strong class imbalance. Bootstrap 95% confidence intervals computed over document-level LOO predictions show overlapping accuracy ranges for both models: BETO uncased obtains 0.876 accuracy [0.845, 0.904] and 0.780 macro-F1 [0.706, 0.834], while SpanBERTa obtains 0.876 accuracy [0.847, 0.904] and 0.774 macro-F1 [0.708, 0.825]. Therefore, differences among the top-performing models should be interpreted cautiously.

Additional runs on raw HTR transcripts, without the normalization pipeline described in Sect. 4.2, showed substantially lower performance across all models.

For BETO uncased, accuracy increased from 0.690 to 0.876 and macro-F1 from 0.630 to 0.780 after normalization. Similar gains were observed for the remaining models, suggesting that reducing surface variability caused by HTR artifacts and historical spelling is important for this corpus.

Table 3 reports per-class F1 scores for the best-performing model (BETO uncased). In particular, PA (Power of Attorney) obtains 0.94, TF (Treaty of Fact) reaches 0.92, LE (Lease) achieves 0.90, LP (Letter of Payment) reaches 0.88, and TE (Testament) also obtains 0.88. These results indicate that categories characterized by stable and recurrent legal expressions are highly robust to transcription noise. Remarkably, even the relatively low-frequency class RI (Risk) (F1 = 1.00, n = 17) achieves perfect performance, suggesting that classification success depends more on semantic distinctiveness than on class frequency alone. In contrast, rare classes with limited support and greater lexical overlap, such as CS, DP, ST, and CN, exhibit higher variance and lower F1 scores.

Table 3. Per-class F1 scores under the LOO protocol using BETO uncased.

Class	PA	LP	DB	LE	TE	SA	RI	CS	DP	ST	CN	TF
F1	0.94	0.88	0.73	0.90	0.88	0.76	1.00	0.55	0.63	0.63	0.55	0.92
Support	240	72	44	32	29	21	17	12	10	9	6	6

The results show that frequent and structurally stable document types are classified reliably despite noisy HTR input. PA, LP, LE, and TE obtain high F1 scores, which suggests that recurrent legal formulae and stable opening expressions remain useful even after automatic transcription. The RI class also achieves perfect performance, indicating that semantic distinctiveness can compensate for low frequency. In contrast, rare classes with limited support and greater lexical overlap, such as CS, DP, ST, and CN, exhibit lower F1 scores. These results suggest that the main difficulty is not only class frequency, but also the degree of lexical and semantic separability between legal categories.

Figure 1 shows the confusion matrix obtained with BETO uncased. Most errors occur between semantically adjacent legal categories, particularly PA, LP, and DB, which share recurrent formulae related to authorization, payment, and obligation. Property-related documents such as SA and LE also exhibit mutual confusion, likely due to overlapping contractual terminology. These patterns suggest that semantic proximity between legal categories is an important factor behind the observed errors. Since manually corrected transcriptions are not available, the specific contribution of HTR noise cannot be directly quantified.

5.2 Long-Document Classification via Overlapping Chunk Aggregation

In this experiment, we evaluate the second strategy described in Sect. 4.2 for handling documents longer than 512 tokens. The experiment is conducted using

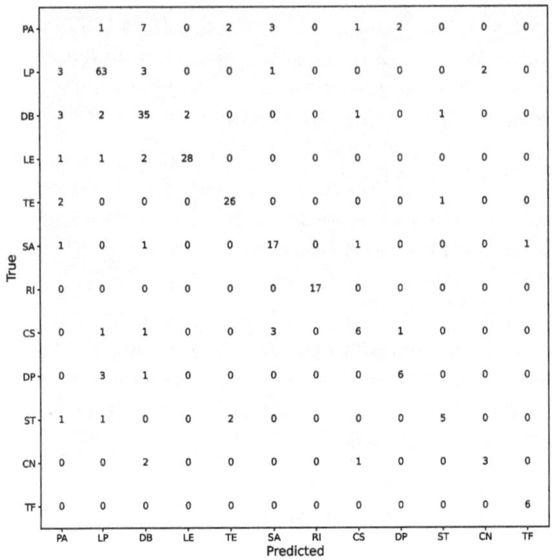

Fig. 1. Confusion matrix under LOO evaluation using BETO uncased.

BETO uncased. Each document is divided into overlapping chunks, each chunk is classified independently, and document-level probabilities are obtained by soft-voting over the chunk predictions. The results are very close to those obtained with simple truncation in terms of accuracy, reaching 0.876. However, macro-F1 decreases from 0.780 to 0.744. Table 4 summarizes the per-class results.

The chunk-based strategy preserves strong performance for frequent and structurally distinctive classes such as PA and RI. However, it does not improve over the simpler truncation strategy and leads to lower macro-F1. This result is informative because notarial deeds are highly formulaic documents: the most discriminative expressions tend to appear in the opening section, where the type of act, parties, and legal formulae are usually introduced.

A possible explanation for the lower macro-F1 is that chunk-level label inheritance introduces noise during training. Each chunk receives the document-level label, even though middle or final fragments may contain generic legal formulae, repeated clauses, or contextual information that is less discriminative of the document type. During inference, averaging probabilities across all chunks may dilute the strong evidence present in the initial segment. Therefore, in this corpus, using the first 512 tokens is not only computationally cheaper but also more robust for low-resource classes.

5.3 Open-Set Classification Methods

In realistic archival scenarios, collections may contain document types that are not represented during training. In such cases, forcing every document into one of the known classes can produce misleading metadata. We therefore evaluate

Table 4. Per-class performance under overlapping chunk aggregation.

Class	Precision	Recall	F1	Support
PA	0.95	0.95	0.95	240
LP	0.86	0.85	0.85	72
DB	0.82	0.82	0.82	44
LE	0.93	0.88	0.90	32
TE	0.84	0.90	0.87	29
SA	0.54	0.95	0.69	21
RI	0.94	0.94	0.94	17
CS	0.60	0.50	0.55	12
DP	1.00	0.60	0.75	10
ST	0.80	0.44	0.57	9
CN	0.75	0.50	0.60	6
TF	0.67	0.33	0.44	6

the two open-set strategies described in Sect. 4.2: confidence-based rejection and explicit unknown-class modeling.

Open-Set Classification via Confidence-Based Rejection

In this experiment, the models are trained using samples from the 12 known classes. However, the test set includes samples from these 12 classes as well as 57 samples from the remaining 29 classes, which are grouped into the Reject class.

This experiment is conducted using the best-performing model identified in Sect. 5.1, namely BETO uncased. For each document, the maximum softmax probability, according to Eq. 1, is used as a confidence score. Documents whose maximum confidence is below a threshold t are rejected. Figure 2 (left) shows the distribution of confidence scores for known and unknown documents. Known documents exhibit a sharply peaked distribution near confidence values close to 1.0, whereas unknown documents display a broader distribution with lower confidence values.

Figure 2 (right) presents the trade-off between rejecting unknown documents and preserving coverage over known classes as the confidence threshold increases. As the threshold grows, the rejection rate for unknown documents increases steadily, while the rejection of known documents remains limited. For instance, at a threshold of $t = 0.70$, only 1.4% of known documents are rejected, while approximately 11.6% of unknown documents are filtered out. Even at higher thresholds, the loss of known documents increases gradually, indicating a stable and well-behaved trade-off. Overall, these results show that confidence-based rejection provides a lightweight selective prediction mechanism for historical document classification. While it does not identify all unknown documents, it

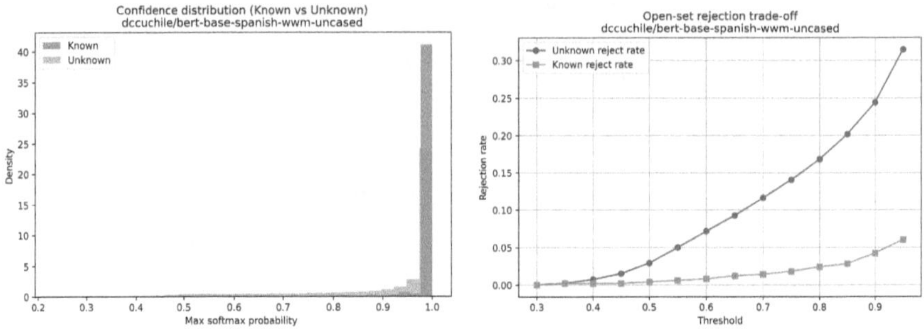

Fig. 2. Left: Distribution of maximum softmax confidence for known and unknown documents. Right: Rejection trade-off for increasing values of the confidence threshold.

reduces some forced misclassifications with minimal impact on known-class coverage.

To further analyze the impact of confidence-based rejection, Table 5 reports the classification performance obtained at this conservative operating point, which prioritizes preserving known-class coverage over aggressive rejection of unknown documents. The results show that the model maintains strong performance on high-frequency classes (e.g., *PA*, *LP*) while lower-frequency categories exhibit greater variability, which is expected given the reduced support.

Table 5. Per-class performance under the selected confidence threshold in the OSC.

Class	Precision	Recall	F1-score	Support
PA	0.9454	0.9375	0.9414	240
LP	0.8481	0.9306	0.8874	72
DB	0.6471	0.7500	0.6947	44
LE	0.9032	0.8750	0.8889	32
TE	0.8710	0.9310	0.9000	29
SA	0.6087	0.6667	0.6364	21
RI	1.0000	0.8235	0.9032	17
CS	0.3000	0.2500	0.2727	12
DP	0.5000	0.5000	0.5000	10
ST	1.0000	0.3333	0.5000	9
CN	0.0000	0.0000	0.0000	6
TF	1.0000	0.8333	0.9091	6

Open-Set Classification via an Explicit Unknown Class

As an alternative to confidence-based rejection, we evaluate an explicit unknown class (UNK). Document types outside the twelve predefined categories are grouped into an additional class, resulting in a 13-class classification problem.

Table 6. Global performance with an explicit UNK class.

Accuracy	Macro-F1	UNK Recall	Known→UNK
0.818	0.715	0.596	0.070

Table 6 summarizes the results obtained with the best-performing model, BETO uncased. The inclusion of the UNK class leads to an overall accuracy of 0.818 and a macro-F1 of 0.715, representing a moderate decrease compared to the 12-class closed-set configuration. This reduction reflects the increased difficulty of the classification task and the introduction of a competing rejection category.

From an open-set perspective, the explicit UNK strategy achieves a recall of approximately 60% on unknown documents, substantially higher than the rejection rates obtained using confidence-based thresholding. However, this gain comes at the cost of increased absorption of known documents into the UNK class: approximately 7% of documents belonging to known categories are misclassified as unknown.

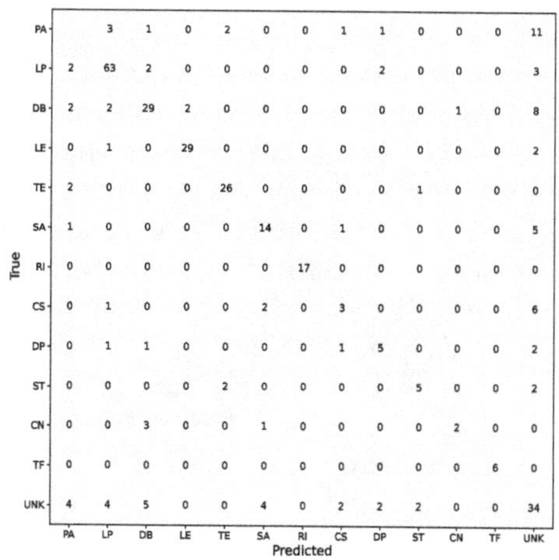

Fig. 3. Confusion matrix obtained under the 13-class configuration.

Figure 3 shows that high-frequency and lexically distinctive classes (PA, LP, TE, RI) remain largely unaffected, while low-support categories (CS, DP, CN) are more frequently absorbed into UNK. This indicates that UNK mainly captures uncertain predictions rather than exclusively novel document types.

Overall, these results highlight a trade-off between selective prediction and supervised rejection. The explicit UNK strategy detects a larger proportion of unknown documents, but it also absorbs some known documents, particularly from low-resource classes. This suggests that the UNK class captures general uncertainty rather than only truly novel document types. In contrast, confidence-based rejection does not require unknown examples during training and better preserves coverage over known classes, but it is more conservative in rejecting unknown documents. The choice between both strategies therefore depends on the availability of representative unknown samples and on whether the archival application prioritizes coverage or rejection.

To contextualize our results, Prieto et al. [16] reported higher performance on the same dataset using probabilistic indexing of untranscribed handwritten images. Their best closed-set configuration achieved 6.2% error, equivalent to 93.8% accuracy, while their open-set configuration with an explicit Reject class achieved 10.5% error, equivalent to 89.5% accuracy. These results confirm that probabilistic indexing remains a strong representation for this collection. In contrast, our approach operates on a single HTR transcript per document and evaluates how much semantic information is preserved after automatic transcription. The two approaches are therefore complementary: probabilistic indexing avoids hard transcription decisions, whereas text-based classification enables direct integration with NLP pipelines once HTR output is available.

6 Conclusions and Future Work

This paper investigates the feasibility of classifying handwritten historical documents using HTR-generated transcriptions and Transformer-based language models. The results show that a text-only pipeline can identify semantic document types in eighteenth-century notarial manuscripts under realistic archival conditions, including noisy transcription, class imbalance, long documents, and unknown classes. Under the LOO protocol, the best-performing model achieves 87.6% accuracy and a macro-F1 score of 0.780, establishing a competitive text-based reference without relying on visual or layout information.

The experiments indicate that the first 512 tokens are sufficient for this corpus, while overlapping chunk aggregation does not improve performance. This suggests that the most discriminative information in notarial deeds is concentrated in their opening formulae, and that assigning document-level labels to all chunks may introduce noise. In the open-set setting, confidence-based rejection and explicit unknown-class modeling show complementary trade-offs between preserving known-class coverage and detecting unknown documents.

The main limitations are the small and imbalanced dataset, its restriction to two eighteenth-century Spanish notarial books, and the lack of manually corrected transcriptions, which prevents a direct analysis of the relationship between

HTR quality and classification performance. Future work will explore cross-archive evaluation, adaptive normalization, stronger open-set methods, multi-modal models combining text with visual and layout information, and automatic deed segmentation.

Acknowledgments. This work was supported by the Generalitat Valenciana under the Santiago Grisolía program, Projecte CIGRIS/2024/129 Ref. CPI25-313 and by the EU4IRI project PID2024-161104OB-C22 funded by MICIU/AEI/10.13039/501100011033 and by FEDER, UE.

References

1. Audebert, N., Herold, C., Slimani, K., Vidal, C.: Multimodal deep networks for text and image-based document classification. In: Machine Learning and Knowledge Discovery in Databases, pp. 427–443 (2020)
2. Bluche, T.: Deep neural networks for large vocabulary handwritten text recognition. Ph.D. thesis, Université Paris-Sud (Paris XI) (2015)
3. Cañete, J., Chaperon, G., Fuentes, R., Ho, J.H., Kang, H., Pérez, J.: Spanish pre-trained BERT model and evaluation data (2023). https://arxiv.org/abs/2308.02976
4. Devlin, J., Chang, M.-W., Lee, K., Toutanova, K.: BERT: pre-training of deep bidirectional transformers for language understanding. In: Proceedings of NAACL-HLT 2019, pp. 4171–4186 (2019). https://doi.org/10.18653/v1/N19-1423
5. Ehrmann, M., Hamdi, A., Pontes, E.L., Romanello, M., Doucet, A.: Named entity recognition and classification in historical documents: a survey. ACM Comput. Surv. **56**(2) (2023). https://doi.org/10.1145/3604931
6. Galal, O., Abdel-Gawad, A.H., Farouk, M.: Rethinking of BERT sentence embedding for text classification. Neural Comput. Appl. **36**, 20245–20258 (2024). https://doi.org/10.1007/s00521-024-10212-3
7. Hamdi, A., Jean-Caurant, A., Sidère, N., Coustaty, M., Doucet, A.: Assessing and minimizing the impact of OCR quality on named entity recognition. In: Digital Libraries for Open Knowledge, pp. 87–101 (2020)
8. Hamza, A., Ren, S., Saeed, U.: ET-network: a novel efficient transformer deep learning model for automated Urdu handwritten text recognition. PLoS ONE **19**(5), 1–21, e0302590 (2024). https://doi.org/10.1371/journal.pone.0302590
9. Jiang, S., Hu, J., Magee, C.L., Luo, J.: Deep learning for technical document classification. IEEE Trans. Eng. Manage. **71**, 1163–1179 (2024)
10. Johnson, K.P., Burns, P.J., Stewart, J., Cook, T., Besnier, C., Mattingly, W.J.B.: The classical language toolkit: an NLP framework for pre-modern languages. In: International Joint Conference on NLP: System Demonstrations, pp. 20–29 (2021)
11. Language Technologies Laboratory, Barcelona Supercomputing Center: BSC-LT/mRoBERTa. https://huggingface.co/BSC-LT/mRoBERTa
12. Li, Q., et al.: A survey on text classification: from traditional to deep learning. ACM Trans. Intell. Syst. Technol. **13**(2) (2022)
13. Liu, Y., et al.: RoBERTa: A Robustly Optimized BERT Pretraining Approach. arXiv preprint arXiv:1907.11692 (2019)
14. Mikolov, T., Chen, K., Corrado, G., Dean, J.: Efficient estimation of word representations in vector space (2013). https://arxiv.org/abs/1301.3781

15. Porta, J., Sancho, J.L., Gómez, J.: Edit transducers for spelling variation in old Spanish. In: Proceedings of the Workshop on Computational Historical Linguistics (NODALIDA 2013), pp. 70–79. Linköping University Electronic Press (2013)
16. Prieto, J.R., Flores, J.J., Vidal, E., Toselli, A.H.: Open set classification of untranscribed handwritten text image documents. Pattern Recogn. Lett. **172**, 113–120 (2023). https://doi.org/10.1016/j.patrec.2023.06.006
17. Puigcerver, J.: Are multidimensional recurrent layers really necessary for handwritten text recognition? In: Proceedings of the International Conference on Document Analysis and Recognition (ICDAR), vol. 1, pp. 67–72 (2017)
18. Quirós, L.: P2pala: page to page layout analysis toolkit (2017). https://github.com/lquirosd/P2PaLA, gitHub repository
19. de la Rosa, J., Ponferrada, E.G., Romero, M., Villegas, P., de Prado Salas, P.G., Grandury, M.: Bertin: efficient pre-training of a Spanish language model using perplexity sampling. Procesamiento del Lenguaje Natural **68**, 13–23 (2022)
20. Schomaker, L.: Retrieval of handwritten lines in historical documents. In: Ninth International Conference on Document Analysis and Recognition (ICDAR 2007), vol. 2, pp. 594–598 (2007). https://doi.org/10.1109/ICDAR.2007.4376984
21. Shi, B., Bai, X., Yao, C.: An end-to-end trainable neural network for image-based sequence recognition and its application to scene text recognition. CoRR abs/1507.05717 (2015)
22. Skim AI Technologies: SpanBERTa base cased Spanish model (2026). https://huggingface.co/skimai/spanberta-base-cased, hugging Face model repository: `skimai/spanberta-base-cased`
23. Sun, C., Qiu, X., Xu, Y., Huang, X.: How to fine-tune bert for text classification? (2020). https://arxiv.org/abs/1905.05583
24. Unicode Consortium: Unicode standard annex #15: Unicode normalization forms (2022). https://www.unicode.org/reports/tr15/, unicode Standard, Version 15.0
25. Vidal, E., et al: The carabela project and manuscript collection: large-scale probabilistic indexing and content-based classification. In: 17th International Conference on Frontiers in Handwriting Recognition (ICFHR), pp. 85–90 (2020). https://doi.org/10.1109/ICFHR2020.2020.00026
26. Voigtlaender, P., Doetsch, P., Ney, H.: Handwriting recognition with large multi-dimensional long short-term memory recurrent neural networks. In: International Conference on Frontiers in Handwriting Recognition (ICFHR), pp. 228–233 (2016)
27. Wu, Y., Schuster, M., Chen, Z., et al.: Google's neural machine translation system: bridging the gap between human and machine translation. In: Conference on Empirical Methods in Natural Language Processing (EMNLP), pp. 493–503 (2016)

Mind the Gaps: A Curated Benchmark for Form Field Detection

Iheb Brini[1(✉)], Omar Moured[2], Hamza Gbada[3], and Elisa Barney[1]

[1] Luleå University of Technology, Luleå, Sweden
ihebbrini.ing@gmail.com, elisa.barney@ltu.se
[2] Karlsruhe Institute of Technology, Karlsruhe, Germany
omar.moured@kit.edu
[3] LATIS-Laboratory of Advanced Technology and Intelligent Systems, National Engineering School of Sousse (ENISo), University of Sousse, Sousse, Tunisia
hamza.gbada@eniso.u-sousse.tn

Abstract. Form Field Detection (FFD) is a fundamental component of document understanding systems, enabling applications ranging from large-scale industrial digitization to accessible form interaction for automated analysis. Unlike conventional object detection tasks, FFD is inherently challenging because fields are often defined by layout structure and whitespace rather than visible foreground content. Existing large-scale datasets frequently rely on heuristic annotation pipelines, resulting in noisy and inconsistent labels that hinder reliable evaluation. In this work, we introduce a carefully curated mini-CommonForms benchmark for FFD with consistent, high-quality annotations, and present a detailed evaluation of state-of-the-art detection approaches. The benchmark is designed to support reproducible research and realistic systems in both document automation and accessibility-oriented systems. Dataset and code are available at https://github.com/moured/mini-commonforms.

Keywords: Form Field Detection · Document Analysis

1 Introduction

Structured Document Understanding (SDU) seeks to extract structured information from document images to support applications such as industrial document digitization, automated workflow processing, and accessible interaction with forms for visually impaired users. While recent advances in Optical Character Recognition (OCR) and document layout analysis have significantly improved the understanding of textual and semantic regions [17,24], accurately identifying form input regions remains a challenging and underexplored problem. Reliable Form Field Detection (FFD) is essential for downstream systems including automatic form completion, document automation platforms, and assistive technologies that enable visually impaired users to navigate and interact with structured documents. FFD can also play a major role in document understanding and paired with OCR to locate relevant information from forms.

F. Shafait et al. (Eds.): DAS 2026, LNCS 16851, pp. 459–473, 2027.
https://doi.org/10.1007/978-3-032-36207-0_27

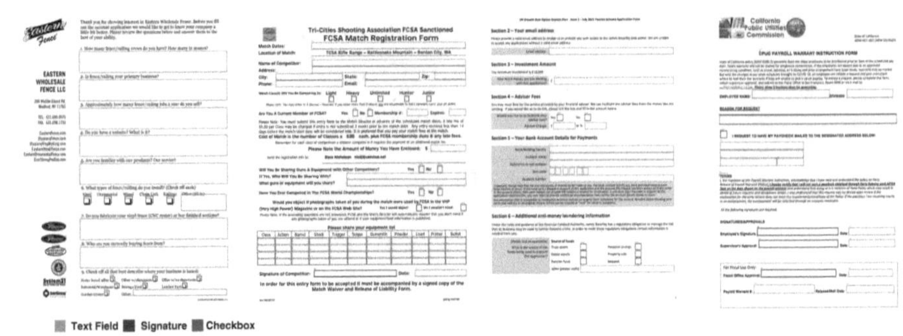

Fig. 1. Sample annotated documents for the Form Field Detection task [5].

We define FFD as the task of localizing form input regions such as text boxes, checkboxes, and signature fields in document images. Unlike conventional object detection, FFD is inherently difficult because many targets correspond to empty or weakly defined regions whose structure is expressed primarily through layout organization, whitespace, alignment, or subtle graphical cues rather than visible foreground content. As shown in Fig. 1, valid fields may appear visually ambiguous, loosely bounded, or only implicitly indicated by surrounding context. Consequently, effective FFD requires modeling structural dependencies beyond textual semantics [1,13].

The importance of dataset curation has been repeatedly demonstrated in document understanding research. A notable example is FUNSD [11], which transformed the generic RVL-CDIP benchmark [9] into a foundational resource for structured form analysis. Starting from noisy, low-resolution document scans annotated only with coarse document categories, the authors manually filtered thousands of samples and introduced dense structural annotations, including word-level bounding boxes, semantic labels, and key-value relations. This careful refinement enabled substantial progress in downstream tasks such as layout reasoning and information extraction, illustrating that high-quality annotations can be more impactful than scale alone for structurally complex document tasks.

This observation is particularly relevant for Form Field Detection (FFD), where subtle layout cues and ambiguous field boundaries make annotation quality critical. Although large-scale datasets such as CommonForms [5] provide broad template diversity, heuristic annotation pipelines often introduce substantial noise and inconsistency. As illustrated in Fig. 2, we observe recurring issues including ① *geometric noise* (imprecise field localization), ② *semantic noise* (missing, duplicated, or ambiguous fields), and ③ *cross-template inconsistencies* caused by heterogeneous layouts and multilingual formatting conventions. Such artifacts hinder reproducible evaluation and limit the reliability of FFD systems deployed in industrial automation and accessibility-oriented applications.

1. We introduce *mini-CommonForms*, a carefully curated and human-validated subset of CommonForms, significantly reducing annotation noise and correct-

ing recurring false positive and false negative labeling errors present in the original dataset.
2. We demonstrate that training on the curated benchmark consistently improves Form Field Detection performance, achieving more than 5% mAP compared to training on the original noisy annotations.
3. We provide the first comprehensive quality analysis of the large-scale CommonForms dataset, identifying recurring annotation issues, structural inconsistencies, and layout-induced failure patterns that impact reliable FFD evaluation and deployment.

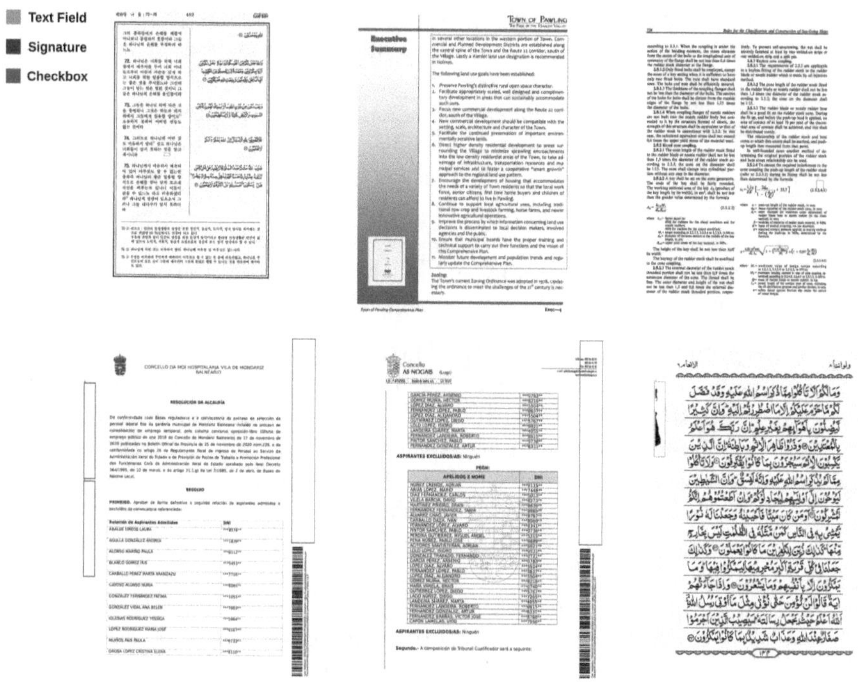

Fig. 2. Examples of annotation gaps observed in the CommonForms dataset. The figure illustrates incorrect or imprecise bounding boxes, semantically meaningless or duplicated annotations, missing fields, inconsistent category, multi-template, and challenges arising from multilingual documents and diverse formatting.

2 Related Work

Transformer-Based Visual Document Understanding. Early works like DocFormer [1] and DocFormerV2 [2] integrated textual, visual, and spatial features through novel self-attention mechanisms to achieve state-of-the-art results on various benchmarks. Similarly, FormNet [13] and its successor FormNetV2 [14] utilized Graph Convolutional Networks (GCNs) and centralized multimodal graph contrastive learning to better capture the structural relationships of tokens in forms. End-to-end approaches such as Dessurt [7] and the OCR-free Donut [12] have further simplified the pipeline by removing the dependency on external OCR engines, generating text autoregressively from document images. While these models excel at general document understanding and key information extraction (KIE), they typically rely on heavy pre-training and often struggle to explicitly model the absence of information, such as empty fields, which are defined by structural gaps rather than textual content.

Benchmark Construction and Curated Subsets. Creating high-quality benchmarks by curating subsets of larger, noisier collections is a well-established practice in SDU research. For instance, FUNSD [11] was created by manually annotating a varied subset of the massive, grayscale RVL-CDIP [9] collection, providing a challenging testbed for spatial understanding. Similarly, XFUND [23] extended this concept to multilingual settings, while DocVQA [15] sourced its varied document images from the extensive UCSF Industry Documents [21] Library. These datasets demonstrate the value of focused, densely annotated subsets for evaluating specific capabilities like key-value extraction and reasoning. Our work follows this precedent by introducing *mini-CommonForms*, a rigorously curated subset of the CommonForms dataset [5].

Domain-Specific Form and Receipt Datasets. Beyond subset-based benchmarks, the community has developed numerous datasets targeting specific form-like domains. SROIE [10] and CORD [16] focus on receipt information extraction, with CORD offering a hierarchical schema for complex list structures. For longer, more complex layouts, the Kleister [19] datasets (NDA and Charity) challenge models to handle multi-page documents. VRDU [22] benchmarking suite addresses the "visually rich" aspect with diverse templates and schemas. Other specialized datasets include Form-NLU [8] for capturing user intent in form filling and WildReceipt [20] for processing receipts in unconstrained environments. While valuable, these datasets prioritize text-rich extraction over the structural detection of empty or implicit fields, which is the primary focus of our research.

3 Mini-CommonForms Dataset

3.1 Analysis of Common Form Dataset

CommonForms [5] is currently the largest publicly available benchmark for Form Field Detection (FFD). It defines three main field categories as shown in Fig. 1: (1) *Text*, which includes regions for written user input; (2) *Choice*, which covers selection elements such as checkboxes and radio buttons; and (3) *Signature*, which refers to areas reserved for signatures, initials, or approval marks. The dataset contains approximately 436k training pages, 33.1k test pages, and 18.2k validation pages, with a total storage size exceeding 163 GB. Its scale and layout diversity make it an attractive resource for large-scale document understanding research. However, our analysis reveals that the raw dataset contains substantial structural and annotation noise that limits its reliability for rigorous FFD evaluation and training.

First, a significant portion of the dataset consists of low-quality document scans, including blurred pages, compression artefacts, scanning shadows, partially cropped forms, and low-resolution historical documents. Since FFD depends heavily on subtle layout cues such as whitespace, alignment, and faint ruling lines, image degradation disproportionately impacts annotation quality and model learning.

Second, the dataset exhibits strong heterogeneity in document language and formatting conventions. CommonForms contains multilingual documents spanning multiple scripts, layouts, and regional form structures. While this diversity is valuable, it also introduces substantial annotation inconsistencies across templates, especially for structurally ambiguous fields and culturally specific form layouts.

Third, despite heuristic collection procedures, we observe numerous non-form or weakly form-related documents within the dataset. These include letters, invoices, advertisements, and generic scanned pages containing little or no fillable structure. In many cases, annotations are either missing entirely or assigned to visually irrelevant regions, introducing substantial semantic noise during training. We additionally identify several recurrent annotation artefacts, including missing annotations for visually obvious fields, duplicated or overlapping field boxes, invalid bounding boxes with negative or degenerate dimensions, severely misaligned field boundaries, and tiny isolated annotations that likely correspond to parsing artefacts rather than meaningful form fields.

Beyond annotation validity, the dataset also exhibits significant statistical irregularities in both bounding box geometry and category distribution. As illustrated in Fig. 3, the bounding box area distribution contains a substantial number of extreme outliers, including numerous annotations with near-imperceptible spatial extent. Many of these regions correspond to isolated marks, OCR artefacts, or annotation noise rather than semantically meaningful form fields.

Figure 2 illustrates representative examples of these failure cases, where tiny bounding boxes are incorrectly assigned within a multilingual religious document that does not correspond to a genuine fillable form and can lead to negative behavior during the training process (Table 1).

Table 1. Comparison of related form document datasets.

Dataset	Year	Task	Images	train/val/test	FFD	Empty	Checkbox	Signature	Curated
FUNSD [11]	2019	Form Understanding	199	149 / – / 50	×	×	×	×	✓
SROIE [10]	2019	Receipt OCR	987	626 / – / 347	×	×	×	×	✓
AutoFormBench [4]	2026	FFD	407	70 / 15 / 15%	✓	✓	✓	×	✓
CommonForms [5]	2025	FFD	486,954	435,698 / 18,195 / 33,061	✓	✓	✓	✓	×
Mini-CommonForms (Ours)	2026	FFD	65,736	45,285 / 7,487 / 12,964	✓	✓	✓	✓	✓

In addition, the dataset suffers from severe category imbalance. As shown in the category statistics, one annotation class dominates the dataset with more

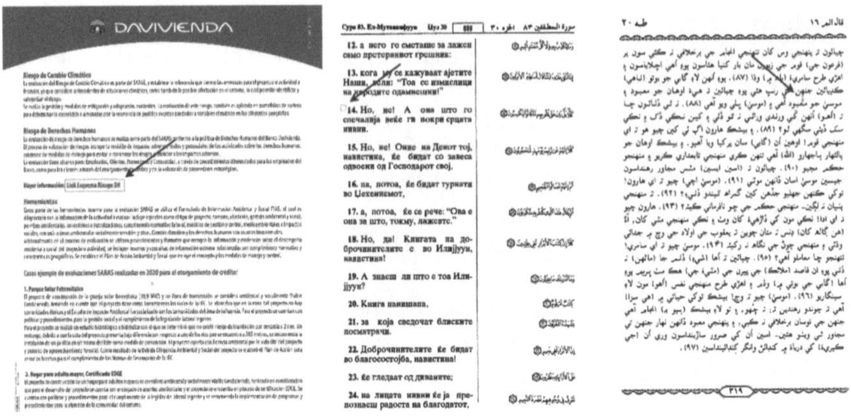

Fig. 3. Example of noisy micro-scale bounding box annotations incorrectly assigned within a multilingual non-form document in the CommonForms dataset. Errors such as annotation artefacts and extreme bounding boxes are common throughout the original dataset.

than 225k instances, while another contains fewer than 2k samples. Such imbalance biases model optimization toward majority classes and reduces robustness for structurally important but underrepresented field types.

These observations highlight a central challenge for large-scale Form Field Detection benchmarks: annotation quality and structural consistency are as critical as dataset scale itself. The combination of noisy labels, non-form documents, invalid geometric annotations, and highly imbalanced category distributions can substantially hinder reproducible evaluation and limit the reliability of systems intended for industrial digitization and accessibility-oriented applications.

3.2 Curation Process

Dataset Cleaning and Verification Pipeline. Based on the original CommonForms COCO annotation, we introduce a multi-stage cleaning and verification pipeline designed to remove annotation artefacts, suppress structural noise, and ensure that retained samples correspond to form documents. The Fig. 4 shows the dataset cleaning and verification pipeline.

Step 1: Geometric Validation. We first remove bounding boxes that violate basic geometric consistency constraints. Specifically, we discard annotations satisfying at least one of the following conditions:

- non-positive width or height,
- minimum side length smaller than 5 pixels,
- extreme aspect ratio outside the interval $[1/50, 50]$,
- coordinates extending outside page boundaries.

These cases typically correspond to corrupted annotations, parsing artefacts, or degenerate regions.

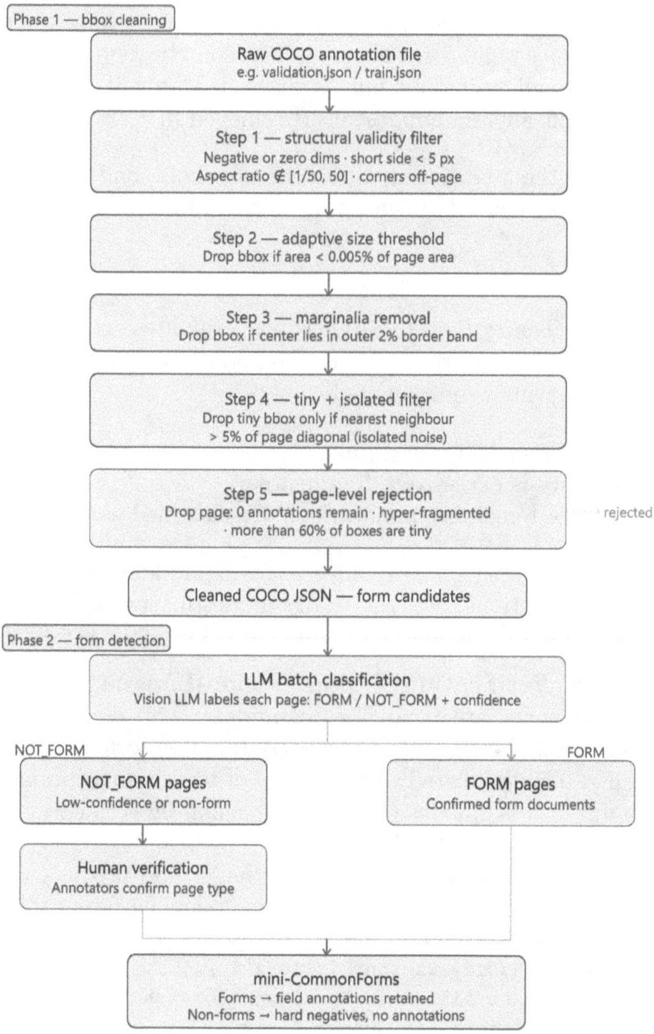

Fig. 4. Dataset cleaning and verification pipeline

Step 2: Adaptive Scale Filtering. To handle differences in document resolution, we apply a page-relative filtering criterion instead of a fixed pixel threshold. A bounding box is removed if: $\frac{\text{bbox area}}{\text{page area}} < 5 \times 10^{-4}$

Step 3: Marginalia Suppression. While document borders often contain scanning artefacts, pagination marks, crop remnants, or decorative elements that are not semantically meaningful form fields we remove annotations whose center lies within the outer 2% border band of the page.

Step 4: Context-Aware Tiny Object Filtering. Small annotations are not systematically removed, since valid form elements can sometimes occupy very limited space. Instead, a small bounding box is discarded only if it is also spatially isolated. More specifically, an annotation is removed if:

- its area is below the adaptive small-object threshold, and
- the distance to its nearest neighboring annotation exceeds 5% of the page diagonal.

Step 5: Page-Level Quality Rejection. Lastly we evaluate each page globally and discard pages exhibiting severe structural degradation. A page is rejected if at least one of the following conditions holds:

- no annotations remain after cleaning,
- annotation density is excessively fragmented,
- more than 60% of remaining annotations are classified as tiny regions.

The impact of this stage is to remove pages dominated by annotation artefacts, fragmented detections, or structurally uninformative layouts.

Document-Type Verification Using Vision-Language Models. The remaining cleaned pages are treated as candidate form documents. However, large-scale web-collected corpora frequently contain non-form documents despite heuristic filtering. This is related to the issues of wrong annotation as depicted in Fig. 3. To address this issue, we introduce a semantic verification stage using a Qwen3.6[1] Vision-Language Model (VLM).

The model receives batches of document images and predicts whether each page corresponds to a form document using the following prompt:

```
You will be shown N images.
For each image, classify whether it is a form, i.e., a document with
fields intended to be filled in by a person, such as a tax form,
application, intake form, or questionnaire.

Respond with exactly N lines, one per image, using the exact
filename:

<filename>: <yes or no>, confidence <0.0 to 1.0>
```

For each image, the VLM produces:

- a binary decision (*form*/*non-form*),
- a confidence score.

The prompt explicitly instructs the model to identify structural indicators of forms, including:

- empty text fields,

[1] https://huggingface.co/Qwen/Qwen3.6-27B.

- checkboxes,
- signature lines,
- fillable tables,
- aligned label-field structures.

Human Verification of Negative Samples. Predictions classified as *non-form* undergo an additional manual Human-in-the loop verification step to reduce false negatives or unknown predictions with low confidence score introduced by the VLM. Documents confirmed as non-form are retained as negative layout examples, while all associated field annotations are removed.

4 Experiments and Results

4.1 Implementation Details

All detection models were trained with MMDetection [6] on 4× NVIDIA A100 GPUs for 24 epochs. We used SGD with momentum 0.9 and weight decay 10^{-4}, and an initial learning rate of 0.01 (auto-scaled with base batch size 16). The learning rate employed a linear warm-up for the first 500 iterations (start factor 10^{-3}), followed by a multi-step schedule with decay factor 0.1 at epochs 16 and 22. Due to the limited number of images we generated from the LLM calls and the its huge cost, we used a subset of the original training set to train these version of models. During training, we used aspect-ratio batch sampling and standard augmentations (random resize and horizontal flip). At inference, images were resized to a scale of 1333×800 preserving aspect ratio.

Table 2. Mini-CommonForms overall and per-class detection performance and evaluation on the original test set from the CommonForms dataset.

Model	Backbone	Parameters	mAP	mAP_{50}	Text		Choice		Signature	
					mAP	mAP_{50}	mAP	mAP_{50}	mAP	mAP_{50}
FFDNet-S	YOLO11	9M	27.5	42.6	52.4	74.5	29.5	51.9	0.6	1.4
FFDNet-L	YOLO11	25M	24.1	36.9	44.6	63.9	25.2	44.0	2.5	2.8
TOOD	ResNet-101	51.02M	**34.6**	51.5	58.9	80.3	**34.9**	61.0	9.7	13.1
DINO	Swin-L	218M	33.7	**54.2**	**60.3**	**83.6**	34.7	**64.3**	9.3	14.6
Def. DETR	ResNet-50	41.075M	30.0	50.7	54.2	79.0	26.3	57.6	9.5	15.4
Cascade R-CNN	ResNeXt-101	127M	28.1	43.1	46.3	70.4	22.8	44.2	**11.9**	**19.7**
Sparse R-CNN	ResNet-101	125.16M	26.9	41.6	46.5	64.5	24.4	45.9	9.7	14.4
Faster R-CNN	ResNeXt-101	99.651M	24.9	40.3	45.3	69.3	18.5	39.3	9.7	16.1
GFL	ResNeXt-101	50.88M	13.0	26.1	30.1	55.5	8.7	22.2	0.3	0.6

4.2 Evaluation Metrics

Layout performance. We follow the standard COCO-style evaluation protocol for object detection and report $mAP_{50:95}$, together with mAP_{50}. In addition to overall scores, we report per-class results for the three Mini-CommonForms categories (*Text*, *Choice*, *Signature*).

4.3 Results

Detection Performance. Table 2 reports the overall and per-class detection performance on Mini-CommonForms. Overall, TOOD achieves the best mAP (34.6), indicating superior robustness under stricter IoU thresholds (0.50:0.95). In contrast, DINO obtains the highest mAP_{50} (54.2), suggesting strong coarse localization performance at the IoU threshold of 0.5. At the class level, *Text* detection is the most mature category across models. DINO achieves the best performance for this class (60.3 mAP / 83.6 mAP_{50}), followed by TOOD. For the *Choice* category, TOOD achieves the highest mAP (34.9), while DINO obtains the highest mAP_{50} (64.3), showing the advantage of strong global reasoning for structured layout elements such as checkboxes. In contrast, for the *Signature* category, Cascade R-CNN attains the highest mAP (11.9) and mAP_{50} (19.7). This indicates that signature detection remains the most challenging class, likely due to its visual variability and lower representation in the dataset. From a model-efficiency perspective, Deformable DETR with 41M parameters provides competitive performance compared to heavier CNN-based two-stage models. However, the highest strict-IoU performance (mAP) is achieved by TOOD, highlighting that precise boundary alignment remains challenging for document-layout detection.

For FFDNet-S and FFDNet-L, which were trained on the original CommonForms dataset, we observe a noticeable drop in overall mAP, mainly due to the very low performance on the Signature class. Figure 5 compares FFDNet-L with our trained Faster R-CNN model. In these examples, signature regions are often predicted as Text fields rather than as Signature fields, which partly explains the degradation in Signature performance. This behavior is also likely related to the strong class imbalance in the original dataset, where the dominant Text category biases the model toward predicting underrepresented classes as Text.

5 Discussion and Ablation Study

As part of our discussion and ablation analysis, we further inspect the behavior of representative detectors. We use two explainability method LRP [3] and Grad-CAM [18] as qualitative diagnostic tools to examine surface-level failure modes.

Figure 6 presents Grad-CAM explanations for representative failure cases, while Fig. 7 shows the corresponding LRP-mixed attributions. In the *Choice* class, Grad-CAM reveals a consistent failure mode across architectures. The detector frequently concentrates on printed label text rather than the checkbox interior, as shown in Fig. 6a and 6b. The model effectively behaves as a

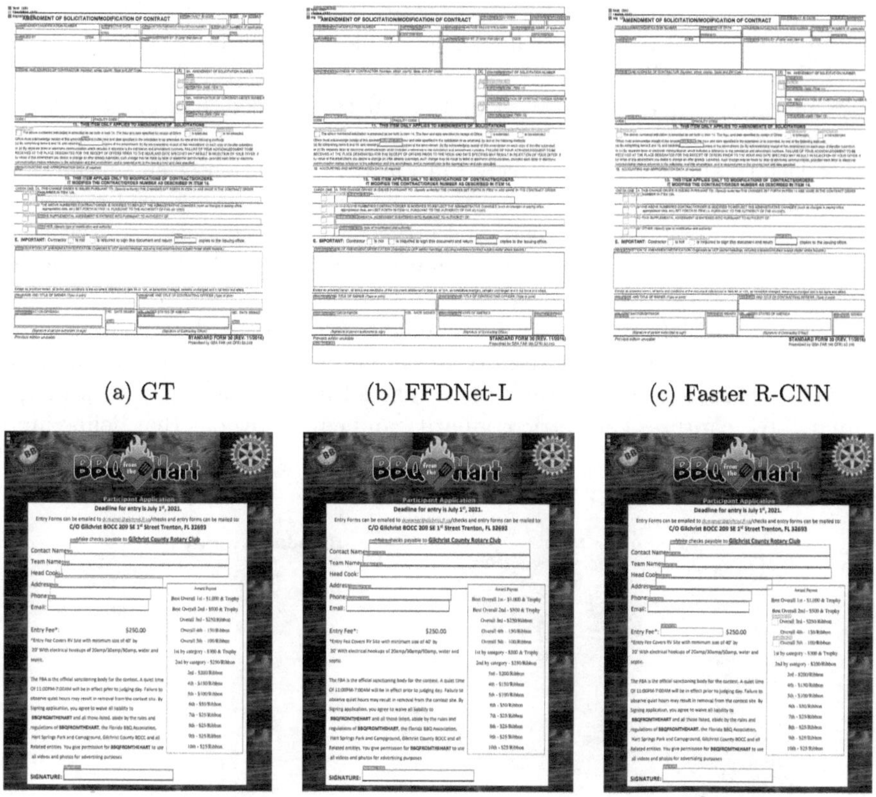

(a) GT (b) FFDNet-L (c) Faster R-CNN

(d) GT (e) FFDNet-L (f) Faster R-CNN

Fig. 5. Qualitative comparison between ground-truth annotations, FFDNet-L from the original commonForms paper, and our trained Faster R-CNN model on mini-commonForms samples.

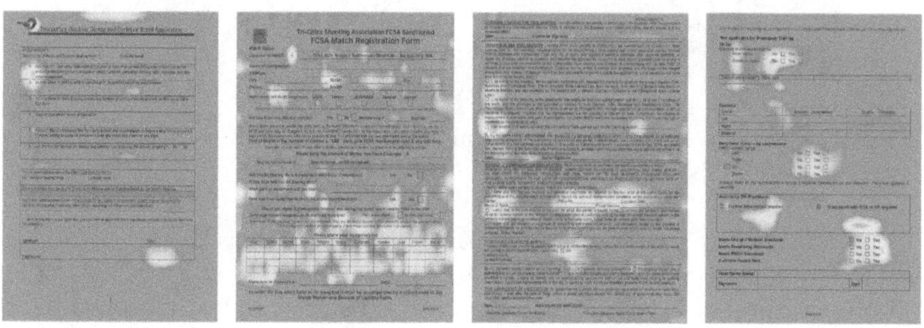

(a) TOOD (b) DINO (c) DETR (d) R-CNN

Fig. 6. Representative failure cases in Mini-CommonForms with Grad-CAM. Misclassified boxes are highlighted in pink. (Color figure online)

key-field detector, anchoring predictions to semantic tokens instead of localizing the empty input region. This behavior is visible not only in transformer-based models but also in CNN-based Cascade R-CNN, as illustrated in Fig. 6d, that indicating a systematic reliance on textual anchors rather than object-internal visual evidence.

For Deformable DETR as depicted in Fig. 6c, attribution often concentrates on the narrow region between label text and predicted box. DINO exhibits a similar pattern, emphasizing relational structures that connect text tokens to nearby box-like regions. In both cases, relevance mass is distributed along spacing and alignment cues rather than within the field interior.

In contrast, LRP-mixed (Fig. 7) highlights whitespace, box contours, and vertical gaps near text. The attribution aligns more closely with structural layout primitives characteristic of form design. Instead of focusing solely on semantic content, LRP captures geometric regularity and empty regions that define input fields.

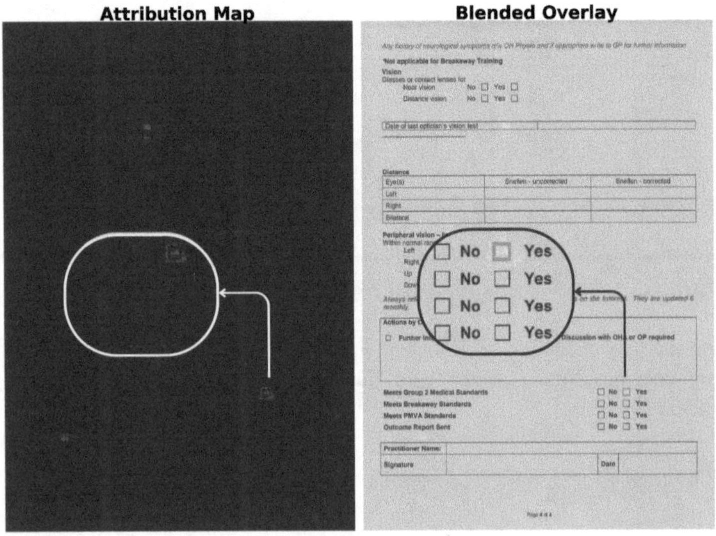

Fig. 7. Failure case in the Mini-CommonForm with LRP on Choice target class.

These qualitative patterns indicate that FFD operates as layout-conditioned reasoning rather than purely object-centric detection.

Unlike natural object detection, form elements are frequently defined by:

- Low-texture or empty interiors,
- Geometric regularity (rectangles, lines),
- Strong spatial dependency on adjacent labels.

Correct predictions, therefore, depend on relational cues such as text–field proximity, alignment consistency, and whitespace distribution. Attribution maps reflect this structural dependency: relevance often extends beyond bounding box interiors toward surrounding layout context.
A consistent architectural trend emerges:

Two-Stage CNN Detectors (e.g., Cascade R-CNN) produce more spatially concentrated attributions and higher localization-oriented scores. Region proposal and RoI refinement encourage object-centered evidence aggregation. However, explanations are denser and less stable under perturbation.

Transformer-Based Detectors (DINO, Deformable DETR) generate smoother and more robust explanations with lower sensitivity and complexity. Global self-attention distributes relevance across spatial tokens, capturing document-level structure but reducing spatial concentration inside bounding boxes.

6 Conclusion

In this work, we introduce mini-CommonForms, a curated compact benchmark derived from CommonForms designed to preserve the characteristics of the original dataset while enabling more reliable evaluation under reduced noise and annotation inconsistencies. We benchmark six representative detectors spanning two-stage CNN, one-stage CNN, and Transformer-based architectures, and we analyze their decisions using instance-level post-hoc explanations. Future FFD systems should better exploit document topology, whitespace, alignment, and key-field relationships to improve robustness on complex real-world forms.

References

1. Appalaraju, S., Jasani, B., Kota, B.U., Xie, Y., Manmatha, R.: Docformer: end-to-end transformer for document understanding. In: Proceedings of the IEEE/CVF International Conference on Computer Vision (2021)
2. Appalaraju, S., Tang, P., Dong, Q., Sankaran, N., Zhou, Y., Manmatha, R.: Docformerv2: local features for document understanding. In: Proceedings of the AAAI Conference on Artificial Intelligence (2024)
3. Bach, S., Binder, A., Montavon, G., Klauschen, F., Müller, K.R., Samek, W.: On pixel-wise explanations for non-linear classifier decisions by layer-wise relevance propagation. PLoS ONE (2015)
4. Baral, G., Zhou, J.: Autoformbench: benchmark dataset for automating form understanding. arXiv preprint arXiv:2603.29832 (2026)
5. Barrow, J.: Commonforms: a large, diverse dataset for form field detection. In: Proceedings of the IEEE/CVF Winter Conference on Applications of Computer Vision (2026)

6. Chen, K., et al.: Mmdetection: open mmlab detection toolbox and benchmark. arxiv 2019. arXiv preprint arXiv:1906.07155 (2019)
7. Davis, B., Morse, B., Price, B., Tensmeyer, C., Wigington, C., Morariu, V.: End-to-end document recognition and understanding with dessurt. In: European Conference on Computer Vision (2022)
8. Ding, Y., et al.: Form-NLU: dataset for the form natural language understanding. In: Proceedings of the 46th International ACM SIGIR Conference on Research and Development in Information Retrieval, pp. 2807–2816 (2023)
9. Harley, A.W., Ufkes, A., Derpanis, K.G.: Evaluation of deep convolutional nets for document image classification and retrieval. In: International Conference on Document Analysis and Recognition (ICDAR) (2015)
10. Huang, Z., et al.: ICDAR2019 competition on scanned receipts OCR and information extraction (SROIE). In: International Conference on Document Analysis and Recognition (ICDAR) (2019)
11. Jaume, G., Ekenel, H.K., Thiran, J.P.: Funsd: a dataset for form understanding in noisy scanned documents. In: 2019 International Conference on Document Analysis and Recognition Workshops (ICDARW), vol. 2, pp. 1–6. IEEE (2019)
12. Kim, G., et al.: OCR-free document understanding transformer. In: European Conference on Computer Vision (2022)
13. Lee, C.Y., et al.: Formnet: structural encoding beyond sequential modeling in form document information extraction. In: Proceedings of the 60th Annual Meeting of the Association for Computational Linguistics (Volume 1: Long Papers) (2022)
14. Lee, C.Y., et al.: Formnetv2: multimodal graph contrastive learning for form document information extraction. In: Proceedings of the 61st Annual Meeting of the Association for Computational Linguistics (Volume 1: Long Papers) (2023)
15. Mathew, M., Karatzas, D., Jawahar, C.: Docvqa: a dataset for VQA on document images. In: Proceedings of the IEEE/CVF Winter Conference on Applications of Computer Vision (2021)
16. Park, S., Shin, S., et al.: Cord: a consolidated receipt dataset for post-OCR parsing. In: NeurIPS Workshop on Document Intelligence (2019)
17. Pfitzmann, B., Auer, C., Dolfi, M., et al.: Doclaynet: a large human-annotated dataset for document-layout segmentation. In: Proceedings of the 28th ACM SIGKDD Conference on Knowledge Discovery and Data Mining (2022)
18. Selvaraju, R.R., Cogswell, M., Das, A., Vedantam, R., Parikh, D., Batra, D.: Grad-cam: visual explanations from deep networks via gradient-based localization. In: Proceedings of the IEEE International Conference on Computer Vision, pp. 618–626 (2017)
19. Stanisławek, T., Graliński, F., et al.: Kleister: key information extraction datasets involving long documents with complex layouts. In: International Conference on Document Analysis and Recognition (ICDAR) (2021)
20. Sun, H., Kuang, Z., et al.: Spatial dual-modality graph device for document information extraction. In: Proceedings of the 29th ACM International Conference on Multimedia (2021)
21. University of California, San Francisco Library: Industry documents library (2026). https://www.industrydocuments.ucsf.edu. Accessed 28 Feb 2026
22. Wang, Z., Zhou, Y., Wei, W., Lee, C.Y., Tata, S.: Vrdu: a benchmark for visually-rich document understanding. In: Proceedings of the 29th ACM SIGKDD Conference on Knowledge Discovery and Data Mining (2023)

23. Xu, Y., et al.: Xfund: a benchmark dataset for multilingual visually rich form understanding. In: Findings of the Association for Computational Linguistics: ACL 2022 (2022)
24. Zhong, X., Tang, J., Yepes, A.J.: Publaynet: largest dataset ever for document layout analysis. In: International Conference on Document Analysis and Recognition (ICDAR) (2019)

SciXplain: Structure-Aware Representation Learning for Scientific Method Diagrams

Zhicheng He[1], Zheyi Deng[2], Junhao Ren[3,4], and Xiaoqing Lyu[5](✉)

[1] National University of Singapore, Singapore, Singapore
[2] Beijing University of Posts and Telecommunications, Beijing, China
[3] Beijing Jiaotong University, Beijing, China
[4] Rochester Institute of Technology, New York, USA
[5] Wangxuan Institute of Computer Technology, Peking University, Beijing, China
lvxiaoqing@pku.edu.cn

Abstract. Scientific method diagrams are a common class of document graphics in research papers, but their meaning depends on topology, geometry, text anchors, and directional relations rather than visual appearance alone. This makes them difficult for vision-language models trained or evaluated mainly through fluent figure descriptions, because such objectives can underemphasize explicit topology and directional dependencies. We introduce SciXplain, a structure-aware representation learning framework for method-diagram understanding. SciXplain distills structural and geometric priors from auxiliary teachers used only during training into a compact visual encoder, enabling a shared decoder to support structural description and diagnostic explanation without requiring a separate graph-extraction module at inference. We further construct SciStruct, a method-diagram benchmark with paragraph context, OCR tokens, region masks, and derived structural signals. Experiments under a fixed relation-evaluation protocol show that SciXplain improves structure-sensitive relation expression over same-split adapted VLM baselines. Teacher ablations support the complementary roles of structural and geometric supervision, while perturbation diagnostics reveal shortcut drift toward paragraph-context-dominated generation. Together, the framework and benchmark support relation-centered analysis of scientific method diagrams (Code is available at https://github.com/hezhicheng2002/SciXplain.).

Keywords: Document Analysis Systems · Scientific Method Diagrams · Structure-aware Representation Learning · Multi-teacher Distillation

1 Introduction

Scientific papers frequently use method diagrams, workflow figures, and model schematics to communicate complex procedures. These figures are document

F. Shafait et al. (Eds.): DAS 2026, LNCS 16851, pp. 474–490, 2027.
https://doi.org/10.1007/978-3-032-36207-0_28

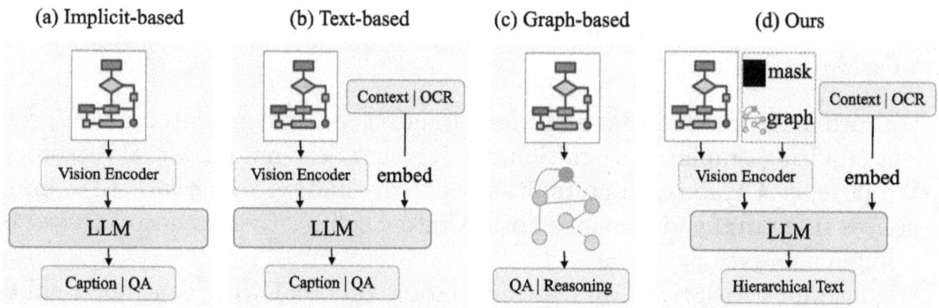

Fig. 1. Structural modeling comparisons for diagram understanding. Compared with implicit-structure, text-augmented, and explicit-graph methods, our approach distills structural and geometric priors into the visual encoder for structure-sensitive generation with a single inference backbone.

graphics: their meaning is carried by connectivity, layout, arrows, OCR labels, and surrounding paragraph context, not by object appearance alone [8,10]. Understanding such figures therefore requires models to express structured relations and not only produce fluent descriptions. This setting is directly relevant to document analysis systems, where graphics recognition, semantic analysis, and document reasoning must be connected within one pipeline.

As illustrated in Fig. 1, existing approaches differ in how they expose structure to the model. Implicit methods encode relations inside dense visual features and directly generate captions or answers, as in scientific captioning models such as SciCap [5]. Text-based methods add paragraph context or OCR tokens to improve alignment and discourse coherence, as in document-oriented models such as mPLUG-PaperOwl [7]; however, in many text-augmented generation settings, component relations are still represented implicitly rather than evaluated as explicit relation expressions. Graph-based methods explicitly parse symbolic structures for downstream reasoning [10], but the parser is often separated from the generation backbone. These paradigms leave a practical gap for method-diagram systems: structure is either hidden inside a large vision-language model or externalized into a separate parser, making it difficult to maintain relation-sensitive generation across structural description and explanation settings.

To address this gap, we propose SciXplain, a structure-aware representation learning framework for scientific method diagrams. Rather than adding an inference-time graph-extraction stage, SciXplain distills node-, role-, edge-, and geometry-level priors into the visual encoder through training-time multi-teacher supervision. The resulting encoder is coupled with a shared multimodal decoder that uses the same visual representation across structural description and diagnostic explanation settings. We also construct SciStruct, a method-diagram benchmark with manually verified figures, paragraph context, OCR tokens, region masks, and Rex-Omni-derived structural signals. The benchmark measures relation preservation with a fixed relation parser and examines shortcut behavior through controlled perturbations. Our primary contributions are summarized as follows:

- We formulate method-diagram understanding as a document-analysis problem requiring graphics recognition, OCR semantics, and relational reasoning.
- We develop a training-time multi-teacher distillation framework that internalizes structural and geometric priors into a compact visual encoder used by a single inference backbone.
- We introduce SciStruct and an evaluation protocol with controlled splits, verified diagram curation, parser-based relation metrics, perturbation diagnostics, and teacher ablations.

2 Related Work

2.1 Scientific Document Graphics Understanding

Early research on scientific figure understanding primarily framed the task as figure captioning, with benchmarks such as SciCap [5,23] and richer variants such as SciCapenter [6]. FlowchartQA [17] and FlowCE [25] then showed that directional and relational cues remain challenging for multimodal models. More recent document-level settings, including mPLUG-PaperOwl [7] and SciER [26], integrate paragraph context and OCR tokens for scientific figures. Recent generation-oriented work such as PaperBanana [30] similarly foregrounds methodology figures within broader academic illustrations. Nevertheless, most of these approaches optimize a single target text, leaving structural alignment across descriptive granularities largely implicit.

2.2 Structure-Aware Vision-to-Text Modeling

Complementing task-driven advances, structure-aware vision-to-text modeling studies how structural priors are incorporated into visual-text representations. Early layout-aware models such as LayoutLM [22] introduced spatial inductive biases for text-rich documents. In diagrammatic contexts, approaches such as Arrow-guided VLM [15] and GenFlowchart [1] injected directional cues or adopted visual textualization, which converts visual structures into textual traces for reasoning [24]. Recent diagram-centric methods such as Chain-of-Region [12] and robustness-oriented scientific diagram analysis [29] emphasize region decomposition and perturbation-sensitive evaluation, while multi-teacher and visual-prior distillation have improved document and image-text understanding [3,14,27]. Across these approaches, structural signals are typically incorporated as auxiliary cues, architectural biases, or intermediate reasoning modules rather than being systematically aligned with the visual representation under a unified generation objective. SciXplain instead targets this representation-level alignment for scientific method diagrams.

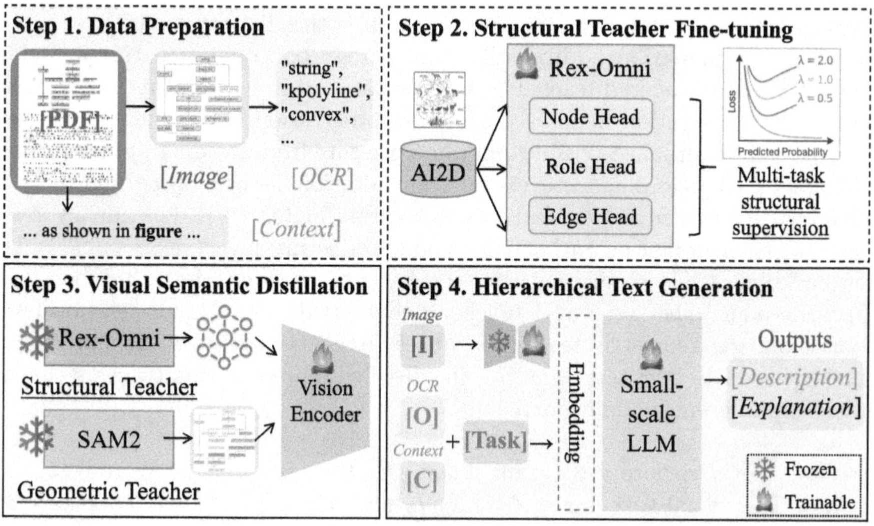

Fig. 2. Overview of SciXplain. **Step 1** is an offline data-preparation stage that builds benchmark and training inputs from scientific PDFs, extracting method-diagram figures, paragraph context, and OCR tokens (see Sect. 4.1). **Step 2** fine-tunes a structural teacher (Rex-Omni [9]) on AI2D [10] with multi-task node-, role-, and edge-level supervision to obtain structure-aware representations (Sect. 3.2). **Step 3** distills structural and geometric priors from frozen teachers into a vision encoder, where the geometric teacher follows the SSSI framework [20] (Sect. 3.3). **Step 4** performs hierarchical text generation using a TinyLLaVA-style decoder-only multimodal backbone [28] through a unified generation interface (Sect. 3.4). In the figure, [I], [O], and [C] denote the diagram image, OCR tokens, and paragraph context, respectively.

3 Method

3.1 Overall Pipeline

Scientific method diagrams convey meaning through topology, geometry, and dense textual anchors, making it difficult for existing vision–language models to maintain consistent structural alignment across descriptive granularities. SciXplain addresses this challenge by internalizing structural and geometric priors within the visual representation and coupling them with a unified hierarchical text-generation interface shared by both generation settings.

Figure 2 illustrates the overall pipeline. The framework has four stages. First, scientific PDFs are converted into structured benchmark inputs in an offline data-preparation stage (Step 1). Second, a structural teacher is fine-tuned under explicit multi-task supervision to model node categories, semantic roles, and directed relations (Step 2). Third, structural and geometric priors are distilled into a shared visual encoder, integrating structure-aware representations (Step 3). Finally, a TinyLLaVA-style decoder-only multimodal backbone generates structure-sensitive textual outputs under a unified hierarchical generation interface (Step 4). At inference, description uses the diagram image and OCR tokens, and explanation additionally uses paragraph context; external region masks, SAM outputs, and teacher predictions are not inputs. In this paper, hierarchical text generation denotes a semantic progression in the generated text: description first verbalizes visible diagram structure from the diagram image and OCR tokens, whereas diagnostic explanation additionally uses paragraph context and internally pooled region features to express higher-level dependency relations under controlled perturbations. The architecture preserves a single inference backbone while separating the structural teacher, geometric teacher, and shared decoder. The structural teacher defines node-, role-, and relation-level supervision; the geometric teacher localizes the same inventory in pixel space; and the shared decoder tests whether the distilled representation transfers across tasks with different modality exposure.

Notation. Let I denote a diagram image, E_θ the visual encoder, and D_ϕ the decoder. The visual tokens are $Z = E_\theta(I)$ and are split into region tokens Z^r and global tokens Z^g. Paragraph context, OCR tokens, and the task indicator are denoted by P, O, and τ, respectively. Teacher masks and student-predicted masks are denoted by M and $\hat{M}$.

Given a diagram image I, the visual encoder E_θ maps the image to a set of visual tokens:

$$Z = E_\theta(I) \in \mathbb{R}^{N \times d}, \tag{1}$$

where N is the number of tokens and d is the embedding dimension. We partition these tokens into region tokens Z^r and global tokens Z^g. Let N_r denote the number of region tokens. Region tokens correspond to localized visual elements and are explicitly supervised to encode structural dependencies such as connectivity and flow direction, while global tokens provide complementary context-level representations of the overall layout. Concretely, teacher masks define the supervised region inventory, and student patch features are pooled with the current student masks to form $Z^r \in \mathbb{R}^{N_r \times d}$. These student masks are aligned to the teacher masks in Sect. 3.3, while Z^g summarizes the full-layout visual context. Thus, $Z = \{Z^r, Z^g\}$ denotes the complete image-conditioned token sequence rather than a separate latent branch [12]. Both generation settings share an identical input interface. Let P denote paragraph context and O OCR tokens. Instead of directly concatenating raw text, we embed each modality into a dedicated structural slot through modality-specific mappings. The final conditioning sequence is written as

$$X(\tau, P, O) = \left[\tau,\ \Phi_{para}(\tilde{P}),\ \Phi_{ocr}(\tilde{O})\right], \tag{2}$$

where τ is the task indicator and $\tilde{P}, \tilde{O}$ denote either observed content or predefined missing-modality markers when a modality is absent. The functions Φ_{para} and Φ_{ocr} map each source into fixed subspaces, ensuring that all tasks operate under a consistent input structure regardless of the available modalities.

Text generation is formulated as conditional language modeling:

$$p_\phi(Y \mid Z, X) = \prod_{t=1}^{T} p_\phi(y_t \mid y_{<t}, Z, X), \tag{3}$$

where Z is computed once per diagram image and reused across description and explanation settings. This formulation ensures that structural visual representations are shared across tasks, while introducing semantic variation only through conditioning signals and the task indicator τ.

3.2 Structural Teacher Fine-Tuning

We fine-tune a Rex-Omni-based structural teacher on AI2D [10] with explicit graph supervision. Annotated nodes and directed edges are parsed into node categories, bounding boxes, semantic roles, and relations. Rex-Omni [9] is the backbone of this structural teacher. The pretrained base is frozen, and task-specific heads are trained for node classification, box regression, role prediction, and relation classification. Optional low-rank adaptation can be applied while keeping most parameters frozen.

Supervision provenance. SciXplain separates teacher preparation, benchmark construction, and inference. AI2D graph annotations are used only to train the structural teacher. SciStruct figures are manually verified as method diagrams, while their node, role, relation, and mask signals are derived from frozen auxiliary teachers and used as training or diagnostic supervision. At inference time, SciXplain receives only the diagram image, OCR tokens, and paragraph context available for the current setting; the structural teacher, geometric teacher, and evaluation parser are not used by the inference model. In this work, "teacher" denotes a frozen auxiliary model that provides feature, relation, or mask targets for encoder-side structural supervision. The overall objective is

$$\mathcal{L} = \lambda_{node}\mathcal{L}_{node} + \lambda_{box}\mathcal{L}_{box} + \lambda_{role}\mathcal{L}_{role} + \lambda_{rel}\mathcal{L}_{rel} + \lambda_{rel\_dir}\mathcal{L}_{dir}. \tag{4}$$

Relation negatives include reversed edges, same-source same-type pairs, and randomly paired nodes labeled as *none*. Although all heads are trained jointly, supervision primarily emphasizes node semantics and relational structure, which are later distilled into the student visual encoder.

3.3 Multi-teacher Structural Distillation

The student visual encoder is trained under supervision from the fine-tuned structural teacher and a geometric teacher following the SSSI framework [20]. The structural teacher provides one region embedding $\hat{z}_i^r$ and teacher-predicted relation labels R^T over detected region pairs, while the geometric teacher provides the corresponding binary teacher mask M_i. We denote the teacher mask set by $M = \{M_i\}_{i=1}^{N_r}$ and the student-predicted mask set by $\hat{M} = \{\hat{M}_i\}_{i=1}^{N_r}$.

The geometric branch follows SSSI [20]: OCR text boxes and region geometry are used to form self-generated box and point prompts, and a SAM-family mask predictor iteratively refines method-diagram region masks without manual prompts. In our implementation, SAM2 [16] is used as the mask predictor. These masks are aligned with the same region inventory used by the structural teacher. Here, "topology-consistent" refers to index consistency with teacher regions and teacher-predicted relations. The objective combines feature alignment, graph supervision, and geometric alignment:

$$\mathcal{L}_{vis} = \mathcal{L}_{feat} + \mathcal{L}_{graph} + \mathcal{L}_{geom}. \tag{5}$$

Feature-level alignment distills intermediate structural representations by minimizing the discrepancy between student and teacher region embeddings:

$$\mathcal{L}_{feat} = \frac{1}{N_r} \sum_{i=1}^{N_r} \|z_i^r - \hat{z}_i^r\|_2^2. \tag{6}$$

Graph-level supervision enforces consistency with teacher-provided relational annotations. Given teacher relation labels R^T and student relation logits $\hat{R}_{pred}$ inferred from Z^r, we apply multi-class cross-entropy:

$$\mathcal{L}_{edge} = \mathrm{CE}(R^T, \hat{R}_{pred}), \tag{7}$$

In addition, node-type and semantic-role classification losses are incorporated:

$$\mathcal{L}_{graph} = \lambda_{node}\mathcal{L}_{node} + \lambda_{role}\mathcal{L}_{role} + \lambda_{edge}\mathcal{L}_{edge}. \tag{8}$$

The geometric term aligns region tokens with spatial layout supervision from the geometric teacher:

$$\mathcal{L}_{geom} = \mathcal{L}_{align}(\hat{M}, M). \tag{9}$$

Operationally, $\hat{M}_i \in [0,1]^{H\times W}$ is produced by a lightweight two-layer upsampling head attached to Z, and $\mathcal{L}_{align}$ is realized as a pixel-wise binary cross-entropy against the teacher mask M_i, averaged over the N_r supervised regions. Each region token z_i^r is pooled from the student patch features using $\hat{M}_i$; at inference, the model uses these soft region maps rather than calling teacher-side graph or mask parsers. After convergence, the encoder E_θ is frozen and serves as a shared structure-aware backbone for downstream generation.

3.4 Structure-Aware Generation with Attention

The decoder is shared across tasks, with a lightweight task-conditioned linear map on region tokens. For each task τ, region tokens are transformed as

$$\tilde{Z}^r = Z^r W_\tau + \mathbf{1} b_\tau, \tag{10}$$

while global tokens remain unchanged. The decoder attends to $\tilde{Z} = \{\tilde{Z}^r, Z^g\}$.

To encourage attention toward localized structural elements, we introduce a bias term in visual cross-attention. Let $s_{t,i}$ denote the pre-softmax attention logit from token y_t to visual token z_i. We adjust it as

$$\tilde{s}_{t,i} = s_{t,i} + \beta^{(k)} \cdot \mathbf{1}[z_i \in \tilde{Z}^r], \tag{11}$$

where $\beta^{(k)} \geq 0$ is the current region-attention bias at training step or checkpoint k, controlling the relative emphasis on region tokens in line with region-level attention biasing for diagrammatic inputs [12,15]. The annealing schedule is reported with the implementation details in Sec. 4.2.

For the explanation task, we further regularize attention over paragraph-conditioned slots. Let $q_{t,j}$ denote the attention logit to textual conditioning token x_j. We define

$$\tilde{q}_{t,j} = q_{t,j} - \gamma \cdot \mathbf{1}[x_j \in \Phi_{para}], \tag{12}$$

where $\gamma \geq 0$ introduces a mild penalty to discourage exclusive reliance on paragraph context. Together, these mechanisms bias the decoder toward jointly utilizing structural visual cues and paragraph context.

3.5 Generation Objective and Inference

The decoder is trained using conditional language modeling:

$$\mathcal{L}_{nll} = -\log p_\phi(Y \mid Z, X). \tag{13}$$

For the explanation task, we introduce a counterfactual visual regularization term to encourage sensitivity to diagram-image-conditional likelihood differences. For each training sample, a similarity-based negative diagram image I' is selected while keeping P, O, and τ fixed. Let $Z = E_\theta(I)$ and $Z' = E_\theta(I')$. We apply a margin-based ranking loss:

$$\mathcal{L}_{cf} = \max\left(0,\, m - \log p_\phi(Y \mid Z, X) + \log p_\phi(Y \mid Z', X)\right), \tag{14}$$

encouraging the likelihood under the correct diagram image to exceed that under a mismatched diagram image. The final generation objective is

$$\mathcal{L}_{gen} = \mathcal{L}_{nll} + \lambda_{cf}\mathcal{L}_{cf}. \tag{15}$$

Explanation is trained jointly in Stage 2 so that the shared decoder can be evaluated under controlled perturbations; we therefore treat the explanation results in Sect. 5.2 as diagnostic evidence rather than a primary performance metric.

At inference time, the diagram image is encoded once to obtain $Z = E_\theta(I)$. Given task indicator τ and available conditioning sources, the decoder generates output according to $p_\phi(Y \mid Z, X)$. Algorithm 1 summarizes this two-stage training and inference flow in the same input/output order used above.

Algorithm 1. Training and Inference of SciXplain

Require: Training data $\mathcal{D}$
Initialize encoder E_θ and decoder D_ϕ
Stage 1: Structural Distillation
for each diagram image I **do**
  Compute $\mathcal{L}_{vis}$ and update θ
end for
Freeze E_θ
Stage 2: Generation Training
for each (I, P, O, τ, Y) **do**
  $Z \leftarrow E_\theta(I)$; compute $\mathcal{L}_{nll}$
  if τ is explanation **then**
    Sample negative diagram image I' and compute $\mathcal{L}_{cf}$
  end if
  Update ϕ using $\mathcal{L}_{gen}$
end for
Inference: encode I once and generate $Y \sim p_\phi(Y \mid Z, X)$

4 Experiments

This section evaluates SciXplain under a fixed benchmark protocol for structure-sensitive description and explanation diagnostics. We first describe the datasets and implementation details, then report parser-based relation metrics, perturbation diagnostics, and teacher ablations.

4.1 Datasets

Publicly Available Datasets. We use two public datasets. AI2D [10] provides 4,903 scientific diagrams with human-annotated structural graphs and geometric relations, and is used only for teacher-side structural supervision under a 9:1 train–test split. MLBCAP [11] contains 7,215 scientific flowcharts with description targets and paragraph context; we repartition it at the figure level under an 8:1:1 split and use only its description field in the reported description setting.

In-house Curated Dataset. SciStruct is a newly curated dataset of 5,246 method diagrams paired with paragraph context, OCR tokens, and offline region-mask supervision. SciPDF Parser is used to recover figures, surrounding paragraphs, source captions, and paper metadata from scientific PDFs. OCR text is extracted with PaddleOCR [4], enabling text-line orientation classification and English text recognition; OCR is preprocessing, not an inference-time engine. After source-caption keyword filtering, two domain researchers double-check candidate figures under a shared standard, retaining method, pipeline, model-schematic, and workflow figures while excluding tables, result plots, chart-only figures, decorative illustrations, and unrelated screenshots; disagreements are adjudicated by the senior researcher. To prevent information leakage, we adopt an article-level 8:1:1 split over 2,500 source articles with no duplicate article or figure

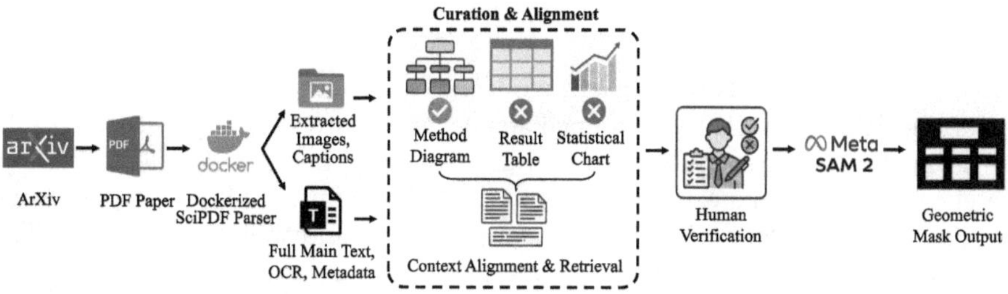

Fig. 3. Curation and derived-supervision workflow of SciStruct. SciPDF Parser recovers figures and paragraph context from scientific PDFs, candidate method diagrams are verified, OCR tokens are extracted, region masks are generated offline, and split-level supervision files are organized for hierarchical text generation.

Table 1. Statistics of the SciXplain method-diagram benchmark. Counts are supervision instances. Context denotes paragraph context; OCR and Region indicate available OCR tokens and offline region-mask supervision.

Setting	Dataset	Train	Val	Test	Context	OCR	Region
Description	MLBCAP	5,772	721	722	–	✓	–
Explanation	SciStruct	4,197	531	518	✓	✓	✓

across splits, yielding 4,197, 531, and 518 figures for training, validation, and testing. Records store article and figure identifiers, figure images, source captions, paragraph context, OCR text, target text fields, and derived supervision files. Region masks are generated offline following SSSI [20] with a frozen SAM-family mask predictor and are used as training supervision, not as inference inputs. Training-side structural signals are produced by the Rex-Omni structural teacher described in Sect. 3.2 and are not inputs to the inference model. Because these signals are derived rather than manually annotated graph labels, we use them as protocol-level supervision and diagnostics rather than definitive human reasoning labels. The released materials will provide split manifests, preprocessing code, the relation-alias inventory, and the deterministic evaluation parser; derived masks and teacher outputs will be included where source-document licensing permits. The overall curation workflow is illustrated in Fig. 3.

Benchmark and Supervision Setting. We organize MLBCAP and SciStruct into a unified method-diagram benchmark with two reported settings: description and explanation. For SciStruct, source captions are used only as document metadata for figure recovery, method-diagram filtering, and paragraph-context extraction; no caption-generation setting is trained or evaluated in the reported experiments. The data sources differ in the visibility of paragraph context, OCR tokens, and region-mask supervision, as summarized in Table 1.

4.2 Implementation Details

SciXplain is trained in two stages. Stage 1 distills structural and geometric priors into a SigLIP2-Large [18] visual encoder for 20,000 steps, using feature, graph, and geometry weights 1.0/8.0/5.0 and gradient clipping at 1.0. Stage 2 freezes the encoder and trains a TinyLLaVA-style decoder-only multimodal backbone for 1,200 AdamW steps with learning rate 1×10^{-5} and weight decay 0.01. The explanation counterfactual loss uses margin $m = 0.25$ and weight $\lambda_{cf} = 1.20$; region attention is enabled only for explanation, with $\beta^{(k)}$ annealed from 1.35 to 0.90 and paragraph penalty $\gamma = 0.05$. Inference uses deterministic beam search with beam size 5, maximum length 176, repetition penalty 1.1, and no-repeat trigram blocking. The main evaluation focuses on description and explanation diagnostics. The reported student-training cost is 13.48 single-GPU hours (10.57 for visual distillation and 2.91 for generation-side training), excluding one-time teacher preparation, OCR, mask generation, and dataset curation. Runs use fixed seed 42 and are reported as single-run point estimates. The final model contains approximately 3.67B parameters; on a 40 GB A100 GPU under half-precision decoding, it uses 9.86 GB peak memory, 25.27 ms per output token, and 1.01–2.07 s per generated sample. The adapted-baseline protocol uses the same split, prompt family, decoding budget, LoRA rank ($r = 8$), and 300 optimizer steps for each baseline.

4.3 Metrics

For description, following recent diagram-evaluation work on structural evaluation [12,29], we emphasize structure-sensitive relation expression rather than surface lexical overlap. Lexical-overlap metrics such as BLEU, ROUGE, METEOR, and CIDEr [19] are not used as primary comparison metrics because they reward reference-level wording agreement, whereas method-diagram understanding depends on whether components, arrows, roles, and dependencies are preserved. All systems are evaluated by the same deterministic relation parser and the same canonical alias inventory; the parser is applied only after outputs are generated and is not tuned per method. Relation Trigger Rate (RTR) measures the proportion of benchmark target relations recovered by the model:

$$\mathrm{RTR} = \frac{|\mathcal{R}_{\text{pred}} \cap \mathcal{R}_{\text{target}}|}{|\mathcal{R}_{\text{target}}|}. \tag{16}$$

Strict Hallucination Rate (SHR) is the fraction of samples containing at least one unsupported relation, and we report $1 - \mathrm{SHR}$. The Grounded Structure Score (GSS) combines both terms:

$$\mathrm{GSS} = \mathrm{RTR} \times (1 - \mathrm{SHR}), \tag{17}$$

where $\mathcal{R}_{\text{pred}}$ is obtained by matching generated relation phrases to the benchmark inventory through canonical aliases, and unmatched relations are counted as unsupported. The parser is a post-hoc evaluator; GSS should therefore be

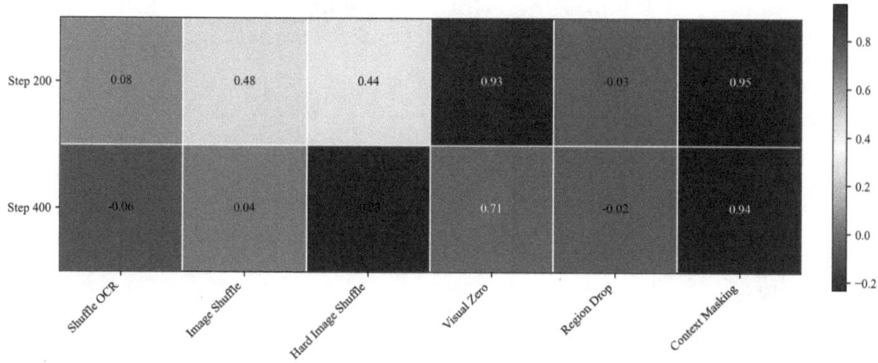

Fig. 4. Explanation sensitivity across training steps. Each cell shows relative CIDEr drop under controlled perturbation; larger values indicate stronger modality sensitivity, while negative values mean no drop.

interpreted as a protocol-level proxy for structure-sensitive relation expression rather than a direct region-localization measure or a substitute for human evaluation.

For explanation analysis, we evaluate robustness under controlled perturbations to OCR tokens, image layout, region cues, or paragraph context. Shuffle OCR shuffles OCR tokens and the packed text-context stream while keeping the diagram image and paragraph field fixed; Image Shuffle keeps paragraph context and OCR tokens fixed but replaces the diagram image with another sample; Hard Image Shuffle uses a lexically similar counterfactual diagram image; Visual Zero removes visual tokens; Region Drop masks a sampled diagram region; and Context Masking removes paragraph context. Here CIDEr is used only as a within-sample relative sensitivity score, not as an absolute text-generation metric. Sensitivity is measured as relative CIDEr drop:

$$\mathrm{Drop}_{rel} = \frac{\mathrm{CIDEr}_{clean} - \mathrm{CIDEr}_{perturbed}}{\mathrm{CIDEr}_{clean}}, \tag{18}$$

where larger drops indicate stronger dependence on the perturbed modality.

To assess the visual encoder independently, we report graph-level distillation metrics: node and role accuracy for categorical and functional prediction, edge accuracy over sampled edge pairs, edge F1 for binary relation existence, and positive-edge accuracy (Pos Acc) over target positive edges (Fig. 4).

5 Results

5.1 Description

Under this protocol, SciXplain scores higher than the adapted baselines on all three parser-based description metrics. Table 2 shows a GSS of 0.8573 for SciXplain, compared with 0.6341 for Qwen3-VL, 0.5992 for InternVL3.5, and 0.7157

Table 2. Same-split adapted description comparison. RTR, 1-SHR, and GSS are parser-based structure-sensitive proxies; best values are bold.

Method	RTR	1-SHR	GSS
InternVL3.5 [21]	0.8161	0.7343	0.5992
Qwen3-VL [2]	0.8297	0.7643	0.6341
LLaVA-1.6 [13]	0.8965	0.7984	0.7157
SciXplain (Ours)	**0.9741**	**0.8801**	**0.8573**

for LLaVA-1.6. The adapted baselines use the same split, prompts, decoding budget, LoRA rank, and parser, but not the same full training budget because SciXplain uses teacher preparation and visual distillation. Because GSS combines target-relation recovery with unsupported-relation suppression, the gain reflects both higher relation-trigger coverage and fewer unsupported relations under the same parser. We interpret the gap as improved structure-sensitive relation expression, not as a direct guarantee of localized visual grounding.

5.2 Explanation Diagnostics

Explanation provides a robustness analysis complementary to the description metrics: the targets are silver proxies, and perturbation-based CIDEr drops measure modality sensitivity rather than reasoning quality. The first two scheduled checkpoints from the same decoder-training run show a clear transition. At step 200, structural perturbations still induce substantial drops; by step 400, image-based effects largely disappear while Context Masking remains dominant. Near-zero Region Drop values reflect local masking, since most visual cues remain in global and unmasked region tokens. This pattern is used to locate shortcut drift, not to certify the generated explanation as a verified reasoning trace. Figure 5 shows the same pattern qualitatively: the step-200 output preserves explicit input–output dependencies, whereas the step-400 output collapses into a shorter title-like phrase.

5.3 Visual Semantics

Table 3 evaluates the visual encoder independently of the language decoder. These metrics test whether distilled tokens retain node, role, and edge information before decoding, rather than whether the generated prose is correct. On SciStruct, the encoder reaches 0.782 node accuracy, 0.785 role accuracy, and 0.765 edge accuracy with edge F1 of 0.718, which indicates competitive structural perception rather than saturated reasoning. On MLBCAP, supervision is sparser, yet edge F1 remains 0.840 on supervised samples. Across both datasets, role prediction is more stable than fine-grained node typing, suggesting that semantic role cues transfer more readily than exact categories.

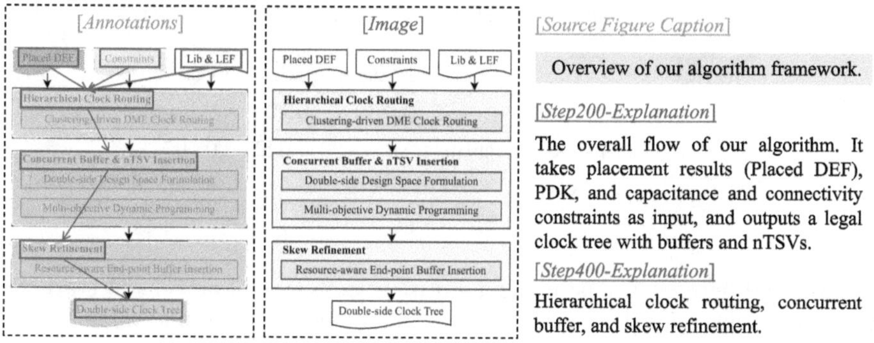

Fig. 5. Representative SciStruct case for explanation diagnostics. The case shows region masks, the source figure/caption, and generated explanations: the step-200 output preserves explicit input–output dependencies, whereas the step-400 output collapses into a shorter title-like phrase.

Table 3. Structural distillation performance. Edge metrics use sampled edge pairs; coverage ratios indicate valid supervision availability. Best values are bold.

Dataset	Coverage Ratio		Node/Role			Edge	
	Struct	Geom	Node Acc	Role Acc	Edge Acc	Edge F1	Pos Acc
SciStruct	**0.890**	**0.875**	**0.782**	0.785	**0.765**	0.718	**0.644**
MLBCAP	0.349	0.283	0.601	**0.791**	0.678	**0.840**	0.597

Table 4 shows that disabling structural-teacher feature alignment lowers edge F1 and positive-edge accuracy, while removing the geometric teacher produces a comparable degradation. This supports the intended role split between structure-aware region representations and mask-aligned spatial supervision.

6 Discussion

The results support the central design choice that explicit structural and geometric priors remain useful when a compact backbone must express method-diagram relations. Under the same-split adaptation protocol, SciXplain obtains higher parser-based relation-expression scores than the adapted VLM baselines with a 3.67B backbone, and teacher ablations support the complementary roles of structural and geometric supervision.

The explanation diagnostics show why robustness analysis is needed for hierarchical text generation. Image-level perturbation sensitivity drops from 0.48 at step 200 to 0.04 at step 400, while Context Masking remains above 0.94, indicating drift toward paragraph-context-dominated generation. Because CIDEr drop is a behavioral sensitivity measure rather than a direct probe of reasoning, we use explanation as diagnostic evidence and rely on RTR, $1 - \text{SHR}$, and GSS for primary description evaluation. Limitations remain: derived structural supervision

Table 4. Teacher ablation on SciStruct. Variants share the Stage-1 schedule.

Setting	Node Acc	Edge F1	Pos Acc
Full	**0.782**	**0.718**	**0.644**
w/o structural-teacher features	0.751	0.642	0.507
w/o geometric teacher	0.742	0.649	0.523

is sparse and heterogeneous; manual curation lacks formal agreement statistics; evaluation relies on a deterministic parser and silver explanation targets; OCR errors can corrupt text anchors; cross-dataset, multi-seed, and cost-normalized comparisons are not reported; and the AI2D-trained teacher may bias the student toward similar diagram styles. Licensing constraints also limit redistribution of some source figures. These limitations delimit the claim to structure-sensitive expression for method diagrams under the proposed protocol, with broader QA, coordinate, region-chain, and human-evaluation alignment left to future work.

7 Conclusion

This paper studies scientific method diagrams as document graphics whose semantics depend on layout, OCR anchors, directional relations, and paragraph context. SciXplain distills structural and geometric supervision from training-time teachers into a compact VLM backbone for structure-sensitive description and diagnostic explanation. On the SciStruct-centered benchmark, it improves parser-based relation-expression scores over same-split adapted VLM baselines, while perturbation diagnostics expose shortcut drift toward context-dominated generation. These results position SciXplain as a relation-centered framework for method-diagram analysis under a bounded, reproducible protocol. Future work should add stronger human-reviewed grounding labels, external-corpus validation, and protocols for broader document graphics.

References

1. Arbaz, A., Fan, H., Ding, J., Qiu, M., Feng, Y.: *GenFlowchart*: parsing and understanding flowchart using generative AI. In: Cao, C., Chen, H., Zhao, L., Arshad, J., Asyhari, T., Wang, Y. (eds.) KSEM 2024. LNCS, vol. 14884, pp. 99–111. Springer, Singapore (2024). https://doi.org/10.1007/978-981-97-5492-2_8
2. Bai, S., et al.: Qwen3-VL technical report. arXiv preprint arXiv:2511.21631 (2025)
3. Ding, Y., et al.: 3MVRD: multimodal multi-task multi-teacher visually-rich form document understanding. In: Findings of the Association for Computational Linguistics: ACL 2024, pp. 15233–15244 (2024)
4. Du, Y., et al.: PP-OCR: a practical ultra lightweight OCR system. arXiv preprint arXiv:2009.09941 (2020)
5. Hsu, T., Giles, C., Huang, T.: SciCap: generating captions for scientific figures. In: Findings of the Association for Computational Linguistics: EMNLP 2021, pp. 3258–3264 (2021)

6. Hsu, T., et al.: SciCapenter: supporting caption composition for scientific figures with machine-generated captions and ratings. In: Extended Abstracts of the CHI Conference on Human Factors in Computing Systems, pp. 1–9 (2024)
7. Hu, A., et al.: mPLUG-PaperOwl: scientific diagram analysis with the multimodal large language model. In: Proceedings of the 32nd ACM International Conference on Multimedia, pp. 6929–6938 (2024)
8. Hu, H., He, Z., Zhou, Y., Zhang, T., Lyu, X.: Multimodal content alignment with LLM for visual presentation of papers. In: Yin, X.C., Karatzas, D., Lopresti, D. (eds.) ICDAR 2025. LNCS, vol. 16025, pp. 238–256. Springer, Cham (2025). https://doi.org/10.1007/978-3-032-04624-6_14
9. Jiang, Q., et al.: Detect anything via next point prediction. arXiv preprint arXiv:2510.12798 (2025)
10. Kembhavi, A., Salvato, M., Kolve, E., Seo, M., Hajishirzi, H., Farhadi, A.: A diagram is worth a dozen images. In: Leibe, B., Matas, J., Sebe, N., Welling, M. (eds.) ECCV 2016. LNCS, vol. 9908, pp. 235–251. Springer, Cham (2016). https://doi.org/10.1007/978-3-319-46493-0_15
11. Kim, J., et al.: Multi-LLM collaborative caption generation in scientific documents. In: Wang, Q., Yin, W., Aich, A., Suh, Y., Peng, K.C. (eds.) AAAI Workshop 2025. CCIS, vol. 2533, pp. 142–160. Springer, Singapore (2025). https://doi.org/10.1007/978-981-96-8912-5_6
12. Li, X., Sun, Y., Cheng, W., Zhu, Y., Chen, H.: Chain-of-region: visual language models need details for diagram analysis. In: The Thirteenth International Conference on Learning Representations (2025)
13. Liu, H., Li, C., Li, Y., Lee, Y.: Improved baselines with visual instruction tuning. In: Proceedings of the IEEE/CVF Conference on Computer Vision and Pattern Recognition, pp. 26296–26306 (2024)
14. Ma, C., Zhang, Y., Tu, M., Zhao, Y., Zhou, Y., Zong, C.: Multi-teacher knowledge distillation for end-to-end text image machine translation. In: Fink, G.A., Jain, R., Kise, K., Zanibbi, R. (eds.) ICDAR 2023. LNCS, vol. 14187, pp. 484–501. Springer, Cham (2023). https://doi.org/10.1007/978-3-031-41676-7_28
15. Omasa, T., Koshihara, R., Morishige, M.: Arrow-guided VLM: enhancing flowchart understanding via arrow direction encoding. arXiv preprint arXiv:2505.07864 (2025)
16. Ravi, N., et al.: SAM 2: Segment anything in images and videos. arXiv preprint arXiv:2408.00714 (2024)
17. Tannert, S., et al.: FlowchartQA: the first large-scale benchmark for reasoning over flowcharts. In: Proceedings of the 1st Workshop on Linguistic Insights from and for Multimodal Language Processing, pp. 34–46 (2023)
18. Tschannen, M., et al.: SigLIP 2: multilingual vision-language encoders with improved semantic understanding, localization, and dense features. arXiv preprint arXiv:2502.14786 (2025)
19. Vedantam, R., Zitnick, C.L., Parikh, D.: CIDEr: consensus-based image description evaluation. In: Proceedings of the IEEE Conference on Computer Vision and Pattern Recognition, pp. 4566–4575 (2015)
20. Wang, T., Zhou, Y., Zhang, T., He, Z., Zhao, Y., Lyu, X.: SSSI: self-prompted segmentation of scientific illustrations. In: Yin, X.C., Karatzas, D., Lopresti, D. (eds.) ICDAR 2025. LNCS, vol. 16026, pp. 347–361. Springer, Cham. (2025). https://doi.org/10.1007/978-3-032-04627-7_20
21. Wang, W., et al.: InternVL3.5: advancing open-source multimodal models in versatility, reasoning, and efficiency. arXiv preprint arXiv:2508.18265 (2025)

22. Xu, Y., Li, M., Cui, L., Huang, S., Wei, F., Zhou, M.: LayoutLM: pre-training of text and layout for document image understanding. In: Proceedings of the ACM SIGKDD International Conference on Knowledge Discovery and Data Mining, pp. 1192–1200 (2020)
23. Yang, Z., Dabre, R., Tanaka, H., Okazaki, N.: SciCap+: a knowledge augmented dataset to study the challenges of scientific figure captioning. J. Nat. Lang. Process. **31**(3), 1140–1165 (2024)
24. Ye, J., Dash, A., Yin, W., Wang, G.: Beyond end-to-end VLMs: leveraging intermediate text representations for superior flowchart understanding. In: Proceedings of NAACL-HLT, pp. 3534–3548 (2025)
25. Zhang, E., Yao, R., Liu, H., Yu, J., Wang, J.: First multi-dimensional evaluation of flowchart comprehension for multimodal large language models. arXiv preprint arXiv:2406.10057 (2024)
26. Zhang, Q., Chen, Z., Pan, H., Caragea, C., Latecki, L., Dragut, E.: SciER: an entity and relation extraction dataset for datasets, methods, and tasks in scientific documents. arXiv preprint arXiv:2410.21155 (2024)
27. Zhang, X., et al.: Powerful teachers matter: text-guided multi-view knowledge distillation with visual prior enhancement. arXiv preprint arXiv:2603.24208 (2026)
28. Zhou, B., et al.: TinyLLaVA: a framework of small-scale large multimodal models. arXiv preprint arXiv:2402.14289 (2024)
29. Zhou, M., Souza, R., Hu, Y., Che, L.: Robust diagram reasoning: a framework for enhancing LVLM performance on visually perturbed scientific diagrams. arXiv preprint arXiv:2508.16972 (2025)
30. Zhu, D., et al.: PaperBanana: automating academic illustration for AI scientists. arXiv preprint arXiv:2601.23265 (2026)

Decoding Romanian Law: A Fine-Tuned LLM Approach

Marinela Adam[1], Alexandru Berindei[1], Darian Onchis[1(✉)], Todor Ivascu[1], and Corina Negrutiu[2]

[1] Faculty of Informatics, West University of Timişoara, Timişoara, Romania
darian.onchis@e-uvt.ro

[2] Faculty of Law, Dragan European University, Lugoj, Romania

Abstract. Romania's complex legal system presents significant accessibility challenges for citizens seeking to understand their rights and obligations. In this paper, we address this challenge by creating a high-quality dataset for Romanian legal Q&A. Our primary contribution is a meticulously curated dataset comprising three complementary components: (1) domain specific questions from judicial entrance examinations, (2) systematically generated Q&A pairs derived from official Romanian legal codes using embeddings and retrieval techniques, and (3) practical scenario-based questions addressing common legal concerns. Building on this foundation, we demonstrate the dataset's utility by fine-tuning three open-source language models (Phi-4, Pansophic, and LLaMA 3.1) using QLoRA techniques, and further enhancing performance through Retrieval-Augmented Generation (RAG) and Retrieval-Augmented Fine-Tuning (RAFT). Our comprehensive evaluation reveals significant improvements across lexical and semantic metrics, with fine-tuned models showing improvements in the structure and grounding of legal responses. This dataset not only addresses a critical gap in Romanian legal AI resources but also provides a replicable methodology for creating similar resources in other low-resource languages and legal systems.

1 Introduction

Large Language Models (LLMs) built on the Transformer architecture have dramatically advanced Natural Language Processing by pre-training on massive text corpora and then fine-tuning for downstream tasks [21,28]. These models can generate fluent, contextually rich text across translation, summarization, dialogue, and more. Yet, when deployed as static systems, they suffer from knowledge staleness and opaque decision pathways, making them ill-suited for domains where accuracy and traceability are essential.

Romanian legal texts pose particular challenges: statutes and codes evolve continuously through parliamentary acts and judicial rulings, legal prose is dense with specialized terminology, and authoritative provisions are scattered across multiple publications [22]. Off-the-shelf keyword search often returns fragmented results, while general-purpose LLMs lack both up-to-date knowledge and the domain expertise to interpret nuanced legal questions accurately.

F. Shafait et al. (Eds.): DAS 2026, LNCS 16851, pp. 491–507, 2027.
https://doi.org/10.1007/978-3-032-36207-0_29

Democratizing access to legal information is vital for empowering citizens, reducing litigation costs, and enabling informed compliance in business and civic life. A transparent, evidence-grounded legal chatbot can help non-specialists understand their rights, guide preliminary research, and serve as a building block for applications such as automated contract review, regulatory monitoring, and legal aid platforms.

In this work, we (i) curate and validate a high-quality Q&A dataset covering key articles of the Romanian civil and penal codes; (ii) adapt three open-source LLMs (Phi-4, Pansophic, LLaMA 3.1) via Quantized Low-Rank Adaptation (QLoRA) fine-tuning, Retrieval-Augmented Generation (RAG) pipelines [8,12], and Retrieval-Augmented Fine-Tuning (RAFT) for more transparency and answer justification [32]; (iii) comprehensively evaluate the models on a test set using metrics for lexical overlap, semantic similarity, readability, coherence, and legal correctness; and (iv) deploy all models through an OpenWebUI interface using Ollama for interactive use.

Our experiments reveal that simple fine-tuning of Phi-4 on Romanian law achieves strong overall performance, boosting n-gram overlap from 0.0051 to 0.0161, cosine similarity from 0.4483 to 0.7343, and legal correctness from 0.5891 to 0.6246. The RAG variant attains the highest factual grounding (LLM accuracy 0.6279, legal correctness 0.6335), while RAFT offers superior citation transparency, highlighting a clear trade-off between completeness and interpretability.

2 Related Work

The intersection of Artificial Intelligence and law offers transformative opportunities to democratize access to legal information and services. Legal systems are characterized by complex language, intricate procedures, and continuously evolving regulations, which can alienate non-experts and impede access to justice [3,8,22]. In Romania, despite the public availability of legal documents, the dispersed nature of legal texts and the specialized knowledge required for interpretation pose significant challenges. These factors motivate the development of more accessible, user-friendly legal resources.

Large language models have emerged as promising tools for bridging these gaps. By leveraging vast pre-training corpora and advanced transformer architectures, LLMs can generate human-like text and distill complex legal jargon into plain language. Legal chatbots powered by LLMs have the potential to guide users through legal procedures and provide personalized legal insights, thereby reducing the need for expensive legal consultations [22]. However, developing effective legal chatbots demands solutions that address both the intrinsic complexity of legal language and the dynamic nature of legal knowledge.

With the development of this technology, competitive open-source models have become increasingly accessible, with hardware requirements that can be met by a larger number of users. One prominent example is LLaMA 3.1 [14], a family of language models developed by Meta and available in various sizes, including an 8-billion-parameter version. These models are designed to support multilinguality, coding, reasoning, and tool usage. The 8B variant is optimized for multilingual dialogue use cases and has been fine-tuned using supervised fine-tuning (SFT) and reinforcement learning with human feedback (RLHF) to align with human preferences for helpfulness and safety.

Another example which is based on the same model but refined more for Romanian language is Pansophic-1-Preview [18], which is an 8-billion parameter large language model fine-tuned from Meta's LLaMA 3.1. Developed by Pansophic, it has been trained on a mix of English and Romanian instructions, making it proficient in both languages. The model utilizes the ChatML prompt format and is aligned with APO-ZERO to enhance its instruction-following capabilities. This model is a strong candidate for our use case because it already has stronger Romanian-language capabilities.

Another interesting option is Phi-4 [15], a 14-billion parameter language model developed by Microsoft. Unlike traditional models that primarily rely on organic data sources, Phi-4 strategically incorporates synthetic data throughout its training process. This approach enhances its capabilities, especially in STEM-focused question-answering tasks, where it has been shown to surpass its teacher model, GPT-4. The model is efficient and has a favorable balance between size and inference time, making it suitable for our purpose.

One significant limitation of static LLMs is their reliance on outdated or incomplete internal knowledge. Retrieval-Augmented Generation (RAG) has been proposed as a solution by integrating real-time information retrieval into the generative process [8,12]. In RAG, a generative model—typically based on architectures such as BART—is conditioned on documents retrieved from an external knowledge base via dense vector search mechanisms (e.g., using Dense Passage Retriever). This approach enables the system to generate outputs that are both contextually relevant and factually grounded, which is particularly critical in legal domains where accuracy and timeliness are paramount. In fact, RAG has achieved remarkable performance: it has reached state-of-the-art scores on benchmarks such as Natural Questions (44.5%), TriviaQA (68.0%), and WebQuestions (45.5%). Moreover, on the MS MARCO Abstractive QA task, RAG outperformed a BART-only baseline by improvements of 2.6 BLEU and 2.6 ROUGE-L points. In a three-way classification task, its label accuracy was within 4.3% of the state-of-the-art without explicit supervision on retrieved evidence, and in Jeopardy Question Generation, RAG was judged more factual than BART in 42.7% of cases compared to only 7.1% for BART.

Building on the RAG paradigm, Retrieval-Augmented Fine-Tuning (RAFT) [32] offers a tailored approach for domain-specific applications. RAFT fine-tunes LLMs on curated, domain-specific corpora while incorporating a retrieval mechanism designed to distinguish between relevant documents and distractors. This dual strategy encourages the model to generate chain-of-thought responses that explicitly cite supporting legal texts, thereby enhancing transparency and trustworthiness. Nonetheless, RAFT's reliance on large, annotated datasets and substantial computational resources poses challenges—especially for low-resource languages such as Romanian.

To alleviate the computational demands of fine-tuning large LLMs, techniques like Quantized Low-Rank Adaptation (QLoRA) have been introduced [7]. QLoRA reduces memory and compute requirements by quantizing most model parameters to 4-bit precision, while only updating a small set of adapter weights during fine-tuning. This strategy not only democratizes the process of fine-tuning large models but also facilitates the deployment of domain-specific systems in resource-constrained environments, a crucial consideration for developing Romanian legal chatbots. For example, QLoRA has reduced the memory requirement for fine-tuning a 65B parameter model from over 780 GB (with 16-bit precision) to under 48 GB, without any degradation in performance. The Guanaco 65B model, trained using QLoRA, achieved 99.3% of ChatGPT's performance on the Vicuna benchmark, and even the 7B version outperformed Alpaca 13B by more than 20 percentage points in memory-constrained setups. Guanaco models have consistently outscored other open-source models such as Vicuna, Alpaca, and Open Assistant, with the 65B model achieving an Elo rating of 1022 surpassing ChatGPT-3.5.

In addition to these model adaptation strategies, several works have focused on specialized retrieval and embedding techniques that enhance semantic search. For instance, [1] present CREA2, an intelligent conversational agent designed for legal dispute resolution in the European Union. Their system leverages similar question retrieval using SBERT embeddings [24] and employs data augmentation via the Parrot framework [19], which is based on T5 [31], to improve robustness in handling diverse query formulations. Similarly, [22] propose an information retrieval approach that modifies the StarSpace algorithm [30] to learn joint embeddings of question-answer pairs, thereby increasing training data availability and retrieval accuracy while reducing computational overhead compared to BERT-based methods.

Another important challenge in retrieval-based systems is the preservation of context when encoding text chunks. Traditional RAG systems sometimes suffer from context loss when text is segmented into isolated chunks. To mitigate this, [2] propose augmenting each text chunk with contextual metadata that situates it within its broader document and enhancing the BM25 ranking algorithm. Their approach, which includes a subsequent re-ranking stage, significantly reduces retrieval failure rates and demonstrates that integrating contextual information with dense embeddings leads to more precise outputs.

3 Methodology

Our methodology comprises four integral and sequentially dependent components: comprehensive dataset curation, strategic model selection, targeted fine-tuning methodologies, evaluation and deployment of the models in a web interface. The code and experimental resources are available at https://anonymous.4open.science/r/LegalAISystem-24C2/README.md.

3.1 Dataset

We began by developing a dataset focused on the Romanian legal system, providing a basis for fine-tuning models with domain-specific legal knowledge. The initial phase of dataset creation targeted foundational legal understanding. We leveraged questions from the entrance examinations of the Institutul National al Magistraturii (INM). These questions, specifically those focusing on civil and penal law, are designed to assess a candidate's grasp of fundamental legal codes and procedures. This provided a core set of examples emphasizing conceptual understanding.

To broaden the dataset's scope and incorporate a wider range of legal text, we then integrated data from the official Romanian Legislative Portal, Portal Legislativ. All articles from this portal were extracted, embedded using OpenAI text-embedding-3-small [17] model and indexed within an Azure Search Index. This indexing serves a dual purpose: it facilitates efficient retrieval of relevant legal articles for future model development, particularly in the context of retrieval-augmented generation, and it provides a comprehensive corpus of Romanian legal text.

As shown in Fig. 1, the collected corpus was large. Therefore, due to resource constraints, we selected a subset of legal codes: four related to civil and penal law, and a fifth containing contextual explanations and interpretations of the Penal Code. Building upon the indexed legislative articles, we proceeded to generate a significant number of question-answer pairs using a large language model (LLM).

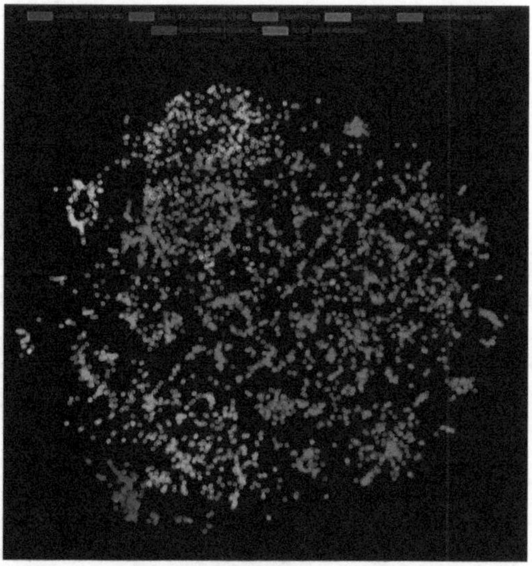

Fig. 1. Embeddings for 7000 articles from the primary Romanian legislation: Civil Code (blue), Penal Procedure (red), Fiscal Code (orange), Traffic Code (yellow), Constitution (turquoise), Labor Code (purple), Fiscal Code (gray) [29] (Color figure online)

These codes were pre-processed to isolate the core articles and any directly associated observations or amendments, ensuring a clean and focused input for the LLM. The Gemini LLMs [26] were then employed to generate questions based on these codes. Crucially, the text of each article served as the context for the generated questions, and the scenarios presented within the questions were designed to reflect real-life situations. This approach aimed to create training data that not only tested factual knowledge but also encouraged the application of legal principles in practical contexts. A detailed visualization of the data acquisition process is provided in Fig. 2. This combination of curated exam questions, indexed legislative text, and LLM-generated question-answer pairs forms a robust and multifaceted dataset for fine-tuning language models on the

Romanian legal system. The resulting dataset provides a basis for training language models specialized in the Romanian legal domain.

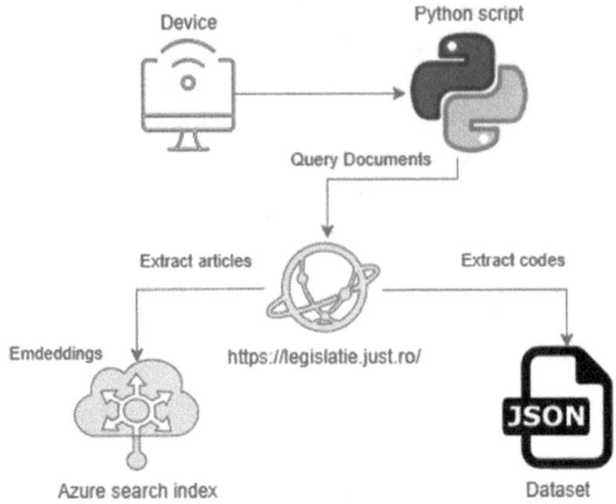

Fig. 2. Dataset acquisition for specific codes

Finally, we added a third dataset component consisting of general questions based on practical real-life situations, extending beyond civil and penal law. This component covers areas such as employment contracts, work-from-home regulations, and environmental authorizations, thereby broadening the dataset's coverage of Romanian law. Combining the three components resulted in a base set of 3,825 questions: 9% from INM examinations, 37% general-purpose questions, and 54% questions generated from specific legal codes (Fig. 3).

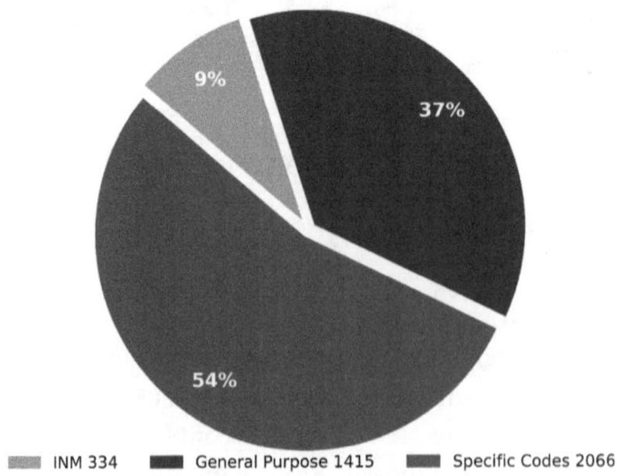

Fig. 3. The distribution of the dataset

The final step in constructing the complete dataset was to generate answers for these questions. For this part we employed a system to obtain qualitative answers. We started with obtaining questions embeddings using OpenAI text-embedding-3-small [17], then for each question we used the previously created Azure search index to find top 3 most related articles to them. We created context for the model, explaining how to answer the questions, handling some situations (abrogated laws) and making it take into account the question and the articles related to it when giving the final answer, that way we targeted to obtain law grounded answer based on the specific legislation applicable in Romania, and we make sure that at the end of the answer, the model also lists the articles used. The detailed process of creating the dataset parts along the final one can be seen in Fig. 4.

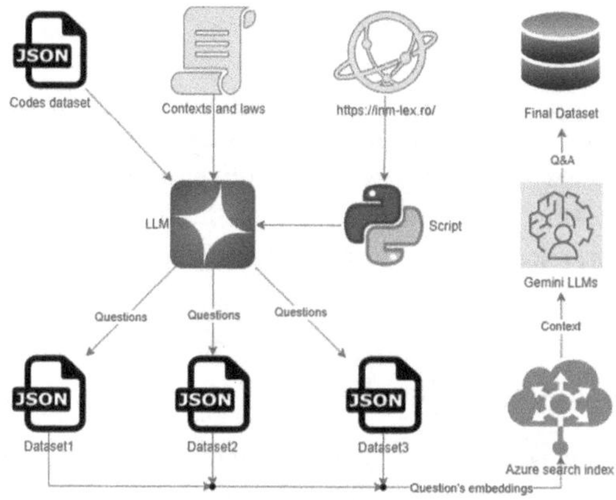

Fig. 4. Dataset composition

3.2 Models

For our benchmark, we fine-tuned three open-source large language models: **Phi-4 (14B)**, **Pansophic-1-Preview-LLaMA-3.1-8B-Instruct**, and **LLaMA 3.1 8B**. Each model has distinct characteristics relevant to Romanian legal question answering. We selected these models based on performance, accessibility, open availability, and the existence of model variants suitable for resource-constrained experimentation.

3.3 Fine-Tuning Strategies

For our fine-tuning experiments with the selected LLMs using our legal dataset with the goal of enhancing the performances in Q&A regarding the Romanian law, we have employed QLoRA.

Because of our hardware resources constraints we decided to use for all three models QLoRA with 4bit quantization for a faster training and more room for experiments. For further optimization we used Unsloth [4] which combined with QLoRA provided a fine-tuning process 5x faster than normally (Fig. 5). We used the suggested parameters with few tweaks.

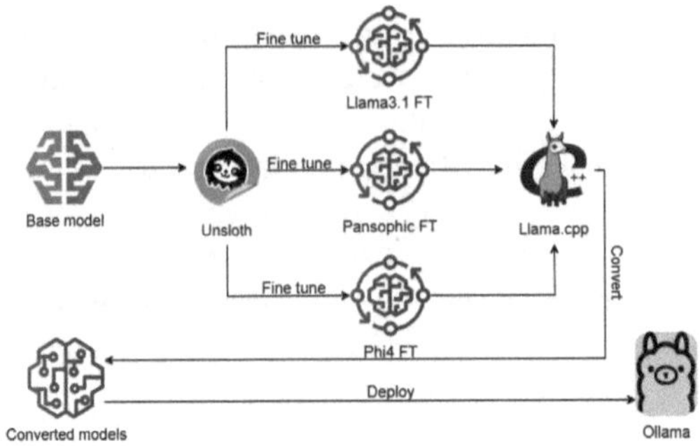

Fig. 5. Fine-tuning process

QLoRA (Quantized Low-Rank Adaptation) is a technique that enables efficient fine-tuning of large language models by using quantization and low-rank adapters. This method significantly reduces memory requirements while maintaining high performance, making it a practical solution for fine-tuning large-scale models on consumer-grade GPUs [7].

3.4 Legal AI Chatbot Implementation

Once we obtained the fine tuned models, we focused on system's reliability and explainability making sure it offers the legal ground for its answers. For obtaining the desired results we employed RAG and RAFT techniques.

RAG [12] is a hybrid approach that combines pre-trained language models with an external retrieval mechanism. Instead of relying solely on parametric memory, RAG dynamically retrieves relevant documents to improve the accuracy and factual consistency of the responses.

RAFT [20] is an enhancement of RAG that integrates retrieval-based learning during the fine-tuning phase. Unlike standard RAG, which retrieves external information only at inference time, RAFT incorporates retrieval mechanisms into the model's training process, allowing it to learn from external sources.

Beyond individual techniques, recent studies emphasize the importance of balancing efficiency and accuracy in legal AI systems. The integration of retrieval-augmented methods with chain-of-thought reasoning, as seen in RAFT, alongside the efficiency gains provided by QLoRA, ensures that legal chatbots are not only accurate and context-aware but also scalable and maintainable.

Collectively, these works illustrate the potential of integrating retrieval-augmented methods, efficient fine-tuning techniques, and advanced semantic search frameworks to build robust legal AI systems. While significant progress has been made in high-resource languages such as English, Italian, and French [1,6,23], the adaptation of these techniques to the Romanian legal domain remains under-explored. Our work builds on this foundation by integrating RAG, RAFT, QLoRA and semantic search advances—as exemplified by DeepSeek [5]—to develop a scalable, efficient, and transparent legal chatbot tailored to the nuances of Romanian law.

To make the system complete, we converted the models to GGUF format using Llama.cpp [9], which facilitates the quantization and the conversion of the models, then we deployed them in Ollama [16]. From that point we can use them in console or as an API endpoint in a web client such as OpenWebUI [27] (Fig. 6).

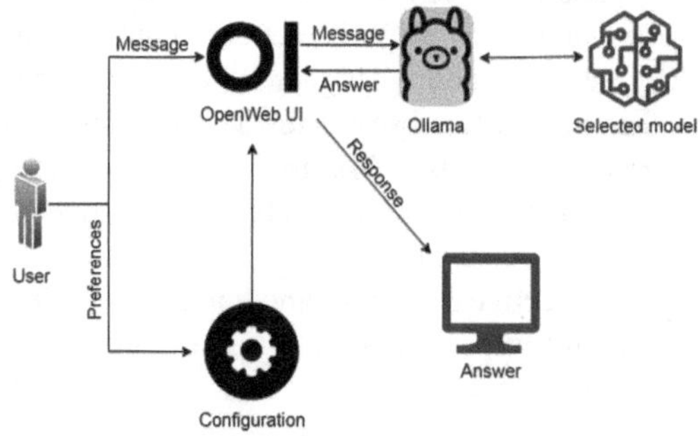

Fig. 6. System workflow

4 Experiments and Results

In this section, we describe how we evaluated and compared the fine-tuned models with the base models, both with and without retrieved context, and present the metrics used in the evaluation. For reproducibility, the code and experimental resources are available in the repository reported in Sect. 3.

Firstly, we did a research on the metrics used for evaluating LLMs and we found out that there is not a single metric to "rule them all" so we have to pick several of them for offering a better overview on how the models perform. As mentioned [25] multi-metric approach is essential to accurately assess and compare the performance of LLMs, particularly as they become more complex and integrated with external data sources or human feedback loops, and we decided to pick as first metrics for generation quality F1, BLEU and ROUGE scores [13], those being one of the most popular general ones. This idea is also supported in the comprehensive survey [11] that focuses on methods of evaluating LLMs and also explains on how metrics used work. Then we decided to move on with a bit more complex ones like N-gram overlap, Cosine Similarity, Flesch Reading Ease, Gunning Fog, SBERT Coherence and LLM as judge (with focus on coherence, accuracy, correctness and completeness). With those metrics we can have a better overview, a clearer image on how each method applied affects the model performance and the final result that user can see.

We split our dataset in train and test sets, and for all metrics calculated we used the test set. The metrics used focus on both simpler and complex aspects of LLMs answers. **F1 score** balances precision and recall to provide a single measure of lexical overlap, ensuring that both missing content (low recall) and extra content (low precision) are penalized. **BLEU score** measures surface-level similarity by quantifying exact matching n-gram sequences, making it a standard metric for tasks like machine translation. **ROUGE scores** evaluates content coverage by measuring how well the generated text captures the key information present in the reference text, which is especially critical for summarization and text generation tasks. **N-gram Overlap** measures surface-level text similarity (lexical precision/recall). **Cosine Similarity** evaluates semantic alignment beyond exact word matches (Word2Vec pretrained model was used here). **Flesch Reading Ease** quantifies text readability (higher is better). Critical for legal text simplification tasks. **Gunning Fog Index** assesses text complexity (lower is better). Aligns with legal clarity requirements. **SBERT Coherence** uses Sentence-BERT embeddings to measure semantic consistency across sentences. **LLM as judge** mimics expert human evaluation at scale, giving a more insightful view of the model performance for the specific task. It was proven as effective evaluation method in both [10, 33] where they used and tested this approach then compared it with human evaluation and showed an agreement of 80 % and also stated that this method is efficient and scalable. For this metrics we decided to split attention as mentioned above on four essential point coherence, accuracy, correctness and completeness. We used Gemini 2.0 Flash [26] as the judge, it was provided with question, context and expected answer and tested model's answer.

In Table 1 we can see the performance of our fine-tuned legal Q&A models evaluated with classic metrics such as F1 score, BLEU, and ROUGE. The significant improvements observed across these metrics indicate that our integration of retrieval-augmented strategies—specifically RAFT—enables the models to more accurately replicate the lexical structure of the reference answers. The higher F1 scores reflect a better balance between precision and recall, suggesting that the models are capturing the essential legal content without introducing extraneous information. Meanwhile, the elevated BLEU scores imply that the models are generating responses with improved n-gram overlap, which in the legal domain is critical for maintaining precise terminology and phrasing. The ROUGE metrics further confirm that the models are effectively covering the key points and information present in the reference texts, ensuring comprehensive content generation.

In Table 2, we delve deeper into the qualitative aspects of the generated responses using advanced metrics. For instance, the N-gram Overlap metric, combined with Cosine Similarity, provides insight into the semantic alignment between the generated responses and the reference answers. The higher Cosine Similarity scores indicate that the fine-tuning process has led to an enhanced understanding of the underlying semantic relationships, enabling the models to capture nuanced legal meanings beyond mere word matching.

Additionally, the readability of the generated text—assessed by Flesch Reading Ease and the Gunning Fog Index—has also improved. Higher Flesch Reading Ease scores coupled with lower Gunning Fog values suggest that the responses are not only technically accurate but also more accessible to non-expert users. This is particularly important in the legal domain, where clarity can bridge the gap between complex legal language and public understanding.

The SBERT Coherence scores further reinforce that our models produce semantically consistent responses across multiple sentences, an essential quality for ensuring

Table 1. Basic Model Performance Metrics

Model	Approach	F1 Score	BLEU Score	ROUGE-1 Precision	ROUGE-L Precision
Phi-4	Baseline	0.137	3.700	0.295	0.133
	RAG	0.075	3.400	0.170	0.092
	Fine-tuned	**0.195**	**10.800**	0.403	0.216
	RAFT	0.188	10.400	0.415	0.236
Panso-phic-1	Baseline	0.134	5.900	0.272	0.135
	RAG	0.140	7.000	0.313	0.163
	Fine-tuned	0.188	8.500	0.430	0.242
	RAFT	0.169	6.500	0.459	**0.286**
LLaMA 3.1	Baseline	0.141	3.400	0.382	0.196
	RAG	0.140	2.400	**0.530**	**0.360**
	Fine-tuned	**0.198**	9.900	0.450	0.252
	RAFT	0.165	5.400	0.493	**0.314**

that multi-part legal arguments remain logically connected. Evaluations via an LLM-as-judge approach—which mimics expert human assessment—corroborate these findings, highlighting improvements in overall response coherence, accuracy, correctness, and completeness.

Collectively, these results validate our approach by demonstrating that the combined use of RAG, RAFT, and QLoRA techniques not only boosts lexical accuracy but also enhances semantic quality and readability. However, our error analysis reveals that the models occasionally falter when confronted with ambiguous queries, suggesting that further refinements in semantic disambiguation are needed. Future work could explore more sophisticated context modeling or interactive clarification mechanisms to better handle such ambiguities.

4.1 Analysis of Methodology Performance

To qualitatively illustrate the benefits of our methodology, we analyze a response from the fine-tuned `Phi-4` model (enhanced with RAFT principles) to a practical legal query. The question, the answer from base model and fine tuned model can be seen in Section ??. The response from fine tuned model presents several crucial improvements, the most notable ones being:

Enhanced Legal Correctness and Grounding: The model accurately identifies Article 117(1)–(2) of the Romanian Code of Civil Procedure as the pertinent legal basis for determining territorial competence. The explicit `LEGAL_GROUND` citation, a direct benefit of RAFT-like mechanisms, provides verifiability and aligns with the higher legal correctness scores observed for RAG variants.

Table 2. Advanced Model Performance Metrics

Model	Base	FT	RAG	RAFT
N-gram Overlap (Higher is better)				
phi4	0.01	0.02	0.01	**0.02**
pansophic	0.02	0.02	**0.03**	0.02
llama3.1	0.01	0.02	0.01	**0.02**
Cosine Similarity (Higher is better)				
phi4	0.45	**0.73**	0.54	0.65
pansophic	**0.78**	0.62	0.67	0.47
llama3.1	0.40	**0.60**	0.23	0.34
Flesch Reading Ease (Higher is better)				
phi4	34.00	38.00	24.00	**39.00**
pansophic	37.00	38.00	32.00	**42.00**
llama3.1	39.00	**41.00**	34.00	**41.00**
Gunning Fog (Lower is better)				
phi4	15.30	**13.30**	16.80	13.70
pansophic	13.90	14.00	14.30	**13.70**
llama3.1	15.10	**13.20**	18.10	13.90
SBERT Coherence (Higher is better)				
phi4	0.28	**0.47**	0.31	0.46
pansophic	0.44	**0.47**	0.43	0.45
llama3.1	0.35	**0.47**	0.25	0.36
LLM Coherence (Higher is better)				
phi4	0.79	**0.80**	0.72	0.68
pansophic	**0.72**	0.70	0.60	0.57
llama3.1	0.63	**0.71**	0.50	0.51
LLM Accuracy (Higher is better)				
phi4	**0.62**	0.61	**0.62**	0.60
pansophic	0.54	**0.55**	0.50	0.50
llama3.1	0.47	**0.53**	0.47	0.45
LLM Correctness (Higher is better)				
phi4	0.59	0.63	**0.64**	0.52
pansophic	0.50	**0.56**	0.51	0.52
llama3.1	0.45	**0.54**	0.48	0.47
LLM Completeness (Higher is better)				
phi4	**0.67**	**0.67**	0.60	0.55
pansophic	**0.60**	0.56	0.47	0.43
llama3.1	0.48	**0.56**	0.38	0.39

Improved Semantic Nuance: The answer correctly distinguishes between two potential legal scenarios: a boundary dispute (*delimitarea terenurilor*) and property encroachment (*încălcare a proprietății*). It then tailors the application of Art. 117(1)–(2) to each scenario, focusing on the location of the "contested immovable" or the "source of prejudice." This showcases the model's deeper semantic understanding fostered by fine-tuning on our domain-specific dataset, leading to improved coherence.

Contextual Problem Solving: The model effectively addresses the user's specific dilemma regarding properties in different localities by guiding them to identify the legally relevant location to determine the competent court. This practical reasoning is a testament to the dataset's inclusion of scenario-based questions and the model's improved ability to apply learned knowledge.

Clarity and Structure: The response is logically structured, first establishing general principles and then applying them to the user's situation. This contributes to better readability, as reflected in improved Flesch Reading Ease and Gunning Fog Index scores, making complex legal information more accessible to non-experts.

Overall, these findings underscore the potential of our fine-tuning methodology to improve legal Q&A performance in a low-resource context, offering a scalable and generalizable framework for advancing legal AI systems.

5 Conclusion

Our fine-tuning methodology for legal Q&A in Romanian demonstrates notable improvements in accuracy and contextual understanding. The integration of retrieval-augmented strategies (RAG and RAFT) significantly enhances response quality by grounding answers in legislative texts. Additionally, QLoRA ensures efficient adaptation of large models without excessive computational overhead. For BLEU, F1 and ROUGE scores the fine-tuned models obtain higher scores than the baseline by a significant margin and we can also see that the RAFT is very sensible to the context received. In case of the metrics that show the fluency and the coherence we can notice that also fine tuning and RAFT have the best results meaning that they can provide an easier to understand and better structured answer. The proposed methodology for legal Q&A in the Romanian language can serve as a guideline for developing similar models in other understudied languages, providing a scalable and adaptable framework for improving legal AI assistants in diverse linguistic contexts.

6 Limitations

While our approach demonstrates promising improvements in Romanian legal question answering, several limitations warrant discussion. First, our dataset is currently limited to select aspects of Romanian civil and penal law. Although carefully curated, it may not fully capture the complexity of the broader legal domain or generalize well to other legal areas. Future work should expand the dataset to cover additional legal subdomains and real-world variations.

Second, our reliance on retrieval-augmented methods (RAG and RAFT) introduces a dependency on the quality and completeness of the external legal corpus. Inaccurate, outdated, or incomplete legal texts could propagate errors into the generated responses, particularly when handling ambiguous queries or cases where relevant documents are missing.

Third, while QLoRA enables efficient fine-tuning on resource-constrained hardware, the quantization and adapter-based fine-tuning approach may incur some performance trade-offs compared to full-precision models. This balance between computational efficiency and model accuracy is especially critical in high-stakes legal contexts.

Fourth, our evaluation is primarily based on automatic metrics and the analysis of a limited number of experimental results. Additional extensive human evaluation and testing on a wider range of legal queries are necessary to fully validate the model's practical utility and reliability.

Lastly, although our approach provides some degree of interpretability through cited legal texts, ensuring consistent and transparent reasoning in the model's outputs remains an open challenge. Further research is needed to enhance the explainability of our system's decision-making process.

6.1 Dataset Quality Validation

Given that 54% of the dataset consists of automatically generated question–answer pairs, we performed a manual quality audit to assess the reliability of the generated content and estimate the prevalence of generation errors. A stratified random sample of 200 Q&A pairs (5.2% of the complete dataset) was selected proportionally from all dataset components, including examination questions, practical legal scenarios, and LLM-generated code-based examples.

The sampled instances were evaluated by a legal domain expert with expertise in Romanian legislation according to a predefined validation protocol. Each Q&A pair was assessed across four dimensions: (i) legal correctness with respect to the applicable legislation, (ii) consistency between the answer and the supporting legal provisions, (iii) completeness of the legal reasoning provided, and (iv) absence of hallucinated, unsupported, or factually incorrect legal references.

The audit identified 35 instances containing at least one material defect, corresponding to an estimated error prevalence of 17.5%. Consequently, 82.5% of the reviewed samples were judged legally valid and consistent with the referenced legislation. The observed defects were primarily associated with incomplete legal reasoning, imprecise interpretation of statutory provisions, or omission of relevant legal details, while severe factual inaccuracies and hallucinated legal references occurred less frequently.

Although the validation was conducted by a single legal expert and therefore does not permit the computation of inter-annotator agreement statistics, the audit provides an empirical estimate of dataset quality and quantifies the risk of error propagation from automatically generated examples into downstream models. Importantly, despite the presence of some defective samples, the resulting models demonstrated consistent improvements across both automatic and LLM-based evaluation metrics, suggesting that the overall signal-to-noise ratio of the dataset remains favorable for supervised fine-tuning. The dataset is available on request from marinela.adam01@e-uvt.ro.

7 Experimental Setup Details

All the experiments were run on a Windows machine using PyTorch and CUDA for acceleration. For fine-tuning the pre-trained models and the evaluation benchmarks we used NVIDIA RTX 4090 GPU mobile along Ryzen 9 7945HX CPU.

7.1 Fine-Tuning Hyperparameters

In Table 3 we can see the hyperparameters used for fine-tuning the models:

Table 3. The hyperparameters used for fine-tuning the models.

Model Configuration	
Model Name	Llama-3.1-8B/Pansophic/Phi-4
Max Sequence Length	1024
Data Type	torch.bfloat16
Load in 4-bit	True
LoRA Configuration	
LoRA Rank (r)	16
Target Modules	q_proj, k_proj, v_proj, o_proj, gate_proj, up_proj, down_proj, lm_head, embed_tokens
LoRA Alpha	16
Use Gradient Checkpointing	unsloth
Random State	3407
Use RSLora	False
Training Configuration	
Per Device Train Batch Size	4
Gradient Accumulation Steps	8
Warmup Steps	5
Learning Rate	2e−4
BF16	True
Save Total Limit	2
Save Steps	5
Optimizer	adamw_8bit
Weight Decay	0.01
LR Scheduler Type	linear
Seed	3407
Packing	False

Acknowledgment. This research is supported by the project "Romanian Hub for Artificial Intelligence - HRIA," in the program PCIDIF/709/PCIDIF_P4/OP1/RSO1.6/PCIDIF_A12 - Action 4.1 - MySMIS no. 351416.

References

1. Amato, F., Fonisto, M., Giacalone, M., Sansone, C.: An intelligent conversational agent for the legal domain. Information **14**(6), 307 (2023)
2. Anthropic: Introducing contextual retrieval (2024). https://www.anthropic.com/news/contextual-retrieval. Accessed 24 Nov 2024
3. Bex, F.J.: Ai, law and beyond. A transdisciplinary ecosystem for the future of AI law. Artif. Intell. Law, 1–18 (2024)
4. Daniel Han, M.H., team, U.: Unsloth (2023). http://github.com/unslothai/unsloth
5. DeepSeek-AI, Guo, D., et al., D.Y.: Deepseek-r1: incentivizing reasoning capability in llms via reinforcement learning (2025). https://arxiv.org/abs/2501.12948
6. Deroy, A., Ghosh, K., Ghosh, S.: Applicability of large language models and generative models for legal case judgement summarization. Artif. Intell. Law, 1–44 (2024)
7. Dettmers, T., Pagnoni, A., Holtzman, A., Zettlemoyer, L.: Qlora: efficient finetuning of quantized LLMs, **52**, 3982–3992. arXiv preprint arXiv:2305.14314 (2023)
8. Finardi, P., et al.: The chronicles of rag: the retriever, the chunk and the generator. arXiv preprint arXiv:2401.07883 (2024)
9. Gerganov, G.: llama.cpp: efficient CPU inference for llama (2023). https://github.com/ggerganov/llama.cpp. Accessed 25 Jan 2025
10. Gu, J., et al.: A survey on LLM-as-a-judge (2025). https://arxiv.org/abs/2411.15594
11. Guo, Z., et al.: Evaluating large language models: a comprehensive survey (2023). https://arxiv.org/abs/2310.19736
12. Lewis, P., et al.: Retrieval-augmented generation for knowledge-intensive NLP tasks. Adv. Neural Inf. Process. Syst. **33**, 9459–9474 (2020)
13. Liu, C.W., Lowe, R., Serban, I.V., Noseworthy, M., Charlin, L., Pineau, J.: How not to evaluate your dialogue system: an empirical study of unsupervised evaluation metrics for dialogue response generation. arXiv preprint arXiv:1603.08023 (2016)
14. MetaAI: Llama 3.1:8b model (2024). https://huggingface.co/meta-llama/Llama-3.1-8B. Accessed 13 Feb 2025
15. Microsoft: Phi-4 technical report (2024). https://www.microsoft.com/en-us/research/publication/phi-4-technical-report/. Accessed 13 Feb 2025
16. Ollama, I.: Ollama: get up and running with large language models locally (2023). https://github.com/ollama/ollama. Accessed 1 Feb 2025
17. OpenAI: New embedding models and API updates (2024). https://openai.com/index/new-embedding-models-and-api-updates/. Accessed 17 Jan 2025
18. Pansophic: Pansophic-1-preview-llama3.1:8b-instruct (2024). https://huggingface.co/pansophic/pansophic-1-preview-LLaMA3.1-8b. Accessed 13 Feb 2025
19. Pellicer, L.F.A.O., Ferreira, T.M., Costa, A.H.R.: Data augmentation techniques in natural language processing. Appl. Soft Comput. **132**, 109803 (2023)
20. Prager, J., Glass, J., Weston, J.: Raft: retrieval-augmented fine-tuning for large language models. arXiv preprint (2023). https://arxiv.org/abs/2307.12901
21. Qin, L., et al.: Large language models meet NLP: a survey. arXiv preprint arXiv:2405.12819 (2024)
22. Queudot, M., Charton, É., Meurs, M.J.: Improving access to justice with legal chatbots. Stats **3**(3), 356–375 (2020)

23. Ragazzi, L., Moro, G., Guidi, S., Frisoni, G.: Lawsuit: a large expert-written summarization dataset of Italian constitutional court verdicts. Artif. Intell. Law, 1–37 (2024)
24. Reimers, N.: Sentence-BERT: sentence embeddings using Siamese BERT-networks. arXiv preprint arXiv:1908.10084 (2019)
25. Sai, A.B., Mohankumar, A.K., Khapra, M.M.: A survey of evaluation metrics used for NLG systems (2020). https://arxiv.org/abs/2008.12009
26. Team, G., et al.: Gemma: open models based on Gemini research and technology. arXiv preprint arXiv:2403.08295 (2024)
27. Team, O.W.: Open Webui: self-hosted offline AI interface (2023). https://github.com/open-webui/open-webui. Accessed 1 Feb 2025
28. Touvron, H., et al.: Llama: open and efficient foundation language models. arXiv preprint arXiv:2302.13971 (2023)
29. Tudor, V., Moldovan, F.: Rag done right - legal AI search case study. Decoding ML on Substack (2024). https://decodingml.substack.com/p/rag-done-right-legal-ai-search-case-study
30. Wu, L., Fisch, A., Chopra, S., Adams, K., Bordes, A., Weston, J.: Starspace: embed all the things! In: Proceedings of the AAAI Conference on Artificial Intelligence, vol. 32 (2018)
31. Xue, L.: mt5: A massively multilingual pre-trained text-to-text transformer. arXiv preprint arXiv:2010.11934 (2020)
32. Zhang, T., et al.: Raft: adapting language model to domain specific rag. arXiv preprint arXiv:2403.10131 (2024)
33. Zheng, L., et al.: Judging LLM-as-a-judge with MT-bench and chatbot arena (2023). https://arxiv.org/abs/2306.05685

Author Index

F. Shafait et al. (Eds.): DAS 2026, LNCS 16851, pp. 509–510, 2027.
https://doi.org/10.1007/978-3-032-36207-0

FSC
www.fsc.org
MIX
Papier | Fördert gute Waldnutzung
FSC® C083411

Kolibri 360 GmbH
Ferdinand-Jühlke-Straße 7
99095 Erfurt, Deutschland
produktsicherheit@kolibri360.de